W9-CRL-476

The Subnuclear Series • Volume 31

Proceedings of the International School of Subnuclear Physics

# FROM SUPERSYMMETRY TO THE ORIGIN OF SPACE-TIME

**Edizioni Scientifiche Galileiane**

Edited by

**Antonino Zichichi**

*European Physical Society*
*Geneva, Switzerland*

*Published by*

World Scientific Publishing Co. Pte. Ltd.

P O Box 128, Farrer Road, Singapore 9128

*USA office:* Suite 1B, 1060 Main Street, River Edge, NJ 07661

*UK office:* 57 Shelton Street, Covent Garden, London WC2H 9HE

**Library of Congress Cataloging-in-Publication Data**

International School of Subnuclear Physics.
From supersymmetry to the origin of space-time : proceedings of the International School of Subnuclear Physics / edited by Antonino Zichichi. -- Edizioni scientifiche Galileiane.
p. cm. -- (The subnuclear series ; v. 31)
Includes bibliographical references
ISBN 9810219172
1. Particles (Nuclear physics) -- Congresses. 2. Electroweak interactions -- Congresses. 3. Superstring theories -- Congresses. 4. Space and time -- Congresses. I. Zichichi, Antonino. II. Title. III. Series.
QC793.I5549 1995
539.7'54--dc20 94-29400
CIP

**British Library Cataloguing-in-Publication Data**
A catalogue record for this book is available from the British Library.



Printed in Singapore.

# CONTENTS

## PREFACE

During July 1993, a group of 80 physicists from 56 laboratories in 15 countries met in Erice for the 31st Course of the International School of Subnuclear Physics. The countries represented by the participants were: Bulgaria, Canada, China, Denmark, France, Germany, Greece, Hungary, Israel, Italy, Netherlands, Norway, Palestine, Peru, Poland, Romania, Russia, Sweden, Taiwan, Turkey, United Kingdom and the United States of America.

The School was sponsored by the European Physical Society (EPS), the Italian Ministry of Public Education (MPI), the Italian Ministry of University and Scientific and Technological Research (MURST), the Sicilian Regional Government (ERS) and the Weizmann Institute of Science.

The purpose of the School was to discuss the main developments in the field of Supersymmetry including the implications on the Origin of Space-Time. The Table of Contents illustrates the details of the programme.

I hope the reader will enjoy the book as much as the students attending the lectures and discussion sessions, which are one of the most attractive features of the School. Thanks to the work of the Scientific Secretaries the discussions have been reproduced as faithfully as possible. At various stages of my work I have enjoyed the collaboration of many friends whose contributions have been extremely important for the School and are highly appreciated. I thank them most warmly. A final acknowledgement to all those in Erice, Bologna and Geneva, who have helped me on so many occasions and to whom I feel very indebted.

Antonino Zichichi

# A Non-Critical String Approach to Black Holes, Time and Quantum Dynamics

**John Ellis**[a,◇], **N.E. Mavromatos**[b,◇] and **D.V. Nanopoulos**[a,c]

**Abstract**

We review our approach to time and quantum dynamics based on non-critical string theory, developing its relationship to previous work on non-equilibrium quantum statistical mechanics and the microscopic arrow of time. We exhibit specific non-factorizing contributions to the $\not{S}$ matrix associated with topological defects on the world sheet, explaining the rôle that the leakage of $W_\infty$ charges plays in the loss of quantum coherence. We stress the analogy with the quantum Hall effect, discuss the violation of $CPT$, and also apply our approach to cosmology.

[a] Theory Division, CERN, CH-1211, Geneva 23, Switzerland
[b] Laboratoire de Physique Thèorique ENSLAPP (URA 14-36 du CNRS, associeè à l' E.N.S de Lyon, et au LAPP (IN2P3-CNRS) d'Annecy-le-Vieux), Chemin de Bellevue, BP 110, F-74941 Annecy-le-Vieux Cedex, France.
On leave from S.E.R.C. Advanced Fellowship, Dept. of Physics (Theoretical Physics), University of Oxford, 1 Keble Road, Oxford OX1 3NP, U.K.
[c] Center for Theoretical Physics, Dept. of Physics, Texas A & M University, College Station, TX 77843-4242, USA and Astroparticle Physics Group, Houston Advanced Research Center (HARC), The Mitchell Campus, Woodlands, TX 77381, USA.

**◇ Lectures presented at :**
**the Erice Summer School, 31st Course: From Superstrings to the Origin of Space-Time, Ettore Majorana Centre, Erice, July 4-12 1993**

# 1 Introduction

String theory is widely heralded as a consistent quantum theory of gravity. As such, it should not only enable us to calculate meaningfully quantum-gravitational corrections to scattering processes in a fixed space-time background, but also to take into account quantum fluctuations in space and time themselves. The way to carry out the first part of this double programme in string theory is well known: calculate higher-genus effects in a given critical string vacuum that describes the appropriate classical background. The way to carry out the second part of the quantum gravity programme in string theory, namely to understand space-time foam, is less evident. One must master a multitude of string vacua and the quantum transitions between them, which necessarily involve non-critical string theory [1]. No-one can do this at present: the best one can do is study some tractable examples of non-critical string models [2] that one hopes are relevant and representative, and abstract from them features that may be generic.

It is to be expected that some shibboleths of conventional physics will be cast down when this programme is carried out. Certainly general relativity must be modified, and perhaps also special relativity and even quantum mechanics. We review in these lectures our work [2, 3, 4] indicating that our understandings of special relativity, quantum mechanics and quantum field theory should indeed be modified. We start from the string black hole solution of Witten [5], and abstract from it general features associated with a renormalization group analysis of the string effective action [6].

There is no arrow of time in conventional quantum field theory, nor in critical string theory. Indeed, it is even possible to formulate critical strings without introducing a time variable at all. However, the time we experience does have an arrow, both microscopically as codified in the second law of thermodynamics, and macroscopically as evidenced in the cosmological Hubble expansion. One of the main thrusts of our work has been to understand the arrow of time in the framework of non-critical string theory, and to relate its microscopic and macroscopic manifestations. As we shall see, an essential feature of this understanding is an apparent modification of quantum mechanics and quantum field theory, entailing the abandonment of the $S$-matrix description of scattering.

Some have long suspected that such a modification might be necessary, in view of the fact that black holes apparently behave thermodynamically in the context of local quantum field theory[7, 8, 9]. The appearance of an event horizon is acompanied by non-zero entropy proportional to its area, and a related non-zero temperature, properties that require a mixed-state quantum treatment. This entails use of the density-matrix formalism, and Hawking has suggested [7] that when space-time foam is taken into account scattering must be formulated as an asymptotic linear transformation from incoming states $\rho^A_{in,B}$ to outgoing states $\rho^C_{out,D}$ :

$$\rho^C_{out,D} = \$^{CB}_{DA}\rho^A_{in,B} \tag{1}$$

where the superscattering matrix $\$^{CB}_{DA}$ does not in general factorize as a product of $S$ and $S^\dagger$ matrix elements [7]

$$\$^{CB}_{DA} \neq S^C_A(S^\dagger)^B_D \tag{2}$$

as in conventional quantum field theory. Correspondingly, the time-evolution of a quantum system cannot be governed simply by the Liouville equation, which integrates to yield just the conventional $S$ matrix, but there should be an extra term in the quantum Liouville equation due to space-time foam, which we may write in the form [10]

$$\partial_t \rho = i[\rho, H] + \not{\delta} H \rho \tag{3}$$

Such a modification of the quantum Liouville equation is characteristic of open quantum-mechanical systems, in which the observed (sub)system is in contact with an unobserved reservoir. It introduces a microscopic arrow of time, with in general dissipation, entropy increase and apparent wave-function collapse in the observed (sub)system. In the quantum-gravitational context discussed here, the unobserved states are associated with non-trivial microscopic event horizons, which are unobservable even in principle. We discuss below how non-trivial contributions to the $\$$ matrix and to $\not{\delta} H$ may arise in string theory, using the general formalism of string backgrounds as $\sigma$-model field theories on the two-dimensional world-sheet. We treat the two-dimensional (spherically-symmetric four-dimensional) string black hole background as an illustrative example in which specific calculations can be performed. We have derived an explicit general form for $\not{\delta} H$ in string theory, expressed in world-sheet $\sigma$-model notation and exemplified by the string black hole model. We have also shown explicitly how this modification of the quantum Liouville equation leads to decoherence and apparent collapse of the wave function, as suggested previously on the basis of more intuitive approaches to quantum gravity.

Before discussing this, however, we discuss in section 2 some relevant features of non-equilibrium quantum statistical mechanics, which is the appropriate framework for the modified quantum Liouville time-evolution equation (3). Specifically, we recall the general formalism of Misra and Prigogine [11, 12], as well as the so-called Lie-admissible approach of ref. [13], which are compatible under certain conditions[14]. In section 3 we recall [2] the general world-sheet $\sigma$-model derivation of the modified quantum Liouville equation (3), and show that it obeys the Lie-admissibility condition of ref. [14]. This follows from the existence of the Zamolodchikov metric in world-sheet $\sigma$-model coupling space. The arrow of time is associated with renormalization group flow in this space. Energy is conserved in the mean [15], as a consequence of renormalizability, which replaces the time translation invariance of conventional target-space field theory. Probability is also conserved, whilst entropy increases monotonically [2]. Section 4 contains a more detailed discussion of time, which is interpreted in our approach as a renormalization group scale identified with a Liouville field [2, 4, 16]. Section 5 reviews the string black hole model and its interpretation in terms of monopoles [17], as well as instantons [18] in this model and their interpretation. Section 6 reviews our previous calculations of specific contributions to the $\not{S}$ matrix and $\not{\delta}H$ due to monopoles and instantons on the world sheet associated with string black holes [17]. In section 7 we discuss in more detail the relation between our string black hole calculations and the approach of ref. [11], underlining in particular the role played by $W$ symmetries. Section 8 reviews the violation of $CPT$ in our formalism, relating it to the analysis of ref. [19] and discussing its possible manifestation in the neutral kaon system. In section 9 we mention applications of our approach to cosmology, with particular mention of the initial singularity, inflation, the time-dependences of the fundamental parameters, and the cosmological constant [16]. Finally, in section 10 we discuss the outlook for our approach.

## 2 Non-Equilibrium Quantum Statistical Mechanics for Pedestrians

In this section we review at an elementary level some relevant features of non-equilibrium quantum statistical mechanics, which we link in the next section with our string-modified quantum Liouville equation. The authors of [11] sought to include time irreversibility and the second law of thermodynamics into a description of microphysics based on the density matrix. It is clear that this aim entails modifications of conventional classical and quantum mechanics, that are to be regarded as incomplete in this context. As a first step in this programme, the authors of [11] introduced a dynamical transformation $\Lambda$, which is not unitary in general, that relates the full density matrix $\rho$ to that for the physically-relevant system, denoted by $\tilde{\rho}$:

$$\tilde{\rho} = \Lambda\rho \tag{4}$$

In our case, we shall consider as $\tilde{\rho}$ the density matrix for the low-mass light string states that can be measured locally, as distinct from extended solitonic string states whose properties can only be characterized by global measurements. We shall refer to $\tilde{\rho}$ as the locally-measurable density matrix. Misra and Prigogine [11] ponted out that the transformation $\Lambda$ should satisfy certain consistency conditions: it should preserve the positivity of $\rho$, it should obey the equations

$$\Lambda \cdot I = I \tag{5}$$

and

$$\int_\Gamma d\mu Tr \Lambda\rho = \int_\Gamma d\mu Tr\rho \tag{6}$$

where $\Gamma$ denotes the phase-space manifold, and the time-evolution operator $U_t = e^{-iLt}$, where $L$ is the Liouville operator, should have the intertwining property

$$\Lambda U_t = W_t^* \Lambda \tag{7}$$

for $t \geq 0$, where $W_t^*$ is an adjoint strongly-irreversible Markov semigroup operator.

Two different possibilities for $\Lambda$ should be distinguished. One is that in which $\Lambda$ has an inverse, and the similarity relation

$$W_t^* = \Lambda U_t \Lambda^{-1} \tag{8}$$

applies for $t \geq 0$. This mathematical invertibility does not mean, however, that physical information is retained, as we shall see later. In the other case, $\Lambda$ has no inverse, and can be regarded as a projection operator onto the physically-relevant states.

In the former case, which is the one that concerns us, as we show in the next section, the locally-measurable density matrix $\tilde{\rho}$ obeys the time-evolution equation

$$i\partial_t \tilde{\rho} = \Phi(L)\tilde{\rho} \tag{9}$$

where

$$\Phi(L) = \Lambda^{-1} L \Lambda \tag{10}$$

with $L$ the Liouvillian. The existence of a function in phase space which varies monotonically with time, called a Lyapounov function, is guaranteed if $\Phi(L)$ obeys the condition [11]

$$i\Phi(L) - i[\Phi(L)]^\dagger \geq 0 \tag{11}$$

which holds if $\Lambda$ has the *star hermiticity* property

$$\Lambda^{-1}(L) = \Lambda^\dagger(-L) \equiv [\Lambda(L)]^* \tag{12}$$

However, this property may not hold in general.

Time is not necessarily irreversible in this framework. The mere existence of a Lyapounov function is not sufficient to guarantee physical time-reversal symmetry breaking. Generalizing equation (8), one can define distinct Markov semigroup operators

$$W_t^{\pm} = \Lambda_{\pm} U_t \Lambda_{\pm}^{-1} \tag{13}$$

for $t \geq 0$ and $t \leq 0$, respectively. Thanks to the time-reversal invariance of the initial unitary evolution operator $U_t$, time is reversible if $\Lambda_+ = \Lambda_-$, but this is not the case in general. In most cases there are physical reasons for distinguishing $\Lambda_+$ and $\Lambda_-$, and thereby determining the arrow of time-reversal symmetry breaking, as is for instance the case in which the initial conditions in one of the Markov processes are such that their preparation requires infinite entropy. It has been shown [11] that a dynamical system of the type described above admits an internal time variable, suitable for the discussion of aging, if $\Lambda$ obeys the conditions (5), (6) and (7), and is sufficiently large in the sense that for any given $\rho$ and any $\epsilon \geq 0$ there is another density matrix $\rho'$ with the property that

$$||\rho - U_t \Lambda \rho'|| < \epsilon \tag{14}$$

i.e., the subspace generated by $\Lambda$ evolves backward in time to generate arbitrarily good approximations to all possible states. In this case, it has been shown [11] that one can introduce a time-evolution operator $T$ conjugate to the Liouville operator $L$:

$$[L, T] = i\hbar \tag{15}$$

The converse of this theorem is also true.

In such a dynamical system, a point in phase space, when evolved backwards in time, may become an arbitrarily complicated and non-local region of phase space, via a sort of inverse butterfly effect. It is not possible to reconstruct the past history of the system unless one measures all components of its final-state density matrix with arbitrarily high precision, which is impracticable. Thus, information is effectively lost during the time-evolution, and hence entropy increases. The system approaches an equilibrium state in which there is no memory of the initial state.

One example of such a system is provided by geodesics in an expanding Universe, described by a Robertson-Walker-Friedmann metric in $3 + 1$ dimensions:

$$ds^2 = dt^2 - R(t)^2 \sum_{i=1}^{3} (dx^i)^2 \tag{16}$$

Test particles move along four-dimensional geodesics that project onto geodesics in the three-dimensional hypersurfaces of simultaneity. Geodesic flow in four dimensions defines a corresponding geodesic flow in three dimensions, and the theorem of [12] tells us that an internal time $T$, or age, can be defined for the system, which is not the cosmological time, but is a monotonic function of it. It has the interesting property that

$$U^*_\lambda T U_\lambda = T + \lambda(t) I \tag{17}$$

where

$$\lambda(t) = \lambda(t_0) + A \int_{t_0}^{t} ds \frac{1}{R(s)} \qquad \textit{for massless particles} \tag{18}$$

is the affine parameter of the projected geodesic flow, and $t$ is the cosmic time. This tells us that the rate of change of $\lambda$ becomes very rapid at early times, and very slow at late times. Thus the rate of approach to equilibrium is huge close to the initial singularity, and very slow in an old Universe. This is an example of apparent indeterminacy in the *a priori* deterministic theory of General Relativity. As we shall see later, a very similar situation arises in our Liouville approach to non-critical string theory.

Thus far, we have not considered in detail the form of the dynamical equations of motion describing the time-evolution of such a system. An appropriate framework is provided by the so-called Lie-admissible formulation of dissipative statistical mechanics, as reviewed in [13]. The starting point that describes the dissipative motion of a single particle is the open version of the Lagrange equation:

$$\frac{d}{dt}\frac{\partial L}{\partial \dot{q}^i} - \frac{\partial L}{\partial q^i} = \mathcal{F}_i(t, q^i, \dot{q}^i) \tag{19}$$

The corresponding first-order Hamilton equations take the form

$$\begin{aligned} \dot{q}^i &= \frac{\partial H}{\partial p_i} \\ \dot{p}_i &= -\frac{\partial H}{\partial q^i} + F_i \qquad : F_i(t, q^i, p_i) = \mathcal{F}_i(t, q^i, \dot{q}^i) \end{aligned} \tag{20}$$

The extension of this formulation to the statistical evolution of the phase-space density function $\rho(q, p, t)$ entails a generalized Liouville equation

$$\frac{\partial \rho}{\partial t} + \{\rho, H\} + F_i \frac{\partial \rho}{\partial p_i} + \rho \frac{\partial F_i}{\partial p_i} = 0 \tag{21}$$

where $\{,\}$ denotes a conventional Poisson bracket. As we shall see in section 3, in the string case of interest to us the last term in this equation is absent, whilst the previous term is very much present.

The treatment of such a system must answer the question how the physical energy operator $E$ can be identified with the generator $H$ of time translations, since $\dot{E} \neq 0$ at the operator level, due to dissipation or, more general, interactions with the "environment", whereas

$$\dot{H} = \{H, H\} = 0 \tag{22}$$

where $\{,\}$ denotes a conventional Poisson bracket, which becomes a commutator $[,]$ in the quantum case. The answer presented in ref. [13] is to modify the Lie-algebraic structure, replacing $\{,\}$ by an object $\{\{,\}\}$ with the property that

$$\dot{H} = \{\{H, H\}\} \neq 0 \tag{23}$$

and analogously $[,]$ becomes $(,)$ in the quantum case.

A generalized product $(,)$ is said to form a Lie-admissible algebra if it is linear, obeys the generalized Jacobi identity

$$\begin{aligned} ((A, B), C) &+ \textit{cyclic permutations} + \\ (C, (B, A)) &+ \textit{cyclic permutations} - \\ (A, (B, C)) &- \textit{cyclic permutations} - \\ ((C, B), A) &- \textit{cyclic permutations} = 0 \end{aligned} \tag{24}$$

and the condition

$$(A,B)-(B,A)=2[A,B] \tag{25}$$

where $[,]$ denotes the conventional Lie product. As an example, consider a dynamical system with $\frac{\partial H}{\partial p}\neq 0$, in which case

$$\dot{A}=\frac{\partial A}{\partial \xi^i}\omega_{ij}\frac{\partial H}{\partial \xi^j}++\frac{\partial A}{\partial \xi_i}F_i \tag{26}$$

where $\xi^i=(q^i,p_j)$, $\omega_{ij}=-\omega_{ji}$ is a convenient notation for the Poisson bracket commutator, and

$$F^i=T^{ij}\frac{\partial H}{\partial \xi^j} \tag{27}$$

with

$$T^{ij}=\begin{pmatrix}0_{n\times n} & 0_{n\times n}\\ 0_{n\times n} & s_{n\times n}\end{pmatrix} \qquad s_{ij}=\delta_{ij}(F_i/(\frac{\partial H}{\partial p_i})) \tag{28}$$

and $0_{n\times n}$ is an $n\times n$ zero matrix. In this case, the time-evolution equation (26) can be written in the form

$$\dot{A}=\frac{\partial A}{\partial q^i}\frac{\partial H}{\partial p_i}-\frac{\partial A}{\partial p_i}\frac{\partial H}{\partial q^i}+\frac{\partial A}{\partial p_i}s_{ij}\frac{\partial H}{\partial p_j}=(A,H) \tag{29}$$

It is easy to check that the product $(,)$ defined in this equation is of the Lie-admissible form defined earlier, and reduces to the conventional Lie product $[,]$ if $s_{ij}=0$.

We are now in a position to compare this Lie-admissible formulation of dissipative statistical mechanics with the previous formulation of [11], based on a non-unitary time-evolution operator $\Lambda$. We recall the form (9) of the time-evolution equation for the locally-measurable density matrix $\tilde{\rho}$ in that formulation, and the star hermiticity condition (11,12). For the type of dynamical system described by (19,20, and 29), we have

$$\Phi=i\sum_{i=1}^{n}(\frac{\partial H}{\partial q^i}\frac{\partial}{\partial p_i}-\frac{\partial H}{\partial p_i}\frac{\partial}{\partial q^i})+\sum_{i,j=1}^{n}\frac{\partial H}{\partial p_i}s_{ij}\frac{\partial}{\partial p_j} \tag{30}$$

which is consistent with star-hermiticity if and only if

$$\mathcal{F}=\mathcal{F}^{\dagger} \quad ; \quad \mathcal{F}\equiv\sum_{i,j=1}^{n}\frac{\partial H}{\partial p_i}s_{ij}\frac{\partial}{\partial p_j} \tag{31}$$

A necessary and sufficient condition that this be satisfied is that the matrix $s_{ij}$ be real and symmetric [14]. Dissipative statistical systems with this property provide examples of the $\Lambda$-transformation theory of [11]. As we shall see in subsequent sections, string theory is one such example.

## 3 Review of String Density Matrix Mechanics

In order to accommodate the mixed states inevitable in a quantum theory of gravity, one must use a density matrix formalism [7, 10]. The asymptotic $\not{S}$ matrix may not have the familiar analyticity properties, and may only exist in a distribution-theoretic sense, as we shall see later. The evolution of the density matrix over finite times is given by a modification (3) of the quantum Liouville equation [10]. The commutator term in (3) would, when integrated alone over time, give the conventional $S$ matrix. The extra term $\not{\delta}H$ leads, when integrated over time, to the non-factorization property $\not{S} \neq SS^{\dagger}$. As seen in section 2, the presence of such an extra term is characteristic of a dissipative quantum-mechanical system such as an open system interacting with an unobserved environmental reservoir [20, 21, 22]. In string density matrix mechanics this term would reflect the mixing of observable particles to unobservable quantum gravitational states, whose information is lost across microscopic event horizons. Such states are delocalized, and can be thought of as remnants of a symmetric (topological) phase of gravity [23], whose breaking might result in the emergence of the ordinary space-time.

How should one set about modelling space-time foam and exploring these possibilities in string theory? In order for the background in which it propagates have a classical space-time interpretation, the string theory must be critical, i.e. must be characterized by a conformal field theory on the world sheet with central charge $c = 26$ (15) for a bosonic (supersymmetric) string. It is well known [24] how to reproduce $S$-matrix elements via the operator product expansion for such a critical-string conformal field theory. One must look beyond this framework if one is to have any chance of locating a non-trivial contribution to the $\not{S}$ matrix. This means looking at quantum fluctuations in the space-time background, and in particular transitions between them. The appropriate space of theories to discuss these is that of two-dimensional field theories on the world-sheet, but without the restriction that these be conformal. This is the space of generalized $\sigma$-models on the world sheet that we have used to derive [3, 4] a form for $\not{\delta}H$ in the framework of non-critical string theory[1].

Such models are characterized by deformations of conformal points, described by vertex operators $V_i$ associated with background fields $g_i$ in the target space in which the string propagates. Mathematical consistency requires the restoration of criticality by turning a deformation into an exactly marginal one by Liouville dressing, thereby leading, in our interpretation, to a time-dependent background. To be more explicit, consider a two-dimensional critical conformal field theory model described by an action $S_0(r)$ on the world sheet, where the $\{r\}$ are matter fields spanning a $D$-dimensional target manifold of Euclidean signature, that we term "space". Consider, now, a deformation

$$S = S_0(r) + g \int d^2 z V_g(r) \tag{32}$$

Here $V_g$ is a (1,1) operator, i.e. its anomalous dimension vanishes, but it is not *exactly marginal* in the sense that the operator-product expansion coefficients $C_{ggg}$ of $V_g$ with itself are non-zero in any renormalization scheme, and hence 'universal' in the Wilsonian sense. The scaling dimension $\alpha_g$ of $V_g$ in the deformed theory (32) is, to $O(g)$ [25],

$$\alpha_g = -gC_{ggg} + \dots \tag{33}$$

Liouville theory [26] requires that scale invariance of the theory (32) be preserved. The non-zero scaling dimension (33) would jeopardize this, but scale invariance is restored if one dresses $V_g$ gravitationally on the world-sheet as

$$\int d^2zgV_g(r) \to \int d^2zge^{\alpha_g\phi}V_g(r) = \int d^2zgV_g(r) - \int d^2zg^2C_{ggg}V_g(r)\phi + \dots \tag{34}$$

where $\phi$ is the Liouville field. The latter acquires dynamics through integration over world-sheet covariant metrics $\gamma_{\alpha\beta}$ after conformal gauge fixing $\gamma_{\alpha\beta} = e^{\phi}\hat{\gamma}_{\alpha\beta}$ in the way discussed in [26, 27]. Scale invariance is guaranteed through the definition of renormalized couplings $g_R$ , given in terms of $g$ through the relation

$$g_R \equiv g - C_{ggg}\phi g^2 + \dots \tag{35}$$

This equation leads to the correct $\beta$-functions for $g_R$

$$\beta_g = -C_{ggg}g_R^2 + \dots \tag{36}$$

implying a renormalization-group scale $\phi$-dependence of $g_R$. The reader might have noticed that above we viewed the Liouville field as a local scale on the world-sheet [2, 4]. Local world-sheet scales have been considered in the past [28, 29], but the crucial difference in our Liouville approach is that this scale is made dynamical by being integrated out in the path-integral. In this way, in Liouville strings the local dynamical scale acquires the interpretation of an additional target coordinate. If the central charge of the matter theory is $c_m > 25$, the signature of the kinetic term of the Liouville coordinate is opposite to that of the matter fields $r$, and thus the Liouville field is interpreted as Minkowski target time [1, 30], as we discuss in more detail in the next section and in ref. [4, 16].

The target-space density matrix in such a framework is viewed as a function of coordinates $g^i$ that parametrize the couplings of the generalized $\sigma$-models on the string world-sheet, and their conjugate momenta $p_i$ : $\rho(g^i, p_i)$. The effective action functional for this dynamical system can be identified with the Zamolodchikov $C$-function [31] $C(g^i)$, whose gradient determines the rate of change of the coordinates (couplings) $g^i$ :

$$\dot{g}^i = \beta^i(g) \qquad : \qquad \frac{\delta C(g)}{\delta g^i} = G_{ij}\beta^j \tag{37}$$

where

$$C[g] = \int dt(p_i\dot{g}^i - E) \tag{38}$$

where $E$ is the Hamiltonian and $G_{ij}$ is the metric in $g$-space,

$$G_{ij}[g] = 2|z|^4 < V_i(z)V_j(0) > \tag{39}$$

with $V_i$ the associated vertex operators corresponding to the background $g^i$ and $< \ldots >$ denotes a $\sigma$-model vacuum expectation value. Here and subsequently dots denote derivatives with respect to the renormalization scale. As described above and in ref. [16] and the next section, we identify the renormalization scale with a Liouville field, which has negative metric in target space because fluctuations make the string supercritical [1], and we identify it with the target time variable. This is the main reason for considering in (38) an 'effective action' and not simply a Lagrangian in target space. This is crucial in our formalism, as a result of the integration over the dynamical Liouville scale in a $\sigma$-model path integral. In this formalism, the simple gradient flow of the off-shell corollary of the $C$-theorem [31, 29, 32] is extended to a non-trivial functional derivative

$$\frac{\delta}{\delta g^i}C[g] = -\frac{d}{dt}\frac{\partial \mathcal{L}(\},\Y,\sqcup)}{\partial \dot{g}^i} + \frac{\partial \mathcal{L}}{\partial g^i} \neq 0 \tag{40}$$

which includes a generalized non-potential force term in the evolution equation for $g^i$ [2] derived from (38). The renormalizability of the world-sheet $\sigma$-model implies

$$\frac{d}{dt}\rho[g^i, p_i, t] = 0 = \frac{\partial}{\partial t}\rho + \dot{g}^i\frac{\partial}{\partial g^i}\rho + \dot{p}_i\frac{\partial}{\partial p_i}\rho \tag{41}$$

From (40) one can then derive straightforwardly a modified Liouville equation (3) with the explicit form [2]

$$\not{\delta} H\rho = G_{ij}\beta^j\frac{\partial \rho}{\partial p_i} = -iG_{ij}[\rho, g^i]\beta^j \tag{42}$$

where the second form holds in the quantum formulation.

We note here the fundamental point that this modification of the quantum Liouville equation obeys the Lie-admissibility condition of ref. [14], because the Zamolodchikov tensor (39) is real and symmetric, and hence defines a metric in coupling constant space. This property is non-trivial, since it does not hold in a a more general renormalization scheme [29], in which the couplings $g^i$ have an arbitrary dependence on the world-sheet coordinates $g^i(\sigma, \tau)$ and not a simple local scale dependence [29]. In such a scheme there exists a local (in renormalization group space) function whose variations with respect to the couplings have off-shell relations with the $\beta$-functions given by matrices that have antisymmetric parts. More explicitly, one can prove the relation [29]:

$$\partial_i\tilde{\beta}^\Phi = \chi_{ij}\beta^j + \partial_j W_i\beta^j - \partial_i W_j\beta^j \tag{43}$$

where $\Phi$ is the dilaton background coupled to the world-sheet curvature, the $g^i$ denote the rest of the backgrounds, existing even on flat world sheets, and $\chi_{ij}$

is symmetric. It is related to divergences of the two-point functions of $V_i$, but its positivity is not evident in this approach. Finally, $\tilde{\beta}^\Phi \equiv \beta^\Phi + W_i\beta^i$, where the $W_i$ are (computable) renormalization counterterms related to total world-sheet derivative terms in the expression for the trace of the stress-tensor. Such terms are crucial for the local scale invariance of the theory. In dimensional regularization the $W_i$ are non-trivial beyond three $\sigma$-model loop order [29]. Such off-shell relations appear only if the $g^i$ are allowed to depend arbitrarily on the world-sheet coordinates, which is not the case in the framework adopted above and in ref. [28] where the $g^i$ depend only on the renormalization scale that we identify with target time [2, 4].

The more general framework incorporates off-shell generalized forces that depend on the conjugate momenta $p_i$. It could be regarded as providing an 'atlas' relating 'charts' or 'patches', in each of which the $g^i$ space is torsion-free. In our interpretation, the Universe observable within our event horizon is contained within one patch, enabling physics everywhere within it to be described by the approach to a unique conformal field theory or critical string vacuum. There may well be other patches beyond our observable horizon, in which a different string vacuum is approached, with the generalized renormalization scheme [29], including torsion (43) describing transitions between the patches. This would provide a physical realization of the rather abstract analysis in ref. [33], where it was argued that the 'classical' coupling constant space should not be simply connected, with each component corresponding to one of our patches.

This 'torsion-free' modification of the quantum Liouville equation, that applies within our patch of the Universe, has other several important properties. One is that the total probability $P = \int dp_l dg^l Tr[\rho(g^i, p_j)]$ is conserved :

$$\dot{P} = \int dp_l dg^l Tr[\frac{\partial}{\partial p_i}(G_{ij}\beta^j\rho)] \tag{44}$$

which can receive contributions only from the boundary of phase space, that must vanish for an isolated system. Secondly, energy is conserved on the average [15]. This can be seen by computing

$$\partial_t << E^n >>= n << (\partial_t E)E^{n-1} >> -i << \beta^i G_{ij}[g^j, E^n] >>=$$
$$n << (\partial_t E)E^{n-1} >> -i << \beta^i G_{ij} E[g^j, E^{n-1}]] >> + << \beta^i G_{ij}\beta^j E^{n-1} >> \tag{45}$$

where $E$ is the Hamiltonian operator, and $<< \ldots >>\equiv Tr[\rho(\ldots)]$. In arriving at this result we took into account the quantization rules in coupling constant space discussed in ref. [2],

$$[g^i, g^j] = 0 \qquad ; \qquad [g^i, p^j] = -i\delta^{ij} \tag{46}$$

as well as the fact that in string $\sigma$-models the 'quantum operators' $\beta^i G_{ij}$ are functionals of the coordinates only $g^i$ and not of the generalized momenta $p^i$. For future use we note that the total time derivative of an operator $\hat{Q}$ is given as usual by

$$\frac{d}{dt}\hat{Q} = -i[\hat{Q}, E] \tag{47}$$

We recall that total time derivatives incorporate both explicit and implicit (via running couplings) renormalization-scale dependence, whilst partial time derivatives incorporate only the explicit dependence.

For the energy conservation law we should take $n = 1$ in (45), in which case we find

$$\frac{\partial}{\partial t} << E >>= \frac{\partial}{\partial t} Tr(E\rho) =<< \partial_t (E - \beta^i G_{ij} \beta^j) >> \tag{48}$$

Using the $C$-theorem results (37,38) [31, 15] and the formalism developed in ref. [2] it is straightforward to arrive at

$$\frac{\partial}{\partial t} << E >>= \frac{\partial}{\partial t}(p_i \beta^i) = 0 \tag{49}$$

due to the renormalizability of the stringy $\sigma$-model. The latter implies that any dependence on the renormalization group scale in the $\beta^i$ functions is implicit through the renormalized couplings. Renormalizability replaces the time-translation invariance of conventional target-space field theory.

This conservation result does not generalize to quantum fluctuations in the energy $\Delta E =<< E^2 >> -(<< E >>)^2$. To get the the energy fluctuations we set $n = 2$,

$$\partial_t << E^2 >>= -i << [\beta^j, E]\beta^i G_{ij} >>=<< \frac{d\beta^j}{dt} \beta^i G_{ij} >> \tag{50}$$

which is non-zero for a non-critical string. This result implies that despite energy conservation the uncertainties in the energy $\Delta E$ depend on the (Liouville) time $t$. This point is relevant to the discussion of uncertainty in section 9 [16].

The entropy $S = -Tr(\rho ln\rho)$ is also not conserved:

$$\dot{S} = (\beta^i G_{ij} \beta^j) S \tag{51}$$

implying a monotonic increase for unitary theories for which $G_{ij}$ is positive definite. We see from (51) that *any* running of *any* coupling will lead to an increase in entropy, and we have interpreted [2] this behaviour in terms of quantum models of friction [21]. The increase (51) in the entropy corresponds to a loss of quantum coherence, which is also known in these models. Note that entropy increases within any 'torsion-free' cosmological 'patch' in coupling space: this is not in general true at the boundaries between patches, where 'torsion' may appear.

The final comment in this brief review of density matrix mechanics is that Ehrenfest's theorem continues to hold. The time evolution of the expectation value of any observable $O(g^i)$ that is a function of the coordinates alone, and not the momenta $p_i$, is given by

$$\frac{\partial}{\partial t} << O(g^i) >>= \frac{\partial}{\partial t} Tr(O(g^i)\rho) = Tr(O(g^i)\dot{\rho})$$
$$= iTr[O(g^i)[\rho, H]] + iTr[g^i, O(g^i)\beta^i G_{ij}\rho] = iTr\{O(g^i)[\rho, H]\} \quad (52)$$

as usual.

It may be helpful to bear in mind a simple two-state system, whose $2 \times 2$ density matrix can be decomposed with respect to the hermitian Pauli $\sigma$-matrix basis [10] $(\mathbf{1}, \sigma_x, \sigma_y, \sigma_z)$, $\rho = \rho_0 \mathbf{1} + \rho \cdot \sigma$. In conventional quantum mechanics with a Hamiltonian $H = \Delta E \sigma_z$, a pure initial state $\rho_{in} = \frac{1}{2}(|1> + |2>)(<1| + <2|)$ evolves unitarily :

$$\rho(t) = \frac{1}{2} \begin{pmatrix} 1 & e^{-i\Delta Et} \\ e^{i\Delta Et} & 1 \end{pmatrix} \quad (53)$$

A generic "open system" modification $\not{\delta} H$ can be expressed as a $4 \times 4$ matrix w.r.t. the coordinates $(0, \sigma_x, \sigma_y, \sigma_z)$. The probability and energy conservation derived above (44,49) tell us that

$$\not{\delta} H_{0\beta} = 0 = \not{\delta} H_{\beta 0}, \qquad \not{\delta} H_{3\beta} = 0 = \not{\delta} H_{\beta 3} \quad (54)$$

respectively. We can therefore write

$$\not{\delta} H_{\alpha\beta} = \begin{pmatrix} 0 & 0 & 0 & 0 \\ 0 & -\alpha & -\beta & 0 \\ 0 & -\beta & -\gamma & 0 \\ 0 & 0 & 0 & 0 \end{pmatrix} \quad (55)$$

where positivity imposes $\alpha, \gamma > 0, \alpha\gamma > \beta^2$. It is easy to see that with the addition of such a term

$$\rho(t) = \frac{1}{2} \begin{pmatrix} 1 & e^{-(\alpha+\gamma)t/2} e^{-i\Delta Et} \\ e^{-(\alpha+\gamma)t/2} e^{i\Delta Et} & 1 \end{pmatrix} \quad (56)$$

which becomes asymptotically a completely mixed state

$$\rho(\infty) = \frac{1}{2} \begin{pmatrix} 1 & 0 \\ 0 & 1 \end{pmatrix} \quad (57)$$

Such evolution towards a completely mixed state is the generic consequence of the monotonic increase (51) in the entropy. The completely-mixed form (57) of the density matrix corresponds to the advertized loss of coherence.

As discussed in more detail in section 8, the above formalism can be adapted, with conceptually minor modifications, to accommodate decays to the neutral $K^0 - \overline{K}^0$ system [10, 15], which is one of the best microscopic laboratories for testing quantum mechanics and its possible modification (3). The coresponding parameters $\alpha, \beta, \gamma$ (55) violate $CPT$. String theory suggests that $CPT$ violation should be considered a generic feature of our density matrix mechanics [15]. The normal field-theoretical proof of the $CPT$ theorem is based on locality, Lorentz invariance and unitarity. Clearly string theory is not local in space-time, and Lorentz invariance may be considered *a derived property of critical string theory that does not hold in our treatment of time as a renormalization scale in non-critical string theory.* Indeed, we have related $CPT$ violation to charge non-conservation on the world-sheet associated with topological fluctuations such as monopoles whose appearance drives the string supercritical [15, 4].

# 4 Time and the Two-Dimensional String Black Hole Model

As an illustration of our approach to non-critical string theory, we now discuss the two-dimensional black hole model of ref. [5]. We regard it as a toy laboratory that gives us insight into the nature of time in string theory and contributes to the physical effects mentioned in the previous section.

The action of the model is

$$S_0 = \frac{k}{2\pi} \int d^2z[\partial r \overline{\partial} r - tanh^2 r \partial t \overline{\partial} t] + \frac{1}{8\pi} \int d^2 z R^{(2)} \Phi(r) \tag{58}$$

where $r$ is a space-like coordinate and $t$ is time-like, $R^{(2)}$ is the scalar curvature, and $\Phi$ is the dilaton field. The customary interpretation of (58) is as a string model with $c = 1$ matter, represented by the $t$ field, interacting with a Liouville mode, represented by the $r$ field, which has $c < 1$ and is correspondingly space-like [1, 30]. As an illustration of the approach outlined in the previous section, however, we re-interpret (58) as a fixed point of the renormalization group flow in the local scale variable $t$. In our interpretation, the "matter" sector is defined by the spatial coordinate $r$, and has central charge $c_m = 25$ when $k = 9/4$ [5]. Thus the model (58) describes a critical string in a dilaton/graviton background. The fact that this is static, i.e. independent of $t$, reflects the fact that one is at a fixed point of the renormalization group flow [4, 16, 34].

We now outline how one can use the machinery of the renormalization group in curved space, with $t$ introduced as a local renormalization scale on the world sheet, to derive the model (58). A detailed technical description is given in [34, 4]. There are two contributions to the kinetic term for $t$ in this approach, one associated with the Jacobian of the path integration over the world-sheet metrics, and the other with fluctuations in the background metric.

To exhibit the former, we first choose the conformal gauge $\gamma_{\alpha\beta} = e^{\rho}\hat{\gamma}_{\alpha\beta}$ [26, 27], where $\rho$ represents the Liouville mode. We will later identify $\rho$ with an appropriate function of $\phi$, thereby making the local scale $\phi$ a dynamical $\sigma$-model field. Ref.[27] contains an explicit computation of the Jacobian using heat-kernel regularization, which yields

$$-\frac{1}{48\pi}[\frac{1}{2}\partial_{\alpha}\rho\partial^{\alpha}\rho + R^{(2)}\rho + \frac{\mu}{\epsilon}e^{\rho} + S'_G] \tag{59}$$

where the counterterms $S'_G$ are needed to remove the non-logarithmic divergences associated with the induced world-sheet cosmological constant term $\frac{\mu}{\epsilon}e^{\rho}$, and depend on the background fields. This procedure reproduces the critical string results of ref. [5] when one identifies the Liouville field $\rho$ with $2\alpha'\phi$. Equation (59) contains a negative (time-like) contribution to the kinetic term for the Liouville (time) field, but this is not the only such contribution, as we now show.

We recall that the renormalization of composite operators in $\sigma$-models formulated on curved world sheets is achieved by allowing an arbitrary dependence of the couplings $g^i$ on the world-sheet variables $z, \bar{z}$ [28, 29]. This induces counterterms of "tachyonic" form, which take the following form in dimensional regularization with $d = 2 - \epsilon$ [29]:

$$\int d^2 z \Lambda_0 \tag{60}$$

where

$$\Lambda_0 = \mu^{-\epsilon}(Z(g)\Lambda + Y(g)) \tag{61}$$

Here $Z(g)$ is a common wave function renormalization that maps target scalars into scalars, $\Lambda$ is a residual renormalization factor, and the remaining counterterms $Y(g)$ can be expanded as power series in $1/\epsilon$, with the the one-loop result giving a simple pole. Simple power-counting yields the following form for $Y(g)$:

$$Y(g) = \partial_{\alpha}g^i \mathcal{G}_{ij}\partial^{\alpha}g^j \tag{62}$$

where $\mathcal{G}_{ij}$ is the analogue of the Zamolodchikov metric [31] in this formalism, which is positive for unitary theories. It is related to the divergent part of the two-point function $< V_i V_j >$ [29] that cannot be absorbed in the conventional renormalization of the operators $V_i$. We need to consider a $\sigma$-model propagating in a graviton background $G_{MN}$, in which case a standard one-loop computation [29] yields the following result for the simple $\epsilon$-pole in $Y$:

$$Y^{(1)} = \frac{\lambda}{16\pi\epsilon}\partial_{\alpha}G_{MN}\partial^{\alpha}G^{MN} \tag{63}$$

where $\lambda \equiv 4\pi\alpha'$ is a loop-counting parameter. We note that the wave-function renormalization $Z(g)$ vanishes at one-loop. In ref. [29] $G_{MN}$ was allowed to depend arbitrarily on the world-sheet variables, and all world-sheet derivatives of the couplings were set to zero at the end of the calculation. In our Liouville mode interpretation, we assume that such dependence occurs only through the local scale $\mu(z,\overline{z})$, so that

$$\partial_\alpha g^i = \hat{\beta}^i \partial_\alpha \phi(z,\bar{z}) \tag{64}$$

where $\hat{\beta}^i = \epsilon g^i + \beta^i(g)$ and $\phi = ln\mu(z,\bar{z})$. Taking the $\epsilon \to 0$ limit, and separating the finite and $O(\frac{1}{\epsilon})$ terms, we obtain for the former

$$O(1)-terms: \qquad ResY^{(1)} = \alpha'^2 R \partial_\alpha \phi \partial^\alpha \phi \tag{65}$$

where $R$ is the scalar curvature in target space, and we have used the fact that the one-loop graviton $\beta$-function is

$$\beta^G_{MN} = \frac{\lambda}{2\pi} R_{MN} \tag{66}$$

The terms without logarithmic divergences,

$$\frac{1}{\epsilon}\beta^i \mathcal{G}^{(1)}_{ij} \beta^j \tag{67}$$

do not contribute to the renormalization group, and can be removed explicitly by target-space metric counterterms

$$S_G = \frac{1}{\epsilon} G_{\phi\phi} \partial_\alpha \phi \partial^\alpha \phi + \delta S(\phi, r) \tag{68}$$

where the coefficients $G_{\phi\phi}$ are fixed by the requirement of cancelling the $\frac{1}{\epsilon}$ terms (67). The $\delta S$ denotes arbitrary finite counterterms, which are invariant under the simultaneous conformal rescalings of the fiducial world-sheet metric, $\hat{\gamma} \to e^\sigma \hat{\gamma}$, and local shifts of the scale $\phi \to \phi - \sigma$. This last requirement arises as in the conventional approach to Liouville gravity [26, 27], where the local renormalization scale $\phi$ is identified with the Liouville mode $\rho$, after appropriate normalization. In our interpretation one is forced to treat the scale $\phi$ simultaneously as the target time coordinate.

In the case of the Minkowski black hole model of ref. [5], the Lorentzian curvature is

$$R = \frac{4}{cosh^2 r} = 4 - 4tanh^2 r, \tag{69}$$

which we substitute into equation (65) to obtain the form of the second contribution to the kinetic term for the Liouville field $\phi$. Combining the world-sheet metric Jacobian term in (59) with the background fluctuation term (65,69), we finally obtain the following terms in the effective action

$$\frac{1}{4\pi\alpha'} \int d^2 z [\partial_\alpha r \partial^\alpha r - tanh^2 r \partial_\alpha \phi \partial^\alpha \phi + dilaton - terms] \tag{70}$$

Thus we recover the critical string $\sigma$-model action (58) for the Minkowski black-hole. Dilaton counterterms are incorporated in a similar way, yielding the dilaton background of [5]. In addition, as standard in stringy $\sigma$-models, one also obtains the necessary counterterms that guarantee target-space diffeomorphism invariance of the Weyl-anomaly cefficients [28]. Details are given in ref. [34].

It should be noticed that the renormalization group yields automatically the Minkowski signature, due to the $c_m = 25$ value of the matter central charge [1, 30]. However, as we remarked in ref. [34, 4], one can also switch over to the Euclidean black hole model, and still maintain the identification of the compact time with some appropriate function of the Liouville scale $\phi$ that takes into account the compactness of $t$ in that case. The formalism of exactly-marginal deformations that turn on matter in the model (58) is better studied in this Euclidean version [35]. In ref. [35] it was argued that the exactly-marginal deformation that turned on a static tachyon background for the black hole of ref. [5] necessarily involved the higher-level topological string modes, that are non-propagating delocalized states, which are interrelated by an infinite-dimensional $W$ symmetry[1]. This is a consequence of the operator product expansion of the tachyon zero-mode operator $\mathcal{F}^c_{-\frac{1}{2},0}$ [35]:

$$\mathcal{F}^c_{-\frac{1}{2},0} \circ \mathcal{F}^c_{-\frac{1}{2},0} = \mathcal{F}^c_{-\frac{1}{2},0} + W^{hw}_{-1,0} + W^{lw}_{-1,0} + \dots \tag{71}$$

where we only exhibit the appropriate holomorphic part for reasons of economy of space. The $W$ operators and the ... denote level-one and higher string states. The corresponding exactly-marginal deformation, constructed by tensoring holomophiic and antiholomorphic parts, is given by [35]

$$L_0^1 \overline{L}_0^1 \propto \mathcal{F}^{c-c}_{\frac{1}{2},0,0} + i(\psi^{++} - \psi^{--}) + \dots \tag{72}$$

where the $\psi$ denote higher-string-level operators [35], and the 'tachyon' operator is given by

$$\mathcal{F}^{c-c}_{\frac{1}{2},0,0}(r) = \frac{1}{coshr} F(\frac{1}{2}, \frac{1}{2}; 1, tanh^2 r) \tag{73}$$

with

$$F(\frac{1}{2}, \frac{1}{2}, 1; tanh^2 r) \simeq \frac{1}{\Gamma^2(\frac{1}{2})} \sum_{n=0}^{\infty} \frac{(\frac{1}{2})_n (\frac{1}{2})_n}{(n!)^2} [2\psi(n+1) - 2\psi(n+\frac{1}{2}) + \\ + \; ln(1+|w|^2)](\sqrt{1+|w|^2}\,)^{-n} \tag{74}$$

There is an additional marginal deformation, dictated by the $SL(2,R)$ symmetry structure [35], which consists of topological string modes only. At large $k$, this operator rescales the black hole metric, as can be seen from its contribution to the

[1] The elevation of this symmetry to target space-time is discussed in more detail in section 7.

action of the deformed Wess-Zumino $\sigma$-model after the gauge field integration [35],

$$g L_0^2 \overline{L_0}^2 \ni \int d^2z \{ \partial r \overline{\partial} r (1 - 2g csch^2 r - 2g sech^2 r) + \partial\theta\overline{\partial}\theta (sinh^2 r + 2g - \frac{(sinh^2 r + 2g)^2}{cosh^2 r + 2g}) \} \tag{75}$$

Changing variables $cosh^2 r + 2g \to cosh^2 r$ in (75) one finds that to $O(g)$ the target space metric is rescaled by an overall constant.

The topological (higher-level) string modes cannot be detected in local scattering experiments, due to their delocalized character. From a formal field-theoretic point of view, such states cannot exist as asymptotic states to define scattering, and also cannot be integrated out in a local path-integral. An 'experimentalist' therefore sees necessarily a truncated matter theory, where the only deformation is the tachyon $\mathcal{F}^c_{-\frac{1}{2},0}$, which is a (1,1) operator in the black hole $\sigma$-model (58), but is not *exactly* marginal. This truncated theory is non-critical, and hence Liouville dressing in the sense of (34) is essential, thereby implying time-dependence of the matter background. Due to the fact that the appropriate exactly-marginal deformation associated with the tachyon in these models involves all the higher-level string states that are interrelated by $W$-symmetry, one can conclude that in this picture the ensuing non-equilibrium time-dependent backgrounds are a consequence of information carried off by the unobserved topological string modes. These states are delocalized modes with definite (target-space) energies and momenta, so a low-energy scattering process involving propagating string degrees of freedom will not have any observable energy violations. This is an apparent physical explanation in this case for the general result (49) of energy conservation on the average in density matrix mechanics. The rôle of the space-time singularity[2] was crucial for this argument. Indeed, in flat target-space matrix models [37] the tachyon zero-mode operator $\mathcal{F}$ is exactly marginal. As we shall argue later on, these flat models can be regarded as an asymptotic ultraviolet limit in time of the Wess-Zumino black hole. Hence, any time-dependence of the matter disappears in the vacuum, leading to equilibrium.

The above 'truncation' procedure can be compared with the $\Lambda$ transformation theory of Misra and Prigogine [11], and its Lie-admissible formulation [13], discussed in section 2. It has been argued in ref. [38] that the 'topological' modes can in principle be observed by either Aharonov-Bohm (global) scattering experiments, in four-dimensional string theories, or via selection rules characterizing tachyon scattering off black-hole backgrounds in the (effective) two-dimensional

[2]We would like to stress that the notion of 'singularity' is clearly a low-energy effective-theory concept. The existence of infinite-dimensional stringy symmetries associated with higher-level string states ($W_\infty$-symmetries [36]) 'smooth out' the singularity, and render the full string theory finite.

($s$-wave four-dimensional) string theory, in an analogous fashion to fermion scattering on a monopole in the Callan-Rubakov effect [39]. This, as well as the coherence-maintaining property of the associated $W_\infty$-algebras that characterize the two-dimensional string theory [36], imply that no information is lost in the $\Lambda$ transformation and the associated $\Lambda$ operator is therefore invertible, as we discuss in more detail in section 7. The Lie-admissible structure then is evident from the work of [14] and the reality of the Zamolodchikov metric tensor $G_{ij}$ (39). The modified Liouville equation of string density matrix mechanics (42) is to be compared to the (quantum version) of the generalized Liouville equation (21) or the equivalent Misra-Prigogine form (30). The $C$-theorem of Zamolodchikov [31], as extended in section 3 to incorporate Liouville strings, provides us with a natural Lyapounov function in the coupling constant (background field) phase space, as discussed in sections 2 and 3.

Having such a Lyapounov function, it is natural to enquire into the irreversibility in target time of the effective theory of locally-measurable observables. To this end, we recall that in Liouville theory a correlation function of $(1,1)$ matter deformations $V_i$ is given by [40]

$$< V_{i_1} \dots V_{i_n} >_\mu = \Gamma(-s) < V_{i_1} \dots V_{i_n} >_{\mu=0} \tag{76}$$

where $s$ is the sum of the appropriate Liouville energies, and $< \dots >_\mu$ denotes a $\sigma$-model average in the presence of an appropriate cosmological constant $\mu$ deformation on the world-sheet[3]. The important point for our discussion is the $\Gamma$-factor $\Gamma(-s)$. For the interesting case of matter scattering off a two-dimensional ($s$-wave four-dimensional) string black hole, the latter is excited to a 'massive' (topological) string state [38] corresponding to a positive integer value for $s = n^+ \in \mathbf{Z}^+$ [4]. In this case, the expression (76) needs regularization. By employing the 'fixed area constraint' [26] one can use an integral representation for $\Gamma(-s)$

$$\Gamma(-s) = \int dA e^{-A} A^{-s-1} \tag{77}$$

where $A$ is the covariant area of the world-sheet. In the case $s = n^+ \in \mathbf{Z}^+$ one can then employ a regularization by analytic continuation, replacing (77) by a contour integral as shown in fig. 1 [42, 4, 3]. This is a well-known method of regularization in conventional field theory, where integrals of form similar to (77) appear in terms of Feynman parameters. We note that it is the same regularization which was also used to prove the equivalence of the Bogolubov-Parasiuk-Hepp-Zimmerman renormalization prescription to the dimensional regularization of 't Hooft [43]. One result of such an analytic continuation is the appearance of imaginary parts in the respective correlation functions, which in our case are interpreted [42, 4, 3] as renormalization group instabilities of the system.

[3] In the case of a black-hole coset model this operator is a 'modified cosmological constant' involving some mixing with appropriate ghost fields parametrising the $SL(2,R)$ string [41].

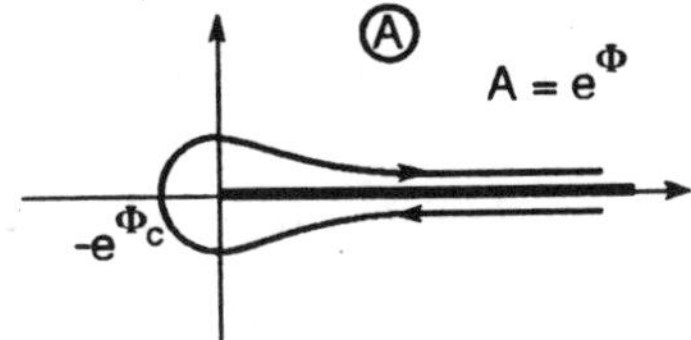


Figure 1: - Contour of integration in the analytically-continued (regularized) version of $\Gamma(-s)$ for $s \in Z^+$. This is known in the literature as the Saalschutz contour, and has been used in conventional quantum field theory to relate dimensional regularization to the Bogoliubov-Parasiuk-Hepp-Zimmermann renormalization method.

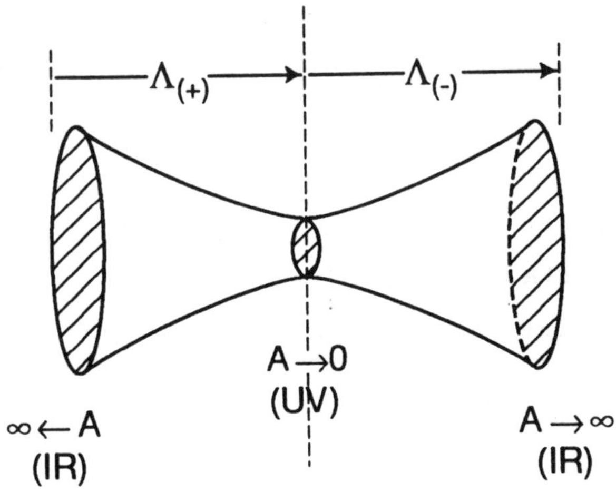


Figure 2: - Schematic repesentation of the evolution of the world-sheet area as the renormalization group scale moves along the contour of fig. 1.

Interpreting the latter as an actual time flow, we then interpret the contour of fig. 1 as implying evolution of the world-sheet area in both (negative and positive) directions of time (c.f. fig. 2), i.e.

$$Infrared\ fixed\ point \rightarrow Ultraviolet\ fixed\ point \rightarrow Infrared\ fixed\ point \quad (78)$$

In each half of the world-sheet diagram of fig. 2, the Zamolodchikov $C$-theorem tells us that we have an irreversible Markov process. According to the analysis of section 2, the physical system will be time-irreversible if the associated $\Lambda$ transformations are not equivalent. It has been argued in ref. [23] that a highly-symmetrical phase of the two-dimensional black hole occurs at the infrared fixed point of the world-sheet renormalization group flow. At that point, the associated $\sigma$-model is a topological theory constructed by twisting [5, 44] an appropriate $N = 2$ supersymmetric black-hole $SL(2, R)$ Wess-Zumino $\sigma$-model. The singularity of a stringy black hole, then, describes a topological degree of freedom. The highly-symmetric phase is interpreted as the state with the most 'appropriate' initial conditions, whose preparation requires finite entropy. This in turn implies a 'bounce' interpretation of the renormalization group flow of fig. 2, in which the infrared fixed point is a 'bounce' point, similar to the corresponding picture in point-like field theory [45]. Thus, the "physical" flow of time is taken to be *opposite* to the conventional renormalization group flow, i.e. from the infrared to the ultraviolet fixed point on the world sheet. In the next section we shall see this explicitly, by using world-sheet instanton calculus to represent, at least qualitatively, the renormalization group flow of the effective target-space theory, providing a concise expression for the effects of the topological modes that are linked to the tachyon modes by $W$ symmetries.

# 5 The String Black Hole Model and its World-Sheet Instantons

In this section we review some calculations we have made of specific contributions to the $\$$ matrix due to topologically non-trivial world-sheet configurations. The action of $SL(2, R)/U(1)$ coset Wess-Zumino model [5] describing a Euclidean black hole can be written in the form

$$S = \frac{k}{4\pi}\int d^2z \frac{1}{1+|w|^2}\partial_\mu \bar{w}\partial^\mu w + \dots \quad (79)$$

where the conventional radial and angular coordinates $(r, \theta)$ are given by $w = sinhre^{-i\theta}$ and the target space $(r, \theta)$ line element is

$$ds^2 = \frac{dwd\overline{w}}{1+w\overline{w}} = dr^2 + tanh^2 r d\theta^2 \quad (80)$$

The Euclidean black hole can be written as a vortex-antivortex pair [17], which is a solution of the following Green function equations on a spherical world sheet:

$$\partial_z \partial_{\bar{z}} X_v = i\pi \frac{q_v}{2}[\delta(z - z_1) - \delta(z - z_2)] \tag{81}$$

The world-sheet can also accommodate monopole-antimonopole pairs [17], which are solutions of:

$$\partial_z \partial_{\bar{z}} X_m = -\frac{q_m \pi}{2}[\delta(z - z_1) - \delta(z - z_2)] \tag{82}$$

These are related to Minkowski black holes with masses $\propto q_m$. Vortex and monopole configurations can both be regarded as sine-Gordon deformations of the effective action for the field $X \equiv \beta^{-\frac{1}{2}}\tilde{X}$, where $\beta^{-1}$ is an effective 'pseudo-temperature': $\beta = \frac{3}{\pi(C-25)}$ in Liouville theory. The partition function [46]

$$\begin{aligned} Z &= \int D\tilde{X} exp(-\beta S_{eff}(\tilde{X})) \\ \beta S_{eff} &= \int d^2z[2\partial\tilde{X}\bar{\partial}\tilde{X} + \frac{1}{4\pi}[\gamma_v \omega^{\frac{\alpha}{2}-2}(2\sqrt{|g(z)|})^{1-\frac{\alpha}{4}} : cos(\sqrt{2\pi\alpha}[\tilde{X}(z) + \tilde{X}(\bar{z})]) : \\ &+ (\gamma_v, \alpha, \tilde{X}(z) + \tilde{X}(\bar{z})) \rightarrow (\gamma_m, \alpha', \tilde{X}(z) - \tilde{X}(\bar{z}))]] \end{aligned} \tag{83}$$

requires for its specification an angular ultraviolet cut-off $\omega$ on the world-sheet. Here $\gamma_{v,m}$ are the fugacities for vortices and spikes respectively, and $\frac{\alpha}{4}$ is the conformal dimension $\Delta$. This deformed sine-Gordon theory has a low-temperature phase in which monopole-antimonopole pairs are bound in dipoles as irrelevant deformations with the conformal dimension

$$\Delta_m = \frac{\alpha_m}{4} = \frac{\pi\beta}{2} q_m^2 > 1 \tag{84}$$

Monopole-antimonopole pairs correspond to the creation and anihilation of a microscopic black hole in the space-time foam.

As shown in ref. [18], the $SL(2,R)/U(1)$ Wess-Zumino coset model describing a Euclidean black hole also has instantons given by the holomorphic function

$$w(z) = \frac{\rho}{z - z_0} \tag{85}$$

with topological charge

$$Q = \frac{1}{\pi} \int d^2z \frac{1}{1 + |w|^2}[\bar{\partial}\bar{w}\partial w - h.c.] = -2ln(a) + \ const \tag{86}$$

where $a$ is an ultraviolet cut-off discussed later. The instanton action on the world-sheet also depends logarithmically on the ultraviolet cut-off. As in the case of the more familiar vortex configuration in the Kosterlitz-Thouless model, this logarithmic divergence does not prevent the instanton from having important dynamical effects. The instanton-anti-instanton vertices take the form [18]

$$V_{I\bar{I}} \propto -\frac{d}{2\pi}\int d^2z \frac{d^2\rho}{|\rho|^4} e^{-S_0} (e^{(\frac{k[\rho\partial\overline{w}+h.c.+\ldots]}{f(|w|)}} + e^{(\frac{k[\rho\partial w+h.c.+\ldots]}{f(|w|)}}) \tag{87}$$

introducing a new term into the effective action. Making a derivative expansion of the instanton vertex and taking the large-$k$ limit, i.e. restricting our attention to instanton sizes $\rho \simeq a$, this new term has the same form as the kinetic term in (79), and hence corresponds to a renormalization of the effective level parameter in the large $k$ limit:

$$k \to k - 2\pi k^2 d' \quad : \quad d' \equiv d \int \frac{d|\rho|}{|\rho|^3} \frac{a^2}{[(\rho/a)^2+1]^{\frac{k}{2}}} \tag{88}$$

If other perturbations are ignored, the instantons are irrelevant deformations and conformal invariance is maintained. However, in the presence of "tachyon" deformations, $T_0 \int d^2z \mathcal{F}^{c,c}_{-\frac{1}{2},0,0}$ in the $SL(2,R)$ notation of ref. [35], there are extra logarithmic infinities in the shift (88), that are visible in the dilute gas and weak-"tachyon"-field approximations. In this case, there is a contribution to the effective action of the form

$$T_0 \int d^2z d^2z' < \mathcal{F}^{c,c}_{-\frac{1}{2},0,0}(z,\bar{z}) V_{I\bar{I}}(z',\bar{z}') > \tag{89}$$

Using the explicit form of the "tachyon" vertex $\mathcal{F}$ (73,74) given by $SL(2,R)$ symmetry [35], it is straightforward to isolate a logarithmically-infinite contribution to the kinetic term in (79), associated with infrared infinities on the world-sheet expressible in terms of the world-sheet area $\Omega/a^2$ [4, 34],

$$\begin{aligned} gT_0 \int d^2z' \int \frac{d\rho}{\rho} (\frac{a^2}{a^2+\rho^2})^{\frac{k}{2}} \int d^2z \frac{1}{|z-z'|^2} \frac{1}{1+|w|^2} \partial_{z'} w(z') \partial_{\bar{z}'} \overline{w}(z') + \ldots \\ \propto gT_0 ln \frac{\Omega}{a^2} \int d^2z' \frac{1}{1+|w|^2} \partial_{z'} w(z') \partial_{\bar{z}'} \overline{w}(z') \end{aligned} \tag{90}$$

Such covariant-scale-dependent contributions can be attributed to Liouville field dynamics, through the "fixed-area constraint" in the Liouville path integral [26, 47]. The zero-mode part can be absorbed in a scale-dependent shift of $k$[4], which for large $k >> 1$ may be assumed to exponentiate:

$$k_R \propto (\frac{\Omega}{a^2})^{(const).\beta^I T_0} \tag{91}$$

where $\beta^I$ is the instanton $\beta$-function [18]. In ref. [4] we gave general arguments and verified to lowest order that instantons represent massive mode effects, enabling us to identify $\beta^I = -\beta^T$, where $\beta^T$ is the renormalization-group $\beta$-function of a matter deformation of the black hole[4]. Notice that in (91) both the infrared and

[4]Notice that this implies that the matter $\beta$-function has to be computed in a non-perturbative way, which is consistent with the exact conformal field theory analysis of ref. [35].

the ultraviolet cut-off scales enter. In the following we shall not distinguish between infrared and ultraviolet cut-offs. The physical scale of the system, which varies along a renormalization group trajectory, is the dimensionless ratio of the two, which is identified with the Liouville field.

The change in $k$ and the associated change in the central charge $c = \frac{3k}{k-2} - 1$ and the black-hole mass $M_{bh} \propto (k-2)^{-\frac{1}{2}}$ do not conflict with any general theorems. An analogous instanton renormalization of $\theta$ (c.f. $k$) has been demonstrated [48] in related $\sigma$-models that describe the Integer Quantum Hall Effect (IQHE), discussed further in section 7. Instanton renormalization of $k$ can also be seen in the Minkowski black hole model of ref. [5], defined on a non-compact manifold $SL(2,R)/O(1,1)$[5]. In our case, as we have seen, the instantons reflect a shift of the central charge between the matter and background sectors of a combined matter + black hole theory, in which the total central charge is unchanged. They correspond to a combination of world-sheet deformation operators in the Wess-Zumino model [5]: the exactly marginal operator $L_0^2\overline{L}_0^2$ and the irrelevant part of the exactly-marginal deformation $L_0^1\overline{L}_0^1$, which involves an infinite sum of massive string operators [35], as we saw in section 4 [see equations (72),(75)]. The fact that the $L_0^2\overline{L}_0^2$ operator rescales the target-space metric by an overall constant, implies that such perturbations have the same effect as the instanton. Thus the instanton represents the effects of massive string modes that are related to each other and to massless excitations by a $W$ symmetry. Matrix elements of the full exactly marginal light matter + instanton operator have no dependence on the ultraviolet cut-off $a$, but the separate matter and instanton parts do depend on $a$, as we have seen above.

Since instantons rescale the target-space metric and the black hole mass, they may also be used to represent black hole decay. This is higher-genus effect in string theory [50], so one should expect that instantons could reflect the contributions of higher genera. This expectation is indeed supported by an explicit computation of instanton effects in a dilute-gas approximation in the presence of dilatons. It is well known [51] in $\sigma$-model perturbation theory theory that modular infinities of resummed world-sheet surfaces may be absorbed as an effective renormalization of lower-genus Riemann surfaces, above and beyond the local renormalization effects at fixed genus. For example, modular infinities for the sphere and torus may be summarized by a logarithmically-divergent contribution to the dilaton background $\Phi$ [52]

$$\Phi = \Phi_R + (\frac{d-26}{6} + \frac{d+2}{4}) ln|\Lambda/a| + O[\frac{1}{a^2}] \tag{92}$$

[5]The $\sigma$-model action of such a theory contains [49], in addition to the action (79), a total-derivative $\theta$-term which can be thought of as a deformation of the black hole by an "antisymmetric tensor" background, which in two dimensions is a discrete mode as a result of the abelian gauge symmetry. Its Euclideanized version has also instanton solutions of the form (85), but with *finite* action, which induce "Liouville"-time-dependent shifts to $k$, prior to matter couplings.

for a string propagating in a $d$-dimensional target space. The first term $(d-26)/6$ in the coefficient of the logarithmic divergence is the world-sheet sphere contribution to the conformal anomaly, whilst the second term is due to the torus. It has the effect of increasing the effective central charge, driving a critical model superciritical, where it is subject to the renormalization group instability discussed in section 4. There is a similar instanton effect in the string black hole model (79) when one considers the dilaton term, which takes the form

$$exp(\frac{1}{8\pi}\int d^2zR^{(2)}[ln(1+|w|^2)+\ldots] \tag{93}$$

where $R^{(2)}$ is the world sheet curvature. It is convenient for technical reasons to use conformal invariance to concentrate the world-sheet curvature at a point $z^*$

$$R^{(2)}=4\pi\chi\delta^{(2)}(z-z^*) \tag{94}$$

where $\chi=2$ for a world sheet with spherical topology. Taking the limit $z^*\to\infty$, and integrating over the location $z_0$ of the instanton one finds the following contribution to the partition function

$$\int\frac{d^2\rho}{\rho^4}(\frac{a^2}{\rho^2})^{\frac{k}{2}}(\frac{\rho^2}{\Lambda^2}) \tag{95}$$

where $\Lambda$ is the infrared cutoff. Since the leading instanton contributions to the path-integral come from $\rho=O(a)$, the ultraviolet cut-off, we find that instantons contribute

$$\int d^2zR^{(2)}ln|\Lambda/a| \tag{96}$$

to the world-sheet effective action. This reproduces the above mention higher-genus effects (92). Such shifts in the dilaton are essential for the consistency of the full string theory within the interpretation of target time as a local renormalization scale.

We conclude this section by discussing the rôle of instantons in the renormalization group flow, in particular to justify the bounce picture discussed in section 4. Close to the infrared fixed point, the system is believed to be topological *both* on the world sheet and in target space-time. A topological $\sigma$-model is described by an appropriate twist of the $N=2$ supersymmetric $SL(2,R)/U(1)$ Wess-Zumino model, under which the fermions of the $N=2$ model become ghosts [5, 44, 23]. The topological model possesses an enhanced symmetry which includes a bosonic $W_\infty\otimes W_\infty$ symmetry. The breaking of $W_\infty\otimes W_\infty$ down to a single $W_\infty$ generates space-time dynamically [23]. This breaking of twisted supersymmetry (or topological BRST symmetry) arises from instanton effects [49], associated with the appearance in the presence of such configurations of logarithmic divergences in correlation functions, whose form has been computed in the dilute-gas approximation. It is essential for the discussion of this effect to include a constant $u$ in the definition of the instanton field [49]

$$w=u+\frac{\rho}{z-z_0} \tag{97}$$

The rôle of the field $u$ is similar to that of the Higgs field in supersymmetric models [53] where the instantons break supersymmetry dynamically. In that case the field $u$ labels vacua of the theory. In the limit where the infrared cutoff $\Lambda$ is large: $\Lambda/a >> u$, in which case the relevant correlator has a double-logarithmic divergence [49]:

$$< O(x_1)O(x_2) >= -8\pi^2 g^I ln\frac{|x_1 - x_2|}{a} ln(\Lambda|x_1 - x_2|) + \ldots \tag{98}$$

where the ... denote subleading terms in the infrared limit. The dependence of the correlator (98) on the distance between points on the world sheet indicates that the world-sheet topological symmetry is broken, with the appearance of a metric. On the other hand, the same correlator vanishes in the ultraviolet limit $\Lambda \simeq a$ and the world-sheet topological symmetry is restored. The ultraviolet fixed point of the flow is a stable conformally-invariant background for the $c = 1$ string, which can be regarded as an appropriate mixing of the $SL(2, R)/U(1)$ coset with ghost fields [54, 55]. In our picture, instantons provide a qualitative description of this mixing.

This picture is supported by a computation of the vacuum energy associated with instanton-anti-instanton configurations in the toy topological model described above. Identifying thevacuum energy with the one-point function of the anti-instanton vertex in an instanton background:

$$E^{I\bar{I}}_{vac} =< V_{\bar{I}} >_I= -g^{\bar{I}}\partial_{x_1} < O(x_1)O(x_2) > |_{x_1 \to x_2} \tag{99}$$

and recalling that the dominant anti-instanton configurations have sizes $\rho \simeq a$, we can use (98) to estimate that in the infrared limit when $\Lambda/a >> u$

$$E_{vac} = 16\pi^2 g^I g^{\bar{I}} \frac{V^{(2)}}{a^2}[ln(\Lambda/a) + O(1)] \tag{100}$$

where $V^{(2)}$ is the world-sheet volume. This logarithmic divergence will be removed in the full theory when all topological string modes are taken into account, but the vacuum energy of the effective theory of locally-measurable observables will be non-zero. On the other hand, the limit $u \to \infty$ yields zero vacuum energy [49], as expected for a theory with unbroken BRST symmetry. According to our previous discussion this limit coincides with the critical $c = 1$ string. Indeed, it has been argued in [55] that the $c = 1$ string theory could be topological. In this case, the topological nature pertains *only* to the world-sheet, the target space theory being described by a flat spacetime, over which the 'tachyon' matter field propagates [55][6]. Thus the $c = 1$ model should be regarded as the stable ground state, to which the false string vacua with broken BRST symmetry flow, justifying the bounce interpretation given in section 4, according to which temporal flow is opposite to the renormalization flow, i.e., from the infrared fixed point to the ultraviolet.

[6]The topological world-sheet character allows for a formal resummation of the world-sheet genera in the $c = 1$ string case [55], and should be contrasted to the situation at the infrared fixed point, where both the world-sheet and the target space theories are argued to be topological [23].

# 6 Valley Contributions to the $\not{S}$ Matrix

We now consider the contributions of monopoles and instantons to $\not{S}$ matrix elements giving transitions between a generic initial-state density matrix $\rho_B^{A(in)}$ and final-state density matrix $\rho_D^{C(out)}$. This is described by an absorptive part of a world-sheet correlation function

$$
\begin{aligned}
&\sum_{X_{out}} {}_{in}< A|D,X >_{out} \; {}_{out}< X,C|B >_{in} = \\
= &\sum_{X_{out}} {}_{in}< 0|T(\phi(z_A)\phi(z_D))|X >_{out} \; {}_{out}< X|\overline{T}(\phi(z_C)\phi(z_B))|0 >_{in} = \\
= &\; {}_{in}< 0|T(\phi(z_A)\phi(z_D))\overline{T}(\phi(z_C)\phi(z_B)|0 >_{in} \qquad (101)
\end{aligned}
$$

Here we have used the optical theorem [56] on the world sheet, which is valid because conventional quantum field theory, and indeed quantum mechanics, remain valid on the world sheet, to replace the sum over unseen states $X$ by unity. Next, we use dilute-gas approximations to estimate the leading monopole-antimonopole and instanton-anti-instanton contributions to the absorptive part (101). We expect these to be dominated [57] by valley configurations in the Euclidean functional integral, so that in a semi-classical approximation

$$\not{S} \propto Abs \int D\phi_c exp(-S_v(\phi_c))F_{kin} \qquad (102)$$

where the integral is over the collective coordinates $\phi_c$ of the valley, whose action is $S_v(\phi_c)$. The function $F_{kin}$ depends on kinematic factors, taking generically the form

$$F_{kin} = exp(E\Delta R) \qquad (103)$$

in the case of a four-point function for large $E\Delta R$, where $E$ is the centre-of-mass energy and $\Delta R$ is the valley separation parameter. This enables us to make a saddle-point approximation to the integral (102), which we then continue back to Minkowski space.

Valley trajectories $\psi_v$ have a homotopic parameter $\mu$ and obey an equation of the form

$$\frac{\partial S_0}{\partial \psi}|_{\psi=\psi_v} = W_\psi(\mu)\frac{\partial \psi_v}{\partial \mu} \qquad (104)$$

where $W_\psi(\mu)$ is a weight function that is positive definite and decays rapidly at large distances [58, 57]. We adapt techniques used in the $O(3)$ $\sigma$-model [59, 60] to find the monopole-antimonopole and instanton-anti-instanton valleys in a reduced version of the $SL(2,R)/U(1)$ model. The separations of the topological defects and anti-defects are well-defined in the presence of conformal symmetry breaking, which is provided in our case by the dilaton field [4]. Valleys can be found by using the analogy [59] between $\mu$ and a 'time' variable for defect-anti-defect scattering. We do not discuss here the details of their construction, but record the results.

The monopole-antimonopole valley function, expressed in terms of the original world-sheet variables, reads

$$w(z,\bar{z}) = \frac{(v-1/v)\bar{z}}{1+|z|^2} \tag{105}$$

where $v$ denotes the separation in the $\sigma$-model framework. Eq. (105) represents a concentric valley, which can then be mapped into an ordinary valley by applying appropriate conformal transformations. The function (105) interpolates between a far-separated monopole-antimonopole pair ($v \to \infty$) and the trivial vacuum ($v = 1$). For large but finite separations the corresponding valley action leads to the action of a monopole-antimonopole pair interacting via dipole interactions. The action of the monopole-antimonopole valley depends on the angular ultraviolet cut-off $w$ introduced in section 5:

$$S_m = 8\pi q^2 ln(2)\sqrt{2}e^{\gamma} + 2\pi q^2 ln\frac{2R}{\omega} + 2\pi q^2 ln[\frac{|z_1 - z_2|}{(4R^2+|z_1|^2)^{\frac{1}{2}}}\frac{4}{(4R^2+|z_2|^2)^{\frac{1}{2}}}] \tag{106}$$

for a monopole and antimonopole pair of equal and opposite charges $q$, which we treat as a collective coordinate over which we must integrate, where $\gamma$ is Euler's constant, the second term in (106) is a logarithmically-divergent self-energy term on a spherical world sheet of radius $R$, and the last term in (106) is a dipole interaction energy. For finite separations $0 < |z_1 - z_2| < \infty$ and very small world-sheets $R = O(a \to 0)$, the action (106) yields

$$S_m = 2\pi q^2 ln\frac{a}{\omega} + finite\ parts \tag{107}$$

where the ultraviolet cut-off dependence is apparent.

To construct the instanton valley, we notice that in the reduced model used for the construction of the monopole valley (105) the solution for an instanton-anti-instanton pair is derived from the corresponding monopole case via a conformal transformation in the $(\mu, ln|z|)$-plane. In ref. [4] we give arguments why this construction is true for finite separations as well, thereby leading to an expression of the instanton valley as an (approximate) conformal transform of the monopole valley (105). The action of the instanton-anti-instanton valley in the large-separation limit of the dilute-gas approximation is

$$S_{I\bar{I}} = kln(1+|\rho|^2/a^2) + O(\frac{\rho\bar{\rho}}{(\Delta R)^2)}) \tag{108}$$

where $\Delta R$ is the separation of an instanton of size $\rho$ and an anti-instanton of size $\bar{\rho}$, and we find a dependence on the ultraviolet cut-off $a$. The actions (107,108) substituted into the general expression (102) make non-trivial contributions to the $\not{S}$ matrix that do not factorize as a product of $S$ and $S^\dagger$ matrix elements, as we shall now see.

In the dilute-gas approximation introduced in section 5, the topologically trivial zero monopole-antimonopole, zero instanton-anti-instanton sector in the unitary sum in (101) provides the usual $S$-matrix description of scattering in a fixed background, with no back reaction of the light matter on the metric. This result is well-known in the conformal field theory approach to critical string theory, and is discussed explicitly in the present context in section 6 of ref. [4]. This $S$-matrix contribution corresponds to the usual Hamiltonian description of quantum mechanics, via the representation $S = 1 + iT : T = \int_{-\infty}^{\infty} dt H(t)$. Any topologically non-trivial contribution to the unitarity sum in (101) goes beyond the usual treatment of conformal field theory in critical strings, and makes a contribution to the non-factorization of the $\not{S}$-matrix : $\not{S} = SS^{\dagger} + \ldots$. Two such contributions that we have identified above come from the monopole-antimonopole and instanton-anti-instanton sectors discussed above, which we expect to be dominated by the valley actions (106) and (108) respectively.

The dependence of the monopole-antimonopole valley action (106) on the ultraviolet cutoff $\omega$, which we identify with the target-space time $t = -ln\omega$, and of the instanton-anti-instanton valley action (108) on the local scale-dependent level parameter $k$ (88, 91) where $t = -lna$, tell us that both valleys contribute to the non-Hamiltonian term in the modified quantum Liouville equation (3), that are proportional to the anomalous dimensions $(\Delta_m - 1)$ (84) of an irrelevant dipole-like monopole-antimonopole pair and $\gamma_0$ (91) of a matter deformation respectively. Integrating up the corresponding modified quantum Liouville equation (3), we find that a generic $\not{S}$ matrix element, defined at finite time $t$ by $\rho(t) = \not{S}(t)\ \rho(0)$, contains a factor

$$\not{S} \simeq e^{-2(\Delta_m - 1)t + \ldots} \tag{109}$$

associated with an irrelevant dipole-like monopole-antimonopole pair, and

$$\not{S} \simeq e^{-2\gamma_0 t + \ldots} \tag{110}$$

from the instanton-anti-instanton valley. Both of these time-dependences apply in limits of far-separated defect and anti-defect.

Before discussing the rôle of topologically non-trivial world-sheet configurations in the suppression of coherence at large times, we review a similar phenomenon in Hall conductors, namely the suppression of spatial correlations by "de-phasons" [61]. As we discuss in more detail in section 7, the two-dimensional black-hole model is analogous to a fractional Hall conductor [17], with the Wess-Zumino level parameter $k$ corresponding to the transverse conductivity. Hall systems generally are described by appropriate $\sigma$-models with Wess-Zumino $\theta$-terms, defined on the two-dimensional space of electron motion [61]. The fields of such $\sigma$-models, which are space-time coordinates in the black-hole case, correspond to electrons propagating in the plane, with the transverse and longitudinal conductivities $\sigma_{\mu\nu}$ corresponding to background fields in the black-hole case. The Wess-Zumino terms are associated with instantons that renormalize non-perturbatively these conductivities[48] :

$$\beta_{\mu\nu} = \frac{d\sigma_{\mu\nu}}{dlnL} \neq 0 \tag{111}$$

where $L$ is an infrared cut-off on the instanton size that serves as a renormalization group scale[48]. The effects of these instantons are seen clearly in the case of a $\sigma$-model defined on a compact manifold $U(m+n)/U(m)\otimes U(n)$, where $m$,$n$ are electron field replicas with the physical case corresponding to $m$ and $n \to 0$. This limit is equivalent to the corresponding limit of the non-compact models $U(m,n)/U(m) \otimes U(n)$, where the corresponding instantons have infinite action when $m,n \neq 0$, and might naively be thought unimportant. However, the physical limit of $m$ and $n \to 0$ is sensitive to instanton effects. Another example of the importance of infinite-action topological solutions is provided by three-dimensional anyonic Chern-Simons theories [62], which are relevant to the fractional quantum Hall effect (FQHE).

We believe that localization in Hall systems is directly related to our problem of quantum coherence. In the IQHE model of ref. [61], impurities are responsible for the localization of the electron wave function in the plane. The localization is achieved formally by representing collectively the effects of impurities on elelctron trajectories via extended, static scattering centres termed "de-phasons", which trap the electron waves into localized states with sizes $O(1/\sqrt{\rho})$, where $\rho$ is the de-phason density. As a result, the electron correlation functions are suppressed at large spatial separations:

$$\propto exp[-(x-y)^2\rho] \tag{112}$$

at zero magnetic field ($\theta = 0$). As the magnetic field is varied so that the transverse conductivity becomes a half-integer (in units of $e^2/h$), corresponding to a discrete value of the instanton angle $\theta = \pi$, the property of the de-phasons to destroy phase coherence between the advanced and the retarded electron propagators is lost. Quantitatively [61], the expectation value of an electron loop that encircles a de-phason, in the presence of a magnetic field, is proportional to $e^{-(x-y)^2\rho cos\frac{\theta}{2}}$. Thus, for $\theta = \pi$ the "effective de-phason density " $\rho cos(\theta/2)$ vanishes, and the electrons delocalize implying a non-zero longitudinal conductivity. This delocalization property is responsible for the transition between two adjacent plateaux of the transverse conductivity in the Hall conductivity diagram [61]. These ideas can be extended to the FQHE [63] via the three-dimensional anyonic Chern-Simons theories mentioned above, which are closer to our black-hole interests[7]. In our case, the massive modes of the $SL(2,R)/U(1)$ black-hole model are the analogues of the de-phasons. As discussed earlier in this section, the instantons in this model renormalize the Wess-Zumino level-parameter $k$ (c.f. $\theta$), changing the mass and size of the black hole. The delocalized phase at $\theta = \pi$ may be identified with the "topological" phase at

[7]In fact, it appears to be the zero-field Hall effect that describes physics at the space-time singularity [17].

the space-time singularity [23], which is an infrared fixed point. The propagating "tachyon" mixes in this limit, as we have discussed above, with the delocalized topological modes of the string that are analogous to the de-phasons. The localization properties are consistent with shrinking of the world-sheet as one approaches the ultraviolet fixed point that corresponds to a flat target space-time where the tachyons are normal localized fields that do not mix with topological modes.

Our formalism for the time evolution of the density matrix is analogous to the Drude model of quantum friction [20, 21], with the massive string modes playing the rôles of 'environmental oscillators'. In the language of world-sheet $\sigma$-model couplings $\{g\}$, the reduced density matrix of the observable states is given, relative to that evaluated in conventional Schrödinger quantum mechanics, by an expression of the form

$$\rho(g,g',t)/\rho_S(g,g',t) \simeq e^{-\eta\int_0^t d\tau \int_{\tau'\simeq\tau} d\tau' \beta^i G_{ij}\beta^j} \simeq e^{-Dt(\mathrm{g}-\mathrm{g}')^2+\dots} \tag{113}$$

where $\eta$ is a calculable proportionality coefficient, and $G_{ij}$ is the Zamolodchikov metric [31] in the space of couplings. In string theory, the identification of the target-space action with the Zamolodchikov $C$-function $C(\{g\})$ [31] enables the Drude exponent to be written in the form

$$\beta^i G_{ij}\beta^j = \partial_t C(\{g\}) \tag{114}$$

which also determines the rate of increase of entropy

$$\dot{S} = \beta^i G_{ij}\beta^j S \tag{115}$$

In the string analogue (114) of the Drude model (113) the rôle of the coordinates in (real) space is played by the $\sigma$-model couplings $g^i$ that are target-space background fields. Relevant for us is the tachyon field $T(X)$, leading us to interpret $(g-g')^2$ in (113) as [4]

$$(g-g')^2 = (T-T')^2 \simeq (\nabla T)^2(X-X')^2 \tag{116}$$

for small target separations $(X, X')$. Equation (116) substituted into (113) gives us a suppression very similar to the IQHE case (112).

The effect of the time-dependences (109,110, 113,116) is to suppress off-diagonal elements in the target configuration space representation [64] of the out-state density matrix :

$$\rho_{out}(x,x') = \hat{\rho}(x)\delta(x-x') \tag{117}$$

This behaviour can be understood intuitively [4, 3] as being related to the apparent shrinking of the string world sheet in target space, which destroys interferences between strings localized at different points in target configuration space, c.f. the de-phasons in the Hall model [61]. This behaviour is generic for string contributions to the space-time foam, which make the theory supercritical locally, inducing renormalization group (target time) flow. The two specific contributions (109,110) to this suppression (117) of space-time coherence that we have identified in this paper correspond (109) to microscopic black hole formation and (110) to the back-reaction of matter on a microscopic black hole, entailing in each case information loss across an event horizon.

# 7 $W$ Symmetries and Non-Hamiltonian Time Evolution

At various places in earlier sections, we have mentioned the infinite-dimensional $W$ algebra underlying the string black hole model, and its rôle in interrelating different solitonic states in the spectrum of the model. In earlier papers [36] we have stressed the rôle of $W$ symmetry in preserving quantum coherence when back-reaction is neglected. Subsequently, we have argued that the coherence of the effective theory of light states is suppressed when back-reaction is taken into account, in particular via the monopole-antimonopole and instanton-anti-instanton contributions discussed in the previous section. In this section we discuss the $W$-transformation properties of these explicit contributions to $\not{S}$ and $\not{\delta}H$, and relate them to the formalisms of refs [11] and [13]. In this way, we link explicitly the loss of coherence to the leakage of $W$ quantum numbers.

As a warm-up, we first present an analogous phenomenon in the theory of the Quantum Hall Effect [65]. The ground state of an integer Quantum Hall conductor with filling fraction $\nu = 1$ is represented by a non-singular wave function $\Psi_0(z_1, z_2, ..., z_N)$ for a system of non-interacting electrons. On the other hand, the ground state of a fractional Quantum Hall conductor is singular:

$$\Psi(z_1, z_2, ..., z_N) = S_p \Psi_0(z_1, z_2, ..., z_N) \tag{118}$$

where $p$ is related to the filling fraction $\nu$ by $\nu = 1/(2p+1)$ and

$$S_p = \Pi_{k<l}(z_k - z_l)^{2p} \tag{119}$$

The prefactor $S_p$ can be regarded as the matrix element of an operator that creates monopoles and vortices on the world sheet:

$$S_p \propto exp(\sum_{k<l}[2pReln(z_k - z_l) + i2pImln(z_k - z_l)]) \tag{120}$$

This representation realizes the plasma picture of [66], and is in direct analogy with the representation of Minkowski and Euclidean black holes as monopoles and vortices on the world sheet which we introduced in section 5.

The important aspect of this analogy for the purposes of this section is the observation that the relation (118) between the IQHE and FQHE ground states can be regarded as an invertible but non-unitary relation between the corresponding density matrices:

$$\rho_p = S_p {S_p}^{\dagger} \rho_0 \tag{121}$$

The relationship between the density matrices $\rho_0$ and $\rho_p$ corresponds to that between the $\rho$ and $\tilde{\rho}$ of [11], as we shall see in more detail shortly.

It has been pointed out that the IQHE and FQHE systems both possess an infinite-dimensional $W$ symmetry associated with the presence of incompressible quantum electron fluids [67], in which the quantum deformation parameter [68] $\frac{1}{k}$ is related to the filling fraction $\nu$ by

$$\nu = \frac{1}{k} \tag{122}$$

Thus the operator $S_p$ can be regarded as inducing a quantum deformation of the $W$ algebra that does not change the classical $W$ charges. The operator $S_p$ induces $\delta$-function Schwinger terms in the operator product expansion, associated with the singularities in $S_p$. They are related to the corresponding term in the operator product expansion for two energy-momentum tensors, which is a measure of the central charge $c$ in the Virasoro algebra, which is related in turn to the level parameter $k$ and hence to the filling fraction $\nu$:

$$c = \frac{3k}{k-2} = \frac{3}{1-2\nu} \tag{123}$$

As we shall see shortly, the world-sheet monopoles and instantons discussed in section 5 play a similar rôle in the black-hole case.

In the case of a string black hole contribution to space-time foam, the rôle of the density matrix $\rho$ of [11] is played by the exact density matrix for external tachyon fields dressed by higher-level operators as discussed in sections 4 and 5: As is well known [69], there is a well-defined $S$-matrix for the scattering of these dressed tachyons off a string black hole, and hence a well-defined Hamiltonian $H$, and the time evolution of the density matrix is given by the conventional Liouville equation using this Hamiltonian.

However, as we have mentioned previously, a realistic scattering experiment does not measure the non-local topological solitonic states created by the operators $L_0^1 \bar{L}_0^1$ etc., and hence deals only with bare tachyonic operators $T$. As we have described in section 5, the scattering of these bare operators exhibits additional renormalization scale (i.e., Liouville field $\phi$, i.e., time) dependence that cannot be absorbed within the usual Hamiltonian/$S$-matrix description. It is the density matrix of this $(T, \phi)$

system that we interpret as the $\tilde{\rho}$ of [11]:

$$\tilde{\rho} = \frac{e^{-\beta\mathcal{H}([\mathcal{F}^{c-c}_{-\frac{1}{2},0,0}]_{r,\phi})}}{Tr[e^{-\beta\mathcal{H}([\mathcal{F}^{c-c}_{-\frac{1}{2},0,0}]_{r,\phi})}]} \tag{124}$$

where $\mathcal{F}^{c-c}_{-\frac{1}{2},0,0}$ is the 'tachyon' deformation defined in (73), and $[\ldots]_\phi$ denotes the appropriate Liouville dressing. It is because of the extra time dependence mentioned above that this density matrix $\tilde{\rho}$ obeys the modified Liouville equation (3) derived in section 3 [2].

The monopole-antimonopole pairs discussed in the previous section contribute to the relation between $\rho$ (72) to $\tilde{\rho}$ (124), by representing the creation and annihilation of black holes, and the instanton-anti-instanton pairs make a contribution to this relation that is associated with rescaling of the target-space metric due to changes in the black hole mass. According to the analysis of section 2, this relation is associated with a loss of quantum coherence, and we have exhibited just such a loss in section 5.

In order to understand the physical origin of this loss of coherence, it is instructive to consider the $W$-transformation properties of these topological defects on the world sheet. To do this, we express the string black hole action in terms of target-space Kruskal-Szekeres coordinates $u \equiv sinhre^{it}$, $\overline{u} = sinhre^{-it}$:

$$S = \frac{k}{4\pi}\int d^2z \frac{\partial u\overline{\partial}\overline{u} + \partial\overline{u}\overline{\partial}u}{1-\overline{u}u} \tag{125}$$

Monopoles generate transformations of the form

$$u \rightarrow ue^{i\alpha}, \bar{u} \rightarrow \bar{u}e^{-i\alpha} \tag{126}$$

It is useful to construct parafermion operators $\psi_{+,-}$ by attaching Dirac strings to these monopoles [70]:

$$\begin{aligned} \psi_+ &= \frac{\partial u}{\sqrt{1-\overline{u}u}}V_+ \\ \psi_- &= \frac{\partial \overline{u}}{\sqrt{1-\overline{u}u}}V_- \\ V_\pm &= exp[\pm\frac{1}{2}\int_C (dzA + d\bar{z}\overline{A})] \\ A &= \frac{u\partial\overline{u} - \overline{u}\partial u}{1-\overline{u}u} \quad ; \quad \overline{A} = \frac{\overline{u}\partial u - u\overline{\partial}\overline{u}}{1-\overline{u}u} \end{aligned} \tag{127}$$

The $W$ currents are then given in terms of these constructs by

$$W_s = \sum_{k=1}^{s-1}(-1)^{s-k-1}A^s_k\partial^{k-1}\psi_+\partial^{s-k-1}\psi_- \tag{128}$$

where $s$ denotes the conformal spin, and

$$A_k^s = \frac{1}{s-1}\begin{pmatrix} s-1 \\ k \end{pmatrix}\begin{pmatrix} s-1 \\ s-k \end{pmatrix} \tag{129}$$

We note that all these $W_\infty$ transformations on the world sheet are generated by $(1,0)$ or $(0,1)$ currents, so the corresponding stress-tensor deformations are $(1,1)$, and hence can surely be elevated to target space [36]. This elevation has been worked out in [70], where it was shown that this elevation preserves the $W_\infty$ structure. One considers variations of the action $S$ under infinitesimal transformations $\delta u$, $\delta \bar{u}$:

$$\delta S = \int (\delta_\epsilon^{(s)} u \frac{\delta S}{\delta u} + \delta_\epsilon^{(s)} \overline{u} \frac{\delta S}{\delta \overline{u}}) \tag{130}$$

where

$$\begin{aligned} \frac{\delta S}{\delta u} &= \frac{\partial \overline{\partial} u}{1 - u\overline{u}} + \frac{\overline{u}\partial u \overline{\partial} u}{(1-u\overline{u})^2} \\ \frac{\delta S}{\delta \overline{u}} &= \frac{\partial \overline{\partial} \overline{u}}{1 - u\overline{u}} + \frac{u\partial \overline{u} \overline{\partial} \overline{u}}{(1-u\overline{u})^2} \end{aligned} \tag{131}$$

Knowing that the $W_\infty$ transformations are generated by chiral fields, we compare (130) with the Ansatz [70]

$$\delta S = \int d^2 z \epsilon \overline{\partial} W_s \tag{132}$$

to find the following infinite set of infinitesimal transformations:

$$\begin{aligned} \delta_\epsilon^{(s)} u \; = \; & \sum_{i=0}^{s-2} B_i^s \partial^i \epsilon \sqrt{1 - u\overline{u}} V_- \partial^{s-i-2} \psi_+ + \\ & \sum_{k=1}^{s-1} \sum_{l=0}^{k-2} (-1)^l A_k^s \begin{pmatrix} k-1 \\ l \end{pmatrix} u \partial^{k-2-l} [\epsilon \partial^l \psi_- \partial^{s-k-1} \psi_+ - \\ & (-1)^s \epsilon \partial^l \psi_+ \partial^{s-k-1} \psi_-] \end{aligned} \tag{133}$$

with

$$B_l^s = \frac{1}{s-1}\begin{pmatrix} 2s-l-2 \\ s \end{pmatrix}\begin{pmatrix} s-1 \\ l \end{pmatrix} \tag{134}$$

The coresponding transformation for $\overline{u}$ is obtained by interchanging $u \longleftrightarrow \overline{u}$ and multiplying by a factor $(-1)^s$.

It is easy to verify directly that these variations satisfy the $W_\infty$ algebra:

$$[\delta_{\epsilon_1}^s, \delta_{\epsilon_2}^{s'}] = \delta_{(s'-1)\epsilon_1'\epsilon_2 - (s-1)\epsilon_1\epsilon_2'}^{s+s'-2} + \cdots \tag{135}$$

where the ... denote lower-spin terms. Hence the $W$ charges

$$Q^s = \int_C dz W^s(z) \tag{136}$$

commute, as they constitute an infinite-dimensional Cartan subalgebra of $W_\infty$ [68]. It is easy to check that these charges are non-zero in general if $u$ and $\bar{u}$ are not holomorphic functions of the world-sheet coordinates: $u \neq u(z)$, etc., which is the case for the string black hole background. Therefore, their values make available an infinite set of quantum numbers (hair) to specify the black hole state. However, the instantons discussed in section 5 have vanishing $W$ quantum numbers, because they are holomorphic maps $u = u(z)$. One finds that

$$\delta^s_\epsilon u = \sum_{l=0}^{s-2} B^s_l \partial^l \epsilon \partial^{s-l-1} u \qquad ; \qquad \delta^s_\epsilon \overline{u} = 0 \tag{137}$$

Note that instantons and anti-instantons behave differently under the $W_\infty$ transformations.

Now we are ready to discuss in the light of these results the monopole and instanton contributions to the loss of coherence. As has already been mentioned, monopoles make contributions to the functional integral representation of $\tilde{\rho}$ that are identical in form to the prefactor $S_p$ in the relation (121) between the IQHE and FQHE ground states, and play the same rôle as the similarity transformation of [11]. We saw in the previous section [see equation (106)] that the action of a monopole-antimonopole pair depends logarithmically on the ultraviolet cutoff, which translates into a time-dependence, and hence a contribution to $\not{\delta} H$ that tends to suppress quantum coherence. We have seen in this section that monopoles possess $W$ hair, by virtue of being non-holomorphic maps of the world sheet into target space. Thus they carry off $W$ quantum numbers, which entails information loss and hence explains the loss of coherence. The action of the instanton-anti-instanton valley exhibits a logarithmic dependence on the ultraviolet cutoff (and hence time-dependence, a contribution to $\not{\delta} H$ and a loss of coherence) only when there is a finite separation between the instanton and anti-instanton. As we have discussed just above, an isolated instanton is represented by a holomorphic map from the world sheet into target space, hence carries no $W$ quantum numbers and does not contribute to decoherence. However, once an instanton and anti-instanton are superposed at finite separation, their configuration can no longer be represented by such a holomorphic map, and they will carry $W$ quantum numbers in general, explaining this loss of coherence.

Thus we have a complete understanding of the relation between our non-critical string formalism and the general similarity transformation theory of [11], including a physical understanding of the resulting loss of coherence due to string black holes in terms of a leakage of $W$ quantum numbers. We expect that a similar discussion could be given for any other topologically non-trivial contributions to the space-time foam in string theory, in which other string symmetries would play the rôle of the black hole's $W_\infty$ algebra, which is but a small part of the enormous full local symmetry of string.

# 8 CPT Violation

It is a basic theorem [71] of quantum field theory that $CPT$ should not be violated, as a consequence of locality, Lorentz invariance and unitarity. String theory is of course based on local quantum field theory on the world sheet, but this is not equivalent to locality in target space-time. Lorentz invariance is a property of critical string theory, but requires re-examination in the context of the non-critical string approach to time that we have espoused here. Thus, even though we do not challenge the unitarity of the effective low-energy theory, not all the conditions needed to derive the $CPT$ theorem are satisfied in our string framework, and it is appropriate to re-open the possibility of $CPT$ violation.

This has often been discussed [19] in the general context of quantum gravity, motivated largely by the likelihood of non-locality, and also specifically in the context of string theory [72, 73, 74]. In this latter case, it was shown that $CPT$ violation was in principle possible if certain world-sheet charges were not conserved [72], but it was also shown that this could not occur in a string model with a flat target space-time [74].

We have taken this analysis one step further, pointing out that space-times that appear singular from the point of view of conventional general relativity, such as a black hole, can be described by topological defects on the world sheet, such as monopoles or vortices in the black-hole case [17]. Moreover, it is well-known that quantum numbers carried by fermionic fields may no longer be conserved in the presence of topological defects, c.f., monopoles and instantons in conventional three-dimensional space. To demonstrate that $CPT$ is violated in any specific elementary-particle system, such as the neutral kaon system, would require a complete string-derived model in which the world-sheet fermion contents of all quarks and leptons were known. Would that we had such a complete model! However, even in its absence, we [15, 75] and Huet and Peskin [76] have argued that the likelihood of $CPT$ violation in string theory makes it worthwhile to re-examine the traditional phenomenology of the neutral kaon and other systems, parametrizing possible $CPT$-violating effects and constraining their magnitudes. Before reviewing that work here, we would like to make contact with other work [19] on $CPT$ violation in the general framework of quantum gravity.

The conventional, strong form of $CPT$ invariance proved in local quantum field theory implies the existence of a $CPT$ operator $\Theta$ that transforms any initial density matrix $\rho_{in}$ into some final-state density matrix $\rho_{out}{}'$:

$$\rho_{out}{}' = \Theta \rho_{in} \tag{138}$$

and correspondingly

$$\rho_{in}' = \Theta^{-1}\rho_{out} \tag{139}$$

where $\rho_{out}$ is related to $\rho_{in}$ by the familiar $\not{S}$ matrix:

$$\rho_{out} = \not{S}\rho_{in} \tag{140}$$

and likewise for $\rho_{out}'$ and $\rho_{in}'$:

$$\rho_{out}' = \not{S}\rho_{in}' \tag{141}$$

The following relation is a trivial consequence of equations (138) to (141):

$$\Theta = \not{S}\Theta^{-1}\not{S} \tag{142}$$

Again, it is a trivial consequence of (142) that the $\not{S}$ matrix must have an inverse :

$$\not{S}^{-1} = \Theta^{-1}\not{S}\Theta^{-1} \tag{143}$$

However, there can be no inverse of the superscattering matrix $\not{S}$ in any framework that allows pure states to evolve into mixed states, as has been argued to be a general necessity in any quantum theory of gravity, and as we have found specifically in our non-critical string approach. Thus, there cannot be a $CPT$ operator with the properties assumed above, and the $CPT$ theorem cannot hold in its strong field-theoretical form.

A weaker form of CPT invariance has been proposed [19], according to which the probability $P(\phi \to \psi)$ that a pure initial state $\phi$ will be observed to become a given pure final state $\psi$ is equal to the probability for the $CPT$-reversed process:

$$P(\psi \to \phi) = P(\theta^{-1}\phi \to \theta\psi) \tag{144}$$

This is automatically true in any theory in which the final-state density matrix is proportional to the unit matrix, as was the case in our simple two-state model in section 3. However, we will not address here the interesting question whether this or some other weak form of $CPT$ invariance holds in string theory. For now, we simply observe that there is no reason for the strong $CPT$ theorem to hold in string theory. To substantiate this claim we review briefly the arguments presented in ref. [15] concerning $CPT$-violation in our string model. Following [72] we consider the target-space $CPT$ operator $\Theta$ (138) as being derived from an appropriate world-sheet operator $\Theta_w$. Any state in target-space can be considered as an eigenstate of the $\sigma$-model Hamiltonian $E$ with eigenvalue $m_i$, the mass of the corresponding particle, i.e.

$$E|m_i, Q_i >= m_i|m_i, Q_i > \tag{145}$$

where $\{Q_i\}$ denotes a set of conserved world-sheet charges that can be elevated to target space,

$$[E, \hat{Q}_i] = 0 \qquad ; \qquad \hat{Q}_i|m_i, Q_i >= Q_i|m_i, Q_i > \tag{146}$$

The $\hat{A}$ denote quantum-mechanical operators on the world-sheet. World-sheet $CPT$ invariance is guaranteed if and only if [72]

$$[E, \Theta_w] = 0 \qquad ; \qquad \hat{Q}_i \Theta_w + \Theta_w \hat{Q}_i = 0 \tag{147}$$

This implies $CPT$-invariance in target space in the following sense [72] : the $CPT$ transform of a state of mass $m_i$ and 'charge' $Q_i$, is $\Theta_w|m_i, Q_i>$. Using (147) we can readily see that it will be an eigenstate of $E$ with mass $m_i$ and 'charge' $-Q_i$. In our case, the existence of valleys of topological defects on the world-sheet spoils the conservation of $Q_i$, and thus (147), as a result of logarithmic divergences, as we have discussed in sections 5 and 6. These imply temporal dependences of the 'charges' $Q_i$, and hence their conservation is spoiled. As a consequence, the above 'proof' of $CPT$ invariance in target space fails.

Although the above picture is rather heuristic, and much more work is required to define the elevation process of the $CPT$ operation from the world sheet to target space in a mathematical rigorous way, however it is certainly suggestive of the kind of $CPT$ violation one should expect in string theory formulated in highly curved space-time backgrounds. It is therefore worthwhile to explore the possibility of its violation, though we also cannot exclude the possibility that the strong $CPT$ theorem might not be violated detectably in any given experiment.

We now describe briefly the formalism [10, 15] for describing the possible modification of quantum mechanics and violation of $CPT$ in the neutral kaon system, which is among the most sensitive microscopic laboratories for studying these possibilities. In the normal quantum-mechanical formalism, the time-evolution of a neutral kaon density matrix is given by

$$\partial_t \rho = -i(H\rho - \rho H^\dagger) \tag{148}$$

where the Hamiltonian takes the following form in the $(K^0, \overline{K}^0)$ basis:

$$H = \begin{pmatrix} (M + \frac{1}{2}\Delta M) - \frac{1}{2}i(\Gamma + \frac{1}{2}\Delta\Gamma) & M^*_{12} - \frac{1}{2}i\Gamma^*_{12} \\ M_{12} - \frac{1}{2}i\Gamma_{12} & (M - \frac{1}{2}\Delta M) - \frac{1}{2}i(\Gamma - \frac{1}{2}\Delta\Gamma) \end{pmatrix} \tag{149}$$

The non-hermiticity of $H$ reflects the process of $K$ decay: an initially-pure state evolving according to (148) and (149) remains pure.

In order to discuss the possible modification of this normal quantum-mechanical evolution, and allow for the possibility of $CPT$ violation, it is convenient to rewrite [15] (148) and (149) in a Pauli $\sigma$-matrix basis [10], introducing components $\rho_\alpha$ of the density matrix:

$$\rho = 1/2\rho_\alpha\sigma_\alpha \tag{150}$$

which evolves according to

$$\partial_t \rho_\alpha = h_{\alpha\beta}\rho_\beta \tag{151}$$

with

$$h_{\alpha\beta} \equiv \begin{pmatrix} Imh_0 & Imh_1 & Imh_2 & Imh_3 \\ Imh_1 & Imh_0 & -Reh_3 & Reh_2 \\ Imh_2 & Reh_3 & Imh_0 & -Reh_1 \\ Imh_3 & -Reh_2 & Reh_1 & Imh_0 \end{pmatrix} \tag{152}$$

It is easy to check that at large times $\rho$ takes the form

$$\rho \simeq e^{-\Gamma_L t} \begin{pmatrix} 1 & \epsilon^* \\ \epsilon & |\epsilon|^2 \end{pmatrix} \tag{153}$$

where $\epsilon$ is given by

$$\epsilon = \frac{\frac{1}{2} i Im\Gamma_{12} - ImM_{12}}{\frac{1}{2}\Delta\Gamma - i\Delta M} \tag{154}$$

in the usual way.

A modification of quantum mechanics of the form discussed in section 3 can be introduced by modifying equation (151) to become

$$\partial_t \rho_\alpha = h_{\alpha\beta}\rho_\beta + \not{h}_{\alpha\beta}\rho_\beta \tag{155}$$

The form of $\not{h}_{\alpha\beta}$ is determined if we assume probability and energy conservation, as proved in the string context in section 3, and that the leading modification conserves strangeness:

$$\not{h}_{\alpha\beta} = \begin{pmatrix} 0 & 0 & 0 & 0 \\ 0 & 0 & 0 & 0 \\ 0 & 0 & -2\alpha & -2\beta \\ 0 & 0 & -2\beta & -2\gamma \end{pmatrix} \tag{156}$$

It is easy to solve the $4 \times 4$ linear matrix equation (155) in the limits of large time:

$$\rho_L \propto \begin{pmatrix} 1 & \frac{-\frac{1}{2}i(Im\Gamma_{12}+2\beta)-ImM_{12}}{\frac{1}{2}\Delta\Gamma+i\Delta M} \\ \frac{\frac{1}{2}i(Im\Gamma_{12}+2\beta)-ImM_{12}}{\frac{1}{2}\Delta\Gamma-i\Delta M} & |\epsilon|^2 + \frac{\gamma}{\Delta\Gamma} - \frac{4\beta ImM_{12}(\Delta M/\Delta\Gamma)+\beta^2}{\frac{1}{4}\Delta\Gamma^2+\Delta M^2} \end{pmatrix} \tag{157}$$

and of short time:

$$\rho_S \propto \begin{pmatrix} |\epsilon|^2 + \frac{\gamma}{|\Delta\Gamma|} - \frac{-4\beta ImM_{12}(\Delta M/\Delta\Gamma)+\beta^2}{\frac{1}{4}\Delta\Gamma^2+\Delta M^2} & \epsilon - \frac{i\beta}{\frac{\Delta\Gamma}{2}-i\Delta M} \\ \epsilon^* + \frac{i\beta}{\frac{\Delta\Gamma}{2}+i\Delta M} & 1 \end{pmatrix} \tag{158}$$

We note that the density matrix (157) for $K_L$ is mixed to the extent that the parameters $\beta$ and $\gamma$ are non-zero. It is also easy to check [15] that the parameters $\alpha$, $\beta$ and $\gamma$ all violate $CPT$, in accord with the general argument of [19], and consistent with the string analysis mentioned earlier in this section.

Experimental observables $O$ can be introduced [10, 15] into this framework as matrices, with their measured values being given by

$$< O >= Tr(O\rho) \tag{159}$$

Examples are the $K$ to $2\pi$ and $3\pi$ decay observables

$$O_{2\pi} = \begin{pmatrix} 0 & 0 \\ 0 & 1 \end{pmatrix} \quad ; \quad O_{3\pi} = (0.22)\begin{pmatrix} 1 & 0 \\ 0 & 0 \end{pmatrix} \tag{160}$$

and the semileptonic decay observables

$$\begin{aligned} O_{\pi^- l^+ \nu} &= \begin{pmatrix} 1 & 1 \\ 1 & 1 \end{pmatrix} \\ O_{\pi^+ l^- \overline{\nu}} &= \begin{pmatrix} 1 & -1 \\ -1 & 1 \end{pmatrix} \end{aligned} \tag{161}$$

A quantity of interest is the difference between the $K_L$ to $2\pi$ and $K_S$ to $3\pi$ decay rates [15]:

$$\delta R \equiv R_{2\pi} - R_{3\pi} = \frac{8\beta}{|\Delta\Gamma|}|\epsilon| sin\phi_\epsilon \tag{162}$$

where $R^L_{2\pi} \equiv Tr(O_{2\pi}\rho_L)$, and $R^S_{3\pi} \equiv Tr(O_{3\pi}\rho_S)/0.22$, and the prefactors are determined by the measured [77] branching ratio for $K_L \to 3\pi^0$. (Strictly speaking, there should be a corresponding prefactor of 0.998 in the formula (160) for the $O_{2\pi}$ observable.)

Using (161), one can calculate the semileptonic decay asymmetry [15]

$$\delta \equiv \frac{\Gamma(\pi^- l^+ \nu) - \Gamma(\pi^+ l^- \overline{\nu})}{\Gamma(\pi^- l^+ \nu) + \Gamma(\pi^+ l^- \overline{\nu})} \tag{163}$$

in the long- and short-lifetime limits:

$$\begin{aligned} \delta_L &= 2Re[\epsilon(1 - \frac{i\beta}{ImM_{12}})] \\ \delta_S &= 2Re[\epsilon(1 + \frac{i\beta}{ImM_{12}})] \end{aligned} \tag{164}$$

The difference between these two values

$$\delta\delta \equiv \delta_L - \delta_S = -\frac{8\beta}{|\Delta\Gamma|}\frac{sin\phi_\epsilon}{\sqrt{1+\tan^2\phi_\epsilon}} = -\frac{8\beta}{|\Delta\Gamma|} sin\phi_\epsilon cos\phi_\epsilon \tag{165}$$

with $tan\phi_\epsilon = (2\Delta M)/\Delta\Gamma$, is a signature of $CPT$ violation that can be explored at the CPLEAR and DA$\phi$NE facilities.

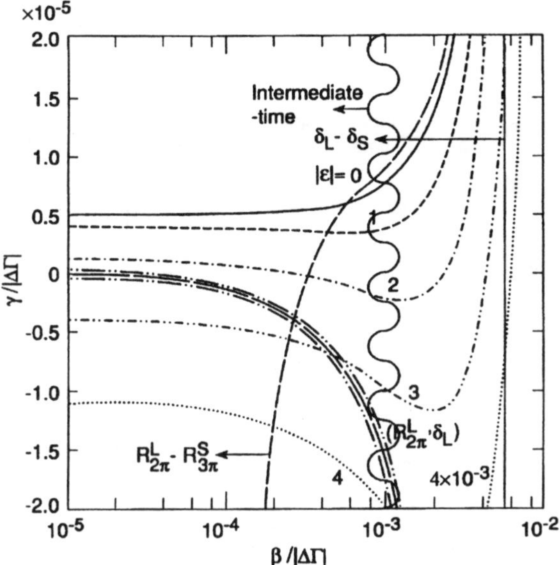


Figure 3: - The $(\beta, \gamma)$ plane on a logarithmic scale for $\beta > 0$. We plot contours of the conventional $CP$-violating parameter $|\epsilon|$, evaluated from the $K_L \rightarrow 2\pi$ decay rate. The dashed-double-dotted band is that allowed at the one-standard-deviation level by the comparison between measurements of the $K_L \rightarrow 2\pi$ decay rate and the $K_L$ semileptonic decay asymmetry $\delta_L$. The dashed line delineates the boundary of the region allowed by the present experimental upper limit on $K_S \rightarrow 3\pi^0$ decays $(R^L_{2\pi} - R^S_{3\pi})$ and a solid line delineates the boundary of the region allowed by a recent preliminary measurement of the $K_S$ semileptonic decay asymmetry $\delta_S$. A wavy line bounds approximately the region of $|\beta|$ which may be prohibited by intermediate-time measurements of $K \rightarrow 2\pi$ decays.

We have used [15] the latest experimental values of $R_{2\pi}$ and $R_{3\pi}$ to bound $\delta R$, and the latest experimental values of $\delta_{L,S}$ to bound $\delta\delta$, expressing the results as contours in the $(\beta, \gamma)$ plane as seen in figure 3. Also shown there are contours of the usual $CP$-violating parameter $\epsilon$, which is given in our case by [15]

$$|\epsilon| = -\frac{2\beta}{|\Delta\Gamma|} sin\phi_\epsilon + \sqrt{\frac{4\beta^2}{|\Delta\Gamma|^2} - \frac{\gamma}{|\Delta\Gamma|} + R^L_{2\pi}} \tag{166}$$

On the basis of this preliminary analysis, it is safe to conclude that

$$|\frac{\beta}{\Delta\Gamma}| \lesssim\ 10^{-4}\ to\ 10^{-3} \quad ; \quad |\frac{\gamma}{\Delta\Gamma}| \lesssim\ 10^{-6}\ to\ 10^{-5} \tag{167}$$

In addition to more precise experimental data, what is also needed is a more complete global fit to all the available experimental data, including those at intermediate times, which are essential for bounding $\alpha$, and may improve our bounds (167) on $\beta$ and $\gamma$ [75, 76] .

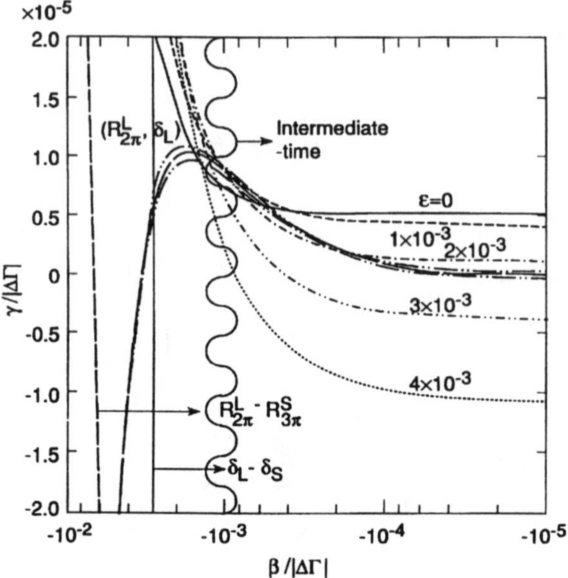


Figure 4: - As in Fig. 3, on a logarithmic scale for $\beta < 0$.

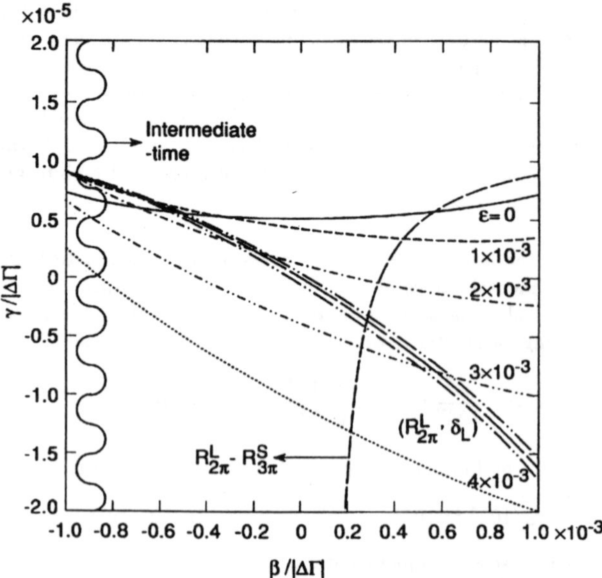


Figure 5: As in Fig. 3, on a linear scale for the neighborhood of $\beta = 0$.

We cannot resist pointing out that the bounds (167) are quite close to

$$O(\Lambda_{QCD}/M_P)m_K \simeq 10^{-19}GeV \tag{168}$$

which is perhaps the largest magnitude that any such $CPT$- and quantum-mechanics-violating parameters could conceivably have. Since any such effects are associated with topological string states that have masses of order $M_P$, we expect them to be suppressed by some power of $1/M_P$. This expectation is supported by the analogy with the Feynman-Vernon model of quantum friction [20], in which coherence is suppressed by some power of the unobserved oscillator mass or frequency. If the $CPT$- and quantum-mechanics-violating parameters discussed in this section are suppressed by just one power of $M_P$, they may be accessible to the next round of experiments with CPLEAR and/or DA$\phi$NE.

# 9 Connection with Cosmology

We complete this review with a discussion of cosmology in the context of our non-critical approach to string theory, commenting on inflation, entropy generation, variations in fundamental parameters, and the cosmological constant [16]. The first explicit cosmological string theory with a time-dependent background was that discovered in [1]. It has a dilaton field that depends linearly on the time variable $t$, and the string black hole can be regarded as a Minkowski rotation of this model, in which the $t$-dependence of the dilaton field is replaced by the corrsponding dependence on the radial coordinate $r$. It was suggested in [1] that the Universe might make quantum transitions between models with different values of the central charge associated with the dilaton/time variable, but this suggestion was not worked out in detail. The mechanism for such quantum transitions via instantons has now been worked out in the black hole case [16, 3, 4]. As discussed in section 5, it leads (91) to a scale-dependent value of the level parameter $k$, corresponding in turn to a time-dependence

$$k(t) \simeq ke^{-4\pi\beta^I T_0 t} \tag{169}$$

in the dilute-gas approximation, where we recall that the corresponding instanton $\beta$-function

$$\beta^I = -(k/2)g^I \tag{170}$$

in the large-$k$ limit is negative, leading to an increase in the effective level parameter $k$, and hence an approach to the flat limit $k \rightarrow \infty$. In the cosmological context, this implies a slowing down in the rate of expansion of the Universe. According to the analogy with the quantum Hall effect [17], this process is analogous to a series of transitions between different conductivity plateaux.

Several questions arise in this picture, including the following. What was the initial state of the Universe? Is there any analogue of inflation in this approach, particularly as regards entropy generation, which is an essential feature of our non-critical treatment of string theory, as seen in equation (51)? Do fundamental physics parameters such as the velocity of light $c$ and Planck's constant $\hbar$ vary during the cosmological expansion [16]? How does the effective cosmological constant relax to zero, as it should do in the flat-space limit?

In partial answer to the first of these questions, we recall that the central charge

$$c = 2(k+1)/(k-2) \tag{171}$$

becomes infinite in the limit $k \to 2$. The dilute gas approximation is not reliable here, but is suggestive that the Universe started from such a limit. In the picture of section 4, as shown in figure 2, this limit would correspond to the infrared limit of the renormalization group flow. In the case of the string black hole, it corresponds to the region close to the core where there is an appropriate description [23] in terms of a twisted $N = 2$ supersymmetric theory, which is equivalent to a topological field theory in which the concepts of space and time break down. Thus we are led to the conclusions that the origin of the Universe is presumably also described by such a topological field theory, and that the concepts of space and time also break down at the beginning of the Universe.

Our simple cosmological scenario provides a qualitative picture of the entropy production rate in the Universe. In our framework, the rate of entropy increase with time is given by [2, 16]

$$\partial_t S = \beta^i G_{ij} \beta^j S \quad ; \quad G_{ij} = 2|z|^4 < V_i V_j > \tag{172}$$

where the unitarity requirement of the world-sheet theory implies the positivity of the Zamolodchikov metric [31] $G_{ij} > 0$. Using the C-theorem [31], especially in its string formulation [78] on the fiducial-metric world-sheet, one may write

$$\beta^i G_{ij} \beta^j = \partial_t C(g) \quad : \quad C(g) = -\frac{1}{12} \int d^D y \sqrt{G} e^{-2\Phi} < TT > + \dots \tag{173}$$

In this expression, the $y$ denote target spatial coordinates, $\Phi$ is the dilaton field, and $T \equiv T_{zz}$ is a component of the world-sheet stress tensor. The ... denote the remaining two-point functions that appear in the Zamolodchikov C-function [31], which involve the trace $\Theta$ of the stress tensor, i.e. $< T\Theta >$ and $< \Theta\Theta >$. Taking into account the off-shell corollary of the C-theorem, $\frac{\delta C(g)}{\delta g^i} = G_{ij}\beta^j$, it can readily be shown [79] that such terms can always be removed by an appropriate renormalization-scheme choice, that is by appropriate redefinitions of the renormalized couplings $g^i$, and hence play no rôle in the physics. Thus, one can solve (172) for the entropy $S$ in terms of the Zamolodchikov C-function

$$S(t) = S_0 e^{-\frac{1}{12} \int_0^t \int d^D y \sqrt{G} e^{-2\Phi} <TT> + \dots} \tag{174}$$

where the minus sign in the exponent indicates the opposite flow of the time $t$ with respect to the renormalization-group flow. Expression (174) reduces a complicated target-space computation of entropy production in an inflationary scenario to a conformal-field-theory computation of two-point functions involving components of the stress tensor of a first-quantized string. We observe from (172) that the rate of entropy increase is maximized on the maximum-$\beta^i$ surface in coupling constant space. At late stages of the inflationary era, i.e. close to the ultraviolet fixed point, the rate of change of $S$ is strongly suppressed, due to the smallness of the $\beta^i$.

In order to discuss the possible variation of fundamental physical parameters during the expansion of the Universe [16], we first recall the relation between a string black hole mass and the level parameter $k$:

$$M/M_{Planck} = \sqrt{\frac{1}{k(t)-2}} e^{const} \tag{175}$$

It is well known that light cones are distorted in the presence of a black hole. Specifically, the exact space-time background metric of the black hole Wess-Zumino model has the following asymptotic form for large $r$:

$$ds^2 = 2(k(t)-2)(dr^2 - \frac{k(t)}{k(t)-2}dt^2) \tag{176}$$

This implies a $k$-dependence of the apparent velocity of light, which becomes a time-dependence as a result of equation (169):

$$c_q = c\sqrt{\frac{k(t)}{k(t)-2}} \tag{177}$$

where $c$ is the usual flat-space velocity. We note that the fact that $c_q \to \infty$ as $k \to 2$, corresponding to the broadening out of the effective light-cone, is consistent with the suggestion made above that the concepts of space and time break down in this limit. Specifically, in a Robertson-Walker-Friedmann universe the horizon distance $d$ in co-moving coordinates over which an observer can look back is [80]

$$d = \int c_q(t) = \int dt \sqrt{\frac{k(t)}{k(t)-2}} \tag{178}$$

which is larger than the naive estimate $d = ct$. Indeed, the horizon distance could even become infinite if $k(t) \to 2$ in a suitable way as $t \to 0$, but this conjecture takes us beyond the dilute-gas approximation where we can compute reliably.

The time-dependence of string physics is also reflected in a computation of the string position-momentum uncertainty relation. Defined appropriately to incorporate curved gravitational backgrounds, this uncertainty can be expressed as [16]

$$(\Delta X \Delta P)_{min} \equiv \hbar_{eff}(t) = \hbar(1 + O(\frac{1}{k(t)}) \tag{179}$$

where $\Delta A = (< A^2 > - < A >^2)^{\frac{1}{2}}$, $< \ldots >$ denotes a $\sigma$-model vacuum expectation value, and $\hbar$ is the critical-string Planck's constant. The string uncertainty relation introduces a minimum length $\lambda_s$, that in our case also decreases with time [16]:

$$\lambda_s(t) \equiv (\frac{\hbar_{eff}(t)\alpha'(t)}{c_q(t)^2})^{\frac{1}{2}} = \lambda_s^0(1 + O(\frac{1}{k(t)})) \tag{180}$$

We mention in passing that the effective string Regge slope is also time-dependent [16]:

$$\frac{\alpha'(t)}{\alpha'^0} = \frac{c_q(t)^2}{c^2} \tag{181}$$

which stems from the relation between $k$ and $\alpha'$ in this model. It is also worth mentioning that this cosmological model exhibits certain Jeans-like instabilities [81] leading to the exponential growth of low-energy string modes at finite $k$ [16], thereby providing a scenario for string-sustained inflation [82].

The non-critical string scenario for the expanding Universe described in the preceeding paragraphs offers the prospect of solving the three basic problems of the standard-model cosmology in a manner reminiscent of conventional inflation [83]. The *horizon problem* could be solved by the enhanced look-back distance (178), and/or the breakdown of the normal concepts of space and time in a transition to a topological phase close to the infrared fixed point. The *flatness* problem could be solved by an epoch of exponential expansion, induced by a Jeans-like instability [16]. The *entropy problem* could be solved by the enhanced rate of entropy production (174) at early times. However, the crucial difference in our approach is that the fundamental scalar field, usually termed the *inflaton*, is replaced by a world-sheet field, the Liouville mode, in our approach. Fluctuations of this field create the renormalization group flow of the system that leads to the generation of propagating matter, in the way described above and in previous works[17, 4]. Of course, this mode is associated with the appearance of a target space scalar, the dilaton, but the latter is part of the metric background. This can be seen clearly in the two-dimensional Wess-Zumino string theory, which may be considered as a prototype for the description of a spherically-symmetric ($s$-wave) four-dimensional Universe [84]. In this model the dilaton belongs to the graviton level-one string multiplet, which is a non-propagating (discrete) string mode, and as such can only exist as a background, in contrast to a massless 'tachyon' mode, which propagates and scatters.

In our approach to the cosmological constant question, we start by considering the following one-loop results for the dilaton and graviton $\beta$-functions in bosonic $\sigma$-models [85, 29]:

$$\begin{aligned} \beta^{\Phi} \equiv \frac{d\Phi}{d\phi} &= -\frac{2}{\alpha'}\frac{\delta c}{3} + \nabla^2\Phi - (\nabla\Phi)^2 \\ \beta^G_{MN} \equiv \frac{dG_{MN}}{d\phi} &= -R_{MN} - 2\nabla_M\nabla_N\Phi \end{aligned} \tag{182}$$

where $\Phi$ is the dilaton field and $\phi$ denotes our covariant Liouville cutoff (c.f. the relative minus sign compared with the notation of ref. [85, 29] where the cutoff is defined with the dimensions of mass), and $\delta c = C - 26$, the 26 coming from the space-time reparametrization ghosts. If the central charge of the theory is not 26, as is the case of non-critical bosonic strings, then a cosmological constant term appears in the target space effective action. The form of this target-space action, whose variations yield the $\beta$-functions (182), reads:

$$\mathcal{I} = \frac{2}{\alpha'} \int d^D y \sqrt{G} e^{-\Phi} \{ \frac{1}{3} \delta c - \alpha' (R + 4(\nabla\Phi)^2 + \ldots) \} \tag{183}$$

where the ... denote other fields in the theory that we shall not use explicitly. We now notice that the effects of the tachyons in our two-dimensional target-space string model amount to a shift of the level parameter $k(\phi)$ with the renormalization group scale. This is the result of the combined effects of tachyon and instanton deformations, the latter representing higher-genus instabilities [50, 52, 16, 4]. The instantons alone, as irrelevant deformations, produce an initial instability by inducing an increase of the central charge, which then flows downhill towards 26 in the presence of relevant matter (tachyon) couplings. Hence there is a running central charge $c(\phi) > 26$, according to the C-theorem [31], that will, in general, imply a non-vanishing, time-varying (running), *positive* cosmological constant, $\Lambda(\phi)$, for the background of (176). Its precise form is determined by consistency with the equations (182).

For simplicity, we assume that the only effect of the dilaton is a constant contribution to the scale anomaly, which is certainly the case of interest. This allows one to decouple $\Phi$ in the field equations obtained from (183). Then the latter read

$$\begin{aligned} \frac{\delta \mathcal{I}}{\delta \Phi} &= \Lambda(\phi) - R \\ \frac{\delta \mathcal{I}}{\delta G_{MN}} &= -R_{MN} + \frac{1}{2} G_{MN} R \end{aligned} \tag{184}$$

In two dimensions the second equation is satisfied identically. Decoupling of the dilaton field also implies that the first equation yields

$$R = \Lambda(\phi) \tag{185}$$

The metric background (176) has a maximal symmetry in its space part. To make the analysis more general, we extend the background to $d = 2 + \epsilon$ dimensions, keeping the maximal symmetry in the spatial part of the metric [8]. Relating time to the Liouville field as we have discussed earlier, we find the following solution for

[8]The $G_{00}$-component depends at most on the time $\phi$ and can be absorbed in a redefinition of the time variable. It will not be of interest to us here.

$\Lambda(\phi)$[9]

$$\Lambda(t) = \frac{\Lambda(0)}{1 + t\frac{\Lambda(0)}{d-1}} \qquad ; \qquad t \equiv -\phi > 0 \tag{186}$$

which for positive $\Lambda$ implies an asymptotically-free cosmological constant $\beta$-function, thereby leading to a vanishing cosmological constant at the ultraviolet fixed point on the world-sheet.

The rate of the decrease of $\Lambda(\phi)$ is determined by its initial value at the infrared fixed point, where we conjectured that the theory makes a transition to a topological (twisted $N = 2$ supersymmetric) $\sigma$-model. It is of great interest to estimate this value in our two-dimensional model. This can be done by noticing that

$$\Lambda(0) = \frac{2}{\alpha'(0)} \frac{c(0) - 26}{3} \tag{187}$$

where $c(0) - 26 = \frac{3k(0)}{k(0)-2} - 27 \simeq \frac{3k(0)}{k(0)-2}$, given that $k(\phi) \to 2$ as $\phi \to 0$. Thus, taking into account (187) one observes that $\Lambda(0)$ is determined by the critical string tension, as it should be, given that $\alpha'_0$ is the only scale in the problem (or equivalently the minimal string length): the result is

$$\Lambda(0) = \frac{2}{\alpha'_0} \tag{188}$$

The latter result implies a really fast decay of the cosmological constant in this model. Notice that the finite initial value of $\Lambda(0)$ implies from (185) no curvature singularity in the Euclidean model at the origin of target space $r = 0$, as is indeed the case of the two-dimensional black-hole model of ref. [5], given that this point is a pure coordinate singularity. The above analysis for the cosmological constant, therefore, applies most likely to singularity-free inflationary universes [87]. It is understood that until the precise behaviour of the running couplings near the infrared fixed point is found, there will always be uncertainties in the above estimates. String perturbation theory is not applicable near the topological phase transition, and the infinities we get in the various running couplings constitute an indication of this. In the complete theory, these infinities should be absent.

# 10 Outlook

We have outlined in these lectures an approach to non-equilibrium quantum statistical mechanics, black holes, time, quantum mechanics and cosmology that is based

[9] We remark that a similar equation has also been considered in ref. [86], but the flow of time in that reference coincides with the renormalization group flow. In such a case, one gets sensible results only for negative initial values of the cosmological constant, contrary to our case where we have a vanishing cosmological constant asymptotically, starting from positive initial values.

on non-critical string theory, with time described by a Liouville field. Our basic aim has been to understand some of the qualitative features of quantum fluctuations in the structure of space-time, and their physical consequences, i.e., to understand foam. We have tried to short-circuit the general ignorance of string field theory by using a sort of mini-superspace approach, exploiting our knowledge of one particular class of such fluctuations, namely string black holes, and arguing that their consequences are likely to be quite general. Specifically, we expect that many other types of space-time fluctuation will tend to suppress quantum coherence in the manner discussed here for the black hole case.

We are aware that many details of these ideas remain to be worked out, and that many questions could be raised. However, we believe that our approach manifests such a high degree of internal consistency, and brings together so many apparently unconnected features of different areas of physics, that it is worthy of further constructive study.

There are several aspects of our work that we have not discussed here in any detail. These include the possible rôle of the type of decoherence discussed here in the transition between microscopic quantum physics and macroscopic classical physics. We have already given some discussion of this, and see a connection with ideas of Penrose [88] that we plan to discuss elsewhere. It also seems that our approach can put the problem of measurement in a new light. These are just some examples of basic physical problems where others have conjectured that quantum gravity may play a rôle. In string theory, we have for the first time a consistent quantum theory of gravity in which these questions can be addressed in a meaningful way. We have started to provide some answers. Some of them may be incomplete or wrong in detail, but we believe we have put our fingers on some important aspects of the truth. We urge the reader to examine our approach with a cool head.

# Acknowledgements

One of us (N.E.M.) thanks C. Ktorides for valuable discussions about his work. The work of N.E.M. is supported by a EC Research Fellowship, Proposal Nr. ERB4001GT922259. The work of D.V.N. is partially supported by DOE grant DE-FG05-91-ER-40633.

## References


[1] I. Antoniadis, C. Bachas, J. Ellis and D.V. Nanopoulos, Phys. Lett. B211 (1988), 393; Nucl. Phys. B328 (1989), 117; Phys. Lett. B257 (1991), 278.

[2] J. Ellis, N.E. Mavromatos and D.V. Nanopoulos, Phys. Lett. B293 (1992), 37.

[3] J. Ellis, N.E. Mavromatos and D.V. Nanopoulos, CERN, ENS-LAPP and Texas A & M Univ. preprint, CERN-TH.6896/93, ENS-LAPP-A426-93, CTP-TAMU-29/93; ACT-09/93 (1993); hep-th/9305116.

[4] J. Ellis, N.E. Mavromatos and D.V. Nanopoulos, CERN, ENS-LAPP and Texas A & M Univ. preprint, CERN-TH.6897/93, ENS-LAPP-A427-93, CTP-TAMU-30/93; ACT-10/93 (1993); hep-th/9305117.

[5] E. Witten, Phys. Rev. D44 (1991), 314.

[6] See, for instance, A.M. Polyakov, *Gauge Fields and Strings* (Harwood, New York 1984).

[7] S. Hawking, Comm. Math. Phys. 87 (1982), 395.

[8] J. Bekenstein, Phys. Rev. D12 (1975), 3077.

[9] S. Hawking, Comm. Math. Phys. 43 (1975), 199.

[10] J. Ellis, J.S. Hagelin, D.V. Nanopoulos and M. Srednicki, Nucl. Phys. B241 (1984), 381.

[11] B. Misra, I. Prigogine and M. Courbage, *Physica* A98 (1979), 1;

I. Prigogine, *Entropy, Time, and Kinetic Description*, in *Order and Fluctuations in Equilibrium and Non-Equilibrium Statistical Mechanics*, ed G. Nicolis *et al.* (Wiley, New York 1981);

B. Misra and I. Prigogine, *Time, Probability and Dynamics*, in *Long-Time Prediction in Dynamics*, ed G. W. Horton, L. E. Reichl and A.G. Szebehely (Wiley, New York 1983).

[12] B. Misra, *Proc. Nat. Acad. Sci. U.S.A.* 75 (1978), 1627.

[13] R. M. Santilli, Hadronic J. 1 (1978), 223 , 574 and 1279 ; *Foundations of Theoretical Mechanics*, Vol. I (1978) and II (1983) (Springer-Verlag, Heidelberg-New York);

For an application of this approach to dissipative statistical systems, which is directly relevant to our work here, see: J. Fronteau, A. Tellez-Arenas and R.M. Santilli, Hadronic J. 3 (1979), 130;

J. Fronteau, Hadronic J. 4 (1981), 742.

[14] J.P. Constantopoulos and C.N. Ktorides, J. Phys. A17 (1984), L29.

[15] J. Ellis, N.E. Mavromatos and D.V. Nanopoulos, Phys. Lett. B293 (1992), 142; CERN and Texas A & M Univ. preprint CERN-TH. 6755/92; ACT-24/92;CTP-TAMU-83/92; hep-th/9212057.

[16] J. Ellis, N.E. Mavromatos and D.V. Nanopoulos, preprint CERN-TH.7000/93, CTP-TAMU 66/93, ENSLAPP-A-445/93, OUTP-93-26P, hep-th/9311148, *Opening lecture at HARC workshop on "Recent Advances in the Superworld", The Woodlands, Texas (USA), April 14-16 1993*, to appear in the proceedings.

[17] J. Ellis, N.E. Mavromatos and D.V. Nanopoulos, Phys. Lett. B289 (1992), 25; *ibid* B296 (1992), 40.

[18] A.V. Yung, Int. J. Mod. Phys. A9 (1994), 591.

[19] R. Wald, Phys. Rev. D21 (1980), 2742.

[20] R.P. Feynman and F.L. Vernon Jr., Ann. Phys. (NY) 94 (1963), 118.

[21] A.O. Caldeira and A.J. Leggett, Ann. Phys. 149 (1983), 374.

[22] Y.R. Shen, Phys. Rev. 155 (1967), 921;

A.S. Davydov and A. A. Serikov, Phys. Stat. Sol. B51 (1972), 57;

B.Ya. Zel'dovich, A.M. Perelomov, and V.S. Popov, Sov. Phys. JETP 28 (1969), 308;

For a recent review see : V. Gorini *et al.*, Rep. Math. Phys. Vol. 13 (1978), 149.

[23] J. Ellis, N.E. Mavromatos and D.V. Nanopoulos, Phys. Lett. B288 (1992), 23.

[24] M.B. Green, J.H. Schwarz and E. Witten, *String Theory*, Vol. I and II (Cambridge Univ. Press 1986).

[25] See, for instance, P. Ginsparg, Nucl. Phys. B295 (1988), 153.

[26] F. David, Mod. Phys. Lett. A3 (1988), 1651;

J. Distler and H. Kawai, Nucl. Phys. B321 (1989), 509.

[27] N.E. Mavromatos and J.L. Miramontes, Mod. Phys. Lett. A4 (1989), 1847.

[28] G. Shore, Nucl. Phys. B286 (1987), 349.

[29] H. Osborn, Nucl. Phys. B294 (1987), 595; *ibid* B308 (1988), 629; Phys. Lett. B222 (1989), 97.

[30] J. Polchinski, Nucl. Phys. B324 (1989), 123;

D.V. Nanopoulos, in *Proc. Int. School of Astroparticle Physics*, HARC-Houston (World Scientific, Singapore, 1991), p. 183.

[31] A.B. Zamolodchikov, JETP Lett. 43 (1986), 730; Sov. J. Nucl. Phys. 46 (1987), 1090.

[32] N.E. Mavromatos, J.L. Miramontes and J.M. Sanchez de Santos, Phys. Rev. D40 (1989), 535.

[33] C. Vafa, Phys. Lett. B212 (1987), 27.

[34] J. Ellis, N.E. Mavromatos and D.V. Nanopoulos, to appear.

[35] S. Chaudhuri and J. Lykken, Nucl. Phys B396 (1993), 270.

[36] J. Ellis, N.E. Mavromatos and D.V. Nanopoulos, Phys. Lett. B267 (1991), 465; *ibid* B272 (1991), 261.

[37] E. Brézin and V.A. Kazakov, Phys. Lett. B236 (1990), 144;

M. Douglas and A. Shenker, Nucl. Phys. B335 (1990), 635;

D. Gross and A.A. Migdal, Phys. Rev. Lett. 64 (1990), 127;

For a recent review see, e.g., I. Klebanov, in *String Theory and Quantum Gravity*, Proc. Trieste Spring School 1991, ed. by J. Harvey et al. (World Scientific, Singapore, 1991), and references therein.

[38] J. Ellis, N.E. Mavromatos and D.V. Nanopoulos, Phys. Lett. B284 (1992), 43.

[39] V.A. Rubakov, Nucl. Phys. B203 (1982), 311;

C. Callan, Nucl. Phys. B212 (1983), 391.

[40] M. Goulian and M. Li, Phys. Rev. Lett. 66 (1991), 2051.

[41] M. Bershadsky and D. Kutasov, Phys. Lett. B266 (1991), 345.

[42] I. Kogan, Phys. Lett. B265 (1991), 269.

[43] T. Roy and A. Roy Chowdhuri, Phys. Rev. D15 (1977), 3768.

[44] T. Eguchi, Mod. Phys. Lett. A7 (1992), 85.

[45] S. Coleman, Phys. Rev. D15 (1977), 2929; 1248 (E) (1977).

C. Callan and S. Coleman, Phys. Rev. D16 91977), 1762.

[46] B.A. Ovrut and S. Thomas, Phys. Rev. D43 (1991), 1314.

[47] D. Kutasov, Mod. Phys. Lett. A7 (1992), 2943.

[48] A. Pruisken, Nucl. Phys. B290[FS20] (1987), 61.

[49] A.V. Yung, Swansea preprint SWAT 94/22 (1994).

[50] J. Ellis, N.E. Mavromatos and D.V. Nanopoulos, Phys. Lett. B276 (1992), 56.

[51] W. Fischler and L. Susskind, Phys. Lett. B171 (1986), 262; *ibid* B171 (1986), 383.

[52] E. Cohen, H. Kluberg-Stern, and R. Peschanski, Nucl. Phys. B328 (499) (1989);
E. Cohen, H. Kluberg-Stern, H. Navelet, and R. Peschanski, Nucl. Phys. B347 (802) (1990).

[53] I. Affleck, M. Dine and N. Seiberg, Nucl. Phys. B241 (1984), 493.

[54] E. Witten, Nucl. Phys. B373 (1992), 187.

[55] S. Mukhi and C. Vafa, Harvard Univ. and Tata Inst. preprint HUTP-93/A002; TIFR/TH/93-01 (1993).

[56] A. Mueller, Phys. Rep. 73 (1981), 237.

[57] V.V. Khoze and A. Ringwald, Nucl. Phys. B355 (1991), 351.

[58] I.I. Balitsky and A.V. Yung, Phys. Lett. B168 (1986), 113;
A.V. Yung, Nucl. Phys. B297 (1988), 47.

[59] N. Dorey, Los Alamos National Lab. preprint, LA-UR-92-1380 (1992), Phys. Rev. D to be published.

[60] J. Bossart and Ch. Wiesendanger, ETH and Univ. of Zürich preprint, ETH-TH/91-42; ZH-TH-32/91 (1991).

[61] A. Pruisken, Nucl. Phys. B235[FS11] (1984), 277;
H. Levine, S. Libby and A. Pruisken, Phys. Rev. Lett. 51 (1983), 1915; Nucl. Phys. B240[FS12] (1984), 30;49;71.

[62] Kimyeong Lee, Nucl. Phys. B373 (1992), 735.

[63] C.A. Lütken and G.G. Ross, Oxford preprint OUTP-92-22P (1992), and references therein.

[64] J. Ellis, S. Mohanty and D.V. Nanopoulos, Phys. Lett. B221 (1989), 113.

[65] M Eliashvili, LAPP (Annecy) preprint ENSLAPP-A-462/94 (1994).

[66] R.B. Laughlin, in *The Quantum Hall Effect*, ed. R.A. Prange and S.M. Girvin (Springer-Verlag New York 1990).

[67] A. Capelli, C. Trugenberger and G. Zemba, Nucl. Phys. B396 (1993), 465 ;
S. Iso, D. Karabali and B. Sakita, Phys. Lett. B296 (1992), 143.

[68] I. Bakas and E. Kiritsis, Int . J. Mod. Phys. A7 (Suppl. 1A) (1992), 55.

[69] R. Dijkgraaf, E. Verlinde and H. Verlinde, Nucl. Phys. B371 (1992), 269.

[70] I. Bakas and E. Kiritsis, ENS (Paris) preprint LPTENS-92-30 (1992).

[71] G. Luders, Ann. Phys. (N.Y.) 2, (1957), 1.

[72] E. Witten, Comm. Math. Phys. 109 (1987), 525.

[73] V. A. Kostelecký and R. Potting, Nucl. Phys. B359 (1991), 545.

[74] H. Sonoda, Nucl. Phys. B326 (1989), 135.

[75] J. Ellis, J. Lopez, N.E. Mavromatos and D.V. Nanopoulos, in preparation ;

J. Lopez, preprint CPT-TAMU 38/93 (1993), talk at *HARC workshop on "Recent Advances in the Superworld", The Woodlands, Texas (USA), April 14-16 1993*, to appear in the proceedings.

[76] P. Huet and M. Peskin, SLAC (Stanford) preprint SLAC-PUB-6454 (1994), hep-ph/9403257.

[77] *Review of Particle Properties*, Particle Data Group, Phys. Rev. D45 (No 11, part II) (1992), 1.

[78] N.E. Mavromatos and J.L. Miramontes, Phys. Lett. B212 (1988), 33;

N.E. Mavromatos, Phys. Rev. D39 (1989), 1659.

[79] N.E. Mavromatos, Mod. Phys. Lett. A3 (1988), 1079.

[80] See, for instance, S. Weinberg *Gravitation and Cosmology* (Wiley, New York 1972 ).

[81] N. Sanchez and G. Veneziano, Nucl. Phys. B333 (1990), 253.

[82] N. Turok, Phys. Rev. Lett. 60 (1988), 549.

[83] See, for instance, A.H. Guth, *Phase Transitions in the Embryo Universe*, Phil. Trans. Roy. Soc. Lond. A307 (1982), 141.

[84] J. Ellis, N. Mavromatos and D.V. Nanopoulos, Phys. Lett. B278 (1992), 246.

[85] A.A. Tseytlin, Phys. Lett. B178 (1986), 34.

[86] I. Kogan, Univ. of British Columbia preprint, UBCTP 91-13 (1991).

[87] I. Antoniadis, G.F.R. Ellis, J. Ellis, C. Kounnas and D.V. Nanopoulos, Phys. Lett. B191 (1987), 393;

J. Barrow and N. Deruelle, Nucl. Phys. B297 (1988), 733.

[88] R. Penrose, *The Emperor's New Mind* (Oxford Univ. Press, 1989).

**CHAIRMAN: J.Ellis**

*Scientific Secretaries: M.Giovannini, B.Zheng*

## DISCUSSION

– *Gourdin:*

I wish to clarify a point concerning the $K^0 - \overline{K}^0$ system. In the normal treatment we use both the superposition principle and the Schrödinger–type time evolution equation.The physical states are linear combinations of the PC eigen states $K_1^0$ and $K_2^0$ with

$$K_s^0 = K_1^0 + \varepsilon_s K_2^0$$

and

$$K_L^0 = K_2^0 + \varepsilon_L K_1^0$$

CPT invariance implies $\varepsilon_S = \varepsilon_L$ and a violation of CPT invariance is exhibited by $\varepsilon_L \neq \varepsilon_S$. In your approach the superposition principle is no longer valid and you describe a violation of CPT with 3 parameters $\alpha\ \beta\ \gamma$. How can we distinguish between these various types of CPT violations?

– *Ellis:*

Your question is asking how it is possible to distinguish between this type of CPT violation and the other type of CPT violation. We start from the evolution equation of the density operator

$$\partial_t \hat{\rho} = -i(\hat{H}\hat{\rho} - \hat{\rho}\hat{H}^t) + \partial \hat{H} \hat{\rho}$$

or in components

$$\partial_t \rho_\alpha = h_{\alpha\beta}\ \rho_\beta + \not{h}_{\alpha\beta}\ \rho_\beta$$

where

$$\not{h}_{\alpha\beta} = \begin{pmatrix} 0 & 0 & 0 & 0 \\ 0 & 0 & 0 & 0 \\ 0 & 0 & -2\alpha & -2\beta \\ 0 & 0 & -2\beta & -2\gamma \end{pmatrix}$$

The hermitian part of the Hamiltonian gives an anti–symmetric contribution (is symplectic). The non-hermitian part is not symplectic and gives you the decay during the evolution. If I introduce CPT violation in the normal way I would have opposite signs in the off-diagonal elements of the Hamiltonian. In our approach we have the same sign $(-2\beta)$. This enables one in principle to distinguish between our form of CPT violation and other kinds of violations: look at the sign of the off-diagonal elements of the Hamiltonian.

– *Morpurgo:*

I have in fact two questions: The first is related to that of Prof. Gourdin and is as follows: in addition to the $K^0 - \overline{K}^0$ system are there other cases that may provide an experimental verification of your modification of quantum mechanics?

The second question is this: If I am not wrong, in the Ghirardi–Rimini–Weber modification of quantum mechanics, macroscopic systems are affected considerably while microscopic ones are affected negligibly. Is the same true in your case? Quantitatively can you give an order of magnitude of how much microscopic systems are affected?

– *Ellis:*

To your first question I can answer that in our first paper in 1983 we found that the effects of modification of quantum mechanics could be detectable also. If you do an experiment of interference of two neutrino beams the amount of loss of phase coherence during the time evolution is comparable with the kaon system.

In answer to your second question: – the Ghirardi–Rimini–Weber modification of quantum mechanics appears to take place at the level of the wave function (i.e. the Schrödinger equation). Our modification is encoded in the quantum Liouville equation, which is the evolution equation for the density matrix. The difference between our approaches and the Ghirardi–Rimini–Weber approach is not completely clear to me. However the mechanism with which the off-diagonal terms of the density matrix collapse in the Ghirardi–Rimini–Weber approach is very similar to our mechanism of collapse even if the reasons for their modification of Q.M. seem very different from ours. About the magnitude of $|\partial H/H|$ I can say that it is likely to be of the order of $\left(\frac{\Lambda_{QCD}}{M_p}\right)^n$ with n $\geq 1$. If you look at the mass difference between $K^0$ and $\overline{K}^0$, it is $O(10^{-15}$ Gev) and the CP violation is $O(10^{-3})$ smaller. So $O\left(\frac{\Lambda_{QCD}}{M_p}\right)^1$ is very close. I might add that if you want our non-Hamiltonian to suppress interference for macroscopic bodies à la Ghirardi–Rimini–Weber you probably need n $= 1$.

– *Zichichi:*

Could you comment about the relationship between space-time and the coupling parameter space and also about the relationship between the increase of entropy and our time?

– *Ellis:*

I was not talking about gauge coupling space. I was talking about a general space of couplings of sigma models. In String Theory we do not know which is the effective field theory which correctly describes physics on the world-sheet. Any individual string vacuum is presumably described by one particular 2-dimensional field theory. In String Theory we have to study the set of possible field theories sitting on the world sheet. The co-ordinates $q_i$ describe a sort of abstract space which is the space of possible 2-dimensional field theories on the world-sheet. We call them $\sigma$ models just for practical reasons. For a black hole the world-sheet is described by a non-linear $\sigma$ model. In other cases things may be different. The dynamics in this "space of theories" is controlled by the Zamolodchikov $C$ function, which can be regarded as a sort of effective action as N.Mavramatos will discuss later on during the school.

– *Lopez:*

How are your microscopic black holes related to the macroscopic black holes?

– *Ellis:*

The connection is not so clear at the moment. Astrophysical observations are very difficult even if the Hawking–Bekenstein predictions are quite direct. If observed, black hole evaporation will not necessarily shed much light on the 2-D black hole physics.

– *Lu:*

Some of the principles of local field theories (i.e. locality and Lorentz invariance) will not be true in String Theory. Could you be more specific about this point?

– *Ellis:*

You have no problems with theories on the world-sheet which are completely local and where conventional quantum field theory and quantum mechanics apply. The problem comes out when you discuss the elevation of the field theories on the world-sheet to space-time. String Theory is not strictly local in the sense of local field theories. In the case of a black hole there is a large multiplicity of states which are related together with W-algebra. These states are non-local states. This modification of locality is very important. On the other hand, Lorentz invariance is something which is not so obvious in non-critical String Theory. If you want to keep Lorentz invariance just because the world is Lorentz invariant I can tell you that the world is not so Lorentz invariant! We live in a Friedmann–Robertson–Walker universe with a certain arrow of time. We claim that it is possible to connect this macroscopic arrow of time with a microscopic arrow of time.

– *Giannakis*

A crucial point in your analysis was the existence of an infinite symmetry. Do you expect that the Coleman–Mandula theorem holds in String Theory and if it does what will be the consequences?

– *Ellis:*

No, the Coleman–Mandula theorem is not applicable here. It applies to local quantum number only. The states here which carry these quantum numbers are extremely extensive states, and the quantum numbers are defined only when one integrates over the whole of space–time. These quantum numbers are not local.

– *Montgomery*

There are existing measurements of the relative phases of $\eta^{00}$ and $\eta^{+-}$ in the Kaon system with an error of one to two degrees and an experiment with an error of 0.5 degrees when its analysis is complete. Has this any relevance to the CPT violation which you discussed?

– *Ellis:*

Yes, it is absolutely and directly relevant. Lopez, Mavromatos, Nanopoulos and myself have a project to do the intermediate-time analysis, including the experiments. It will involve $\alpha, \beta$ and $\gamma$ not only $\beta$ and $\gamma$.

– *Hoang:*

Is the $B-\overline{B}$ system also suitable for testing the modified quantum mechanics?

– *Ellis:*

Unfortunately the $B-\overline{B}$ system is not so good because the mixing in the $B-\overline{B}$ system is much larger and CP violation in the mass matrix is not so accessible.

– *Budzynski:*

Are there any a priori reasons to expect the conservation of energy for matter interacting with gravity when quantum gravitational effects become important?

– *Ellis:*

Yes, we still expect the conservation of energy on the average because a) experimentally it exists and b) there is time translation invariance. I am happy to say that we have been able to prove that in String Theory we do have energy conservation as in normal quantum mechanics. Also, energy fluctuations are as in conventional mechanics. Thus an objection by Banks, Peskin and Susskind to our formulism has been overcome.

– *Kaplunovsky:*

Your modified quantum mechanics is non-linear at least in the level of wave functions. But there is an experiment with extremely high precision showing that quantum mechanics is linear. How does this fit together?

– *Ellis:*

Our modification of quantum mechanics is non-linear, though superpositions do vanish. The Ghirard–Rimini–Weber formulism really is non–linear. The experiment you refer to tests a different type of non-linearity. It would in principle be interesting to study this also, but we have not done so yet.

**CHAIRMAN: N.Mavromatos**

*Scientific Secretaries: M.Cavaglia, N.Sarlis, R.Wegrzyn*

## DISCUSSION

– *Arnowitt:*

There are many kinds of time; cosmological, thermodynamical, Newtonian (which is reversible). Which time or times are you discussing?

– *Mavromatos:*

In strings propagating in non-singular space-time there is no arrow of time, the latter being defined as a $\sigma$–model field. In our approach of strings propagating in singular foamy backgrounds, time is "thermodynamic" in the sense of being associated with loss of information across event horizons. The latter is carried by the global string modes.

– *Arnowitt:*

The topological modes could not be measured in the laboratory because they are not localizable. Are they normalizable which is required by quantum physics? Or is this a breakdown of quantum mechanics?

– *Mavramatos:*

Once I truncate the string theory to individual nodes the wave functions are not normalizable. I would put it differently, there is no concept of state vector and wave function in this effective theory, and the light energy world is necessarily described by a density matrix, as in "open" quantum mechanical systems.

– *Arnowitt:*

You said that Planck's constant is a function of time. Is it decreasing or increasing i.e. is the universe becoming more or less quantum mechanical?

– *Mavramatos:*

Planck's constant in our approach seems to decrease exponentially with time in contrast to Ghirardi Rimini Weber's approach to the theory of measurement, where the uncertainty for position and momenta seem to increase with time.

– *Lopez:*

What is the difference between a $C = 26$ bosonic string with background charges and a $D = 4$ bosonic string with additional conformal charge (e.g. free fermion)?

– *Mavramatos:*

For critical strings the number of fields of the $\sigma$–model is different. For non-critical (e.g.2–D) strings we use fewer fields but the cost we pay is the introduction of non–trivial backgrounds.

– *Lopez:*

Why do you only consider bosonic strings? What about superstrings? How will this change things?

– *Mavromatos:*

Bosonic strings are easier, but one could also consider superstrings. I do not expect to find a big difference as far as the nature of time is concerned . I am taking this question as an opportunity to mention that N=2 supersymmetric $\sigma$–models have been considered. In this context, in order to describe (after twisting) a topological phase of string theory which after some symmetry breaking leads to singular (2 dimensional) space times.

– *Nanopoulos:*

In the case of exact superstring solutions in curved space-time, the difference between a bosonic string and a superstring is not relevant.

– *Lopez:*

What drives the Huble expansion?

– *Mavramatos:*

It is the gravitational friction caused by the interaction of light-matter with the global modes. This, we believe, drives the Huble expansion, in our definition of time.

– *Giovannini:*

I have two questions.

a) Let's go to the interference Gedanken experiment of a string turning around a black hole in two dimensions. In two dimensions the Ahavonov–Bohm effect does not hold, but it seems that you use 2–D black hole physics and 4–D effect. Could you tell me in which is the dimensionality in which you work?

b) You stated that the singularity structure of 2D black hole physics is the same as in 4 dimensions. Could you state this as a theorem?

– *Mavramatos:*

a) The Ahavonov–Bohm measurement refers to the s–wave 4D black hole case (spherically symmetric, 4D black hole). Our 2D string framework can be used to describe the singularity structure of a 4D s–wave black hole which we believe is the same as in 2D string theories. There are attempts to show this either at an effective field theory level (work done by Ellis, Nanopoulos and myself) or at the string level ( Strominger, Giddings and Horowitz). There is a similarity of our approach with the Callan–Dubakov effect which describes the scattering of fermions of spherically symmetric monopoles in 4D. The phenomenon leads to s–wave fermion number non–conservation which can be described effectively in the framework of 2D Yang–Mills theories.

b) No, I can't state this as a theorem. There is a paper of Strominger, Giddings and Horowitz which gives you one example in this direction. But in any case I can't state this as an established result. At the moment I have no general argument but just intuition.

– *Nanopoulos:*

In string theory it seems that the space–time singularity corresponds to a topological theory on the world sheet. Because a topological theory (by definition) does not use a specific metric, the notion of dimension of space is irrelevant, so it may be that the 2–D black hole may define a "universality" class for any dimension.

– *Cadoni:*

In the conventional black hole physics a macroscopic black hole will lose its mass through Hawking radiation leading to a naked singularity. My question is whether there is in your approach a mechanism which prevents complete evaporation of a black hole.

– *Mavramatos:*

In our model it is argued that the black hole evaporates until the Planck mass is reached. This acts as a "cut–off" of black hole masses, in the sense that the microscopic (Planck mass) black hole becomes indistinguishable from the rest of the (virtual) black holes in the foam.

– *McPherson:*

My understanding of observable CPT violation in your model is that CPT is conserved in some total sense, but since we only observe part of the total with the rest remaining hidden, we happen to observe CPT violation in low energy phenomena. Is this correct?

– *Mavromatos:*

Basically, yes. CPT is conserved in the complete string theory but the effective theory applicable to low energy phenomena need not conserve CPT.

– *Giannakis:*

Does the RG flow which defines time for you take place in the space of 2D field theories?

– *Mavramatos:*

The RG flow takes place in the space of 2D conformal field theories.

– *Giannakis:*

Do you expect that the relation $Q = M$ of an extreme Reissner–Noraström type black hole will be modified in your framework because of the existence of the W–charges?

– *Mavramatos:*

The 2–D black hole is an extreme black hole as far as its mass is concerned. Due to the W–charges it is possible that a generalised extremity condition holds, but this is not known yet.

– *Kawall:*

How do you exclude the existence of white hole singularities which would add information to the universe?

– *Mavramatos:*

Our theory implies, in general, CPT violation in the low-energy world, and this prevents the formulation of white holes. If CPT is not violated then they could exist in principle.

– *Lu:*

I am confused about a remark you made this morning. You mentioned that decoupling theories were violated in several places, such as in $\pi^0 \rightarrow 2\gamma$ decay. But $\pi^0 \rightarrow 2\gamma$ was understood via anomaly. Why does the decoupling theorem fail in that case?

– *Mavromatos:*

All I wanted to remark on is an analogy of the case at hand with the anomaly problem. There is a way to understand anomalies in general, not only for the case $\pi^0 \rightarrow 2\gamma$, by means of failures of the decoupling theorem, which implies that physics at the ultraviolet cut-off scale affects the low-energy phenomena. Anomalies are associated with certain loop graphs that contain exceptional momenta, and hence they violate one of the basic assumptions of the theorem, that of scaling of momenta.

– *Stoilov:*

In 2D Quantum Gravity there are physical states in which the ghost number is not zero. Could you comment on their significance for a field model obtained from non-critical string?

– *Mavramatos:*

Allow me to be a little more technical. In 2D String Theory the cohomology of the physical states involves not only standard ghost number states but also non-standard ones at adjacent values of the ghost number (Witten, Lian-Zuckermann). These states form a (ground) ring under OPE and exist in the 2D String Black Hole case as well as in the flat space-time case. These states are associated with (1,0) and (0,1) conformal currents on the world sheet which in turn induce a world-sheet (global) $W_\infty$ algebra. These currents induce deformations of the world-sheet theory, which in turn can be interpreted as target space symmetries. The resulting symmetry algebra is a local $W_\infty$ algebra in space-time, which is the essential property of the string black holes according to this talk. The global string modes that carry the information across the event horizons are closely associated with the $W_\infty$ algebra. In the case of black holes it is the $W_\infty$ target symmetry that is believed to be responsible for the quantum coherence of the full string theory. Upon truncating the theory to its effective light-mode part the $W_\infty$ symmetry, that mixes various string levels, is broken and coherence "seems" to be lost at the effective theory level.

– *Hoang:*

I am asking you a question as someone not involved with string theory. You stressed several times that the renormalisability is extremely important for your model. I want to know whether the concept of renormalisability in string theory is different than in QFT.

– *Mavramatos:*

Renormalising $\sigma$–models on the world-sheet can be done by the normal procedure of renormalisation of 2D QFT. However, because the world-sheet is curved a local renormalisation scale on the world-sheet has to be introduced. This is the difference from the usual flat-space renormalisation in QFT.

# STATUS OF ELECTROWEAK THEORY

Paul Langacker
University of Pennsylvania
Department of Physics
Philadelphia, Pennsylvania, USA 19104-6396

Abstract

The status of precision tests of the standard electroweak model is reviewed, as are implications for supersymmetric grand unification.

## Introduction

- The Standard Model and Its Extension
- The Two Paths, Unification or Compositeness
- The $Z$, the $W$, and the Weak Neutral Current
- Experimental Results
- Standard Model Tests and $m_t$
- Supersymmetry
- (Supersymmetric) Grand Unification

# 1 The Standard Model and Its Extensions

The standard $SU_3 \times SU_2 \times U_1$ model is almost certainly the correct description of nature to an excellent first approximation down to a distance scale of $10^{-16}$ cm, except possibly in the Higgs sector. Despite the great success of the standard model, however, it is clear that there must be new physics that underlies it. This is not so much because there is anything wrong or inconsistent with the standard model, but rather because it leaves so many questions unanswered. Amongst the difficulties are:

- The minimal version of the standard model, with massless neutrinos and including classical general relatively, contains 21 arbitrary parameters which must be determined from experiment.

- The standard model has a complicated gauge structure, and there is no fundamental explanation of charge quantization, *i.e.*, of why the atom is electrically neutral. One possible extension that could explain this is grand unification.

- The fermion mass spectrum is extremely complicated. In the standard model this is put in by hand; there is no explanation for the existence of fermion families or for their masses or mixings. Possible extensions that might bear on this question include compositeness and string theories.

- The Higgs mechanism needed to break the electroweak symmetry has a severe problem involving quadratic divergences of the Higgs mass-squared parameter. These yield higher order corrections which tend to change the Higgs mass from the scale $M_H \sim M_W \sim 100$ GeV needed for electroweak symmetry breaking to the very much larger Planck scale, $M_P \sim 10^{19}$ GeV. One could always cancel these corrections against the bare value, but that involves an unnatural fine-tuning between two things that do not seem to be related. Possible extensions of the standard model suggested by this include (extended) technicolor, in which the Higgs is not an elementary particle, or supersymmetry (SUSY), in which there are cancellations between diagrams involving fermions and bosons.

- The strong $CP$ problem involves a new CP-violating term which can be introduced into the QCD Lagrangian or which is generated by electroweak $CP$ violation. This tends to generate a neutron electric dipole moment $\sim 10^8$ too large unless fine-tunings are invoked. Possible solutions include the Peccei-Quinn symmetry, spontaneous $CP$ violation, or a massless $u$ quark.

- There is no quantum gravity in the standard model, and attempts to add quantum gravity by hand lead to terrible divergences. Furthermore, the symmetry breaking needed for the electroweak theory generates a vacuum energy which can be interpreted as a cosmological constant some 50 orders of magnitude larger than is allowed experimentally, $\Lambda_{\rm SSB} \sim 10^{50}\Lambda_{\rm exp}$. Again, one can have a fine-tuned cancellation between the induced and primordial (bare) values, but this appears extremely unnatural. Possible solutions include supergravity, which at least brings quantum gravity into the game but does not shed light on renormalizability or on the cosmological constant, or superstring theory, which appears to lead to a finite theory of quantum gravity. Its implications for the cosmological constant are still uncertain.

## 2 The Two Paths, Unification or Compositeness

There are many ideas for possible extensions of the standard model. However, most fall into two broad categories: unification or compositeness (or, as I like to say, the Bang or the Whimper).

The Bang scenario is the idea that there is a grand desert up to some grand unification (GUT) or Planck scale. The Bang models are intrinsically perturbative, and are the domain of elementary Higgs fields, SUSY, GUTs, and string theories. The well-known success of the coupling constant unification in the minimal supersymmetric extension of the standard model at least hints that this may be the right scenario. If nature chooses this route it (almost) requires the existence of supersymmetry as observable new physics, and this in turn implies the existence of a light scalar particle which acts like the standard model Higgs but which has a relatively small mass, $M_H <$ 150 GeV. If this sort of approach is correct there are relatively few types of new physics which are possible without messing up the successful coupling constant unification (barring the possibility of cancellations between two new effects). The few safe types of physics include new $Z'$ bosons, new sequential or mirror families of fermions, or exotic multiplets of fermions. The latter include $SU_2$ singlet quarks, $D_{\mathrm{L,R}}$, which have the charge and color of the $d$ quark, and mirror doublet lepton families, $\binom{E^0}{D^-}_{\mathrm{L,R}}$, in which both left and right-handed fields participate in normal weak interactions. Finally, one can add gauge singlet particles such as Higgs or neutrino supermultiplets.

Tests of this scenario include the existence of the light Higgs fields, which should be observable at LEP or at the LHC, and the existence of the superpartners or sparticles, which can be observed at hadron colliders. In addition, there is at least the possibility of observing flavor changing neutral currents (FCNC) or possibly proton decay. Within the precision experiments there is a possibility of observing the indirect effects of $Z'$ bosons or exotic fermions. However, the major prediction of most of the models in this class of extensions for precision experiments is the *absence* of other deviations from the standard model predictions.

The other general scenario is the compositeness (Whimper) scenario. This is the notion that there are onion-like layers of compositeness. It is intrinsically non-perturbative and is the domain of dynamical symmetry breaking or composite fermions. Models of this sort have the generic prediction of the existence of FCNC. The absence of observed FCNC is a severe problem, and in my opinion argue strongly against this scenario. Certainly, it has obstructed the construction of realistic models. If one somehow manages to evade these problems the predictions for the future include anomalous $WW \to WW$ cross sections and triple-gauge vertices which can be observed at hadron and $e^+e^-$ colliders, and the existence of new bound state particles. In the precision experiments one should be able to observe 4-fermion operators, anomalous values of the $Z \to b\bar{b}$ decay rate, and anomalous values of the $\rho_0$ or the $S$, $T$, $U$ parameters.

## 3 The $Z$, the $W$, and the Weak Neutral Current

The weak neutral current and properties of the weak gauge bosons have always been the primary test of the unification part of the electroweak $SU_2 \times U_1$ model. These tests have gone through a number of phases [1].

1. During the mid-1970's the weak neutral current was a successful prediction of the $SU_2 \times U_1$ model, as was verified by the discovery of neutral currents by the Gargamelle Collaboration in 1973 and their subsequent confirmation at Fermilab, in such processes as $\nu N \to \nu X$, $\nu p \to \nu p$, $\nu N \to \nu \pi X$, and $\nu e \to \nu e$.

2. During the late 1970's there was a second generation of experiments, typically at the 10% precision level, which allowed model independent determinations of parameters which could be compared to the predictions of arbitrary gauge theories. The parameters of a number of reactions were successfully determined, with the conclusion that $SU_2 \times U_1$ and not some entirely separate model was indeed correct to first approximation.

3. A third generation of neutral current experiments in the 1980's, as well as the direct observation of the $W$ and $Z$ particles and their masses by the UA1 and UA2 collaborations at CERN in 1983, led to much more precise tests of the standard model, stringent limits on the possible existence of certain types of new physics, and rough tests of the radiative corrections. These experiments were typically at the 1 – 5% precision level. They included purely weak neutral current processes, such as $\nu N \to \nu X$, $\nu p \to \nu p$, $\nu N \to \nu \pi X$, and $\nu e \to \nu e$; weak-electromagnetic interference experiments, such as the famous polarized asymmetry experiment $e^{\uparrow\downarrow} D \to eX$ at SLAC, atomic parity violation, and characteristic forward-backward asymmetries in $e^+e^- \to e^+e^-$, $\mu^+\mu^-$, $\tau^+\tau^-$, $c\bar{c}$, $b\bar{b}$ and $q\bar{q}$ observed at PEP, PETRA, and TRISTAN; and the masses of the $W$ and $Z$ bosons. The result of this generation of experiments was again that the standard model is correct within the experimental precision, that the top quark exists (because of measurements of the weak interactions of its partner, the bottom quark), the upper limit $m_t < 200$ GeV, and a value $\sin^2 \theta_W \sim 0.230 \pm 0.007$ GeV for the weak angle.

4. A fourth generation of ultrahigh precision $Z$-pole experiments began in 1989 at LEP and more recently at the SLC. Many of these experiments are at the $\ll 1\%$ level, and they include such properties of the $Z$ as its mass, various total and partial widths, the effective number of neutrinos, the forward-backward asymmetry for leptons and the polarization of the $\tau$: $M_Z$, $\Gamma_Z$, $\Gamma_{\ell\ell}$, $\Gamma_{\rm had}$, $N_\nu$, $A_{\rm FB}$, and $A_{\rm pol}(\tau)$. At later stages one will observe polarization asymmetries and various non-$Z$-pole observables, such as a much more precise $W$ mass and greatly improved measurements of atomic parity violation and deep inelastic neutrino scattering. As of this writing, all of these measurements are in superb agreement with the standard model, not only testing it precisely but also making predictions for the top quark mass and severely limiting possible extensions of the standard model, especially those associated with the Whimper-type scenario.

# 4 Experimental Results

Recent high precision measurements of $Z$ pole observables by the ALEPH, DELPHI, L3, and OPAL [2, 3, 4] collaborations at LEP and SLD at the SLC [5], the $W$ mass by CDF [6] and UA2 [7], atomic parity violation in cesium [8, 9], neutrino-electron scattering by CHARM II [10], and other weak neutral current observables [11, 12], as well as the direct lower bounds $m_t > 91$ GeV (CDF [13]) and $M_H > 60$ GeV (LEP average [14]) and the determination $\alpha_s(M_Z) = 0.12 \pm 0.01$ from $Z$-pole and low-energy observables [15] allow precise tests of the standard electroweak model and searches for certain types of new physics. In this article, which is an update of previously presented analyses [16]-[20], I review the status of the standard model tests and parameters, the coupling constant predictions in ordinary and supersymmetric grand unified theories (GUTs), and the implications for a variety of types of possible new physics.

Many of the recent results are summarized in Table 1. The LEP results are averages by D. Schaile of the four LEP experiments as of March, 1993 [4], which includes nearly final results for the 1991 LEP run and contains a proper treatment of common systematic errors [3]. $M_Z$ is now known to the incredible precision of better than 0.01%. This was achieved by the method of resonant depolarization, in which the (calculable) energies at which the small ($\sim$ 10%) transverse polarization of the leptons is destroyed by an oscillating $B$ field was used to calibrate the energy of the LEP beams. The method is so precise that the tidal effects of the moon, which cause the size of the LEP ring to change by a few parts in $10^8$ and thus change the energy by $\sim$ 8 $MeV$, had to be measured and corrected for[1]. $\Gamma_Z, \Gamma_{l\bar{l}}, \Gamma_{had}$, $\Gamma_{b\bar{b}}$, and $\Gamma_{inv}$ refer respectively to the total, leptonic (average of $e, \mu, \tau$), hadronic, $b\bar{b}$, and invisible $Z$ widths; $R \equiv \Gamma_{had}/\Gamma_{l\bar{l}}$; $\sigma_p^h = 12\pi\Gamma_{e\bar{e}}\Gamma_{had}/M_Z^2\Gamma_Z^2$ is the hadronic cross section on the pole; and $N_\nu \equiv \Gamma_{inv}/\Gamma_{\nu\bar{\nu}}$ is the number of light neutrino flavors. A number of asymmetries have also been measured. $A_{FB}(f)$ is the forward-backward asymmetry for $e^+e^- \rightarrow f\bar{f}$; $A_{pol}(\tau)$ is the polarization of a final $\tau$ ($L$ is positive), while $A_e(P_\tau)$ is essentially the forward-backward asymmetry in the polarization; $A_{LR}$ is the polarization asymmetry, which has recently been measured for the first time by the SLD collaboration at the SLC [5]. All of the asymmetries are Born contributions, from which various QED, QCD, interference, and box contributions have been removed by the experimenters. Finally, $\bar{g}_A, \bar{g}_V$ are effective Born couplings, related, for example, to $\Gamma_{l\bar{l}}$ and $A_{FB}(\mu)$ by[2]

$$\Gamma_{l\bar{l}} = \frac{G_F M_Z^3}{6\sqrt{2}\pi}(\bar{g}_A^2 + \bar{g}_V^2) \qquad A_{FB}(\mu) = \frac{3\ \bar{g}_V^2\ \bar{g}_A^2}{(\bar{g}_V^2 + \bar{g}_A^2)^2}. \tag{1}$$

Similarly,

$$A_{LR} = \frac{2\ \bar{g}_V\ \bar{g}_A}{\bar{g}_V^2 + \bar{g}_A^2}, \tag{2}$$

with the same expression for $A_{pol}(\tau)$ and $A_e(P_\tau)$.

Of the $Z$-pole observables only $M_Z$, $\Gamma_Z, R$, $\sigma_p^h$, $\Gamma_{b\bar{b}}$, $A_{FB}(\mu)$, $A_{pol}(\tau)$, $A_e(P_\tau)$, $A_{FB}(b)$ (which is corrected for $b\bar{b}$ oscillations), $A_{FB}(c)$, and $A_{LR}$ are used in the analysis. $\Gamma_Z$, $R$, and $\sigma_p^h$ are used rather than the more physically transparent $\Gamma_Z$, $\Gamma_{l\bar{l}}$, and $\Gamma_{had}$ because the former are closer to what is actually measured and are relatively weakly correlated. (The combined LEP values [4] for the correlations are used.) $\bar{s}_W^2$ ($A_{FB}(q)$), which is the effective weak angle obtained from the charge asymmetry in hadronic decays, is not used because the results have only been presented assuming the validity of the standard model. The other LEP observables are not independent but are displayed for completeness.

Recent measurements of the $W$ mass and weak neutral current data are also displayed in Table 1. $Q_W(Cs)$ is the effective charge of the parity-violating interaction in cesium [8], while $g_{V,A}^e$ are the coefficients of the vector and axial electron currents

---

[1]This is the first experiment in which all four interactions were important simultaneously!

[2]I assume lepton universality, throughout. This is strongly supported by the LEP data.

| Quantity | Value | standard model |
|---|---|---|
| $M_Z$ (GeV) | $91.187 \pm 0.007$ | input |
| $\Gamma_Z$ (GeV) | $2.491 \pm 0.007$ | $2.490 \pm 0.001 \pm 0.005 \pm [0.006]$ |
| $R = \Gamma_{had}/\Gamma_{l\bar{l}}$ | $20.87 \pm 0.07$ | $20.78 \pm 0.01 \pm 0.01 \pm [0.07]$ |
| $\sigma_p^h(nb)$ | $41.33 \pm 0.18$ | $41.42 \pm 0.01 \pm 0.01 \pm [0.06]$ |
| $\Gamma_{b\bar{b}}$ (MeV) | $373 \pm 9$ | $375.9 \pm 0.2 \pm 0.5 \pm [1.3]$ |
| $A_{FB}(\mu)$ | $0.0152 \pm 0.0027$ | $0.0141 \pm 0.0005 \pm 0.0010$ |
| $A_{pol}(\tau)$ | $0.140 \pm 0.018$ | $0.137 \pm 0.002 \pm 0.005$ |
| $A_e(P_\tau)$ | $0.134 \pm 0.030$ | $0.137 \pm 0.002 \pm 0.005$ |
| $A_{FB}(b)$ | $0.093 \pm 0.012$ | $0.096 \pm 0.002 \pm 0.003$ |
| $A_{FB}(c)$ | $0.072 \pm 0.027$ | $0.068 \pm 0.001 \pm 0.003$ |
| $A_{LR}$ | $0.100 \pm 0.044$ | $0.137 \pm 0.002 \pm 0.005$ |
| $\Gamma_{l\bar{l}}$ (MeV) | $83.43 \pm 0.29$ | $83.66 \pm 0.02 \pm 0.13$ |
| $\Gamma_{had}$ (MeV) | $1741.2 \pm 6.6$ | $1739 \pm 1 \pm 4 \pm [6]$ |
| $\Gamma_{inv}$ (MeV) | $499.5 \pm 5.6$ | $500.4 \pm 0.1 \pm 0.9$ |
| $N_\nu$ | $3.004 \pm 0.035$ | 3 |
| $\bar{g}_A$ | $-0.4999 \pm 0.0009$ | $-0.5$ |
| $\bar{g}_V$ | $-0.0351 \pm 0.0025$ | $-0.0344 \pm 0.0006 \pm 0.0013$ |
| $\bar{s}_W^2$ $(A_{FB}(q))$ | $0.2329 \pm 0.0031$ | $0.2328 \pm 0.0003 \pm 0.0007 \pm$ ? |
| $M_W$ (GeV) | $79.91 \pm 0.39$ | $80.18 \pm 0.02 \pm 0.13$ |
| $M_W/M_Z$ | $0.8813 \pm 0.0041$ | $0.8793 \pm 0.0002 \pm 0.0014$ |
| $Q_W(Cs)$ | $-71.04 \pm 1.58 \pm [0.88]$ | $-73.20 \pm 0.07 \pm 0.02$ |
| $g_A^e(\nu e \to \nu e)$ | $-0.503 \pm 0.017$ | $-0.505 \pm 0 \pm 0.001$ |
| $g_V^e(\nu e \to \nu e)$ | $-0.025 \pm 0.020$ | $-0.036 \pm 0.001 \pm 0.001$ |
| $\sin^2\theta_W$ | $0.2242 \pm 0.0042 \pm [0.0047]$ | $0.2269 \pm 0.0003 \pm 0.0025$ |

Table 1: Experimental values for LEP [2, 3, 4] and SLC [5] observables, $M_W$ [6], $M_W/M_Z$ [7], the weak charge in cesium $Q_W$ [8, 9], the parameters $g_{V,A}^e$ relevant to $\nu_\mu e$ scattering from CHARM II [10], and $\sin^2\theta_W \equiv 1 - M_W^2/M_Z^2$ from CCFR [11], compared with the standard model predictions for $M_Z = 91.187 \pm 0.007$ GeV, $m_t = 150^{+19}_{-24}$ GeV, and 60 GeV $< M_H <$ 1 TeV. Only the first eleven $Z$-pole observables are independent. The ? for the $\bar{s}_W^2$ $(A_{FB}(q))$ prediction refers to the scheme dependence. The two errors for $Q_W(Cs)$ and $\sin^2\theta_W$ are experimental and theoretical (in brackets). The first error in the predictions is from the uncertainties in $M_Z$ and $\Delta r$, the second is from $m_t$ and $M_H$, and the third (in brackets) is the theoretical QCD uncertainty for $\alpha_s(M_Z) = 0.12 \pm 0.01$ [15] . The older neutral current quantities described in [12] are also used in the analysis.

| Quantity | Experiment | SM | Topless | Mirror | Vector |
|---|---|---|---|---|---|
| $\Gamma_{b\bar{b}}$ (MeV) | 373 ±9 | 376 | 24 | 376 | 728 |
| $A_{FB}(b)$ | 0.093 ±0.012 | 0.096 | 0 | −0.096 | 0 |

Table 2: Predictions of the standard model (SM), topless models, a mirror model with $(t\ b)_R$ in a doublet, and a vector model with left and right-handed doublets, for $\Gamma_{b\bar{b}}$ and $A_{FB}(b)$, compared with the experimental values.

in the effective four-fermi interaction for $\overset{(-)}{\nu}_\mu e \to \overset{(-)}{\nu}_\mu e$ as obtained by CHARM II [10]. The preliminary value of the on-shell weak angle $\sin^2\theta_W \equiv 1 - M_W^2/M_Z^2 = 0.2242 \pm 0.0042 \pm [0.0047]$ obtained from deep inelastic neutrino scattering from CCFR [11] at Fermilab is in reasonable agreement with the earlier CERN values $0.228 \pm 0.005 \pm [0.005]$ [21], and $0.236 \pm 0.005 \pm [0.005]$ [22], though the central value is somewhat lower. The errors in brackets are theoretical. They are dominated by the $c$-quark threshold in the charged current scattering used to normalize the neutral current process, and are strongly correlated between the experiments. Older neutral current results, included in the analysis, are described in [12].

The standard model predictions for each quantity other than $M_Z$ are also shown. These are computed using $M_Z = 91.187 \pm 0.007$ GeV as input, using the range of $m_t$ determined from the global fit and 60 GeV $< M_H <$ 1 TeV. The agreement is excellent.

The $b$ observables $\Gamma_{b\bar{b}}$ and $A_{FB}(b)$ are especially important because the predictions depend on the $SU_2$ assignments of the $b$. In Table 2 the experimental values are compared with topless models and other alternatives with $V+A$ currents. It is seen that the data uniquely picks out the standard model from these alternatives [23]. This conclusion is strenghtened by a recent detailed analysis by Schaile and Zerwas [24] of LEP and lower energy data, which yields

$$t_{3L}(b) = -0.490^{+0.015}_{-0.012} \qquad t_{3R}(b) = -0.028 \pm 0.056 \tag{3}$$

for the third component of the weak isospin of the $b_{L,R}$, respectively, in agreement with the standard model expectations of $-1/2$ and $0$ – *i.e.*, topless models are excluded and the $b_L$ must be in a weak doublet with the $t_L$.

# 5 Standard Model Tests and $m_t$

Results will be presented in the $\overline{\text{MS}}$ [25] and on-shell [26] schemes. I use the radiative corrections calculated by Degrassi *et al.* [27] for the $W$ and $Z$ masses, those of Hollik [28] for the $Z$ widths, and generalized Born expressions for the Born contributions to the asymmetries. The latter are obtained from the data, *e.g.*, by using the program ZFITTER [29]. The calculations in [27]-[29] are in excellent agreement with each other and with those in [30]. Radiative corrections to low energy neutral current processes are described in [12].

In the standard model

$$\begin{aligned} M_Z^2 &= \frac{A_0^2}{\hat{\rho}\hat{c}^2\hat{s}^2(1-\Delta\hat{r}_W)} = \frac{A_0^2}{c^2 s^2(1-\Delta r)} \\ M_W^2 &= \hat{\rho}\hat{c}^2 M_Z^2 = c^2 M_Z^2 \end{aligned} \tag{4}$$

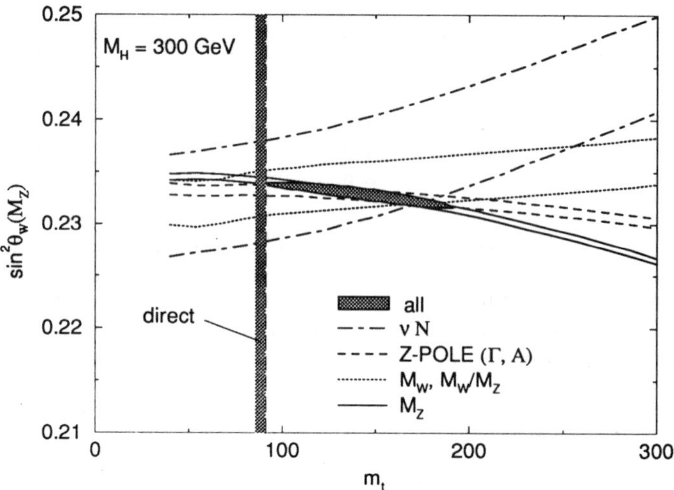


Figure 1: One $\sigma$ uncertainty in $\sin^2\hat{\theta}_W(M_Z)$ as a function of $m_t$ for $\sin^2\hat{\theta}_W(M_Z)$ determined from various inputs for $M_H = 300$ GeV. The direct lower limit $m_t > 91$ GeV and the 90% CL fit to all data are also shown.

where $A_0^2 = \pi\alpha/\sqrt{2}G_F = (37.2803 \text{ GeV})^2, \hat{s}^2 \equiv \sin^2\hat{\theta}_W(M_Z)$ refers to the weak angle in the $\overline{\text{MS}}$ scheme [25], $s^2 \equiv \sin^2\theta_W = 1 - M_W^2/M_Z^2$ refers to the on-shell scheme [26], $\hat{c}^2 \equiv 1 - \hat{s}^2$, and $c^2 \equiv 1 - s^2$; $\Delta\hat{r}_W, \hat{\rho} - 1$, and $\Delta r$ are radiative correction parameters. As is well known [31], $\hat{\rho} \sim 1 + \Delta\rho_t$, where

$$\Delta\rho_t = \frac{3G_F m_t^2}{8\sqrt{2}\pi^2} \simeq 0.0031 \left(\frac{m_t}{100\text{GeV}}\right)^2, \tag{5}$$

has a strong $m_t$ dependence, while $\Delta r \simeq \Delta r_0 - \Delta\rho_t/\tan^2\theta_W$ is even more sensitive. $\Delta\hat{r}_W \sim \Delta r_0 \sim 1 - \alpha/\alpha(M_Z) \sim 0.07$ has no quadratic $m_t$ dependence. There is additional logarithmic dependence on $m_t$ and $M_H$ in $\hat{\rho}, \Delta\hat{r}_W$, and $\Delta r$, as well as $O(\alpha)$ effects associated with low energy physics. These effects are important and are fully incorporated in the analysis, but will not be displayed here.

Gluonic corrections to $\Delta\rho_t$ of order $-\alpha\alpha_s m_t^2/M_Z^2$ can be important for large $m_t$ [32, 33]. The leading perturbative term is [32] $-2\alpha_s(m_t)(\pi^2+3)/9\pi \sim -0.10$ times the expression in (5). These corrections, which increase the predicted value of $m_t$ by about 5%, are included in the analysis[3].

The most accurate determination of $\sin^2\hat{\theta}_W(M_Z)$ and $\sin^2\theta_W$ are from $M_Z = 91.187 \pm 0.007$ GeV. The values are shown in Table 3 for $m_t = 100$ and 200 GeV and $M_H = 300$ GeV, and also for the global best fit range for $m_t$ and $M_H$. It is apparent that the extracted value of $\sin^2\theta_W$ depends strongly on $m_t$, while $\sin^2\hat{\theta}_W(M_Z)$ is considerably less sensitive due to the smaller coefficient of the quadratic $m_t$ term in

[3]These terms were omitted from previous analyses (e.g., [16]), because of uncertainties in both the magnitude and sign of important nonperturbative effects [33]. However, a careful new analysis [34] indicates that the perturbative esimate is an excellent approximation. A future global analysis will include small additional corrections.

| $m_t$ (GeV) | $\sin^2\theta_W$ | $\sin^2\hat{\theta}_W(M_Z)$ |
|---|---|---|
| 100 | $0.2322 \pm 0.0003$ | $0.2340 \pm 0.0003$ |
| 200 | $0.2204 \pm 0.0003$ | $0.2311 \pm 0.0003$ |
| $150^{+19}_{-24}$ | $0.2269 \pm 0.0025$ | $0.2328 \pm 0.0007$ |

Table 3: Values of $\sin^2\theta_W = 1 - M_W^2/M_Z^2$ and $\sin^2\hat{\theta}_W(M_Z)$ obtained from $M_Z = 91.187 \pm 0.007$ GeV, assuming $(m_t, M_H) = (100, 300)$ and $(200, 300)$ GeV. In the last row $m_t = 150^{+19}_{-24}$ GeV (obtained from the global fit to all data) and 60 GeV$< M_H <$ 1000 GeV.

$\hat{\rho}$ than in $\Delta r$. For fixed $m_t$ and $M_H$ the uncertainty of $\pm 0.0003$ in $\sin^2\theta_W$ has two components: the experimental error from $\Delta M_Z$ is only $\pm 0.0001$, while the theoretical error (from the uncertainty of $\pm 0.0009$ in $\Delta r$ from low energy hadronic contributions) is larger, $\pm 0.0003$. The $\pm 1\sigma$ limits on $\sin^2\hat{\theta}_W(M_Z)$ as a function of $m_t$ are shown in Figure 1.

The ratio $M_W/M_Z = 0.8813 \pm 0.0041$ determined by UA2 [7] and $M_W = 79.91 \pm 0.39$ GeV from CDF [6] determine the values of $\sin^2\hat{\theta}_W(M_Z)$ shown in Table 4. From Figure 1 it is apparent that $M_Z$, $M_W$, and $M_W/M_Z$ together imply an upper limit of $O(200$ GeV$)$ on $m_t$. A simultaneous fit of $M_Z$, $M_W$, and $M_W/M_Z$ to $\sin^2\hat{\theta}_W(M_Z)$, and $m_t$ yields $m_t = 145^{+42}_{-49} \pm 16$ GeV, where the second uncertainty is from $M_H$ in the range 60-1000 GeV, with a central value of 300 GeV. The 90(95)% CL upper limits on $m_t$ are 211 (223) GeV (Table 5). The upper limits are for $M_H = 1000$ GeV, which gives the weakest constraint. The value of $\sin^2\hat{\theta}_W(M_Z)$, including the uncertainties from $m_t$ and $M_H$, is also given in Table 5.

The partial width for $\Gamma \to f\bar{f}$ is given by [28, 30],

$$\Gamma_{f\bar{f}} = \frac{C_f \hat{\rho} G_F M_Z^3}{6\sqrt{2}\pi}\left(|a_f|^2 + |v_f|^2\right). \tag{6}$$

The axial and vector couplings are

$$\begin{aligned} a_f &= t_{3L}(f) = \pm\frac{1}{2} \\ v_f &= t_{3L}(f) - 2\sin^2\hat{\theta}_W(M_Z) q_f, \end{aligned} \tag{7}$$

where $t_{3L}(f)$ and $q_f$ are respectively the third component of weak isospin and electric charge of fermion $f$; $\hat{\rho}$ is dominated by the $m_t$ term (cf (5). The coefficient comes about by rewriting the tree-level formula

$$\frac{g^2(M_Z)M_Z}{8\cos^2\theta_W} = \frac{G_F M_Z^3}{\sqrt{2}}. \tag{8}$$

Expressing the width in this way incorporates the bulk of the radiative corrections, except for the large $m_t$ dependence in $\hat{\rho}$. Additional small radiative corrections are included but not displayed here. The factor in front incorporates the color factor and QED and QCD corrections:

$$C_f = \begin{cases} 1 + \frac{3\alpha}{4\pi} q_f^2 & , \text{ leptons} \\ 3\left(1 + \frac{3\alpha}{4\pi} q_f^2\right)\left(1 + \frac{\alpha_s}{\pi} + 1.405\frac{\alpha_s^2}{\pi^2}\right) & , \text{ quarks} \end{cases} \tag{9}$$

where the range $\alpha_s(M_Z) \simeq 0.12 \pm 0.01$ from $Z$-decay event topologies and other data [15] is used.

| Data | $\sin^2\hat\theta_W(M_Z)$ $m_t = 100$ | $m_t = 200$ | $m_t = 150^{+19}_{-24}$ |
|---|---|---|---|
| $M_Z$ | 0.2340 ±0.0003 | 0.2311 ±0.0003 | 0.2328 ±0.0007 |
| $M_W, \frac{M_W}{M_Z}$ | 0.2331 ±0.0022 | 0.2345 ±0.0022 | 0.2339 ±0.0022 |
| $\Gamma_Z, R, \sigma_p^h$ | 0.2333 ±0.0006 | 0.2319 ±0.0006 | 0.2327 ±0.0006 |
| $\Gamma_Z$ | 0.2332 ±0.0006 | 0.2320 ±0.0006 | 0.2327 ±0.0007 |
| $\Gamma_{l\bar{l}}$ | 0.2340 ±0.0007 | 0.2326 ±0.0007 | 0.2333 ±0.0008 |
| $\Gamma_{b\bar{b}}$ | 0.236 ±0.004 | 0.232 ±0.004 | 0.234 ±0.004 |
| $\Gamma_Z/M_Z$ | 0.2329 ±0.0009 | 0.2323 ±0.0009 | 0.2326 ±0.0009 |
| $\Gamma_{l\bar{l}}/M_Z$ | 0.2341 ±0.0010 | 0.2332 ±0.0010 | 0.2336 ±0.0010 |
| $\Gamma_Z/M_Z^3$ | 0.230 ±0.003 | 0.236 ±0.003 | 0.232 ±0.004 |
| $\Gamma_{l\bar{l}}/M_Z^3, R$ | 0.231 ±0.004 | 0.234 ±0.003 | 0.232 ±0.004 |
| $A_{FB}(\mu)$ | 0.232 ±0.002 | 0.232 ±0.002 | 0.232 ±0.002 |
| $A_{pol}(\tau)$ | 0.232 ±0.002 | 0.232 ±0.002 | 0.232 ±0.002 |
| $A_{FB}(b)$ | 0.233 ±0.002 | 0.233 ±0.002 | 0.233 ±0.002 |
| $A_{LR}$ | 0.238 ±0.006 | 0.238 ±0.006 | 0.238 ±0.006 |
| $\nu N \to \nu X$ | 0.233 ±0.005 | 0.238 ±0.005 | 0.235 ±0.005 |
| $\nu p \to \nu p$ | 0.212 ±0.032 | 0.212 ±0.031 | 0.212 ±0.032 |
| $\nu_\mu e \to \nu_\mu e$ | 0.232 ±0.009 | 0.231 ±0.009 | 0.232 ±0.009 |
| $e^{\uparrow\downarrow} D \to eX$ | 0.222 ±0.018 | 0.223 ±0.018 | 0.222 ±0.018 |
| atomic parity | 0.224 ±0.008 | 0.221 ±0.008 | 0.223 ±0.008 |
| All | 0.2337 ±0.0003 | 0.2314 ±0.0003 | 0.2328 ±0.0007 |

Table 4: Values of $\sin^2\hat\theta_W(M_Z)$ obtained from various inputs, for $(m_t,\ M_H) = (100, 300)$ and $(200, 300)$ GeV. In the last column $m_t = 150^{+19+15}_{-24-20}$ GeV, from the global fit, correlated with 60 GeV $< M_H <$ 1000 GeV. For $\nu N \to \nu X$ the uncertainty includes 0.003 (experiment) and 0.005 (theory). For atomic parity, the experimental and theoretical components of the error are 0.007 and 0.004 respectively.

| data | $\sin^2\hat\theta_W(M_Z)$ | $m_t$ (GeV) | $m_t^{max}$ (GeV) |
|---|---|---|---|
| $M_Z, M_W, M_W/M_Z$ | $0.2329 \pm 0.0014$ | $145^{+42}_{-49} \pm 16$ | 211(223) |
| $Z$-POLE | $0.2328 \pm 0.0008$ | $150 \pm 27 \pm 18$ | 198 (206) |
| $Z$-POLE $+M_W, M_W/M_Z$ | $0.2328 \pm 0.0007$ | $150^{+21}_{-26} \pm 18$ | 193(200) |
| $M_Z, \nu N$ | $0.2333^{+0.0011}_{-0.0016}$ | $132^{+54}_{-40} \pm 19$ | 210 (223) |
| All | $0.2328 \pm 0.0007$ | $150^{+19+15}_{-24-20}$ | 190 (197) |

Table 5: Values of $\sin^2\hat\theta_W(M_Z)$ and $m_t$ obtained for various data sets. $Z$-POLE refers to the first 11 constraints in Table 1 (with correlations). The $\sin^2\hat\theta_W(M_Z)$ error includes $m_t$ and $M_H$. The first error for $m_t$ includes experimental and theoretical uncertainties for $M_H = 300$ GeV. The second error is the variation for $M_H \to 60$ GeV (−) and $M_H \to 1000$ GeV (+). The last column lists the upper limits on $m_t$ at 90 (95)% CL for $M_H = 1000$ GeV, which gives the weakest upper limit. The direct CDF constraint $m_t > 91$ GeV is included.

(6) is written neglecting the fermion masses. In practice, fermion mass corrections [28] must be applied for $\Gamma_{b\bar{b}}$. They are also included in the following for $\Gamma_{c\bar{c}}$ and $\Gamma_{\tau\bar{\tau}}$, though the effects are small. There are significant correlations between the experimental values of the various total and partial $Z$ widths, which must be included in a global analysis.

The vertex corrections for $\Gamma_{b\bar{b}}$ depend strongly on $m_t$ and must be included as an extra correction [35]. For fixed $M_Z$ the $b\bar{b}$ width actually decreases with $m_t$, while the other modes all increase (because of the $\hat{\rho}$ factor). This gives a means of separating $\hat{\rho}(m_t)$ from such new physics as nonstandard Higgs representations by comparing $\Gamma_{b\bar{b}}$ or $\Gamma_Z$ with the other data [17].

The standard model predictions for $\Gamma_Z$, $\Gamma_{\ell^+\ell^-}(\ell = e, \mu, \text{ or } \tau)$, $R \equiv \Gamma_{had}/\Gamma_{\ell^+\ell^-}$, and the invisible width $\Gamma_{inv}$ as a function of $m_t$ are compared with the experimental results in Figures 2 and 3. ($\sin^2\hat{\theta}_W(M_Z)$ in $v_f$ is obtained from $M_Z$). One sees that the agreement is excellent for $m_t$ in the 100 – 200 GeV range. The results of fits to the $Z$ widths are listed in Tables 4 and 5. The $R$ ratio, which is insensitive to $m_t$, is slightly above the standard model prediction, though only at the $1\sigma$ level. As will be discussed, $R$ favors a slightly higher value of $\alpha_s(M_Z)$than the value obtained from event topologies and low energy data.

The invisible width in Figure 3 is clearly in agreement with $N_\nu = 3$ but not $N_\nu = 4$. In fact, the result [4] $N_\nu = 3.004 \pm 0.035$ not only eliminates extra fermion families with $m_\nu \ll M_Z/2$, but also supersymmetric models with light sneutrinos ($\Delta N_\nu = 0.5$) and models with triplet ($\Delta N_\nu = 2$) or doublet ($\Delta N_\nu = 0.5$) Majorons [36]. $N_\nu$ does not include sterile ($SU_2$-singlet) neutrinos. However, the complementary bound $N'_\nu < 3.3$ (95% CL) from nucleosynthesis [37] *does* include sterile neutrinos for a wide range of masses and mixings, provided their mass is less than $\sim 20$ MeV.

One can obtain precise ($\Delta = O(\pm 0.0007)$) values of $\sin^2\hat{\theta}_W(M_Z)$ from $\Gamma_Z$ and $\Gamma_{\ell^+\ell^-}$ (Table 4). The major sensitivity is through the $M_Z^3$ factor in (6) rather than from the vertices (*i.e.*, the $v_f$). It is useful to also obtain the $\sin^2\hat{\theta}_W(M_Z)$ from the vertices. Values can be obtained from the "reduced widths" $\Gamma_Z/M_Z^3$, $\Gamma_{\ell^+\ell^-}/M_Z^3$, and $R$. As can be seen in Table 4, the $\sin^2\hat{\theta}_W(M_Z)$ sensitivity from $\Gamma_Z/M_Z^3$ and the combination ($\Gamma_{\ell^+\ell^-}/M_Z^3, R$) is around $\pm 0.004$ ($\Gamma_{\ell^+\ell^-}/M_Z^3$ and $R$ individually give large asymmetric errors). Yet another determination of $\sin^2\hat{\theta}_W(M_Z)$ comes from $\Gamma_Z/M_Z$ and $\Gamma_{l^+l^-}/M_Z$. As can be seen in Table 4 the values obtained are insensitive to $m_t$. This can be understood from (4) and (6), from which one sees that $\Gamma_{f\bar{f}}/M_Z$ has no quadratic $m_t$ dependence (except $f = b$). Of course, the various values of $\sin^2\hat{\theta}_W$ obtained from the $\Gamma$'s are not all independent.

At tree level the asymmetries can be written

$$A_{FB}(f) \simeq 3\eta_e\eta_f, \tag{10}$$

and

$$A_{pol}(\tau) \simeq 2\eta_\tau, \tag{11}$$

where

$$\eta_f \equiv \frac{v_f a_f}{v_f^2 + a_f^2}, \tag{12}$$

and $v_f$ and $a_f$ are the tree-level vector and axial couplings in (7). These expressions are an excellent first approximation even in the presence of higher-order corrections, provided that $v_f$ is expressed in terms of $\sin^2\hat{\theta}_W(M_Z)$ , *i.e.*, one identifies $v_f$ and $a_f$

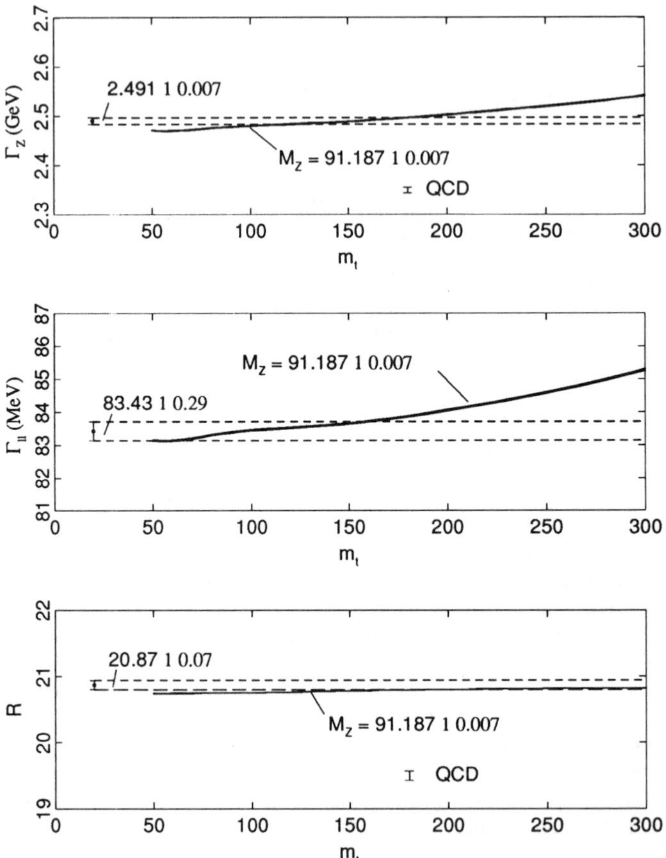


Figure 2: Theoretical predictions for $\Gamma_Z, \Gamma_{\ell^+\ell^-}$, and $R = \Gamma_{had}/\Gamma_{\ell^+\ell^-}$ in the standard model as a function of $m_t$, compared with the experimental results. The $M_H$ dependence is too small to see on the scale of the graph. The QCD uncertainties in $\Gamma_Z$ and $R$ are indicated.

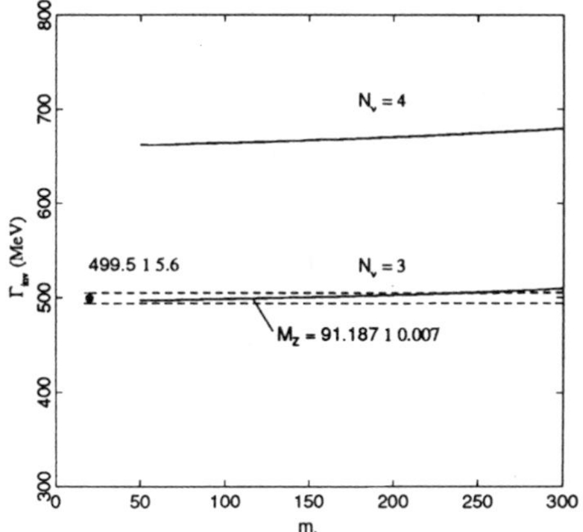


Figure 3: Theoretical prediction for $\Gamma_{inv}$ in the standard model with $N_\nu = 3$ and 4, compared with the experimental value.

with the effective Born couplings $\bar{g}_V$ and $\bar{g}_A$. $A_{FB}(b) = 0.093 \pm 0.012$ has been corrected for $b\bar{b}$ oscillations [2], using

$$A_{FB}(b) = \frac{A_{FB}^{obs}(f)}{1 - 2\chi}, \tag{13}$$

where $\chi = 0.126 \pm 0.012$ is the oscillation probability at the $Z$-pole. $Zb\bar{b}$ vertex corrections can be added to $A_{FB}(b)$ but are negligible numerically. The predictions for $A_{FB}(\mu)$, $A_{pol}(\tau)$, and $A_{FB}(b)$ are compared with the experimental data in Figure 4. Again, the agreement is excellent.

The results for $\sin^2\hat{\theta}_W$ obtained from a variety of low energy neutral current processes are listed in Table 4. The values obtained from atomic parity violation, $e^{\uparrow\downarrow}D$, and $\nu e$ and $\nu p$ elastic scattering are consistent with the value obtained from $M_Z$. They all have a similar dependence on $m_t$ as the $M_Z$ value and therefore do not significantly constrain $m_t$. They are, however, quite important in searches for new physics.

On the other hand, the value of the on-shell $\sin^2\theta_W$ obtained from deep inelastic $\nu N$ scattering [38] is insensitive to $m_t$. As can be seen in Table 4 and Figure 1 the corresponding $\sin^2\hat{\theta}_W(M_Z)$ increases rapidly with $m_t$. From Table 5 deep inelastic $\nu N$ scattering (combined with $M_Z$) gives $m_t < 210(223)$ GeV at 90(95)% CL. These are somewhat weaker than previous limits (193 (207) GeV) [16] due to the inclusion of the new CCFR result [11], with its slightly lower value for $\sin^2\theta_W$ (+6 GeV) and due to the inclusion of $O(\alpha\alpha_s m_t^2)$ radiative corrections (+11 GeV).

The results of global fits to all data are shown in Tables 4 and 5. All results include full statistical and systematic uncertainties in the experimental data as well as all of the important correlations. In particular, one obtains the prediction[4]

$$m_t = 150^{+19+15}_{-24-20} \text{ GeV}, \tag{14}$$

[4]This is in excellent agreement with the result $148^{+18+17}_{-20-19}$ of Schaile [4].

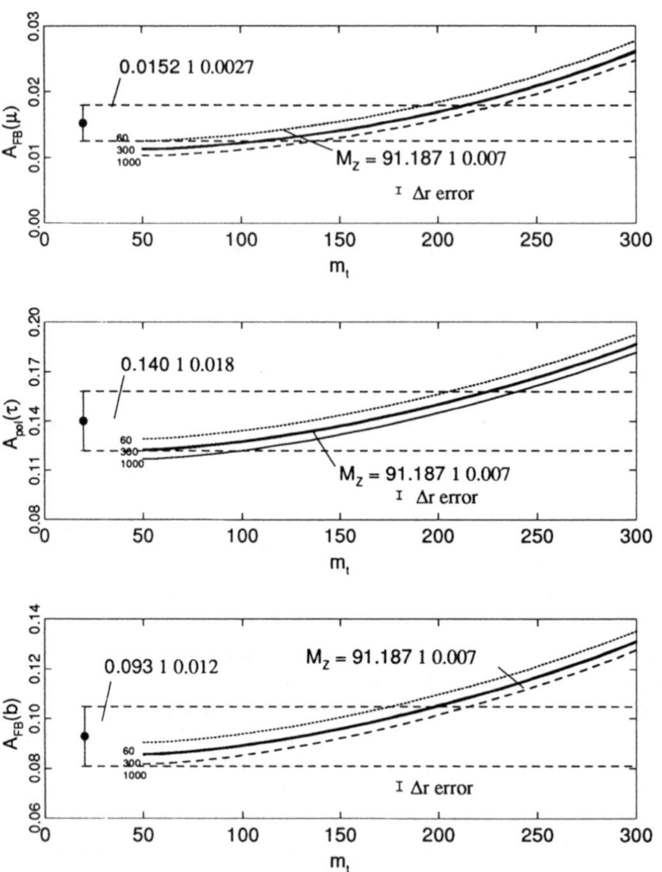


Figure 4: Theoretical prediction for $A_{FB}(\mu)$, $A_{pol}(\tau)$, and $A_{FB}(b)$ in the standard model as a function of $m_t$ for $M_H = 60$ (dotted line), 300 (solid), and 1000 (dashed) GeV, compared with the experimental values. The theoretical uncertainties from $\Delta\Delta r = \pm 0.0009$ are also indicated.

where the central value assumes $M_H = 300$ GeV. The second error is from the Higgs mass, assuming 60 GeV $< M_H <$ 1000 GeV. The $m_t$ and $M_H$ dependences are strongly correlated. The relation between the two in the radiative corrections is not universal, but a reasonable interpolation of the $M_H$dependence is

$$m_t(\text{GeV}) = 150^{+19}_{-24} + 12.5 \ln(M_H/300\text{GeV}). \tag{15}$$

Alternately, we can allow $M_H$ to be a free parameter in the range 60 -1000 GeV, with the result that $m_t = 131^{+47}_{-28}$ GeV, with the lower central value occurring because the best fit is for $M_H = 60$ GeV.

The upper limit on $m_t$ is

$$m_t < \begin{cases} 190 \text{ GeV}, & 90\% \quad CL \\ 197 \text{ GeV}, & 95\% \quad CL \\ 208 \text{ GeV}, & 99\% \quad CL \end{cases} , \tag{16}$$

which occurs for $M_H = 1000$ GeV. For $M_H = 60(300)$ GeV, the 90% CL limit is 158 (175) GeV and the 95% CL limit is 165 (182) GeV. The upper and lower limits on $m_t$ are shown as a function of $M_H$ in Figure 5. The values of $\sin^2\hat{\theta}_W$ and $m_t$ and the $m_t$ limits for various subsets of the data are given in Table 5. The $\chi^2$ distribution as a function of $m_t$ is shown in Figure 6 for $M_H = 60$, $M_Z$, 300, and 1000 GeV. The fit is excellent[5], with a $\chi^2/df$ of $168/206 \sim 0.82$ for $m_t = 150$, $M_H = 300$ GeV.

The result in (14) is very close to the value $149^{+21}_{-27} \pm 16$ obtained about 1 year ago. The agreement is somewhat fortuitous: the new 1991 LEP and other data lower the prediction by $\sim 9$ GeV, but this is compensated by the inclusion of $O(\alpha\alpha_s m_t^2)$ radiative corrections (+8 GeV) and the use of 300 (rather than 250) GeV as the central $M_H$ value (+2 GeV).

The prediction in (14) is for the minimal standard model. The corresponding results in the minimal supersymmetric extension (MSSM) will be discussed in Section 6.

The data also yield an indirect *lower* limit on $m_t$ (Figure 1). For $M_H = 60$ GeV one obtains $m_t > 95(83)$ GeV at 90(95)% CL. The corresponding limits are 118(108) GeV for $M_H = 300$ GeV and 138(129) GeV for $M_H = 1000$ GeV. The lower bound is comparable to the direct CDF limit $m_t > 91$ GeV (95% CL) [13]. However, it is more general in that it applies even for nonstandard $t$ decay modes, for which the direct lower limit is $\sim 60$ GeV.

The data will not significantly constrain $M_H$ until $m_t$ is known separately. At present the best fit occurs for lower values of $M_H$, but the change in $\chi^2$ between $M_H$ = 60 and 1000 GeV is only 0.6. From Figure 6 is it obvious that if $m_t$ is measured directly to within 5-10 GeV it may be possible to constrain $M_H$, particularly if $m_t$ is in the lower part of the allowed range. This is further illustrated in Figure 7, in which are displayed the 68 and 90% CL $M_H$ ranges that could be obtained from present data if $m_t$ were known to 10 GeV.

Assuming the standard model, one therefore concludes 91 GeV $< m_t <$ 197 GeV at 95% CL. In most cases, the effect of new physics is to *strengthen* the upper bound rather than weaken it. The obvious question is, why is $m_t$ so large (or why are the other

[5]In fact, the fit is *too* good: there is only a 2% probability of obtaining a $\chi^2$ this low or lower for random errors. This has always been the case for precision neutral current and $Z$-pole experiments [12]. The most likely explanation is a tendency for experimenters to overestimate systematic errors.

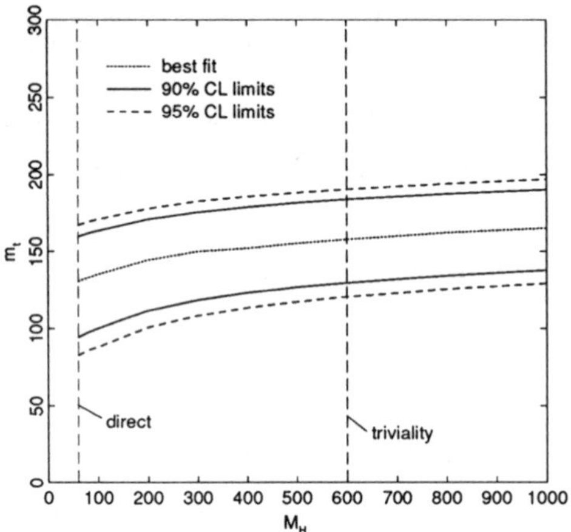


Figure 5: Best fit value for $m_t$ and upper and lower limits as a function of $M_H$. The direct lower limit $M_H > 60$ GeV [14] and the approximate triviality limit [39] $M_H <$ 600 GeV are also indicated. The latter becomes $M_H < 200$ GeV if one requires that the standard model holds up to the Planck scale.

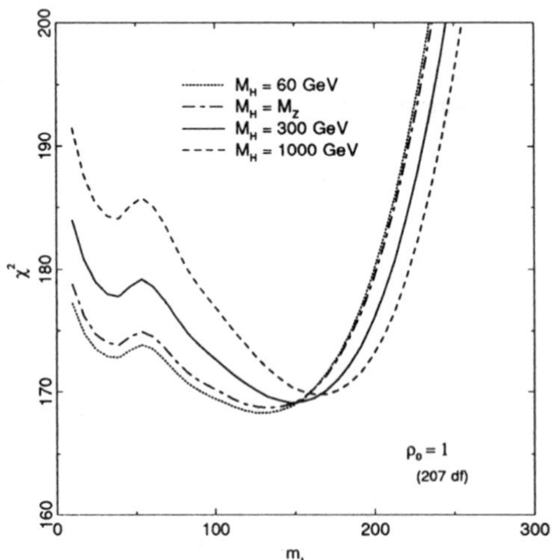


Figure 6: $\chi^2$ distribution for all data (207 df) in the standard model as a function of $m_t$, for $M_H = 60$, $M_Z$, 300, and 1000 GeV. The direct constraint $m_t > 91$ GeV is *not* included.

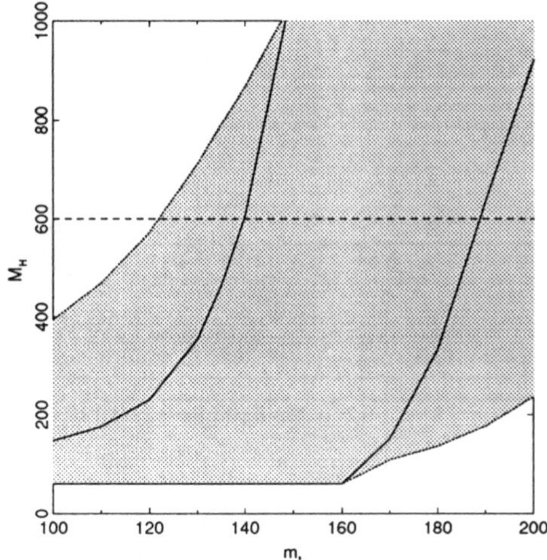


Figure 7: 68 and 90% CL $M_H$ ranges that could be obtained from present data if $m_t$ were known by direct measurement to $\pm$ 10 GeV as a function of the central value of $m_t$.

fermion masses so small)? Note that the value of $m_t$ considered here is the position of the pole in the $t$ propagator (not the running mass). It should coincide (with a theoretical ambiguity of $\sim 5$ GeV) with the kinematic mass relevant for the production of the $t$ quark at hadron colliders.

For the weak angle one obtains

$$\begin{aligned} \sin^2\theta_W &= 0.2267 \pm 0.0024 \\ \sin^2\hat{\theta}_W(M_Z) &= 0.2328 \pm 0.0007, \end{aligned} \tag{17}$$

where the uncertainty is mainly from $m_t$. The small uncertainty from $M_H$ in the range 60 - 1000 GeV is included in the errors in (17). Of course, $\sin^2\hat{\theta}_W(M_Z)$ is much less sensitive to $m_t$ and $M_H$ than $\sin^2\theta_W$. All of the values obtained from individual observables are in excellent agreement with (17). In particular, the $\sin^2\hat{\theta}_W$ values obtained assuming $m_t = 150^{+19}_{-24}$ GeV and 60 GeV $< M_H <$ 1000 GeV are shown in Table 4 and in Figure 8. The agreement is remarkable.

One can also extract the radiative correction parameter $\Delta r$ (eqn. (4)). One finds

$$\Delta r = 0.049 \pm 0.008 \tag{18}$$

compared to the expectation $0.0626 \pm 0.0009(0.0273)$ for $m_t = 100(200)$, $M_H = 300$. Similarly, in the $\overline{\text{MS}}$ scheme, one finds

$$\Delta\hat{r}_W = 0.069 \pm 0.006, \tag{19}$$

compared with the expectation $0.0696 \pm 0.0009(0.0723)$.

The hadronic $Z$ width depends on the value of $\alpha_s(M_Z)$. The quoted results use the value $0.12 \pm 0.01$ obtained from $Z$-decay event topologies and low energy data [15]. One can also obtain a value of $\alpha_s(M_Z)$from the hadronic widths, and in particular from

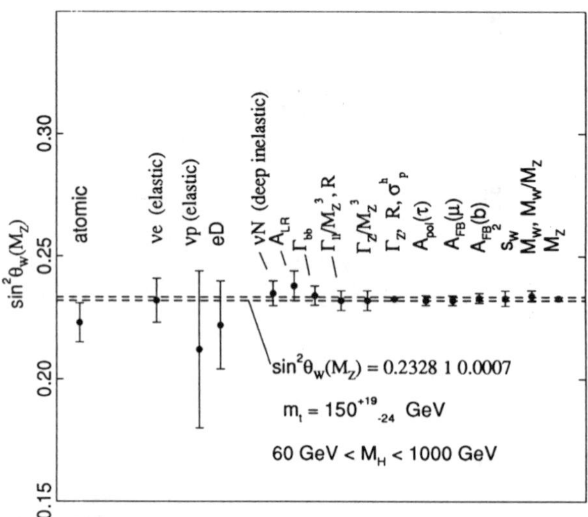


Figure 8: $\sin^2 \hat{\theta}_W(M_Z)$ obtained from various observables assuming $m_t = 150^{+19}_{-24}$ GeV, $60 < M_H < 1000$ GeV.

$R$, which is insensitive to $m_t$. A fit to all $Z$-pole and other data (but not including the constraint $\alpha_s(M_Z)= 0.12 \pm 0.01$) to $\alpha_s(M_Z)$, $\sin^2 \hat{\theta}_W(M_Z)$ , and $m_t$ yields $\alpha_s(M_Z)= 0.130 \pm 0.009$, which is consistent but slightly above the other determinations. From $M_Z, \Gamma_Z, R, \sigma_p^h$, and $\Gamma_{b\bar{b}}$ only, one finds the higher value $0.135 \pm 0.011$. When $\alpha_s(M_Z)= 0.12 \pm 0.01$ is included as a separate constraint in the fit to all data one obtains the average $\alpha_s(M_Z) = 0.126 \pm 0.007$. These values are listed in Table 6, along with the most important low energy determinations [15]. There is a slight tendency for higher values from the $Z$-pole data, but given the uncertainties (which are usually dominated by theoretical errors) there is no real discrepancy[6].

The value of $\alpha_s(M_Z)$ from the precision experiments is strongly anticorrelated

[6]The situation is somewhat aggravated by a recent determination [40] $\alpha_s(M_Z) = 0.105$ from the charmonium spectrum, which used (quenched) lattice methods. The claimed uncertainty is $\pm 0.004$.

| $\alpha_s(M_Z)$ | source |
|---|---|
| $0.130 \pm 0.009$ | precision $Z$-pole and low energy |
| $0.135 \pm 0.011$ | $M_Z, \Gamma_Z, R, \sigma_p^h, \Gamma_{b\bar{b}}$ |
| $0.123 \pm 0.005$ | event topologies [15] |
| $0.118 \pm 0.005$ | $\tau$ decays [15] |
| $0.112 \pm 0.005$ | deep inelastic scattering (DIS) [15] |
| $0.113 \pm 0.006$ | $\Upsilon$, $J/\psi$ [15] |
| $0.12 \pm 0.01$ | event topologies, $\tau$, DIS, $\Upsilon$, $J/\psi$ |
| $0.126 \pm 0.007$ | combined |

Table 6: Values of $\alpha_s(M_Z)$from indirect precision data, event topologies, low energy data, and all data.

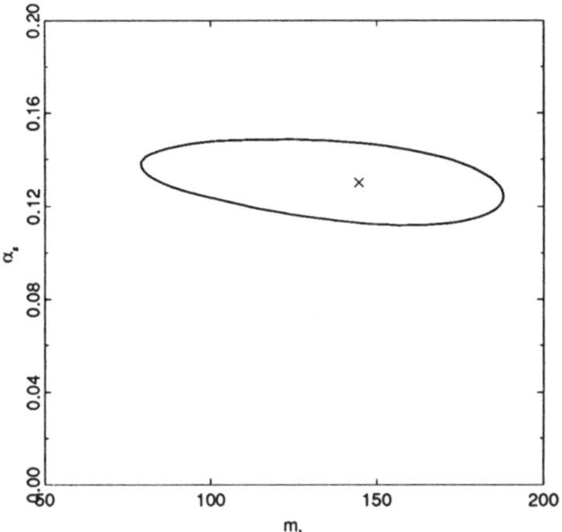


Figure 9: 90% CL allowed region in $\alpha_s(M_Z)$and $m_t$ from a combined fit to precision $Z$-pole and other data (but not including event topology and low energy determinations of $\alpha_s(M_Z)$).

with $m_t$, as can be seen in Figure 9. In particular, larger $m_t$ corresponds to smaller $\alpha_s(M_Z)$, in better agreement with the low energy data.

# 6 Supersymmetry

In the minimal supersymmetric extension of the standard model (MSSM), for most of the allowed parameter range the superpartner and second Higgs doublet masses are much larger than $M_Z$. In that case, the only significant effect on the analysis of precision experiments is in the Higgs sector [41]. There is a light ($M < 150$ GeV) scalar which acts like a light standard model Higgs (as far as radiative corrections are concerned), and the other Higgs particles and superpartners do not contribute significantly. Thus, for the MSSM we will take 60 GeV $< M_H <$ 150 GeV with a central value of $M_Z$, yielding:

$$\text{MSSM}: \qquad m_t = 134^{+23}_{-28} \pm 5 \text{ GeV}. \tag{20}$$

For $M_H$ a free parameter in the range 60 -150 GeV, one obtains $m_t = 131^{+31}_{-28}$ GeV, with the best fit for $M_H = 60$ GeV. The corresponding value of the weak angle is

$$\text{MSSM}: \qquad \sin^2\hat{\theta}_W(M_Z) = 0.2326 \pm 0.0006, \tag{21}$$

which should be compared to (17).

There could be additional effects for the small allowed region in which some of the new particles have masses between the direct limits and $\sim M_Z$. In particular, light charginos or light charged Higgs fields could contribute significantly to the $Zb\bar{b}$ vertex [42], while a large $\tilde{b}$ - $\tilde{t}$ splitting would lead to a smaller prediction for $m_t$(i.e., $m_t^2 \rightarrow m_t^2 + (m_{\tilde{t}} - m_{\tilde{b}})^2 f(m_{\tilde{t}}, m_{\tilde{b}})$ in $\Delta\rho_t$, where $f \geq 1$.) However, these effects and the contributions of the new particles to $S$, $T$, and $U$ are all negligible for masses $\gg M_Z$[41].

Thus, for most of parameter space the only effects of supersymmetry on the precision observables are: (a) a light Higgs, and (b) the *absence* of any deviation from the standard model predictions.

# 7 (Supersymmetric) Grand Unification

It is interesting to compare the value of $\sin^2\hat{\theta}_W(M_Z)$ in (17) or (21) with the only models available which predict it, namely grand unified theories [43, 44, 45, 46].

In a grand unified theory there is only one underlying gauge coupling constant, and when the low energy couplings are extrapolated to high energy they are expected to (approximately) meet at the unification scale $M_X$ above which symmetry breaking can be neglected. We define the couplings $g_s = g_3$, $g = g_2$, and $g' = \sqrt{3/5}g_1$ of the standard model $SU_3 \times SU_2 \times U_1$ group, and the fine-structure constants $\alpha_i = g_i^2/4\pi$. The extra factor in the definition of $g_1$ is a normalization condition [47]. The couplings are expected to meet only if the corresponding group generators are normalized in the same way. However, the standard model generators are conventionally normalized as $\mathrm{Tr}(Q_s^2) = \mathrm{Tr}(Q_2^2) = 5/3\mathrm{Tr}(Y/2)^2$, so the factor $\sqrt{3/5}$ is needed to compensate. Thus,

$$\sin^2\theta_W = \frac{g'^2}{g^2+g'^2} = \frac{g_1^2}{\frac{5}{3}g_2^2+g_1^2} \underset{g_1=g_2}{\longrightarrow} \frac{3}{8}. \tag{22}$$

One expects $\sin^2\theta_W = 3/8$ at the unification scale [47] for which $g_1 = g_2$.

To test the unification, one starts with the couplings at $M_Z$, which are now very well known from the LEP and low energy data. Using as inputs $\alpha^{-1}(M_Z) = 127.9 \pm 0.2$ [27], $\alpha_s(M_Z) = 0.12 \pm 0.01$, and[7] $\sin^2\hat{\theta}_W(M_Z) = 0.2324 \pm 0.0006$, one obtains

$$\begin{aligned}
\alpha_1^{-1}(M_Z) &\equiv \frac{3}{5}\alpha^{-1}(M_Z)\cos^2\hat{\theta}_W(M_Z) = 58.9 \pm 0.1 \\
\alpha_2^{-1}(M_Z) &\equiv \alpha^{-1}(M_Z)\sin^2\hat{\theta}_W(M_Z) = 29.7 \pm 0.1 \\
\alpha_3^{-1}(M_Z) &\equiv \alpha_s^{-1}(M_Z) = 8.3 \pm 0.7.
\end{aligned} \tag{23}$$

These may be extrapolated to high energy using the two-loop renormalization group equations

$$\frac{d\alpha_i^{-1}}{d\ln\mu} = -\frac{b_i}{2\pi} - \sum_{j=1}^{3}\frac{b_{ij}\alpha_j}{8\pi^2}. \tag{24}$$

The 1-loop coefficients are

$$b_i = \begin{pmatrix} 0 \\ -\frac{22}{3} \\ -11 \end{pmatrix} + F\begin{pmatrix} \frac{4}{3} \\ \frac{4}{3} \\ \frac{4}{3} \end{pmatrix} + N_H\begin{pmatrix} \frac{1}{10} \\ \frac{1}{6} \\ 0 \end{pmatrix}, \tag{25}$$

assuming the standard model. $F$ is the number of fermion families and $N_H$ is the number of Higgs doublets. In the MSSM,

$$b_i = \begin{pmatrix} 0 \\ -6 \\ -9 \end{pmatrix} + F\begin{pmatrix} 2 \\ 2 \\ 2 \end{pmatrix} + N_H\begin{pmatrix} \frac{3}{10} \\ \frac{1}{2} \\ 0 \end{pmatrix}, \tag{26}$$

[7]This is a slightly older value for $\sin^2\hat{\theta}_W(M_Z)$ than those in (17) or (21), chosen to agree with the analysis in [20].

where the difference is due to the additional particles in the loops. The 2-loop coefficients can be found in [45]. Equation 24 can be integrated to yield

$$\alpha_i^{-1}(\mu) = \alpha_i^{-1}(M_Z) - \frac{b_i}{2\pi}\ln\left(\frac{\mu}{M_Z}\right) + \sum_{j=1}^{3}\frac{b_{ij}}{4\pi b_j}\ln\left[\frac{\alpha_j^{-1}(\mu)}{\alpha_j^{-1}(M_Z)}\right], \tag{27}$$

for an arbitrary scale $\mu$. To first approximation one can neglect the last (2-loop) term, in which case the inverse coupling constant varies linearly with $\ln\mu$. However, the 2-loop terms must be kept in the final analysis. In a grand unified theory one expects that the three couplings will approximately meet at $M_X$,

$$\alpha_i^{-1}(M_X) = \alpha_G^{-1}(M_X) - \Delta_i. \tag{28}$$

The $\Delta_i$ are small corrections [48] associated with the low energy threshold (*i.e.*, $m_t$ and the new sparticles and Higgs not degenerate with $M_Z$), the high scale thresholds ($m_{\rm heavy} \neq M_X$), or with non-renormalizable operators.

The running couplings in the standard model are shown in Figure 10a, ignoring threshold corrections. They clearly do not meet at a point, thus ruling out simple grand unified theories such as $SU_5, SO_{10}$, or $E_6$ which break in a single step to the standard model [44]. Of course, such models are also excluded by the non-observation of proton decay, but this independent evidence is welcome.

On the other hand, in the minimal supersymmetric extension of the standard model the couplings do meet within the experimental uncertainties [46]. This is illustrated in Figure 10b for the case in which all of the new particles have a common mass $M_{SUSY} = M_Z$. Almost identical curves are obtained for larger $M_{SUSY}$, such as 1 TeV. (In practice, the splittings between the sparticle masses are more important than the average value [48].) The unification scale $M_X$ is sufficiently large ($> 10^{16}$ GeV) that proton decay by dimension$-6$ operators is adequately suppressed, although there may still be a problem with dimension$-5$ operators [49]. This success is encouraging for supersymmetric grand unified theories such as $SUSY\text{-}SU_5$ or $SUSY\text{-}SO_{10}$.

To display the theoretical uncertainties, it is convenient to use $\alpha(M_Z)$ and $\alpha_s(M_Z)$ to predict $\sin^2\hat{\theta}_W(M_Z)$ . Using $\alpha^{-1}(M_Z) = 127.9 \pm 0.2$ and $\alpha_s(M_Z)= 0.12 \pm 0.01$ one predicts

$$\begin{aligned} \sin^2\hat{\theta}_W(M_Z) &= 0.2334 \pm 0.0025 \pm 0.0025 \text{ (MSSM)}, \\ \sin^2\hat{\theta}_W(M_Z) &= 0.2100 \pm 0.0025 \pm 0.0007 \text{ (SM)}, \end{aligned} \tag{29}$$

where the first uncertaintly is from $\alpha_s$ and $\alpha^{-1}$, and the second is an estimate of theoretical uncertainties from the superspectrum, high-scale thresholds, and possible non-renormalizable operators [20]. The MSSM prediction is in excellent agreement with the experimental value 0.2326 $\pm$ 0.0006, while the SM prediction is in conflict with the data. These results are displayed in Figure 11.

Because of the large uncertainty in $\alpha_s(M_Z)$, it is convenient to invert the logic and use the precisely known $\alpha^{-1}$ and $\sin^2\hat{\theta}_W(M_Z)$ to predict $\alpha_s(M_Z)$:

$$\begin{aligned} \alpha_s(M_Z) &= 0.125 \pm 0.002 \pm 0.009 \text{ (MSSM)}, \\ \alpha_s(M_Z) &= 0.072 \pm 0.001 \pm 0.001 \text{ (SM)}, \end{aligned} \tag{30}$$

where again the second error is theoretical. It is seen that the SUSY case is in excellent agreement with the experimental $\alpha_s(M_Z)$= 0.12 $\pm$ 0.01, while the simplest ordinary

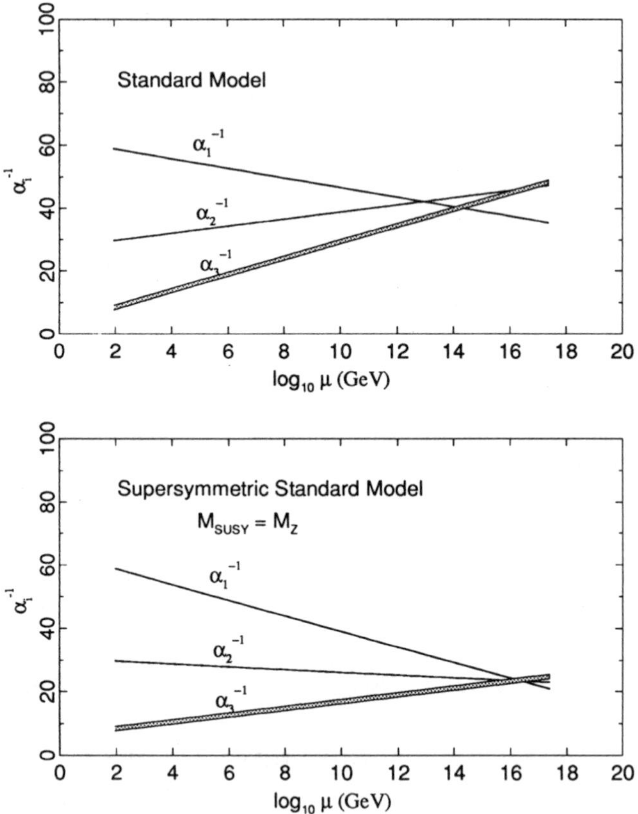


Figure 10: Running couplings in (a) the standard model (SM) and (b) in the minimal supersymmetric extension of the standard model (MSSM) with two Higgs doublets for $M_{SUSY} = M_Z$, from [20]. The corresponding figure for $M_{SUSY} = 1$ TeV is almost identical. It is seen that the couplings unify at $\simeq 10^{16}$ GeV in the MSSM. The effects of threshold uncertainties are seen in Figures 11 and 12.

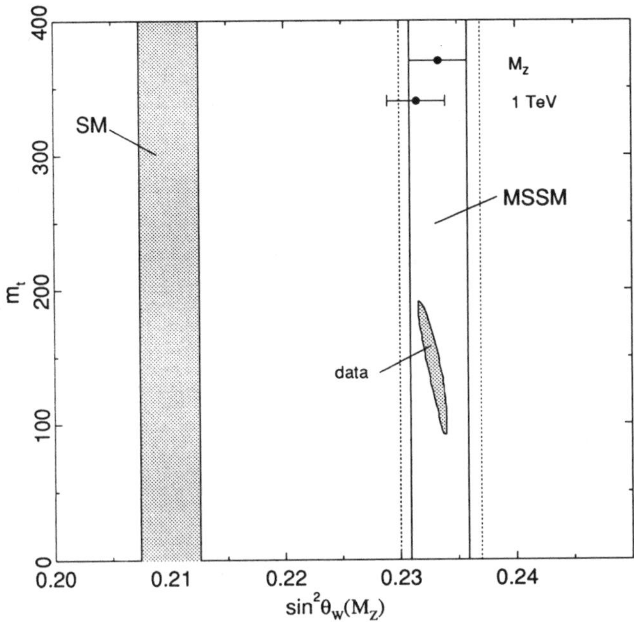


Figure 11: 90% CL regions in $\sin^2\hat{\theta}_W(M_Z)$ vs $m_t$, compared with the predictions of ordinary and SUSY–GUTs, from [20]. The smaller ranges of uncertainties are from $\alpha(M_Z)$ and $\alpha_s(M_Z)$ only, while the larger range includes the various low and high scale uncertainties, added in quadrature. The predictions for degenerate SUSY masses at $M_Z$ and 1 TeV are shown for comparison.

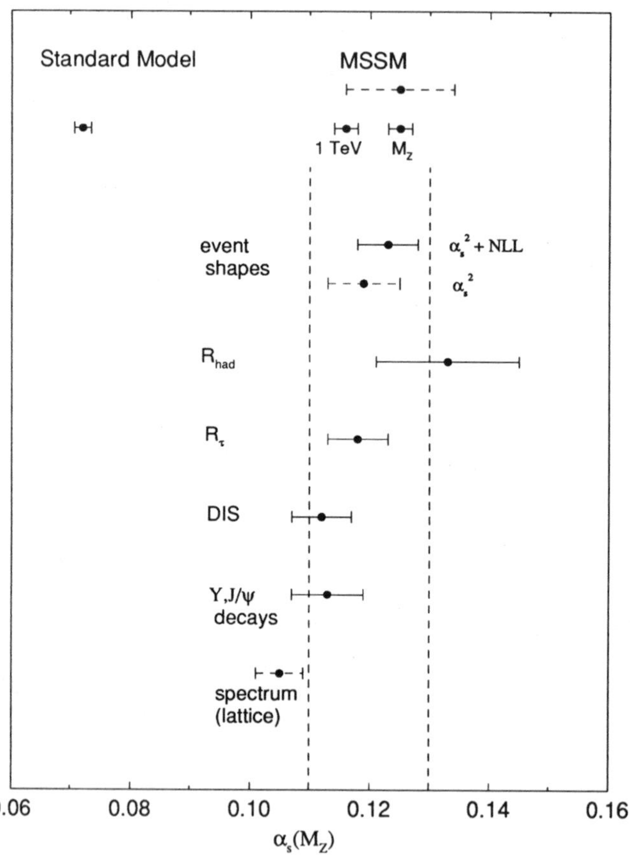


Figure 12: Predictions for $\alpha_s(M_Z)$ from $\alpha^{-1}$ and $\sin^2\hat{\theta}_W(M_Z)$ in ordinary and SUSY GUTs. The dashed lines represent the experimental range $0.12 \pm 0.01$. In the SUSY (MSSM) case the error bar includes the theoretical uncertainties, added in quadrature. The smaller error bars are for degenerate sparticle masses at $M_Z$ and 1 TeV. From [20].

GUTs are excluded. The unification slightly prefers larger values of $\alpha_s(M_Z)$, as suggested by the $Z$-pole data, but the theoretical uncertainties are comparable to the error on the observed $\alpha_s(M_Z)$ (which is also dominated by theory). These theory uncertainties are mainly irreducible, so for this purpose there is no need to improve the precision of the $\alpha_s(M_Z)$ measurement. The $\alpha_s(M_Z)$ predictions are displayed in Figure 12.

The success of the coupling constant unification, which is insensitive to the gauge group and the number of complete families, provides a hint that supersymmetric grand unification (or some superstring imitator) may be on the right track. Of course it is possible that the success may be an accident. Similarly, there are many more complicated schemes which could yield coupling constant unification, such as those involving a large group breaking in two or more stages to the standard model, or those with ad hoc new representations split into light and heavy components. However, the MSSM is the only scheme in which the unification is a prediction rather than being achieved by adjusting new parameters or representations. Perhaps the coupling constants may indeed prove to be the "first harbinger of supersymmetry" [50].

Unless the apparent coupling constant unification is an accident, there are stringent restrictions on the types of new "SUSY-safe" physics which do not drastically disturb the predictions (unless, of course, one allows two large effects to cancel). These are: supersymmetry (required), additional heavy $Z'$ gauge bosons or additional gauge groups which commute with the standard model, additional complete ordinary or mirror fermion families and their superpartners (the neutrinos would have to be very heavy because of the LEP limit $N_\nu = 3.004 \pm 0.035$), or complete exotic fermion supermultiplets such as occur in some $E_6$ models.

# 8 Conclusions

- There is no evidence for any deviation from the standard model.
- $\overline{\text{MS}}$ : $\sin^2 \hat{\theta}_W(M_Z) = 0.2328 \pm 0.0007$
- On-shell: $\sin^2 \theta_W \equiv 1 - {M_W}^2/{M_Z}^2 = 0.2267 \pm 0.0024$, where the uncertainties are mainly from $m_t$.
- In the standard model one predicts: $m_t = 150^{+19+15}_{-24-20}$ GeV, where the central value assumes $M_H$= 300 GeV and the second uncertainty is for $M_H \rightarrow$ 60 GeV $(-)$ or 1 TeV $(+)$.
- In the MSSM $m_t = 134^{+23}_{-28} \pm 5$ GeV, where the difference is due the light Higgs scalar expected in the MSSM.
- Precision data yield the 95% CL constraints

$$83 \text{ GeV} < m_t < 197 \text{ GeV}, \tag{31}$$

  where the lower (upper) limits are for $M_H = 60$ (1000) GeV. The lower limit is valid for any decay mode, and is to be compared with the direct CDF limit $m_t > 91$ GeV, which assumes canonical decays.
- There is no significant constraint on $M_H$ until $m_t$ is known independently.

- Precision $Z$-pole and low-energy data yield the indirect result $\alpha_s(M_Z) = 0.130 \pm 0.009$, in reasonable agreement with the value $0.12 \pm 0.01$ obtained from jet event topologies and low energy direct determinations.
- The low energy couplings are in excellent agreement with the predictions of supersymmetric grand unification, but not with the simplest (and most predictive) non-supersymmetric grand unified theories.
- The precision data place stringent limits on many types of new physics into the TeV range.
- In the future, precision electroweak experiments will be a useful complement to high energy colliders.

## References

[1] The history is described in *From Weak Neutral Currents to the* $(W)/Z$ *and Beyond*, ed. D. Cline and A. Mann, AIP, 1993.

[2] For a recent review, see L. Rolandi, *XXVI ICHEP 1992*, Dallas, August 1992, CERN-PPE/92-175.

[3] The LEP Collaborations, Phys. Lett. **B276**, 247 (1992).

[4] D. Schaile, to be published in *Precision Tests of the Standard Electroweak Model*, (World, Singapore, 1993), ed. P. Langacker.

[5] SLD: K. Abe *et al.*, Phys. Rev. Lett. **70**, 2515 (1993).

[6] CDF: F. Abe *et al.*, Phys. Rev. Lett. **65**, 2243 (1990).

[7] UA2: J. Alitti *et al.*, Phys. Lett. **B276**, 354 (1992). See also H. Plothow-Besch, *Lepton Photon and High Energy Physics Conference (LP-HEP 91)*, Geneva, July, 1991.

[8] Cesium: M. C. Noecker *et al.*, Phys. Rev. Lett. **61**, 310 (1988).

[9] Atomic theory: S.A. Blundell, W.R. Johnson, and J. Sapirstein, Phys. Rev. Lett. **65**, 1411 (1990); V. A. Dzuba *et al.*, Phys. Lett. **141A**, 147 (1989).

[10] CHARM II: D. Geiregat *et al.*, Phys. Lett. **B259**, 499 (1991); S. Cocco, *XXVIIth Rencontres de Moriond*, March 1992.

[11] T. Bolton, *XXVIIth Rencontres de Moriond*, March 1992.

[12] For recent reviews, see U. Amaldi *et al.*, Phys. Rev. **D36**, 1385 (1987); P. Langacker and M. Luo, Phys. Rev. **D44**, 817 (1991); P. Langacker, in *TeV Physics*, ed. T. Huang *et al.*, (Gordon and Breach, Philadelphia, 1991), p. 53; *Review of Particle Properties*, Phys. Rev. **D45**, III-59,VII-159 (1992).

[13] CDF: F. Abe *et al.*, Phys. Rev. **D43**, 664 (1991).

[14] T. Mori, *XXVI ICHEP 1992*, Dallas, August 1992.

[15] This is an approximate range based on the data reviewed in S. Bethke and S. Catani, CERN TH.6484/92, and T. Hebbeker, LP-HEP 91 (Aachen PITHA 91/17).

[16] P. Langacker, in *Electroweak Physics Beyond the Standard Model*, ed. J. W. F. Valle and J. Velasco (World, Singapore, 1991), p. 75. and [12].

[17] P. Langacker and M. Luo, Phys. Rev. **D44**, 817 (1991).

[18] P. Langacker and M. Luo, Phys. Rev. **D45**, 278 (1992).

[19] D. Kennedy and P. Langacker, Phys. Rev. Lett. **65**, 2967 (1990); Phys. Rev. **D44**, 1591 (1991).

[20] P. Langacker and N. Polonsky, Phys. Rev. **D47**, 4028 (1993), UPR-0556T.

[21] CDHS: H. Abramowicz *et al.*, Phys. Rev. Lett. **57**, 298 (1986), A. Blondel *et al.*, Zeit. Phys. **C45**, 361 (1990).

[22] CHARM: J. V. Allaby *et al.*, Zeit. Phys. **C36**, 611 (1987).

[23] Earlier indirect arguments for the existence of the $t$ quark are summarized in P. Langacker, Comm. Nucl. Part. Sci. **19**, 1 (1989).

[24] D. Schaile and P. M. Zerwas, Phys. Rev. **D45**, 3262 (1992).

[25] A. Sirlin, Phys. Lett. **B232**, 123 (1989); S. Fanchiotti and A. Sirlin, Phys. Rev. **D41**, 319 (1990); A. Sirlin, Nucl. Phys. **B332**, 20 (1990), and references therein.

[26] A. Sirlin, Phys. Rev. **D22**, 971 (1980); **D29**, 89 (1984); W. Marciano and A. Sirlin, Phys. Rev. **D22**, 2695 (1980).

[27] G. Degrassi, S. Fanchiotti and A. Sirlin, Nucl. Phys. **B351**, 49 (1991).

[28] W. Hollik, Fortsch Phys. **38**, 165 (1990).

[29] *ZFITTER*, D. Bardin *et al.*, CERN-TH-6443-92.

[30] G. Degrassi and A. Sirlin, Nucl. Phys. **B352**, 342 (1991).

[31] M. Veltman, Nucl. Phys. **B123**, 89 (1977); M. Chanowitz, M. A. Furman, and I. Hinchliffe, Phys. Lett. **B78**, 285 (1978).

[32] A. Djouadi and C. Verzegnassi, Phys. Lett. **B195**, 265 (1987); A. Djouadi, Nuovo Cim. **100A**, 357 (1988).

[33] B. A. Kniehl, J. H. Kühn, and R. G. Stuart, Phys. Lett. **B214**, 621 (1988); B. A. Kniehl and J. H. Kühn, Nucl. Phys. **B329**, 547 (1990); B. A. Kniehl, Nucl. Phys. **B347**, 86 (1990); F. Halzen and B. A. Kniehl, Nucl. Phys. **B353**, 567 (1991); B. A. Kniehl and A. Sirlin, Nucl. Phys. **B371**, 141 (1992).

[34] S. Fanchiotti, B. Kniehl, and A. Sirlin, Phys. Rev. **D48**, 307 (1993).

[35] W. Beenakker and W. Hollik, Z. Phys. **C40**, 141 (1988); A. A. Akhundov *et al.*, Nucl. Phys. **B276**, 1 (1986); J. Bernabeu, A. Pich, and A. Santamaria, Nucl. Phys. **B363**, 326 (1991).

[36] For a review, see P. Langacker, in *Testing the Standard Model*, ed. M. Cvetic and P. Langacker (World, Singapore, 1991), p. 863.

[37] K. A. Olive *et al.,* Phys. Lett. **B236**, 454 (1990); T. P. Walker *et al.,* Astrophys. J. **376**, 51 (1991).

[38] Theoretical uncertainties are discussed in [12].

[39] For a review, see J. F. Gunion *et al., The Higgs Hunter's Guide* (Addison-Wesley, Redwood City, 1990).

[40] A. X. El-Khadra *et al.,* Phys. Rev. Lett. **69**, 729 (1992).

[41] R. Barbieri *et al.,* Nucl. Phys. **B341**, 309 (1990); J. Ellis, G. L. Fogli, and E. Lisi, Nucl. Phys. **B393**, 3 (1993); H. E. Haber, Santa Cruz SCIPP 93/06.

[42] A. Djouadi *et al.,* Nucl. Phys. **B349**, 48 (1991); M. Boulware and D. Finnell, Phys. Rev. **D44**, 2054 (1991).

[43] H. Georgi and S. L. Glashow, Phys. Rev. Lett. 32, 438 (1974).

[44] For reviews, see P. Langacker, Phys. Rep. **C72**, 185 (1981); *Ninth Workshop on Grand Unification*, ed. R. Barloutaud (World, Singapore, 1988) p3; G. G. Ross, *Grand Unified Theories* (Benjamin, 1985).

[45] L. E. Ibáñez and G. G. Ross, Phys. Lett. **105B**, 439 (1981); S. Dimopoulos, S. Raby, and F. Wilczek, Phys. Rev. **D24**, 1681 (1981); D.R.T. Jones, Phys. Rev. **D25**, 581 (1982); M.B. Einhorn and D.R.T. Jones, Nucl. Phys. **B196**, 475 (1982); W. J. Marciano and G. Senjanovic, Phys. Rev. **D25**, 3092 (1982).

[46] Recent studies of the implications of the couplings for grand unification include: P. Langacker and M. Luo, [17]; U. Amaldi, W. de Boer and H. Fürstenau, Phys. Lett. **B260**, 447 (1991). J. Ellis, S. Kelley and D. V. Nanopoulos, Phys. Lett. **B249**, 441 (1990). F. Anselmo, L. Cifarelli, A. Peterman, and A. Zichichi, Nuo. Cim. **104A**, 1817 (1991); P. Langacker and N. Polonsky, [20].

[47] H. Georgi, H. R. Quinn and S. Weinberg, Phys. Rev. Lett. 33, 451 (1974).

[48] See [20]; L. J. Hall and U. Sarid, Phys. Rev. Lett. **70**, 2673 (1993); M. Carena, S. Pokorski, and C. Wagner, Nucl. Phys. **B406**, 59 (1993); and references theirin.

[49] P. Nath and R. Arnowitt, Phys. Rev. **D38**, 1479 (1988), and references therein.

[50] U. Amaldi *et al.*, [12].

[51] For reviews and analyses, see J. E. Kim *et al.*, Rev. Mod. Phys. **53**, 211 (1981); U. Amaldi *et al.*, [12]; G. Costa *et al.,* Nucl. Phys. **B297**, 244 (1988); G.L. Fogli and D. Haidt, Zeit. Phys. **C40**, 379 (1988); P. Langacker and M. Luo, [17]; Rev. Part. Properties, Phys. Rev. **D45**, III-59 (1992); F. Perrier, in *Precision Tests of the Standard Electroweak Model.* For recent results, see [11, 21, 22].

[52] See [10] and J. Panman, in *Precision Tests of the Standard Electroweak Model.*

[53] LANL: R.C. Allen *et al.*, Phys. Rev. Lett. **64**, 1330 (1990); Phys. Rev. **D47**, 11 (1993).

[54] F. Reines *et al.*, Phys. Rev. Lett. **37**, 315 (1976).

[55] C. Y. Prescott *et al.*, Phys. Lett. **84B**, 524 (1979).

[56] Mainz: W. Heil *et al.*, Nucl. Phys. **B327**, 1 (1989); Bates: P.A. Souder *et al.*, Phys. Rev. Lett. **65**, 694 (1990).

[57] BCDMS: A. Argento *et al.*, Phys. Lett. **120B**, 245 (1983); **140B**, 142 (1984).

[58] M. A. Bouchiat *et al.*, Phys. Lett. **134B**, 463 (1984).

[59] Boulder: M.C. Noecker *et al.*, [8]; for a review, see B. P. Masterson and C. Wieman, in *Precision Tests of the Standard Electroweak Model.*

[60] For references, see [61].

[61] P. Langacker, M. Luo, and A.K. Mann, Rev. Mod. Phys. **64**, 87 (1992).

[62] P. Langacker, Phys. Lett. **B256**, 277 (1991).

[63] C. Kiesling, *Standard Theory of Electroweak Interactions*, (Springer, Berlin, 1988); R. Marshall, Z. Phys. **C43**, 607 (1989); Y. Mori *et al.*, Phys. Lett. **B218**, 499 (1989); D. Haidt, in *Precision Tests of the Standard Electroweak Model.*

[64] Some recent analyses of extra $Z$ bosons include P. Langacker and M. Luo [18]; L. S. Durkin and P. Langacker, Phys. Lett. **166B**, 436 (1986); U. Amaldi *et al.*, [12]; the sensitivity of future precision experiments is discussed in [61].

[65] G. Altarelli, *et al.*, Phys. Lett. **B263**, 459 (1991), *ibid.* **B261**, 146 (1991); J. Layssac, F. M. Renard, and C. Verzegnassi, Z. Phys. **C53**, 97 (1992); F. M. Renard and C. Verzegnassi, Phys. Lett. **B260**, 225 (1991).

[66] K. T. Mahanthappa and P. K. Mohapatra, Phys. Rev. D **43**, 3093 (1991); P. Langacker, Phys. Lett. **B256**, 277 (1991).

[67] F. del Aguila, W. Hollik J. M. Moreno, and M. Quiros, Nucl. Phys. B372, 3 (1992); F. del Aguila, J. M. Moreno, and M. Quiros, Nucl. Phys. **B361**, 45 (1991); Phys. Lett. **B254**, 479 (1991); M. C. Gonzalez-Garcia and J. W. F. Valle, Phys. Lett. **B259**, 365 (1991).

[68] M. Peskin and T. Takeuchi, Phys. Rev. Lett. **65**, 964 (1990); M. Golden and L. Randall, Nucl. Phys. **B361**, 3 (1991).

[69] W. Marciano and J. Rosner, Phys. Rev. Lett. **65**, 2963 (1990); D. Kennedy and P. Langacker, [19]; G. Altarelli and R. Barbieri, Phys. Lett. **B253**, 161 (1990); B. Holdom and J. Terning, Phys. Lett. **B247**, 88 (1990); B. W. Lynn, M. E. Peskin, and R. G. Stuart, in *Physics at LEP*, CERN 86-02, Vol. I, p. 90; G. Altarelli, R. Barbieri, and S. Jadach, Nucl. Phys. **B369**, 3 (1992); H. Georgi, Nucl. Phys. **B363**, 301 (1991); M. J. Dugan and L. Randall, Phys. Lett. **B264**, 154 (1991); E. Gates and J. Terning, Phys. Rev. Lett. **67**, 1840 (1991).

[70] P. Langacker and D. London, Phys. Rev. **D38**, 886 (1988); J. Maalampi and M. Roos, Phys. Rep. **186**, 53 (1990); E. Nardi and E. Roulet, Phys. Lett. **B248**, 139 (1990); E. Nardi, E. Roulet, and D. Tommasini, Nucl. Phys. **B386**, 239 (1992), Phys. Rev. **D46**, 3040 (1992). For recent reviews, see E. Nardi, Michigan UM-TH 93-14; D. London, in *Precision Tests of the Standard Electroweak Model.*

[71] P. Langacker and R. Rückl, unpublished.

[72] A. Blondel, A. Djouadi, and C. Verzegnassi, Phys. Lett. **B293**, 253 (1992); G. Altarelli, R. Barbieri, and F. Caravaglios, Nucl. Phys. **B405**, 3 (1993).

[73] R. S. Chivukula, B. Selipsky, and E. H. Simmons, Phys. Rev. Lett. **69**, 575 (1992); R. S. Chivukula *et al.*, Phys. Lett. **B311**, 157 (1993).

**CHAIRMAN: P.Langacker**

*Scientific Secretaries:A. Hoang, M.Lu, M.Stoilov*

## DISCUSSION

– *Gourdin:*

At the beginning of your talk you pointed out the role of FCNC. May I add that in the Standard Model the FCNC are absent at the tree level and they can be computed at the one loop level. Predictions can be made and we have here an ideal way to detect new physics if the observed FCNC effects are larger than those predicted by the Standard Model. The recent qualitative result of Cornell observing $K^* \rightarrow \pi + \gamma$ seems to be quite consistent with the Standard Model estimate and this first experimeental observation of Penguin diagrams is very interesting. More interesting perhaps is the upper limit $BR(b \rightarrow s + \gamma) < 5.4 \times 10^{-4}$ (90% CL) which shows that in this case New Physics is not important.

– *Langacker:*

I certainly agree with that. I believe that in supersymmetric extensions of the Standard Model there are various diagrams involving Winos and Higgses and other things which could contribute possibly even at the current experimental limit. However, since there are a number of possible diagrams and the possibility of cancellations, I do not think that one can make any stringent exclusions based on that. But I think it is very important for the future.

– *Gourdin:*

At the end of your talk you discussed the converging point of the $SU(2), U(1), SU(3)_c$ coupling constants in a Grand Unified Theory. This naïve idea of convergence assumes a desert type situation between $10^{16} GeV$ and $10^2 GeV$. More complicated situations can occur as for instance a cascade of intermediate breaking between the Grand Unified group G and Standard Model group $SU(2) \times U(1) \times SU(3)_c$. The choice of the group G is relevant for the problem of crossing and RGE's describing the evolution of the coupling constants depend on the group G and the intermediate mechanism. The extension to Supersymmetry will not change this situation and the choice of the Supersymmetry group and the nature of subsequent breakings are very important. In this sense the results obtained concerning the crossing points are model dependent and they cannot be considered, for instance as a proof of Supersymmetry. In my opinion there are a lot of models more or less complicated for which the crossing at the same point can be obtained. Some are true and some are not. Finally nature will decide.

– *Langacker:*

Yes, I agree with everything you have said but I do want to emphasize one aspect. The minimal SUSY extension of the SM is the only model in which, to zeroth order, one gets agreement without adjusting any new inputs. It's true that if you take the masses of the supersymmetric particles to an enormous scale like $10^4$ TeV you would mess up the agreement. But I think that's a little bit unfair criticism because nobody has ever advocated a supersymmetric breaking scale up there anyway. As long as the mass scale of SUSY partners is in the TeV range, as everyone has suggested, and if you don't add anything else to the theory, then you get a reasonable agreement within uncertainties. There are many other models in which you can adjust something to get agreement, such as in the SO(10) model without SUSY. You can get agreement if you fit the value of the intermediate scale. That's not a prediction, you are fitting a parameter. You can also have many models involving new multiplets split into light and heavy sectors but again one has no predictions; one is just choosing the representation to fit the data.

– *Kawall:*

You suggest that in the ongoing g–2 experiment there will be a large theoretical uncertainty in the predictions due to light hadrons in the vacuum polarisation loop. Experimentally, what can one measure which is sensitive to these contributions apart from the g–2 experiment itself?

– *Langacker:*

We need more precise data on $e^+e^- \to$ hadrons, say in the 2 GeV region. That's input to the calculations. Unfortunately nobody has done those experiments precisely enough.

– *Montgomery:*

Does anyone know the status of the low energy $e^+e^- \to$ hadrons experiment at Novosibirsk? If it's not going, how much would such an experiment cost?

– *Langacker:*

The g–2 experiment costs close to twenty million dollars. Furthermore this may become one of the largest residual errors associated with the LEP data. In my opinion it is very important. It's not clear that it has to be done at Novosibirsk. I would be happy to advocate it. It may be that the Beijing people can contribute also.

– *Ellis:*

I would like to pose the previous question in a more precise way. What are the relative importances or weights, at present, of the experimental errors in different energy regions, say around 1 GeV where DAFNE could measure, around 2 GeV which you mentioned, the charm threshold which Beijing could measure, and even the region below and around the $b$ threshold? How would these relative weights change with a precise measurement of g–2, assuming no new physics play a role?

– *Langacker:*

I think the most important region is that around 1 GeV. The second most important region is just below the charm threshold. But I can't be more precise at this moment.

– *Anselm:*

I would like to comment on the possibility which may arise if there is no light enough Higgs boson (e.g. $m_H > 1$ TeV). A new strong interaction would appear with many properties similar to those for the usual strong interaction. For instance, particles may lay on Regge trajectories with typical slopes of order 1 TeV ($\alpha' \sim G_F$). So one can expect, for example, an excited electron with the mass $\approx$ 1 TeV and $J = 3/2$, and similarly for say, W-bosons. There are some interesting specific features of this interaction connected with the fact of possible massless states (neutrinos, electrons etc.) in the t–channel when high energy asymptotic behaviour is considered. But the most spectacular manifestation is the possible Reggeization of leptons, quarks etc.

– *Vysotsky:*

Why did you mention that there is no reason for neutrality of the atoms in the Standard Model?

– *Langacker:*

Gauge anomaly calculations give only two constraints on U(1)–charge assignments of particles and there is no guarantee of electrical neutrality. Now, if we assume that every family has the same charge structure and some other extra assumptions, then one can pin down the charges. I personally never found that too convincing, because a priori there is no reason in that scheme to assume the same charge structure of each family.

– *Hoang:*

I have two questions:

1) You told us that the neutrino number is $3.004 \pm 0.0035$. In which way does it make sense to determine the number of neutrinos with a higher accuracy than that of an integer?

2) You mentioned that in the on-shell definition of $\sin^2\theta$ the theory becomes artificially dependent on $m_t$. Can you be precise in what you mean by "artificially"?

– *Langacker:*

1) This number of neutrinos is an effective number that really counts anything into which the $Z$ can decay and which cannot be explicitly seen in the detector. So it counts the number of the ordinary neutrinos which have masses much less than the $Z$. But now suppose that there was a neutrino which had a mass of 40 GeV. Then the $Z$ decay into that neutrino would be allowed, but kinematically suppressed. So that neutrino would contribute a fraction of a unit to the number of neutrinos in the decay. There are other possible things like a light scalar neutrino in SUSY which counts one half. Certain types of Majorons other than the standard Gelmini-Roncadelli triplet Majoron also count 1/2. In this sense to determine the number of neutrinos as precisely as possible is extremely important.

2) The point is that $\sin^2\theta_w$ enters the theory for example in the $Z_f\bar{f}$-vertex. At tree level this vertex has the following structure:

$$t_{3,L}(1-\gamma_5)\gamma^\mu - 2\sin^2\theta_W^o Q\gamma^\mu \tag{1}$$

where $t_{3,L}$ is the third $SU(2)_L$ generator and $Q$ is the electrical charge and the index "o" denotes the unrenormalised quantity. This vertex doesn't know anything about $m_t$. The reason that the top mass enters all these effects so strongly is because it appears in the vacuum polarisation diagrams of the $Z$ and the $W$. This causes a shift of the $W$ mass wih respect to the $Z$ mass. In particular, when one computes the ratio $M_W^2/M_Z^2$ its higher order physical value is different from its tree level value and is dependent on the top quark mass. So this ratio really depends in some significant way on $m_t$. Suppose I defined the renormalised $\sin^2\theta_W$ in all orders of perturbation theory as

$$\sin^2\theta_W = 1 - M_W^2/M_Z^2 \tag{2}$$

which is in fact the on-shell definition. The very presence of the large top quark mass means that this defined quantity has a value significantly different than it would have had in the absence of the top quark. But the vertex still doesn't know anything about the top mass. So if I choose to rewrite the vertex in terms of the $\sin^2\theta_W$ as defined in (2). I have to rewrite the vertex term (1) in the following form:

$$t_{3,L}(1-\gamma_5)\gamma^\mu - 2\sin^2\theta_W KQ$$

where $K$ is a function of $m_t$. This function $K$ has to be introduced in order to cancel the artificial $m_t$ dependence induced into $\sin^2\theta_W$ by definition (2). In the same way you find form factors quadratically dependent on the top quark mass all over your theory, even though the relevant quantity doesn't know anything about the top quark mass. This method is correct and you get correct results. But to me it seems to be an unnecessarily complicated approach.

– *Piccinini:*

In the values quoted for the $Z$ parameters there are always symmetric errors while for the top mass there is always an asymmetric error. I would also expect an asymmetric error for the $Z$ parameters.

– *Langacker:*

The aymmetric error in $m_t$ is mainly because it is determined by experiment as $m_t^2$, whereas the $Z$ parameters are closer to gaussian distributions. I think this is the simplest answer. At some level there are always asymmetric errors. If the asymmetry is small enough people usually take the average and give a symmetric error.

– *Forshaw:*

How direct is the evidence for the existence of the top quark from the measurement of the b-quark forward-backward asymmetry?

– *Langacker:*

The answer is that there are many measurements of the weak interactions of the b quark.For example, the charged current decay sector of the b quark strongly suggested, though to my mind not quite convincingly proved, that the top quark existed. Suppose you just had two generations of doublets and the $b_L$ and all the right hand particles were singlets. Then, if the only interactions are the standard model ones, $b$ can decay only by mixing with $s$ and with $d$ quarks. If $b$ is in a doublet you have a GIM–like mechanism and then the mixing does not induce any flavor changing neutral currents. But if you mix a singlet with a doublet then you necessarily violate the GIM–mechanism and there will be for example

$$b_L \to s_L + e^+ + e^-$$

decay. People at Cornell looked for such decays and several years ago they said this mechanism was excluded and therefore the top must exist. That's a perfectly valid argument except that it has one loophole. Even in the Standard Model the decay is suppressed because even then you still have to rely on the mixing angles so that $b$ goes to $c$ via a charged current. The mixing angle is very small, so the decay is strongly suppressed. One could at least imagine that some new interactions mediate the decay in a topless model to get around that problem. So it's not a really compelling argument. In recent years however, there have been measurements in the neutral current sector of the $b$ quark, starting at PETRA. Currently at LEP we have measurements at the partial widths of

$$Z \to b\bar{b}$$

and the forward-backward asymmetry in the same process as I showed this morning. There is perfect agreement between experimental numbers (373 $\pm$ 9 MeV) and the Standard Model predictions (376 MeV) for $\Gamma_{b\bar{b}}$. If you have a topless model then $\Gamma_{b\bar{b}}$ must be 24 MeV and $A_{FB}(b) = 0$. So, to my mind, this establishes the fact that the topless models are excluded. The difference betwen these arguments and old decay ones is that this is a full strength interaction and could not be mimicked by a superweak interaction. In order to have something else come in you have to have a new interaction just as strong as the normal weak one is. If you believe in topless models you have to argue that 24 is consistent with $373 \pm 9$. This is of course not so. Other models such as mirror models and vector doublet models are also excluded. So I think this is a pretty compelling argument.

– *Zichichi:*

When you talk about a topless model, does this number 24 take into account that $b$ can indeed go to $c$?

– *Langacker:*

In the topless models you still have Z coupling to $b$ but it is a vector coupling, so you do not have any forward/backward asymmetry.

– *Zichichi:*

This I understand but in the topless model do you have $b \to c$ transition?

– *Langacker:*

One should assume that in a topless model there has to be a small mixing between $b$ and either $s$ or $d$ quarks in order to account for the decay of $b$ (unless there is a new interaction). But this mixing angle is very small and doesn't significantly affect the predictions.

– *Zichichi:*

Suppose we do not know anything about Kobayashi–Maskawa angles, and suppose by experimental measurements the $b \to c$ transition is high. Then this number $\Gamma_{b\bar{b}}$ would not be 24.

– *Langacker:*

That is true.

– *Zichichi:*

This is not a theoretical number – it comes from experimental knowledge.

– *Langacker:*

Yes, you are correct. I have put in the experimental fact that the mixing is small.

– *Zichichi:*

Which means that you have already used the information that there is a $t$ quark. The quantity $\Gamma_{b\bar{b}} = 24$ MeV for the "topless" model is obtained using the experimentally known K–M angles for $b \to c$ transition. So it should be called "topless" with hidden top. My point is that if top were not there, the experimental number that you use for $\theta(b \to c)$ would not be too small. The small value of the $\theta$ has the "hidden" information that $t$ is there. If $t$ is not there $\theta(b \to c)$ would be large. I forbid you to use the experimental knowledge if you want to make a topless model.

– *Langacker:*

I disagree with what you are saying because if I formulate the simple topless model – no new interactions – I could do this in a generalised way, which means I have to introduce some arbitrary unknown mixing angles between $b$ and $s$ (and with $d$ also). Then I could write predictions for the width and for the $b$ lifetime. Now I do have a theoretical prediction that relates two experimental quantities in terms of the mixing angle. I think I can make one measurement of the lifetime to determine the mixing angle and I can then predict the other quantity, and it does not agree with experiment. I do not see anything wrong with this procedure.

– *Morpurgo:*

Just for my clarification; I believe that the top was necessary because in its absence there would arise infinities from anomalies but is there a way out of this so that one can have a topless model? In other words can one have a finite modified model without a top?

– *Langacker:*

This argument has been a little bit misused. You can have another way of getting rid of the anomalies in such a model. If you took $b_L$ and $b_R$ , both singlets, and simultaneously said that $\tau_L$ and $\tau_R$ were singlets then you would have in the third family a vector-like family without any anomaly. Such a model is not phenomenologically viable, but it illustrates that the anomaly argument by itself is not compelling.

– *Vysotsky:*

Topless theories have existed for 20 years already. It was the weak interaction theory without the charm quark.

– *Abu Leil:*

How dependent is the measurement of $M_Z$ on radiative corrections?

– *Langacker:*

It is very dependent especially with initial state QED radiation, but it is believed it is under theoretical control.

– *Khoze:*

Can the topless model survive the existing experimental data on $B^0 - \overline{B^0}$ oscillations?

– *Langacker:*

As I have already mentioned, the topless model contradicts the data from CLEO. If we forget about CLEO constraints, these FCNC will contribute to $b \to d$ as well as $b \to s$ processes which can induce $B^0 - \overline{B^0}$ oscillations. That would occur either in the $B_d$ system or $B_s$ system or both. There was one analysis which concluded that there could be generically greater oscillations than experimentally observed for the obvious reason that this is a tree process. This is potentially another problem.

– *Beneke:*

If for some particular quantity you observe strong $m_t$ dependence of radiative corrections in one renormalisation scheme and smaller dependence in another, does that mean you should worry about the two loop corrections in both schemes?

– *Langacker:*

The two loop effects are significant. If they are left out the $m_t$ prediction would have changed by 5 to 10 GeV. In the on-shell scheme in which you have a larger and somewhat artificial $m_t$ dependence, I don't think it's an enhanced effect but I haven't seen any particular discussion of that. Two loop effects are incorporated in $W$ and $Z$ masses where they are important, but not in (for example) electron–neutrino scattering.

– *Vassilevskaya:*

Does the more precise measurement of the $\rho$ - parameter of electroweak theory permit new physics beyond the standard model on the mass scale below 1 TeV?

– *Langacker:*

Yes, the $\rho$ parameter in the generalised $SU(2) \times U(1)$ model has a tree level contribution if there are Higgs fields that are higher dimensional representations than doublets. It's always been hard to disentangle this $\rho_0$ from $m_t$ dependence because except for the $Zb\bar{b}$ vertex all the quadratic dependences appear in a universal way with $\rho_0$. Now, however, with the better measurement of the $Z\ b\bar{b}$-vertex, one can separate them and the conclusion is that this $\rho_0$ is unity to within around one percent. This means that the VEVs of the new Higgs triplets are typically less than of the order 5 - 10 GeV, compared to the VEV of 246 GeV in the doublet. Now there has been a lot of talk in the last three years about radiative corrections, loop corrections that act like this $\rho$ parameter. If there is a fourth family that is non-degenerate, or if there are new scalar multiplets that are non-degenerate, or if there are heavy degenerate multiplets which are chiral, then they will affect the radiative corrections. These are what are called S, T and U on the other side of the Atlantic and $\epsilon_l, \epsilon_2, \epsilon_3$ on this side. The so-called T parameter($\epsilon_1$) is the analogue of loop corrections due to say a non-degenerate fourth family. It does the same thing as $\rho_0$ does from VEVs. Now, these limits tell us about the splittings. They don't tell us anything about whether there is a degenerate TeV fermion doublet. However, if you have fermion doublets which are chiral, then even if they are degenerate, they do lead to effects on the relation between the Z mass and the strength of the neutral current interactions. That's the so-called S parameter or $\epsilon_3$ parameter. Technicolor theories typically predict such multiplets. There have been a lot of arguments that at least the simplest extended technicolor models are eliminated. These are examples but they do have difficulties.

– *Vassilevskaya:*

What is your opinion on the possibility of the existence of a fourth fermion generation in the framework of the standard theory?

– *Langacker:*

I certainly cannot rule it out. There are only two direct experimental constraints. One is from the invisible width of the Z (from the total width), that there could not be the fourth family unless the associated neutrino is heavier than 40 to 45 GeV. The other is from the $S$ parameter. The fourth family could be tolerated, but probably not the fifth or the sixth family.

– *Vassilevskaya:*

It is known that the feasible phenomena of lepton mixing similar to the quark one, appears quite naturally if the neutrinos have a non–degenerate mass spectrum and in itself, does not go beyond the standard EW theory. Are there other explanations of lepton mixing if, for example, the neutrinos have a degenerate mass spectrum and a heavy neutral lepton exists?

– *Langacker:*

Well would there be other ways to get lepton mixing? If you just added neutrino masses to the theory and for some reason they were exactly degenerate, you would still have individual lepton family number conservation. You could invent other new interactions, such as ones involving new scalars, which could lead to lepton mixing even in that case. Also if one neutrino is very heavy and the others are degenerate one can still have mixing effects because the light states produced in weak processes are non-orthogonal. However my personal prejudice is that if neutrinos do have masses there is no particular reason to believe they are degenerate.

– *Piccinnini:*

It is a commonly adopted procedure to test the standard model by using the Z parameters and then derive the values of $M_Z$, $m_t$, $\alpha_s, m_H$ etc. Don't you think that it could be a more correct procedure to fit the measured cross sections and asymetries directly?

– *Langacker:*

There are a couple of issues here. One is that it is necessary to verify the standard model by testing its consistency and measuring parameters like the Z mass and width. The other separate issue is trying to allow more general parametrisations to test various possible extensions of the standard model. These two require different kinds of fit. Each type of new physics may have different effects and one may look for them through some particular windows. The LEP data is precise but it is blind to some regions where low energy experiments such as atomic parity may be more sensitive to new physics. There is no ideal way to look for all new physics, one just has to try to work through them all.

Now back to your question of whether it's better to work with outputs of analysis of data such as widths or to stick with direct observations. There is no easy answer. Before LEP was turned on, I used to try to stay with original data because I thought it was best. But it is hard to work exclusively with raw data. The LEP Monte Carlo and QED radiative corrections are so complicated it's impossible to work with raw data and one must use processed data instead unless you are one of the members of a LEP collaboration. Even for a LEP experimentalist it would be hard to use the data from all four experiments or to interface with other types of experiments.

– *Lu:*

From my understanding, the Higgs sector is the least understood part in the Standard Model. Given the fact that the SSC may not be built, is there any hope of testing it experimentally?

– *Langacker:*

Yes. I am personally prejudiced to the supersymmetric extension of the Standard Model. In that scenario, the Higgs mass is probably lighter than 150 GeV, maybe much lighter. You do not need the SSC to find a light Higgs like that. LEP200 could have a good shot at it if it is very light.

– *Lu:*

Do you really believe in SUSY?

– *Langacker:*

I take what nature gives me. But my personal prejudice is that it is what fits together best at present, given first that it fits the experimental data and second that it has some beautiful theoretical underpinning which might be related to Superstring theories. I don't say this is right, of course, but it looks more promising to me than other theories so I think there is a hope.

# Why it seems too early to report on status of QCD

Yu.L.Dokshitzer

St.Petersburg Nuclear Physics Institute, Russia
and University of Lund, Sweden

## 0 Preface

We have a kind of unspoken agreement with Professor Zichichi: He names my future lecture, which title appears then in the program of the Erice School, I feel free to change it (sometimes completely) when it comes to delivering the lecture. This should explain the genesis of the strange title above. My 1993 talk has been announced as *Status of QCD.* I will try instead to convince you that QCD is alive: it barely has a "Status", rather "Dynamus", if you take my meaning.

## 1 Light Quarks and Colour Confinement

Recently V.N. Gribov [1,2] has proposed a breathtaking solution of the QCD confinement problem. In the Gribov scenario the colour confinement in the real world is strongly determined by existence of very light (practically massless $u$ and $d$) quarks. These lead to a radical change of the perturbative vacuum in the region between $1/\lambda$ and $1/m$, the Compton wavelength of the light quark, in a close analogy with the phenomenon of "super-charged" ions in QED [3].

In QED, when the electric charge of a nucleus exceeds some critical value ($Z_{crit} = 137$ for a point-like charge), light fermions in the vacuum start to "fall on the centre" creating stationary states with negative electron energy, $\epsilon < -m$. This causes instability of the perturbative vacuum. As a result, nuclei with $Z > Z_{crit}$ decay: $Z \to (Z-1) + e^{+}$.

In QCD the Coulomb-like attraction between fermions leads to a similar falling on the centre. Any coloured particle in QCD acquires a spatial colour charge distribution due to gluonic vacuum polarisation. The "super-critical" phenomena develop when the size of the volume $r_0 \equiv \lambda^{-1}$ in which the total charge $\alpha_s(\lambda)$ exceeds some critical value $\alpha_{\rm crit} \approx 0.6$, is much smaller than the light quark Compton wavelength $m^{-1} \sim$ (5–10 MeV)$^{-1}$. Contrary to the QED case where the nuclear charge would decrease by one unit, in the QCD context this results in producing a *colourless* bound state with negative total energy which, in the end of the day, causes instability of any coloured state. It is important to notice that in the QCD context this happens not only for a light quark in an external field of a heavy source (as in the above QED example) but for interaction between light quarks as well.

To give you an insight of the "mathematics" that backs up this picture, here is the differential equation for the fermion (electron/quark) Green function in the momentum space that sums up the most singular effects due to soft boson (photon/gluon) interaction [1]

$$\frac{\partial^2}{\partial q_\mu \partial q_\mu} G^{-1}(q) = \frac{\alpha(q^2)}{\pi} \left(\frac{\partial G^{-1}(q)}{\partial q_\lambda}\right) G(q) \left(\frac{\partial G^{-1}(q)}{\partial q_\lambda}\right) \\ + \text{ less infrared singular } \mathcal{O}\left(\alpha^2\right) \text{ terms.} \tag{1}$$

In principle, the whole PT series expansion may be constructed for the right hand side in terms of exact Green functions (and their momentum derivatives). In particular, with account of the first subleading terms (1) becomes an integro-differential equation and its r.h.s. may be represented in the following graphic form (with black dots standing for the momentum gradient of the inverse propagator[1])

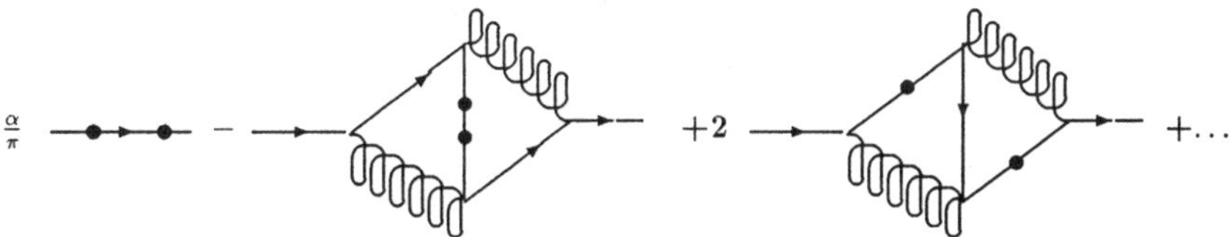


Starting from

$$C_F \frac{\alpha_s}{\pi} > C_F \frac{\alpha_{crit}}{\pi} = \sqrt{1 - \frac{2}{3}} \approx 0.18 \tag{2}$$

the *non-linear* equation (1) has two types of solutions corresponding to

1. fast increase of fermion masses (the one that resembles the properties of a spontaneous symmetry breaking solution) and
2. the phase transition of a new type with two extra *inversely populated* zones in the Dirac sea.

The second (confinement) solution corresponds to the appearance of "super-bound" $q\bar{q}$ states with *negative* total energy, which consist of strongly interacting light flavour quarks with *positive* kinetic energies. In the familiar picture of the Dirac sea all negative kinetic energy fermion states are occupied and the positive are empty. Now, to the contrary, one encounters some positive energy quarks sitting in the true vacuum, tightly bound into "super-critical" colourless pairs. An excitation of a negative energy electron state, thanks to P.A.M. Dirac, corresponds to a physical positron. Heating up a super-bound quark pair would produce a positive energy meson that contains, strangely enough, a pair of *negative* kinetic energy quarks $q_{(-)}$. (To make the picture even more confusing let me mention that these two guys interact *repulsively* which interaction gives rise to positive meson mass. What actually keeps them together is the Fermi principle: The negative energy states *outside* the "bag" are occupied.)

[1]actually, the exact zero-momentum boson emission vertex, due to the Ward identity

## 1.1 Scalars: Conventional and Novel

These unusual vacuum excitations as "novel" mesons should be there in the hadron spectrum of the theory, coexisting with the "normal" hadrons built up of good old positive energy constituent quarks $q_{(+)}$. The theory is not yet developed enough to give a definite quantitative description of such a coexistence. Nevertheless, on a qualitative level some key features may be anticipated [4]:

- The lowest "novel" states are expected to be scalar mesons ($J^{PC} = 0^{++}$), hereafter — VS for "vacuum scalar(s)";
- The most natural mass range would be $M_{\rm VS} \sim \lambda \approx 1 GeV$ corresponding to the characteristic scale where the coupling hits the critical value, $\alpha_s(\lambda) \gtrsim \alpha_{\rm crit}$;
- VS should be spatially compact states with a typical size $r_{\rm VS} \sim \lambda^{-1} \ll 1\,fm$;
- Splitting between the masses of isotopic singlet and triplet VS mesons should be small;
- Compact $q_{(-)}$ states should be strongly decoupled from the "normal" hadron world: Both hadronic and $\gamma\gamma$ decays of VS must be suppressed relative to corresponding widths of their "normal" opposite members with the same quantum numbers.

The anticipated decoupling between "normal" and "novel" states, the mass degeneracy and the suppressed $\gamma\gamma$ decay of the "novel" mesons follow basically from one and the same physical observation: The decay of a scalar (pseudoscalar) state into gluons/photons is helicity violating.

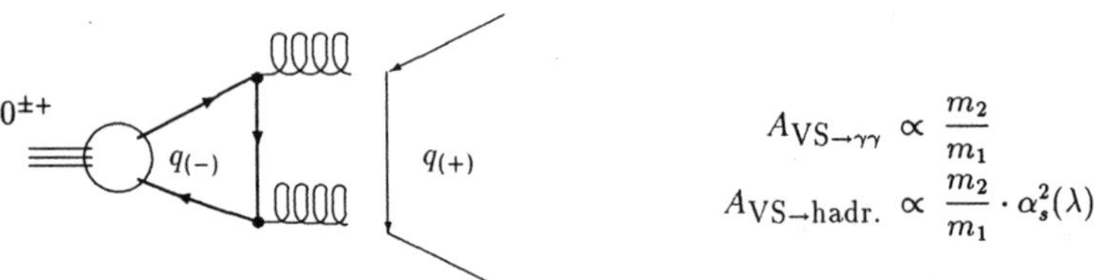


$$A_{\rm VS\to\gamma\gamma} \propto \frac{m_2}{m_1}$$
$$A_{\rm VS\to hadr.} \propto \frac{m_2}{m_1} \cdot \alpha_s^2(\lambda)$$

As a result, both the two-photon decay amplitude and the amplitude for the $q_{(-)} \leftrightarrow q_{(+)}$ transition via gluons acquire the suppression factor $m_2/m_1$ where $m_1$ and $m_2$ are, respectively, the effective mass of the "normal" $q_{(+)}$ and the "novel" $q_{(-)}$ quark states. Since the latter form the compact colourless systems, their effective mass should be closer to the original current value, and so is has to be smaller (formally, *parametrically* smaller) than the usual constituent light quark mass scale,

$$m_1 \approx \tfrac{1}{3} GeV\,, \quad m_2 \ll m_1\,.$$

The ratio of the two-photon decay widths of the isoscalar and isovector VS's would be the most immediate test of the entire hypothesis: Given the standard quark composition (though for the "non-standard" negative energy quarks!),

$$\begin{aligned} \mathrm{VS}_{I=0} &= \qquad \tfrac{1}{\sqrt{2}}\left(u\bar{u} + d\bar{d}\right) \\ \mathrm{VS}_{I=1} &= u\bar{d}\,,\ \tfrac{1}{\sqrt{2}}\left(u\bar{u} - d\bar{d}\right)\,,\ d\bar{u}\,, \end{aligned} \tag{3}$$

this ratio should closely follow the most simple quark charge counting rule

$$\frac{\Gamma(\mathrm{VS}_{I=0} \to \gamma\gamma)}{\Gamma(\mathrm{VS}_{I=1} \to \gamma\gamma)} = \left(\frac{e_u^2 + e_d^2}{e_u^2 - e_d^2}\right)^2 = \frac{25}{9}\,. \tag{4}$$

At the same time, the ***absolute*** value of the decay width $\Gamma_{2\gamma}$ has to be suppressed, compared to "normal" mesons, as $[m_2/m_1]^2$.

Similar factor is responsible for the small mass splitting between $\mathrm{VS}_{I=0}$ and $\mathrm{VS}_{I=1}$. Mass of the former state should have been affected (lowered) by the mixture with the two-gluon (multi-gluon) state. It is helicity violation suppression that makes this mixing tiny.

Let us stress once again that it is an existence of quarks with *very small* current mass $m \ll \lambda \sim 1GeV$ that are essential to the theory. Given this fact, one would expect the SU(3) flavour symmetry to be badly broken with respect to such "novel" states: The strange quark $s$ apparently has too large a current mass to participate in colour blanching in the way $u$ and $d$ do. Therefore a search for non-standard scalar mesons with quite unusual properties (nearly degenerate in mass, relatively narrow etc.) that fit into SU(2) flavour symmetry but fail to have any strange partners would be a crucial and most clear phenomenological test of the Gribov light-quark-confinement picture.

## 1.2 This Messy Scalar World

Even if your occupation is called "theoretical physics", you are familiar with Feynman diagrams and feel sick when facing the HERA kinematics, it is sometimes very inspiring to have a look into Ref.[5] to discover what the real world looks like. So did I, and it came as a surprise to learn that well established scalar mesons that show no respect to $\mathrm{SU}(3)_{\mathrm{flav.}}$ do exist. They are there in the Particle Data Book, ready and packed, a long-standing problem for celebrated hadron zoology based on the standard constituent quark picture.

Let us have a look at Table 1 in which I attempted to collect the PDG mesons together with the recent evidences for scalar excitations. Such a compilation is subject to rapid aging due to active involvement of a number of excellent experimental collaborations in the field of meson spectroscopy.

PDG (p.VII.193) names "the well-established scalars" 1992: $f_0(975)$, $f_0(1400)$, $f_0(1590)$ and $f_0(1710)$. However even those would be apparently too many to embody into the quark model (QM) picture which can manage [15] 2 ground states and 1 radially excited state below 1.8GeV.

The PDG casts some suspicion upon $f_0(1590)$ and $f_0(1710)$ as "non-$q\bar{q}$ Candidates" (p.VII.193). The $f_0(1560)$ that has been seen by Crystal Barrel in the $\eta\eta$ mode and recently confirmed in the $\pi\pi$ channel, is most likely one and the same particle with the puzzling GAMS

| name | seen in | by | Mass | Width | Decay via |
|---|---|---|---|---|---|
| $f_0(975)$ | $pp, \pi p, \psi, \Upsilon, e^+e^-$ | see [5] | $974.1 \pm 2.5$ | $47 \pm 9$ | $\pi\pi : KK \approx 4:1$ |
| $f_0(1240)$ | phase in $\pi^- p$ | MPS [5] | 1240±23 | 140±23 | $K_s^0 K_s^0$ |
| $f_0(1250)$ | $\gamma\gamma$ | CrBall [6] | 1250 | 268±70 | $\pi^0\pi^0$ |
| $f_0(1345)$ | $\bar{p}p$ { see [7] }<br>$\bar{n}p$ | CrBarr<br>OBELIX | $1347 \pm 38$<br>$1345 \pm 45$ | 300–400 | $\pi^+\pi^-2\pi^0$ [8]<br>$2(\pi^+\pi^-)$ [9] |
| $f_0(1400)$ | ph. in $\pi^- p$, $pp$, $\psi$ | see [5] | $\sim 1400$ | 150–400 | $\pi\pi \gg K\bar{K} \gg \eta\eta$ |
| $f_0(1525)$ | $K^- p \to K\bar{K}\Lambda$ | LASS [10] | $\approx 1525$ | $\sim 90$ | $K\bar{K}$ |
| $f_0(1560)$ | $\bar{p}p$ | CrBarr [7] | $\approx 1560$<br>$\sim 1500$ | $\approx 245$<br>100–150 | $\eta\eta$ [11]<br>$\pi^0\pi^0$ [12] |
| $f_0(1590)$ | $\pi^- p$, $\pi^- N$ | GAMS [13] | $1587 \pm 11$ | $175 \pm 19$ | $\eta\eta', \eta\eta, 4\pi^0, \pi\pi, K\bar{K}$ |
| $f_0(1710)$ | $\pi^- p$, $pp$, $\psi$, $e^+e^-$ | see [5] | $1709 \pm 5$ | $146 \pm 12$ | $K\bar{K}, \pi\pi, \eta\eta$ |
| $X(1740)$ | $\pi^- p \to \eta\eta N^*$ | GAMS [14] | $1744 \pm 15$ | $< 80$ | $\eta\eta, \eta\eta'$, $\pi^0\pi^0$ |

Table 1: Scalar Zoo 1993

$f_0(1590)$. The latter has been found to have the very unusual decay pattern dominated by $\eta\eta'$ decay mode,

$$\eta\eta' : \eta\eta : 4\pi^0 : \pi\pi : K\bar{K} \;=\; 3:1:1:(<1):(<1).$$

Still, as far as the two different experiments agree at least on having seen the state, it is a good part of the story. However, in a recent coupled channel fit Lindenbaum and Longacre have claimed [16] to "generate the $f_0(1590)$ effect decaying in $\eta\eta$ and $\eta\eta'$ as a sum of the $f_0(1400)$ and the $f_0(1720)$". Moreover, no evidence for this state has appeared from the first high-statistics analysis of the $\eta\eta$ system produced in $\bar{p}p$ annihilation at FNAL performed by the WA76 collaboration [17]. At the same time, WA76 has observed the $\eta$-rich structure

$$M = 1748 \pm 10\, MeV\;; \quad \Gamma = 264 \pm 25\, MeV,$$

which might be in accord with $f_0(1710)$ and/or with another GAMS particle $X(1740)$, $J^P = (0?, 2?)^+$. This is just one of many examples showing how far from relaxation the present day situation with scalar mesons is.

## 1.3 The $f_0/a_0$ Mystery

The structure of the Table 1 does not mean that the activity in the mass region 1200–1400 MeV is necessarily due to a single state: There may be more than one meson in this range. Before we leave this controversial subject it would suffice to make the following general observation:

- One way or another, there is a good candidate for the "normal" broad $^3P_0$ meson that one expects to have in the 1GeV region.

The latter is a scalar state known as $f_0(1400)$. A recent analysis of $\pi\pi$ and $K\bar{K}$ processes [18] finds that it is better described as "a very broad $f_0(1000)$ of width around 700 MeV". (The CELLO and the MARK-II collaborations have also reported a broad S-wave structure in the mass range $M \approx 1100\,MeV$, for refs. see [6].) This makes it thereby even better a candidate for the QM vacancy [15]:

$$M\left(1^3P_0\right) = 1090\,, \quad \Gamma \approx 850\,MeV.$$

Following [4] we hereafter denote this state as $f(1000/1400)$ . Its $\gamma\gamma$ width is consistent with it being the conventional $^3P_0$ [19,20].

The isovector partner of $f(1000/1400)$ expected in the same mass region should have the large width as well, $\Gamma \approx 400\,MeV$. Fresh evidence for this conventional scalar has come from the Crystal Barrel experiment [12]. Such a state,

$$a_0(\text{'1450'})\ : \qquad M \sim 1450\,MeV\,, \quad \Gamma \sim 250\,MeV,$$

was seen decaying into $\pi^0\eta$ and has been reported to be "obviously necessary to describe the data" [7]. Notice, that this observation seems to be in a reasonable agreement with the old GAMS observation [5],

$$a_0(1320)\ : \quad M = 1322 \pm 30\,MeV\,, \ \Gamma = 130 \pm 30\,MeV.$$

We come to the conclusion that the *lightest* scalars obviously do not belong to the standard QM classification. These are the two very narrow mesons, the isosinglet state $f_0(975)$ from the above Table and the accompanying isotriplet meson $a_0(980)$,

$$\begin{array}{llll} f_0(975)\,, \ I^G(J^{PC}) = 0^+(0^{++}): & M = 974.1 \pm 2.5\,, \ \Gamma = 47 \pm 9\,MeV\,, & \pi\pi : K\bar{K} \approx 4:1\,; \\ a_0(980)\,, \ I^G(J^{PC}) = 1^-(0^{++}): & M = 982.7 \pm 2.0\,, \ \Gamma = 57 \pm 11 MeV\,, & \pi\eta : K\bar{K} > 1\,. \end{array}$$

These mesons are roughly 10 times narrower than expected. Their two-photon decay widths are also up to an order of magnitude smaller than for conventional $q\bar{q}$ $^3P_0$ states [21,19]. At the same time, the $\gamma\gamma$ *branching ratio* (of order of $10^{-5}$) is consistent with the $q\bar{q}$ interpretation, suggesting that a common suppression is at work for the $\gamma\gamma$ and total widths. Moreover, the $f_0(1000/1400)$ state is strongly decoupled from the narrow $f_0(975)$ even though they have superficially the same quantum numbers.

Given these facts, it is tempting to identify the $f_0(975)$ and $a_0(980)$ as compact nearly degenerate systems of (negative kinetic energy) $u$ and $d$ flavours

$$\begin{aligned} f_0(975) &= \quad \tfrac{1}{\sqrt{2}}\left(u\bar{u} + d\bar{d}\right) \\ a_0(980) &= u\bar{d}\,, \ \tfrac{1}{\sqrt{2}}\left(u\bar{u} - d\bar{d}\right)\,, \ d\bar{u}\,. \end{aligned}$$

## 1.4 Coupling to "Normal" Hadrons

At leading order the $f_0(975)$, $f(1000/1400)$ are $q_{(-)}$ and $q_{(+)}$ excitations respectively, and hence are orthogonal states. As we have mentioned above, the wavefunction mixing contains the helicity suppression factor $[m_2/m_1]$ (and, strictly speaking, an extra power of small $\alpha_s(\lambda)$ in the tunnelling amplitude) and thus should be smaller than would be expected were the $f_0(975)$ a glueball, for example. Hence, an extreme decoupling seems most natural in this picture.

**VS Hadron Width.** In reality the normal hadrons will have some admixture of the "novel" quark states and vice versa. As an educated guess[4] one might attempt to estimate the phenomenological magnitude of the "normal"/"novel" mixture by comparing the S wave decays of two P-state mesons: The "novel" $f_0 \to \pi\pi$ and the "normal" axial vector meson decay $a_1 \to \rho\pi$:

$$\frac{\Gamma(f_0(975) \to \pi\pi)}{\Gamma(a_1(1260) \to \rho\pi)} \sim \left[\alpha_s^2 \frac{m_2}{m_1}\right]^2 \approx \frac{40\, MeV}{400\, MeV} = [1/10] \; . \tag{5}$$

If the $f(1000/1400)$ were established as the "normal" $^3P_0$ state, then the $f_0(1000/1400) \to \pi\pi$ would be a more direct comparison; note that this would give essentially the same result as eq.(5).

**VS Mass Splitting.** The $f_0 - a_0$ mass splitting due to (helicity violating) $q\bar{q} \to gg$ annihilation is governed by the same suppression parameter and can be estimated as

$$M(a_0) - M(f_0) \sim \frac{m_1^2}{2M} \cdot \left[\alpha_s^2 \frac{m_2}{m_1}\right]^2 \approx \frac{(\frac{1}{3}\, GeV)^2}{2\, GeV} \cdot \left[\alpha_s^2 \frac{m_2}{m_1}\right]^2 . \tag{6}$$

The suppression of $q_{(-)} \to q_{(+)}$ transitions expressed in eq.(5) implies that

$$M(a_0) - M(f_0) \approx \tfrac{1}{20} GeV \cdot [1/10] \, , \tag{7}$$

in accord with the data[5]. Such a rough estimate demonstrates that the picture appears to have some consistency.

**Suppressed 2-Photon Width.** The characteristic mass ratio $m_2/m_1$ may be extracted also from the two-photon $f_0$ decay. To this end one may compare the corresponding width[5] with 8/9 of the total photon width of the $\eta'(958)$ (due to its nonstrange component).

$$\Gamma(f_0 \to \gamma\gamma) = 0.56 \pm 0.11\, keV \; ;$$
$$\Gamma(\eta_n' \to \gamma\gamma) \approx 4\, keV \, .$$

A naïveestimate disregarding the phase space factors would give

$$\frac{\Gamma(f_0 \to \gamma\gamma)}{\Gamma(\eta_n' \to \gamma\gamma)} \approx \left[\frac{m_2}{m_1}\right]^2 \approx 1/7 \, . \tag{8}$$

This number is in a reasonable agreement with the hadronic estimate (5) which was more accurate since both decays there were S wave dominated.

All the above semi-quantitative considerations point at the welcome result that $m_2 < m_1$; if we adopt the value $m_1 = \frac{1}{3} GeV$ for the constituent light quark mass, then the conservative "upper limit" for the "novel" mass would be $m_1 \lesssim 150\, MeV$.

**Hadron Decays Producing VS.** The tunnelling between $q_{(-)}$ and $q_{(+)}$ states that suppresses the VS's hadronic width will also suppress its production in decays of conventional $q_{(+)}$ hadrons. If A is such a hadron with a "normal" hadronic width of order 150 MeV or more, then by analogy with eq.(5) one may expect that (apart from phase space effects)

$$\Gamma(A \to f_0(975) + X) \approx \frac{1}{10}\,\Gamma(A \to f_0(1000/1400) + X)\,.$$

As an example, among the decay products of the conventional tensor meson $\pi_2(1670)$, whose width is 250 MeV, the $f_0(1000/1400) + \pi$ channel is seen with a branching ratio at the level of ten percent whereas there is no reported sighting of $f_0(975) + \pi$. At the same time, in decay processes where selection rules and/or phase space suppress the total width to be of order tens of MeV or less, the $f_0(975)$ and $a_0(980)$ may show up (the "normal competition" has been removed).

It is curious that it is the *narrow* mesons only that are known to decay into $f_0/a_0$ (though this may be due to lack of search as much as lack of signal for the scalars).

An extreme case is the "onium" where the small width is due to short distance physics and this will make the production of $f_0/a_0$ (as VS) particularly favourable.

1. $f_0$ is seen recoiling against $\omega$ and $\phi$ in the two-body $J/\psi$ decays (with the branchings close to those for the $\omega\eta'$ and $\phi\eta'$ modes [5]);

2. $f_0$ in the $\Upsilon$ region has been recently measured by ARGUS [22]:

$$\frac{f_0(975)}{\rho^0(770)} = \begin{cases} (7.2 \pm 1.8)\% & \text{continuum}\,, \\ (11.7 \pm 3.0)\% & \Upsilon\,. \end{cases} \tag{9}$$

A comparable relative yield of $f_0$ of about 12% has been observed in $e^+e^-$ annihilation at $Z^0$:

$$\frac{f_0(975)}{\rho^0(770)} = \frac{0.10 \pm 0.04}{0.83 \pm 0.14} \qquad \left(\text{DELPHI}^{[23]}\right)\,.$$

Apart from "oniums" there are only two "clear" cases of VS production in hadron decays:

1. The first is the axial vector meson $f_1(1285)$ whose width of 25 MeV is due to G-parity suppressing the would-be-dominant decays. In this case the $a_0(980)$ (at least) is a prominent two-body channel:

$$f_1(1285) \to a_0(980)\pi\,, \qquad \Gamma(\eta\pi\pi)/\,\Gamma = (50 \pm 5)\%\,, \quad \frac{\Gamma(a_0\pi)}{\Gamma(\eta\pi\pi)} = (74 \pm 12)\%$$

2. Another remarkable source of VS is $\eta(1420)$ — one of the 2 pseudoscalar states resolved by MARK-III [24] beneath the $\iota/\eta(1440)^2$:

[2]for other references see also "Note on the $\eta(1440)$" in PDG [5], p.VII.42

| name | in | Mass | Width | Dominant Decay |
|---|---|---|---|---|
| $\eta(1420)$ | $J/\psi \to \gamma K\bar{K}\pi$ | $1416 \pm 10$ | $54 \pm {}^{39}_{32}$ | $a_0(980)\pi$ |
| | $J/\psi \to \gamma\eta\pi\pi$ | $1400 \pm 6$ | $45 \pm 13$ | |
| $\eta(1490)$ | $J/\psi \to \gamma K\bar{K}\pi$ | $1490 \pm {}^{14}_{18}$ | $91 \pm {}^{69}_{49}$ | $K^*(892)\bar{K}$ |

Table 2: $\iota/\eta(1440)$ split by MARK-III [25].

**Looking for Partnership.** The dominance of the $a_0\pi$ mode in the $\eta(1420)$ decay[3], suggests that this might be as well the "novel" (pseudoscalar) $q_{(-)}$ state decaying via $\pi$ emission to its scalar partner. Such a hypothesis was advocated in [4]. By analogy with the $\iota/\eta(1440)$ system, it has also been suggested that the broad ill-defined structure called $\pi(1300)$ may in fact be a combination of a broad "conventional" and a narrow "novel" state which are strongly decoupled from one another. If this is the case, one would expect that the narrow component decays in a manner analogous to the $\eta(1420)$, namely by the chain $\pi(1300) \to \pi f_0(975)$ and with a comparable width (some 50 MeV).

## 1.5 VS or a $K\bar{K}$ Molecule?

Puzzling properties of the $f_0/a_0$ system obviously call for a non-standard interpretation. The only viable hypothesis on the market today is that those are the $K\bar{K}$ bound states, *i.e.*, diffuse "$K\bar{K}$ molecules" as suggested by Weinstein and Isgur [27]. An attractive feature of this picture is that it naturally embodies the relation $m(f_0, a_0) \approx 2m(K)$, which presently looks like a mere coincidence within the VS scenario.

As far as experimental data on the S wave $\pi\pi$ and $K\bar{K}$ systems in concerned, the very fact that $f_0$ is sitting just at the $K\bar{K}$ threshold makes theoretical analysis particularly difficult. A subtle two-channel description is required involving the experimental information from $\pi\pi(K\bar{K}) \to \pi\pi(K\bar{K})$ scattering and central di-meson production in high energy collisions, decay reactions like $D_s\pi(\pi\pi)$ and $J/\psi \to \phi\pi\pi$, $J/\psi \to \phi K\bar{K}$. The $J/\psi$ data are especially constraining. Morgan and Pennington found that the latter require such a structure of the S-matrix (two nearby poles on both unphysical sheets of the amplitude) which makes $f_0(975)$ "most probably not a $K\bar{K}$ molecule" [28].

To be on the safe side, let us leave it as an open problem and concentrate on apparent differences between the $K\bar{K}$ and the VS pictures. These are really striking: The two scenarios look quite "orthogonal".

In the first place, the molecule picture genetically links the $f_0/a_0$ to strange quarks which are just strangers for the VS interpretation. This makes the "s-enriched" objects/processes a particularly good laboratory. As an example, a high "analyzing power" is expected [27,29] for the branching ratios $\phi \to \gamma f_0$, $\phi \to \gamma a_0$, an important part of the foreseeable physical program at DAΦNE [30].

[3]the results of the ASTERIX experiment [26] on the $\iota/\eta(1440)$ production in $p\bar{p}$ annihilation at LEAR also point at the large branching ratio $a_0\pi : K^*\bar{K} = 4:1$; see [25].

Secondly, what actually makes the $f_0/a_0$ narrow in the $K\bar{K}$ picture is the weak binding: Typical size of the $K\bar{K}$ wavefunction is much larger than the normal hadron size, so large as to make quark inter-collisions (annihilation) improbable. To the contrary, the VS mesons are believed to be more compact than the "normal" hadrons. This difference should be seen when the data on $f_0/a_0$ production on nuclei become available. For a loosely bound $K\bar{K}$ molecule the $K^-$ has a high chance of disappearance inside the nucleus through $K^-p \to \pi\Lambda$ whereas the $K^+$ continues to propagate in essentially the same direction as the initial scalar. Meantime, a compact VS state has a good probability to propagate and depart from the nucleus intact with the result that equal numbers of $K^+, K^-$ will be observed from the decay $f_0 \to K^+K^-$. The A dependence and other kinematic dependencies of scalar production on nuclei should be able to distinguish between dynamical models rather clearly [31,32,4].

The following observation is due to Georgio Parisi [33]. If $a_0(980)$ were a $K\bar{K}$ molecule, one would expect $\eta(1420)$ to be of similar nature, namely, the $K^*\bar{K}$ molecular state. Within such a picture the observed decay channel $\eta(1420) \to a_0(980)\pi$ looks quite natural. The "bad news" is, however, *non-observation* of a direct $\eta(1420) \to K^*\bar{K}$ decay mode which would have to be comparable to $a_0\pi$ (if not dominant) in the $K\bar{K}$ scenario. What we have instead according to MARK III (see Table 2), is two clearly separate pseudoscalar states: $\eta(1420)$ decaying *predominantly* into $a_0\pi$ and $\eta(1490)$ seen exclusively in the $K^*(892)\bar{K}$ channel.

Finally, let us come back to the ratio of the two-photon $f_0/a_0$ widths given by (4) in the VS scenario. This is to be contrasted with the diffuse $K\bar{K}$ molecular states in which case the $\gamma\gamma$ coupling will be dominated by the $K\bar{K}$ loop [29] and the widths be equal. The available experimental information is spoiled by the poorly determined branching $a_0 \to \eta\pi$ and is as yet indecisive:

$$\frac{\Gamma(f_0 \to \gamma\gamma)}{\Gamma(a_0 \to \gamma\gamma)} = \frac{21 \pm 4}{9 \pm 3} \cdot B(a_0 \to \eta\pi) = \begin{cases} 25/9\,, & \text{VS} \\ 1\,, & K\bar{K} \end{cases} \tag{10}$$

Information on the $a_0 \to \eta\pi$ coupling from LEAR [34], Fermilab [35] and from the impending $\phi \to \gamma a_0 \to \gamma\eta\pi$ at DAFNE [30,29] will be important in helping to clarify this crucial datum.

## 2 Heavy Quarks and Infrared Regular $\alpha_s$

In the second part of the lecture we shall discuss another provocative subject, the notion of "infrared finite" QCD coupling. As we all know, the perturbative response of the QCD vacuum to implanting a colour charge is "anti-screening": The effective quark-gluon interaction strength, conveniently described by the *running coupling*, starts to increase with distance and blows up at about 1 fm. Formally, $\alpha_s(k^2)$ acquires an "infrared pole" at finite momentum,

$$\left(\tfrac{11}{6}C_A - \tfrac{2}{3}T_R n_f\right) \cdot \frac{\alpha_s(k^2)}{2\pi} \approx \frac{1}{\ln(k^2/\Lambda^2)}\,; \qquad \alpha_s\left(k^2 \to \Lambda^2\right) \to \infty\,. \tag{11}$$

Such a catastrophic behaviour can not be managed within the perturbative approach. It only indicates that starting from the microscopic quark-gluon Lagrangian *and* the non-interacting (perturbative) Dirac vacuum one gets an inconsistent Field Theory which exhibits infrared instability. This perturbative instability was (and still remains) the main indicator in favour of *confinement*, namely, the desired property of QCD as a Field Theory that its physical spectrum consists of colourless objects, hadrons.

## 2.1 Philosophy behind Infrared Finite $\alpha_s$

It is clear that in the True Theory there is no place for infinities: Actual interaction at large distances may become strong, that is $\alpha_s \sim 1$, but never infinite. This is too trivial a statement, however, to be of any practical help. In the meantime, while waiting for the True Theory to become available, one may try to put regular $\alpha_s$ to work. Generally speaking, it is not easy to justify the very notion of $\alpha_s(Q^2)$ at small $Q^2$ where the PT quark-gluon language seems to be hardly applicable at all. Nevertheless, as we shall discuss below, in many practical applications one is actually forced to plug into theoretical analysis, one way or another, an infrared finite coupling.

Let us invoke a classical example that demonstrates, in the first place, the *necessity* of such a "regularization" and, in the same time, the *insensitivity* of the final results to an arbitrary modification of $\alpha_s$ in small momentum region.

This is the PT analysis of $P_\perp$ distributions of massive lepton pairs, $W$ and $Z$ bosons produced in hadron-hadron collisions [36,37]. A colourless massive object is produced in a hard collision of two partons. It is the large invariant mass $P^2$ ($W$, $Z$ mass or invariant mass of the lepton pair) that determines the hardness of the total production cross section. As a result, $d\sigma/dP^2$ is given by the product of the distributions of partons inside two colliding hadrons, $D(x_i, P^2)$, *measured* at scale $P^2$. These distributions are identical, in the leading approximation, to those measured in DIS, in a perfect accord with the original parton model Drell-Yan conjecture. The total Drell-Yan cross section is affected by QCD to a minor extent via logarithmic scaling violations hidden in the parton distributions.

The situation changes drastically, however, when one turns to differential $P_\perp$ spectra. Here multiple gluon radiation effects crucially modify the answer. Physically, the fact that colliding quarks are not point-like objects but characters of a Field Theory screenplay becomes essential: The double logarithmic QCD form factors come on stage.

In the Born approximation it is one additional parton (mainly gluon) that balances the $P_\perp$ of the $W/Z/\mu^+\mu^-$ pair. The corresponding first order quark annihilation cross section is singular in $P_\perp$:

$$P^4 \frac{d\sigma}{dP^2\, dP_\perp^2} \propto c\frac{\alpha_s}{P_\perp^2} \ln\frac{P^2}{P_\perp^2}\,; \qquad P_\perp^2 \ll P^2\,.$$

After proper resummation of multiple gluon radiation effects the double logarithmic quark form factors appear [36] that strongly damp this singularity. Operationally, the following exponentiated structure emerges,

$$c\frac{\alpha_s}{P_\perp^2} \ln\frac{P^2}{P_\perp^2} \Longrightarrow \int_0^\infty \frac{d^2\vec{b}}{(2\pi)^2}\, e^{-i\vec{b}\vec{P}_\perp} \cdot \exp\left\{ c\int_0^P \frac{d^2\vec{k}_\perp}{k_\perp^2}\, \alpha_s(k_\perp^2) \ln\frac{P^2}{k_\perp^2} \left[ e^{i\vec{b}\vec{k}_\perp} - 1 \right] \right\}. \tag{12}$$

Representation in terms of the impact parameter $\vec{b}$ takes care of the fact that the accompanying gluons on aggregate compensate $\vec{P}_\perp$ of the Drell-Yan object. The exponent in the exponent is due to real gluon emission, the unit subtraction corresponds to virtual gluon effects.

Singular behaviour of the differential Born cross section $d\sigma/dP_\perp^{-2} \propto P_\perp^{-2}$ originates from large impact parameters (as compared to the characteristic hardness $P^{-2}$): $b^2 \sim P_\perp^{-2} \gg P^{-2}$. The exponential profile factor in the right hand side of (12) makes, however, the $b$ integration convergent by suppressing large impact parameter region. As a result, $d\sigma/dP_\perp^2 \to$ const in the limit $P_\perp \to 0$. (Differential spectrum flattens off[37].) The shape of the transverse momentum distribution, stays under PT control. In particular, its height and width vary with $P^2$ in a completely predictable (and predicted) way.

This is an example of a very strong (and practically successful) PT QCD prediction, free from phenomenological inputs[4] but one: we have to ensure that the integrand in the exponent of (12) is defined for all $k_\perp^2$ down to 0. (Notice, that cutting off the $k_\perp^2$ integration at some $k_{min}^2$ would induce non-physical oscillations in the answer.) As long as controllable large momentum behaviour of the running coupling remains unaffected, one is free to chose ones preferable model for $\alpha_s$ near the origin. The simplest choice would be to introduce something like an "effective gluon mass" to (11),

$$\frac{1}{\ln(k^2/\Lambda^2)} \Longrightarrow \frac{1}{\ln(k^2/\Lambda^2 + a)}$$

By doing so one observes that huge variations in the value of $a$ does not produce any visible effect in the answer. This observation one may interpret as *Infrared Stability* of the PT description of the Drell-Yan phenomenon.

In the rest of the lecture we shall discuss another elaborated PT prediction, namely, inclusive energy spectra of leading hadrons in heavy quark initiated jets[38–42]. Here we shall also face the problem of "regularizing" $\alpha_s$. In this context the latter can be related to an effective measure of the intensity of accompanying particle production at the confinement stage of jet evolution when a heavy quark $Q$ gets confined inside a heavy flavoured hadron $H_Q$. (In the course of this confinement a *finite* number of light hadrons are produced in addition to the *W-dependent* particle yield due to PT-controlled gluon bremsstrahlung at the first stage of a hard process of creation of a $Q\overline{Q}$ pair.)

Some characteristics (*e.g.*, scaling violation effects) will remain *Infrared Stable* with respect to our ignorance about small momenta, similarly to the above example. Other quantities (such as absolute magnitude of $c$ and $b$ energy losses) will show quite substantial $\alpha_s$ dependence. The latter we shall exploit to extract some quantitative information about effective $\alpha_s$.

## 2.2 "Leading Particle Effect" in Heavy Quark Fragmentation

In the standard "evolution" approach, when calculating the energy spectrum of a heavy flavoured hadron $H_Q(x_H)$, one would convolute the PT quark distribution $Q(1) \to Q(x)$ with a phenomenological fragmentation function responsible for the hadronization of the

[4]such as non-PT "intrinsic" parton transverse momenta

heavy quark, $Q(x) \to H_Q(x_H)$. Realistic fragmentation functions [43] exhibit a parton-model-motivated maximum at

$$1 - \frac{x_H}{x} \sim \frac{\text{const}}{M} . \tag{13}$$

Account of gluon radiation at the PT-stage of evolution would then soften the hadron spectra building up an *infrared stable* radiative tail and would induce scaling violations which broaden the maximum and shift its position to larger values of $1-x_H$ with increasing $W$.

Such an approach has been successfully applied by Paolo Nason and collaborators [44], who have studied effects of multiple soft gluon radiation and have been looking for realistic fragmentation functions to describe the present day situation with heavy particle spectra, and to make reliable predictions for the future.

The so-called Peterson fragmentation function [43] is often used in the literature to describe the $Q \to H_Q$ transition,

$$C_Q(y) = N\frac{\sqrt{\epsilon_Q}}{y}\left[\frac{1-y}{y} + \frac{\epsilon_Q}{1-y}\right]^{-2} , \qquad y = x_H/x , \tag{14}$$

where $N$ is the normalization constant of order 1,

$$\int_0^1 dy\, C_Q(y) = 1 ; \quad N = \frac{4}{\pi}\left[1 + \mathcal{O}\left(\epsilon_Q^{1/2}\right)\right] .$$

The pick-up hadronization picture predicts that the small parameter $\epsilon_Q$ in (14) should scale with heavy quark mass as

$$\epsilon_Q \approx \left(\frac{m_q}{M}\right)^2 \propto M^{-2}, \tag{15}$$

(where $m_q$ is the quantity of the order of constituent light quark mass), in accordance with (13). A sharp peak of the energy distribution at large $y$,

$$1 - y_{max} \propto \sqrt{\epsilon_Q} \propto M^{-1} \ll 1 ,$$

manifests the "leading particle effect" in the heavy quark fragmentation [45].

In the meantime, the PT gluon radiation effects are capable of reproducing the shape of the Peterson fragmentation function *provided* one feels brave enough to continue the PT description down to small gluon momenta [38–42]: The Sudakov form factor suppression of the quasi-elastic region $x \to 1$ results in a distribution which is qualitatively similar to the parton model motivated expression (14).

An emphasis of the rôle of PT dynamics has been successfully tested in studies of light hadron distributions in QCD jets [41,46,47]. As far as heavy quark spectrum is concerned, one may hope that a similar "pure" PT approach is dual to the sum over all possible hadronic excitations (LPHD). Therefore, without invoking any fragmentation function at hadronization stage, one could pretend to describe the energy fraction distribution averaged over heavy-flavoured hadron states. Such a mixture naturally appears, *e.g.*, when studying inclusive hard leptons from heavy quark initiated jets.

Though the pure PT treatment might look rather naïve, it at least is free from the problem of "double counting" which one faces when trying to combine the effects of PT and hadronization stages.

## 2.3 Inclusive Heavy Quark Spectra in PT QCD

Some cumbersome formulae are in order that describe the inclusive energy spectrum of a heavy quark $Q$ produced in $e^+e^- \to Q(x) + \bar{Q}+$ light partons [38,39] as a function of the energy fraction $x$, the total $e^+e^-$ annihilation energy $W$ and the heavy quark mass $M$. It is conveniently written in terms of the Mellin transform

$$D(x; W, M) = \int \frac{dj}{2\pi i}\, x^{-j}\, \exp\left\{ \frac{1}{v} \int_{2m}^{1} dz \left[ z^{j-1} - 1 \right] \frac{C_F}{2\pi}\, \mathcal{R}(z; W, M) \right\} \tag{16a}$$

where the contour in the complex moment plane runs parallel to the imaginary $j$-axis at $Re\, j \geq 1$. The PT expression for the "radiator" $\mathcal{R}$ reads [39]

$$\begin{aligned}\mathcal{R} = \int_{\kappa^2}^{Q^2} \frac{dt}{t} &\left[ \left( \frac{2(z-2m^2)}{1-z} + \zeta(1-z) \right) \alpha_s(t) - (1-z)\alpha_s'(t) + \frac{\alpha_s^2(t)}{2\pi}\, \gamma^{(2)}(z) \right] \\ - \beta &\left\{ \frac{2z}{1-z} \left( \alpha_s(Q^2) + \alpha_s(\kappa^2) \right) - \zeta\, \frac{z(z-2m^2)}{2(1-z)} \left[ \frac{1-z}{1-z+m^2} \right]^2 \alpha_s(Q^2) \right\}\end{aligned} \tag{16b}$$

Here the following notations have been used:

$$\begin{aligned} v &= \sqrt{1-4m^2}\,, & \beta(z) &= \sqrt{1-\frac{4m^2}{z^2}}\,; \qquad \left( m \equiv \frac{M}{W} \leq \tfrac{1}{2} \right) \\ Q^2 &= W^2 \frac{(1-z)^2}{1-z+m^2}\, z_0\,, & \kappa^2 &= M^2 \frac{(1-z)^2}{z_0}\,; \\ z_0 &= \tfrac{1}{2}\left( z - 2m^2 + \sqrt{z^2-4m^2} \right)\,; & \zeta &= (1+2m^2)^{-1} \quad \text{(Vector Current)} \end{aligned}$$

In the relativistic approximation, $m \ll \frac{1}{2}$, the radiator (16b) may be simplified as follows,

$$\begin{aligned}\mathcal{R} = \int_{\kappa^2}^{Q^2} \frac{dt}{t} &\left[ \frac{1+z^2}{1-z} \alpha_s(t) - (1-z)\alpha_s'(t) + \frac{\alpha_s^2(t)}{2\pi}\, \gamma^{(2)}(z) \right] \\ - \frac{2z}{1-z} &\left[ \alpha_s(Q^2) + \alpha_s(\kappa^2) \right] + \frac{z^2}{2(1-z)}\, \alpha_s(Q^2)\end{aligned} \tag{17}$$

$$Q^2 \approx z(1-z)W^2\,, \qquad \kappa^2 \approx \frac{(1-z)^2}{z} M^2\,. \tag{18}$$

$Q^2$ and $\kappa^2$ are the two characteristic scales of the problem. Generally speaking, the first one determines the scaling violation pattern in the inclusive spectra, while the second parameter takes care of the mass dependence of the answer. The latter is responsible, in particular, for an essential difference between $c$- and $b$-quark spectra.

In the derivation of the PT QCD prediction (16) potentially copious production of $Q\overline{Q}$ pairs has been disregarded [5]. Bearing this subtlety in mind, the accuracy of (16) may be expressed as

$$D(x; W, M) \cdot \left[ 1 + \mathcal{O}\left( \alpha_s^2 \right) \right]. \tag{19}$$

[5] apart from the integral effect of the unregistered pairs embodied into the running coupling

This expansion is *uniform* in $x$, that is, the $\alpha_s^2$ correction does not blow up (neither as a power, nor logarithmically) when $(1-x) \to 0$.

The following list displays essential properties of the answer:

- it embodies the exact first order result [48] $D^{(1)} = \alpha_s \cdot f(x;m)$;
- it has the correct threshold behaviour at $W-2M \lesssim M$;
- in the relativistic limit $m \ll \frac{1}{2}$, it accounts for all significant logarithmically enhanced contributions in high orders, including
    1. running coupling effects,
    2. the two-loop anomalous dimension [49] and
    3. the proper coefficient function with exponentiated Sudakov-type logs, which are essential in the quasi-elastic kinematics, $(1-x) \ll 1$;
- it takes into full account the controllable dependence on the heavy quark mass that makes possible a comparison between the spectra of $b$ and (directly produced) $c$ quarks.

Evaluating typical moments $j$ that are essential in (16),

$$\left\langle j^{-1} \right\rangle \sim (1-x) \sim \langle 1-z \rangle \ ,$$

we observe that when $x$ is taken as close to unity as

$$(1-x) \lesssim \frac{\Lambda}{M} \ , \tag{20}$$

the $\kappa^2(z)$ in (18) leaves the PT-controlled momentum region, $\kappa^2 \gg \Lambda^2$. This qualitative estimate shows that the quark spectrum appears to be "infrared sensitive" in the kinematical region (20): A large quark mass $M \gg \Lambda$ does not guarantee an Infrared Stability of the quark distribution at large $x$. This is a direct consequence of the "Dead Cone" physics [41,42]: A finite quark mass suppresses emission of gluons at small angles, $\Theta \le \Theta_0 \equiv 2M/W$, but not necessarily the gluons with small *transverse momenta*. So, when the compensation between the real and virtual *soft* gluon radiation gets destroyed in the "quasi-elastic" region (20), the non-PT confinement physics inevitably enters the game.

## 2.4 Effects of Confinement as seen through PT eyes

One may single out effects of the non-PT momentum region by splitting the radiator into two pieces corresponding to large and small transverse momenta ($t \approx k_\perp^2$ for $x$ close to 1),

$$\mathcal{R} \ = \ \mathcal{R}\left[k_\perp^2 > \mu^2\right] \ + \ \mathcal{R}^{(C)}\left[k_\perp^2 \le \mu^2\right]. \tag{21a}$$

Correspondingly, the spectrum in the moment representation factorizes into the product

$$D_j = D_j\left[k_\perp^2 > \mu^2\right] \ \times \ D_j^{(C)} . \tag{21b}$$

This formal separation becomes informative if one is allowed to choose the boundary value $\mu$ well *below* the quark mass scale. Within such a choice only $z$ close to 1 in (16b) would contribute to $\mathcal{R}^{(C)}$. Keeping track of the leading $\mu/M$ effects, it is straightforward to derive the following formal expression for the "confinement" part,

$$\ln D_j^{(C)} \approx -2C_F \int_0^{\mu} \frac{dk}{k} \frac{\alpha_s(k^2)}{\pi} \ln\left[1 + \frac{k}{M}(j-1)\right]. \tag{22}$$

For moderate moments $j \sim 1$ (that is, for $x$ not specifically close to 1) this expression may be simplified to give

$$D_j^{(C)} \approx \exp\left\{-2C_F \frac{\mu_\alpha}{M}(j-1)\right\} \tag{23a}$$

$$\text{with} \qquad \mu_\alpha \equiv \int_0^{\mu} dk \, \frac{\alpha_s(k^2)}{\pi}. \tag{23b}$$

The PT-motivated expressions (22), (23) for the *non*-PT fragmentation function have any sense at all "if and only if" the running coupling is regular at finite momenta, that is free from a formal infrared singularity (11). This is how the notion of "infrared regular coupling" comes on stage in the context of heavy quark inclusive spectra.

It is worthwhile to remember that neither the Peterson function nor our PT-motivated "confinement" distribution are unambiguously defined objects. The former as an "input" for the evolution is by itself contaminated by gluon radiation effects at the hard scale $t \sim M \gg \Lambda$. On the other hand, $D^{(C)}[k_\perp^2 \leq \mu^2]$ crucially depends on an arbitrarily introduced separation scale $\mu$ that disappears only in the product of the factors responsible for "PT" and "non-PT" stages (21b). Bearing this in mind, one may nevertheless speak of a direct correspondence between these two quantities, namely, $C(x)$ in the $j$ representation and $D_j^{(C)}$ as given by (22), *provided* an "infrared finite $\alpha_s$" (whatever it means) is implanted into the PT analysis.

The formal asymptote $D(x \to 1)$ depends on details of the behaviour of $\alpha_s$ near the origin, eventually, on convergence of the characteristic integral

$$\Xi_0 \equiv \int_0^{\mu} \frac{dk}{k} \frac{\alpha_s(k^2)}{\pi}. \tag{24}$$

If $\alpha_s$ vanished at the origin, then $\Xi_0 < \infty$ and it would be a simple exercise to derive by evaluating the inverse Mellin transform the power-like behaviour,

$$D(x) \propto (1-x)^{-1+2C_F\Xi_0}. \tag{25a}$$

For $\alpha_s(0) \neq 0$ one obtains instead a stronger suppression,

$$D(x) \propto (1-x)^{-1} \cdot \exp\left\{-C_F \frac{\alpha_s(0)}{\pi} \ln^2(1-x)\right\}. \tag{25b}$$

In practice, in both cases the gross features of the popular Peterson fragmentation function are easily reproduced.

An important message comes from comparing the *mean energy losses* that occur at the hadronization stage. The mean $x$ corresponds to the second moment $D_{j=2}$. Our "confinement" distribution (23) immediately leads to

$$1 - \langle x \rangle^{(C)} = 1 - \exp\left\{-2C_F \frac{\mu_\alpha}{M}\right\} \approx 2C_F \frac{\mu_\alpha}{M} ,$$

which justifies the expected [45] scaling law[6]

$$1 - \langle x \rangle_{fragm} \sim \sqrt{\epsilon_Q} \propto M^{-1} .$$

We conclude that instead of convoluting a phenomenological fragmentation function with the $W$-dependent "safe" evolutionary quark distribution one may try to use consistently the pure PT-motivated description that would place no artificial separator between the two stages of the hadroproduction. Introducing the unknown behaviour of the effective long-distance interaction strength $\alpha_s(t)$ at, say, $t \lesssim 1 GeV^2$ constitutes the price one has to pay for enlarging the "domain of applicability" of the PT approach.

From the first sight, one does not gain much profit by substituting the non-PT function $\alpha_s(< 1GeV^2)$ for another unknown — the phenomenological fragmentation function $C_Q(x)$. There is however an essential physical difference between the two approaches: $\alpha_s$ should be looked upon as a *universal* process independent quantity. Therefore, in particular, quite substantial differences between inclusive spectra of c- and b- flavoured hadrons should stay under complete control according to the explicit quark mass dependence embodied in the PT formulae.

## 2.5 Mean Energy Losses: Numerical Results

The first comparison of PT predictions with the experimentally measured mean energy losses for $c$ and $b$ quarks at different $e^+e^-$ annihilation energies [51] has demonstrated high consistency of the PT approach [40]. For lack of anything better to do, we played around with various plausible shapes of $\alpha_s(t)$ at small scales and have arrived at the following conclusions:

1. The scaling violations in $\langle x_{c,b} \rangle (W)$, that is the ratios

   $$\frac{\langle x_c \rangle (W)}{\langle x_c \rangle (W_0)} , \quad \frac{\langle x_b \rangle (W)}{\langle x_b \rangle (W_0)}$$

   viewed as a function of $W$, prove to be insensitive to our ignorance about the small momentum region in the effective coupling.

2. The same kind of stability has been observed with respect to the scaled positions of the *maxima* in the energy spectra,

   $$\frac{x_{max}(W)}{x_{max}(W_0)} .$$

[6]Similar behaviour was advocated recently by R.L. Jaffe and L. Randall [50] who have exploited the difference between the hadron and the heavy quark masses as a small expansion parameter.

3. The ***absolute*** values of $c$ and $b$ energy losses, in particular, the ratios

$$\frac{\langle x_c \rangle (W)}{\langle x_b \rangle (W)}$$

appear to be strongly "confinement-dependent".

4. This "confinement-dependence" reduces to an integral quantity $\mu_\alpha$, defined above in (23b), as the only relevant non-PT parameter. A highly consistent ***simultaneous*** description of $\langle x_{c,b} \rangle (W)$ is achieved by constraining the value of the integral

$$\int_0^{1GeV} dk \, \frac{\alpha_s(k^2)}{\pi} \approx 0.2\, GeV. \tag{26}$$

Given a limited analyzing power of energy losses, the quality of the fit does not depend otherwise on a particular shape of $\alpha_s(k^2)$ near the origin.

This first comparison was based on a simplified (relativistic) version of the radiator (16b) derived in [38] that did not take into account the two-loop effects in the anomalous dimension ($\gamma^{(2)}$ term in (16b)). Therefore the value of the QCD scale parameter extracted from the fit were formulated in terms of *effective* $\Lambda_{HQ}$ that could not be directly compared to the corresponding scale parameter of, say, the standard $\overline{MS}$ scheme:

$$\begin{aligned} \Lambda^{(3)}_{eff,\mathrm{HQ}} &= 440 \pm 150\, MeV\,; \\ \Lambda^{(5)}_{eff,\mathrm{HQ}} &= 310 \pm {}^{140}_{110}\;\; MeV. \end{aligned} \tag{27a}$$

As shown in [39], the bulk of the second loop effects may be absorbed into a finite renormalization of the coupling that determines the intensity of soft gluon radiation. Translation becomes straightforward and results in

$$\begin{aligned} \Lambda^{(3)}_{\overline{MS}} &= 265 \pm 90\, MeV\,; \\ \Lambda^{(5)}_{\overline{MS}} &= 200 \pm {}^{90}_{70}\;\; MeV. \end{aligned} \tag{27b}$$

The corresponding relation reads[7]

$$\alpha_s(\Lambda_{eff}) \equiv \alpha_s(\Lambda_{\overline{MS}}) + \frac{\alpha_s^2}{2\pi} \cdot \left[ C_A \left( \frac{67}{18} - \frac{\pi^2}{6} \right) - \frac{10}{9} n_f T_R \right]. \tag{28}$$

---

[7]identical physically motivated redefinition of the QCD scale has been suggested earlier by S. Catani, G. Marchesini and B. Webber; $\Lambda_{\mathrm{MC}}$ of Ref. [52].

## 2.6 On Hidden Beauty of the Second Loop

Two-loop refinement of PT QCD predictions inevitably involves very difficult calculations and produces results that look cumbersome and absolutely unintuitive. Let me try to convince you that such a sad conclusion is not entirely true: There is a hidden beauty (and relative simplicity) around the corner, that becomes transparent when formal calculus is augmented with a bit of physical sense[8].

Let us take (28) as an example. Those of you who are familiar with the basics of the QCD parton picture (for a recent review see, *e.g.*, [46]) are aware of the fact that it is the total gluon splitting probability that makes the QCD coupling run. Indeed, in the "leading approximation" a gluon may produce either a quark pair or two gluons. As you know, the corresponding "decay probabilities" as a function of offspring energy $x$ read

$$
\begin{aligned}
P_G^F(x) &= x^2 + (1-x)^2\,; \\
P_G^G(x) &= 2\left[\frac{1-x}{x} + \frac{x}{1-x} + x(1-x)\right] = \frac{2}{x(1-x)} + \left[-4 + 2x(1-x)\right].
\end{aligned} \tag{29}
$$

The total gluon splitting probability contains characteristic logarithmic integral $d\theta/\theta \propto dk_\perp/k_\perp$ ("collinear" or "mass" singularity) multiplied by the factor

$$
\int_0^1 dx\left[\tfrac{1}{2!}C_A P_G^G(x) + n_f T_R\, P_G^F(x)\right] = \int_0^1 dx\left[C_A\, P_G^{\rm soft}(x) + C_A\, P_G^{\rm hard}(x) + n_f T_R\, P_G^F(x)\right]. \tag{30}
$$

Here we made use of the symmetry property $P_G^G(x) = P_G^G(1-x)$ to replace

$$
\int_0^1 dx\, \tfrac{1}{2!}\, P_G^G(x) = \int_0^1 dx\; x\cdot P_G^G(x)
$$

and introduced the notation

$$
\begin{aligned}
x\,P_G^G(x) &\equiv P_G^{\rm soft}(x) + P_G^{\rm hard}(x)\,; \\
P_G^{\rm soft}(x) &= \frac{2}{1-x}\,; \qquad P_G^{\rm hard}(x) = -4x + 2x^2 - 2x^3\,.
\end{aligned} \tag{31}
$$

The infrared divergent term $P_G^{\rm soft}$ in (30) goes away in the end of the day, while the remaining finite terms produce the first Gell-Mann-Low coefficient $\beta_2$:

$$
\int_0^1 dx\left[C_A\, P_G^{\rm hard}(x) + n_f T_R\, P_G^F(x)\right] = C_A\left(-2 + \tfrac{2}{3} - \tfrac{2}{4}\right) + n_f T_R\left(\tfrac{1}{3} + \tfrac{1}{3}\right) = -\frac{11}{6}C_A + \frac{2}{3}n_f T_R.
$$

Now, without further ado, here is the message:

$$
\begin{aligned}
\int_0^1 dx\; P_G^{\rm hard}(x)\cdot \ln x &= \left(4 - \tfrac{2}{4} + \tfrac{2}{9}\right) = \frac{67}{18}\,; \\
\tfrac{1}{2}\int_0^1 dx\; P_G^{\rm soft}(x)\cdot \ln x &= -\frac{\pi^2}{6}\,.
\end{aligned}
$$

---

[8]For an excellent example of such an improvement see [53] and Refs. therein.

It looks *as if* the second loop effects in (28) were actually driven by the basic first loop parton splitting. Sleep it over and try to explain the miracle yourself. Just in case you feel curious about the $n_f$ piece of the second loop correction in (28), here are the hints:

1.
$$-\frac{10}{9}n_fT_R = \left(-\frac{5}{3}\right)\cdot\frac{2}{3}n_fT_R\,;$$

2. The exact first-loop fermion polarization operator that "goes into denominator" of the running coupling,

$$\left[\frac{\alpha_s(Q^2)}{2\pi}\right]^{-1} = \left[\frac{\alpha_s(\mu^2)}{2\pi}\right]^{-1} + \beta_{\mathrm{glue}}\ln\frac{Q^2}{\mu^2} + \frac{2T_R}{3}\sum_{f=1}^{n_f}\Pi(Q,m_f)\,,$$

has the ultraviolet asymptote known from the mid-fifties

$$\Pi(Q,m) = \ln\frac{Q^2}{m^2} - \frac{5}{3}\,, \quad Q \gg m\,.$$

# 3 Conclusion

A comparative PT-motivated study of the $c$- and $b$-quark energy losses in $e^+e^-$ annihilation has demonstrated consistency of the hypothesis of the universal infrared finite $\alpha_s$. The same notion of the infrared finite effective coupling can be tried for a good many interesting problems in the light quark sector as well. An incomplete list of such phenomena for which the PT analysis has been carried out recently to next-to-leading order, includes transverse [54] and longitudinal momentum distributions [55] in hadron-initiated processes, the energy-energy correlation [56], the thrust [57] and the heavy jet mass distribution [58] in $e^+e^-$ annihilation *etc.*

The presence of the exponential of the characteristic integral over gluon momenta which emerges after all-order resummation of logarithmically enhanced contributions [36,59] is a common feature of the corresponding PT expressions (*cf.* (12), (16)). It is straightforward to derive quantitative PT-motivated predictions by implementing the universal $\alpha_s$ in these integrals and, at the same time, by getting rid of non-PT "hadronization" effects which are usually taken into account by convoluting the PT distributions with phenomenological fragmentation functions (initial parton distributions for the case of hadron-initiated processes).

Carrying out this laborious but promising program one should get a valuable information about confinement physics as seen through the eyes of the integrated influence of the large-distance hadron production upon inclusive particle distributions and/or event characteristics.

To this end the results of the analysis of heavy quark energy losses displayed above, in particular, (26) should be looked upon as a first hint for more detailed studies including the light quark phenomena listed above.

Now we find ourselves in a position to link together two seemingly uncorrelated parts of this lecture.

We started with the Gribov confinement scenario [1,2] and discussed direct phenomenological consequences of the picture which can be anticipated given the present state of arts [4]. The Gribov theory demonstrates how colour confinement can be achieved in a field theory of light fermions interacting with comparatively small effective coupling. In this context the phenomenological observation we have discussed in the second part of the lecture,

$$\int_0^{1GeV} d\kappa \; \frac{\alpha_s(\kappa^2)}{\pi} \;\approx\; 0.2\, GeV\;,$$

is in line with the light quark confinement picture. In the first place, the characteristic coupling does exceed the critical value (2) that is necessary for super-critical binding to occur. On the other hand, numerical value of the "couplant" $\alpha_s/\pi$ remains sufficiently small to justify, at least qualitatively, the perturbative expansion (1).

We have seen that gluon momenta below 1GeV play a crucial rôle in determining the ***absolute*** values of quark energy losses. In this region the momentum dependence of the effective coupling is poorly known not because of the limited knowledge of higher order effects in the perturbative $\beta$-function but rather because of an essentially different physical phenomenon that enters the game, the one that is usually referred to as confinement. From this point of view the notion of the infrared finite $\alpha_s$ discussed above differs from the one that emerges when the "Principle of Minimal Sensitivity" [60] is applied to perturbative three-loop results for the $e^+e^-$ annihilation cross section [61,62] and the $\tau$-lepton hadronic width [61]. Nevertheless it is worth mentioning that the value of the couplant $\alpha_s(0)/\pi = 0.26$ obtained in [62] agrees numerically with our estimate. (In the model with $\alpha_s$ flattening off at small momenta one derives from the energy loss pattern $\alpha_s(0)/\pi = 0.22 - 0.25$.)

In the section "Phenomenological Virtues of a Frozen Couplant" of the recent paper [63] Mattingly and Stevenson have assembled an impressive list of practical applications which consistently point to $\alpha_s/\pi = 0.2 \rightarrow 0.3$ as a reasonable magnitude of the "long-range" QCD interaction strength. These applications range from rather naïve estimates of hadron-hadron cross sections and form factors to the well elaborated Godfrey-Isgur relativized QCD quarkonium model that describes quite successfully particle spectroscopy from pions all the way up to the $\Upsilon$ family.

Phenomenological verification of the fact that the effective QCD coupling (whatever it means) stays numerically small would produce a great impact for enlarging the domain of applicability of perturbative ideology to physics of hadrons and their interactions.

## References

[1] V.N. Gribov, Lund preprint LU-TP 91–7, March 1991.

[2] V.N. Gribov, in preparation.

[3] I. Pomeranchuk and Ya. Smorodinsky, Journ. Fiz. USSR 9 (1945) 97;
W. Pieper and W. Greiner, Z. Phys. 218 (1969) 327;
Ya.B. Zeldovich and V.S. Popov, Uspekhi Fiz. Nauk 105 (1971) 4.

[4] F.E. Close, Yu.L. Dokshitzer, V.N. Gribov, V.A. Khoze and M.G. Ryskin, *Phys.Lett.* 319B (1993) 291.

[5] Particle Data Group, *Phys.Rev.* D45 (1992) S1.

[6] J.K. Bienlein, CRYSTALL BALL Contribution to the $9^{th}$ Int. Workshop on Photon-Photon Collisions, La-Jolla, March 1992, preprint DESY 92–083, June 1992.

[7] Kay Königsmann, Invited talk at the $13^{th}$ Int. Conf. on Particles and Nuclei (PANIC), Perugia, 28 June – 3 July 1993; preprint CERN-PPE/93–182, October 1993.

[8] C. Amsler *et al.*, to be published in Phys. Lett.

[9] C. Guaraldo, Proc. of the Hadron 93 Conference (Como, Italy, 1993).

[10] D. Aston *et al.*, *Nucl.Phys.* B301 (1988) 525.

[11] C. Amsler *et al.*, *Phys.Lett.* 291B (1992) 347.

[12] M. Doser, Proc. of the Hadron 93 Conference (Como, Italy, 1993).

[13] F. Binon *et al.*, *Nuovo Cimento* 80 (1984) 363;
D. Alde *et al.*, *Nucl.Phys.* B269 (1986) 485; *Phys.Lett.* 201B (1988) 160.

[14] D. Alde *et al.*, *Phys.Lett.* 284B (1992) 457.

[15] S. Godfrey and N. Isgur, *Phys.Rev.* D32 (1985) 189.

[16] S.J. Lindenbaum and R.S. Longacre, preprint BNL–45878 (1991);
S.J. Lindenbaum, to be published in: Proc. of Future Directions in Nuclear and Particle Physics at Multi-GeVHadron Beam Facilities; preprint BNL–49416, April 1993;

[17] T.A. Armstrong *et al.*, *Phys.Lett.* 307B (1993) 394.

[18] D. Morgan and M.R. Pennington, *Phys.Rev.* D48 (1993) 1185.

[19] Z.P. Li, F.E. Close and T. Barnes, *Phys.Rev.* D43 (1991) 2161;
F.E. Close and Z.P. Li, *Zeit.Phys.* C54 (1992) 147.

[20] D. Morgan and M.R. Pennington,*Zeit.Phys.* C48 (1990) 623.

[21] T. Barnes, *Phys.Lett.* 165B (1985) 434.

[22] ARGUS Collaboration, preprint DESY 92–174, December 1992.

[23] DELPHI Collaboration, preprint CERN-PPE/92–183, October 1992.

[24] Z. Bai *et al.*, *Phys.Rev.Lett.* 65 (1990) 2507.

[25] Kay Königsmann, Invited talk at the XI Int. Conf. on Physics in Collision, Colmar, 20–22 June 1991; preprint CERN-PPE/91–160, August 1991.

[26] K.D. Duch *et al.*, *Zeit.Phys.* C45 (1989) 223.

[27] J. Weinstein and N. Isgur, *Phys.Rev.Lett.* 48 (1982) 659; *Phys.Rev.* D27 (1983) 588; *ibid.* 41 (1990) 2236.

[28] D. Morgan and M.R. Pennington, Contribution to Hadron '91: Testing the Nature of the $f_0(S^*)$, preprint RAL–91–070, October 1991.

[29] F.E. Close, N. Isgur and S. Kumano, *Nucl.Phys.* B389 (1993) 513.

[30] F.E. Close, in: Proc. of Workshop on Physics and Detectors for DAFNE, Frascati, April 1991, ed G.Pancheri, p. 310.

[31] W.R. Gibbs and W.B. Kaufmann, Proc. of "Physics with Light Mesons", Los Alamos report LA-11184-C (1987) 28 ;
C. Alexander and T. Sato, *Phys.Rev.* C36 (1987) 1732.

[32] H. Lipkin, *Phys.Lett.* 124B (1983) 509.

[33] G. Parisi, private communication.

[34] Crystal Barrel Collaboration at LEAR,*Phys.Lett.* 291B (1992) 347; *ibid.* 294 (1992) 451.

[35] G. Smith, Proc. of Dallas HEP Conference (1992).

[36] Yu.L. Dokshitzer, D.I. Dyakonov and S.I. Troyan, *Phys.Lett.* 84B (1979) 234; *Phys.Rep.* C58 (1980) 270.

[37] G.Parisi and R.Petronzio, *Nucl.Phys.* B154 (1979) 427;
B.R. Webber and P. Rakow, *Nucl.Phys.* B187 (1981) 254;
G Altarelli, R.K. Ellis, M. Greco and G. Martinelli, *Nucl.Phys.* B246 (1984) 12.

[38] Yu.L. Dokshitzer, V.A. Khoze, S.I.Troyan, "Specific Features of Heavy Quark Production. I. Leading quarks", Lund preprint LU TP 92–10, March 1992.

[39] Yu.L. Dokshitzer, V.A. Khoze, S.I.Troyan, "Specific Features of Heavy Quark Production. II. LPHD approach to heavy particle spectra", Lund preprint LU TP 92–13; to be published.

[40] Yu.L. Dokshitzer, in: *Perturbative QCD and Hadronic Interactions*, Proc. of the 27th Recontres de Moriond, ed. J.Tran Thanh Van, Editions Frontières, Gif-sur-Yvette, 1992, p.259.

[41] Yu.L. Dokshitzer, Light and Heavy Quark Jets in Perturbative QCD, in: *Proceedings of the International School of Subnuclear Physics*, Erice, 1990.

[42] Yu.L. Dokshitzer, V.A. Khoze, and S.I. Troyan, *J. Phys. G: Nucl. Part. Phys.* 17 (1991) 1481; *ibid.* p. 1602, in: *Report of the soft QCD working group.*

[43] C. Peterson *et al.*, *Phys.Rev.* D27 (1983) 105; for review see, *e.g.*,
M. Bosman *et al.*, Heavy Flavours, in: Proc. of the Workshop on $Z$ physics at LEP, CERN Report 89–08, ed. G. Altarelli, R. Kleiss and C. Verzegnassi, volume 1, p.267, 1989.

[44] B. Mele and P. Nason, *Phys.Lett.* 245B (1990) 635; *Nucl.Phys.* B361 (1991) 626;
G. Colangelo and P. Nason, *Phys.Lett.* 285B (1992) 167.

[45] Ya.I. Azimov, L.L. Frankfurt and V.A. Khoze, preprint LNPI–222, 1976;
M. Suzuki, *Phys.Lett.* 71B (1977) 139;
J. Bjorken, *Phys.Rev.* D17 (1978) 171.

[46] Yu.L. Dokshitzer, V.A. Khoze, A.H.Mueller and S.I. Troyan, *Basics of Perturbative QCD*, ed. Tran Than Van, Editions Frontières, Gif/Yvette, 1991.

[47] Yu.L. Dokshitzer, V.A. Khoze and S.I. Troyan, *Zeit.Phys.* C55 (1992) 107.

[48] V.N. Baier and V.A. Khoze, *Sov.Phys.JETP* 21 (1965) 629; preprint NSU, Novosibirsk, 1964.

[49] G. Curci, W. Furmanski and R. Petronzio, *Nucl.Phys.* B175 (1980) 27;
G. Altarelli, R.K. Ellis, G. Martinelli and S.Y. Pi, *Nucl.Phys.* B160 (1979) 301;
W. Furmanski and R. Petronzio, *Phys.Lett.* 97B (1980) 437, *Zeit.Phys.* C11 (1982) 293;
J. Kalinowski, K. Konishi, P.N. Scharbach and T.R. Taylor, *Nucl.Phys.* B181 (1981) 253;
E.G. Floratos, C. Kounnas and R. Lacaze, *Nucl.Phys.* B192 (1981) 417;
for a review see G. Altarelli, *Phys.Rep.* 81 (1982) 1.

[50] R.L. Jaffe and L. Randall, "Heavy Quark Fragmentation into Heavy Mesons", preprint HEP-PH/9306201, May 1993.

[51] P. Mättig, Talk at the $4^{th}$ Int. Symp. on Heavy Flavour Physics; Bonn–HE–91–19, 1991;
T. Behnke, Talk at the XXVI Int. Conf. on High Energy Physics, Dallas, Texas, August 6–12, 1992.

[52] S. Catani, G. Marchesini and B.R. Webber, *Nucl.Phys.* B349 (1991) 635.

[53] Stanley J. Brodsky and Hung Jung Lu, "Commensurate Scale Relations: Relating Observables in QCD without Renormalization Scale or Scheme Ambiguity", preprint SLAC-PUB–6389, November, 1993.

[54] J. Kodaira and L. Trentadue, *Phys.Lett.* 112B (1982) 66;
C.T.H. Davies, J. Stirling and B.R. Webber, *Nucl.Phys.* B256 (1985) 413;
J.C. Collins, D.E. Soper and G. Sterman, *Nucl.Phys.* B250 (1985) 199;
S. Catani, E. d'Emilio and L. Trentadue, *Phys.Lett.* 211B (1988) 335.

[55] G. Sterman, *Nucl.Phys.* B281 (1987) 310;
S. Catani and L. Trentadue, *Phys.Lett.* 217B (1989) 539; *Nucl.Phys.* B327 (1989) 323; *Nucl.Phys.* B353 (1991) 183.

[56] J. Kodaira and L. Trentadue, *Phys.Lett.* 123B (1982) 335; preprint SLAC-PUB-2934 (1982).

[57] S. Catani, G. Turnock, B.R. Webber and L. Trentadue, *Phys.Lett.* 263B (1991) 491.

[58] S. Catani, G. Turnock and B.R. Webber, *Phys.Lett.* 272B (1991) 368.

[59] A. Bassetto, M. Ciafaloni and G. Marchesini, *Nucl.Phys.* B163 (1980) 477;
G. Curci and M. Greco, *Phys.Lett.* 92B (1980) 175.

[60] P.M. Stevenson, *Phys.Rev.* D23 (1981) 2916.

[61] J. Chyla, A. Kataev, S.A. Larin, *Phys.Lett.* 267B (1991) 269.

[62] A.C. Mattingly and P.M. Stevenson, *Phys.Rev.Lett.* 69 (1992) 1320

[63] A.C. Mattingly and P.M. Stevenson, "Optimization of $R_{e^+e^-}$ and "Freezing" of the QCD Couplant at Low Energies", preprint DE-FG05-92ER40717-7.

**CHAIRMAN: Y.Dokshitzer**

*Scientific Secretaries: G. Abu Leil, C.Acerbi, J.R. Forshaw*

## DISCUSSION

– *Forshaw:*

If Gribov's theory of confinement is correct, then what are the implications for lattice QCD?

– *Dokshitzer:*

If the light quarks are responsible for the major part of what we call confinement in the real world you may have to forget about lattice QCD.

– *Forshaw:*

So there is a serious conflict here?

– *Dokshitzer:*

Yes, it is a very serious conflict, at least for you British people who are spending money!

– *Forshaw:*

In the decay $q_{(+)} \to M_{(+)} + q_{(-)}$ what is the meson? Can you explain in the context of the decay $e^+e^- \to D\overline{D}f_0$ ?

– *Dokshitzer:*

A collective $q_{(-)}$ current connects the $q_{(+)}$ and $q_{(-)}$ produced in $e^+e^- \to q\overline{q}$. The excitation of the physical vacuum due to this current results in the production of hadrons; 'normal' and 'novel' mesons in particular. In a typical adiabatic hadronization process the production ratio of the normal/novel states is phenomenologically about 10:1. The yield of the latter states is expected to increase if one selects events with finite but larger than typical isolation between neighbouring hadrons.

– *Forshaw:*

I am a little confused. Do you say that a $q_{(+)}$ decays into a $q_{(-)}$ plus a vacuum scalar?

– *Dokshitzer:*

If you start with the perturbative vacuum, then the picture is very clear. You produce a quark which starts to develop an increasing field around which there appear new empty states in the vacuum – which must be filled in. When you do that you get a negative energy meson and positive energy quark – this cannot be a physical process, which means that your perturbative vacuum is not the correct state to perturb about. We have to start with a physical vacuum in which there are positive kinetic energy quarks bound (into pairs) with total negative energy. The QCD vacuum therefore consists of the Dirac sea plus a new zone in which quarks have positive kinetic energy which, under the interaction, are bound together. So, in the hadronization process, all possible excitations of the physical vacuum can be produced, i.e. both normal hadrons and 'vacuum scalars'.

– *Forshaw:*

So the D meson that has been created (in the process $e^+e^- \to D\overline{D}f_0$ (or $a_0$)) is made up with a negative energy quark – is this right?

– *Dokshitzer:*

Yes and no! 'No' because the sign of kinetic energy is not a conserved quantity. Hence, although it is a $q_{(-)}$ that forms the $f_0$ (or $a_0$) state, its partner (which participates in the $D$ ($\overline{D}$) wavefunction) could be either a $q_{(-)}$ or $q_{(+)}$. And 'yes' because the system of heavy (positive energy) quark and a light companion may well have good reason to bind with the latter in the negative kinetic energy state. The Green function is a complex object which contains all possible states. In principle even normal mesons would have to have some admixture of negative energy states. Therefore the D mesons will have these negative energy quarks around. What you have to do is solve the problem of the D meson wavefunction before one can discuss the importance of the negative energy component of the quark's Green function.

– *Acerbi:*

1) Could you please explain in more detail why, in your explanation, confinement is essentially a $SU(2)_f$ phenomenon?

2) Do you expect heavy quark coloured states to be observable?

– *Dokshitzer:*

1) In this game there are two separate subjects namely how the coupling increases due to the gluon self-interaction and how this anti-screening (infrared problem) is getting stopped by the existence of light quarks. If you start from the pure gluon sector, then at some point (which is parametrized by $\Lambda_{QCD}$), the coupling is formally infinite. When this effective coupling gets larger than some fixed value (about 0.2) the supercritical phenomenon starts to develop, i.e. we require that $\frac{\alpha_s}{\pi}\log(\frac{\lambda}{m}) \simeq 1$ (where $\lambda \simeq 1$ GeV, and $m \simeq 10$ MeV is the light quark current mass). Thus in the presence of the large ratio $(\lambda/m)$, supercritical binding (and thus confinement) occurs whilst the coupling remains small. Once this has

happened the infrared instability of the theory is stopped and the coupling no longer rises sufficiently to allow the heavier quarks to participate, i.e. the effective interaction strength is never sufficient to push the $s\bar{s}$ pairs into the vacuum.

2) A hypothetical world populated with only heavy quarks could be very different. In such a case I would not strongly oppose the popular scenario of increasing (with distance) colour forces between the quarks. At the same time, spontaneous colour symmetry breaking might very well be a preferable way to regularise the infra-red instability: gluons could acquire masses and coloured states would become obervable. A third possibility is that the coupling could rise to a value where the lightest heavy quarks could populate the vacuum (just as in the scenario I have been discussing) only now the coupling would be too large to be called perturbative. Which of these routes nature would choose is a question that may be quantitatively studied and answered – although this is not the world in which we live.

– *Zichichi:*

It is difficult for me to believe that confinement is due to light quarks. So I have a question: would a world made of heavy quarks be without confinement? How can the quantity "mass" play such a fundamental role in the important phenomenon of confinement?

– *Dokshitzer:*

Consider pure gluodynamics – people speak of colourless gluon states. But in principle the answer could be completely different – it could be a field theory with no states at all. This is exactly what happens near the phase transitions in solid state physics, i.e. Green functions have power asymptotics, $G(\kappa) \sim \kappa^{-\gamma}$ and there are no asymptotic particle states.

There is no known field theoretical way to bind massless or nearly massless objects, other than fermions: the crucial thing being Fermi statistics.

– *Abu Leil:*

Can you please explain why the negative energy mesons are expected to be scalars and pseudoscalars?

– *Dokshitzer:*

I wish I could! To answer this question one has to consider the interaction of a $q$ and $\bar{q}$ which have been produced by some current. The aim is to understand the dependence of the binding energy upon the quantum numbers of the pair. This is a difficult problem. However, some information on the $q\bar{q}$ wavefunction may be extracted by utilizing the similarity between this vertex function and the derivative, $d/dm(G^{-1})$, of the quark Green function. One may then deduce that it is the scalar/pseudoscalar $q\bar{q}$ pairs which first develop supercritical binding, i.e. are the first to penetrate the vacuum. This is the statement made by V.Gribov and is still unpublished.

– *Gourdin:*

Is the SU(2) flavour symmetry between u and d the isotopic symmetry?

– *Dokshitzer:*

Yes it is.

– *Gourdin:*

As in $e^+e^- \to D\overline{D}(a_0, f_0)$ can we also have $\to B\overline{B}(a_0, f_0)$ and $\to K\overline{K}(a_0, f_0)$?

– *Dokshitzer:*

Yes. So all of the things I have been talking about are not only relevant to today's experiments but also yesterday's. Because now we are struggling for simplicity, we are trying to go to hard processes in which we can control as much as possible. But in fact the real interesting physics is still hidden in low energy phenomena.

– *Gourdin:*

So there are two classes of quarks the u and d on one side and all the others are called "heavy" quarks.

– *Dokshitzer:*

Yes.

–*Gourdin:*

What is the most recent estimate from experiment of $\Lambda_{\overline{MS}}$ ?

– *Dokshitzer:*

It depends a little on what you mean by $\overline{MS}$ (there are different $\overline{MS}$'s) but basically it is $250 \pm 100$ MeV (from many different phenomena).

There is one thing I would like to mention, which Prof. Zichichi will be very happy about: the fact that $\alpha_s$ measured from LEP is slowly increasing with time. When you take into account not just phenomena in which you can keep track of first and second order but also those in which all orders become important, the typical $\alpha_s$ is not 0.118 but rather 0.125. So from the most elaborate perturbative calculations $\alpha_s$ is going to be larger than the current world average.

– *Shabelski:*

Why is there no $s\underline{s}$ component in the two mesons which appear in e.g. $e^+e^- \to D\overline{D}(a_0, f_0)$? I think, such a component should appear from the mixture with the usual $s\underline{s}$ scalar in the p-wave representation of SU(6).

– *Dokshitzer:*

These kind of states are strongly decoupled since there is a suppression of order 10 in this channel. Certainly, each particle will be a mixture, according to quantum mechanics. In fact, it is known experimentally that we have two kinds of mesons: the normal $f_0(1400)$ and this new $f_0(975)$, and that they practically do not couple to each other. The two $f_0$s have the same quantum numbers so, in principle, the $f_0(1400)$ could decay into the $f_0(975)$ with the emission of $\pi\pi$ for example. But it does not!

– *Khoze:*

I would like to make a comment, connected to Prof. Gourdin's question. I am not so sure that the $f_0$ (or $a_0$) produced in the process $e^+e^- \to K\overline{K}(a_0, f_0)$ can be identified as a bound state of negative energy $q$ and $\overline{q}$: I prefer to have small distances.

– *Dokshitzer:*

If you are serious, then I think you should accept responsibility for the process in which you produce $K\overline{K}$ and a meson.

– *Khoze:*

It was a comment.

– *Anselm:*

Can you clearly define what is the difference between these "novel states" and, say, ordinary mesons? After all, by freezing the coupling constant at large distances you change the interaction. You also told us that it is not very essential how exactly it is changed. It does not seem to be much different from the usually accepted picture. The only thing which you added is that it may be (just "may be", since you don't know the numbers) that $SU(3)_f$ is badly broken down to $SU(2)_f$. But this appears not to be enough to distinguish the "novel" states from the usual ones.

– *Dokshitzer:*

Let me deal with the infra-red behaviour of the coupling first. In this picture an attempt is made to account for confinement on purely perturbative grounds. If you agree that I can use perturbation theory for $\alpha_s/\pi \sim 0.2$, then confinement is going to be perturbative. There is no sign of any linearly increasing potential at all. You expect that the coupling should not rise once you are at distances larger then those determining the onset of the supercritical phenomenon. With regard to the novel states, they really are going to be very different from the normal hadrons. If (and of course this is a large 'if') these novel states exist, then they have to be compact. Which would mean, for example, that they not only decouple from normal hadrons but also behave very differently when you produce them on nuclear targets; there would be no absorption. So there are very different and striking properties which distinguish such states, for example, from those of ordinary non-relativistic quark models. Also, this would be particularly easy to check as it is in the region of 1 GeV physics.

– *Beneke:*

My question is of a more formal nature and concerns the concept of an infrared stable coupling. You are discussing the coupling as if it were a unique object. However in QCD, the coupling is not a quantity which has a direct physical meaning. In fact, it is almost completely ambiguous. In view of this can you explain what kind of entity your infrared stable coupling is and how it connects to the renormalization scheme ambiguities in the definition of the coupling in the perturbative regime?

– *Dokshitzer:*

I think you need not worry about the problems of large momenta, because in this region there are no modifications. The region of small momenta (i.e. $< 1$ GeV) has nothing to do with the problem of scheme dependence in the $\beta$-function etc. Nevertheless there is a justifiable way to define the coupling.

– *Vassilevskaya*

What is the experimental status of $\Lambda_{QCD}$ at present?

– *Dokshitzer:*

In $e^+e^-$ and deep inelastic scattering there has been, until recently, a slight discrepancy between the values of $\Lambda_{QCD}$ obtained from these two hard processes. Namely, $\Lambda_{QCD} : DIS \sim \Lambda_{QCD}^{e^+e^-}/2$, but taking into account the intrinsic uncertainty in perturbative QCD (in my view of order 15%) I wouldn't call it a discrepancy. It may be due to the making of an inappropriate comparison between the two channels.

– *Vassilevskaya*

Is it possible to confirm that the confinement problem is solved in the framework of QCD?

– *Dokshitzer:*

I wouldn't say so.

– *Vassilevskaya:*

What is the role of instantons in understanding the confinement mechanism?

– *Dokshitzer:*

I don't think instantons have anything to do with confinement. Principally because they are Euclidean solutions and have nothing to do with Minkowski space.

– *Zichichi:*

I would like to confirm the point that the low energy (spacelike) data lead to a lower value of $\Lambda_{QCD}$ then the higher energy (timelike) data. We should remember that we do not have data on timelike and spacelike processes at comparable energies. My feeling is that the low energy data suffer from a psychological bias, in that everyone wanted to see a running $\alpha_s$ that was not there; leading to a low value of $\Lambda_{QCD}$. This is a problem that has still not received a satisfactory answer and is an example of how theoretical predictions can have a terrible influence on experimental results.

– *Dokshitzer:*

Yes. It looks like we haven't been too clever in our analysis of the spacelike and timelike data, as a result of the different hardness scales involved, e.g. which can be as large as $Q^2/x$ in deep inelastic scattering, to be compared with $Q^2$ in $e^+e^-$. One can hope to do a better job and rescue the Gribov-Lipatov relation by modifying the appropriate hardness scale in the evolution equations.

– *Acerbi:*

I am a little confused about negative energy states. If energy conservation holds, then, the radiation of such a state would appear to violate energy conservation.

– *Dokshitzer:*

Indeed, no such negative energy states are expected (they populate the vacuum). What one sees is positive energy mesons with negative energy quarks sitting inside.

– *Montgomery:*

I would just like to clarify Prof. Zichichi's view of deep inelastic. If you recall, in 1970 there was scaling, then there were the first DIS $\mu$ and $\nu$ experiments which had $\Lambda_{QCD} = 0.7$ GeV, then $\Lambda_{QCD} = 0.5$ GeV. Then $\mu$-experiments at CERN found $\Lambda_{QCD} = 0.1$ GeV (at which point the proton lifetime died). Now the value has settled around 0.3 MeV. So I don't think the experimentalists have shown any prejudice regarding the value of $\alpha_s$.

# THE NON-RELATIVISTIC QUARK MODEL FROM QCD - AND RELATED TOPICS

G.Morpurgo

Istituto di Fisica dell'Universita' and INFN
Via Dodecaneso 33 -16146 Genova (Italy)

## 1.A SUMMARY

I will show how, starting from QCD -I mean the exact QCD Hamiltonian- one can obtain an exact parametrization of many physical quantities (baryon magnetic moments, hadron masses,.........etc.). It turns out that this ***exact*** parametrization is unexpectedly simple. Its structure, derived from a fully relativistic theory, and thus relativistic (though non-covariant), is similar to that of the Non-Relativistic Quark Model (NRQM). Comparing the exact general parametrization and the results of the NRQM clarifies the reason of the quantitative success of the latter. This is the main purpose of these lectures.

As a byproduct some new results are derived: For instance, 1) A new octet-decuplet baryon mass formula is obtained, correct to a part per thousand; 2) New electromagnetic baryon mass relationships are derived (also well satisfied), as well as a new formula for the magnetic moments of the Δ's; 3) One can understand why the Gell Mann-Okubo mass formula and the Coleman-Glashow one work so well; etc.

The above summary refers to the first part of the title: "NRQM from QCD". The second part "and related topics" will deal with something that I had not in mind when, in 1989, started this work. It emerged naturally from it and has to do with the question if an effective Lagrangian including pions and other local fields as explicit degrees of freedom can be equivalent to a QCD Lagrangian, that is produce the same results.

This will be dealt in the last part of these lectures, where I also discuss briefly the unpleasant proliferation of models in low energy hadron physics, a sort of Babel tower. A few references intended to give an idea of the development of this field appear in Sect.15 (Early history, Ref.1-5; Renaissance, Refs.6-8; Modern period, Refs.9-10). In these lectures I deal only with problems of hadrons composed of light quarks.

## 2.THE IDEA OF PARAMETRIZATION

The NRQM gives not only a general qualitatitive description of hadron spectroscopy but also a fair quantitative agreement with the data. This aspect is perhaps less known. It is the aspect that struck me more since the early days (and that led me to a long experimental search for quarks[5]).

The Magnetic Moments

To exemplify this point I describe the situation of the magnetic moments of the lowest octet baryons. Consider the 2-parameter expression (with parameters $\mu$,a or equivalently $\mu$,A $[A=\mu(1-a)]$):

$$\underset{\sim}{M}=\mu\sum\left[(2/3)\underset{\sim}{\sigma}^{\mathcal{P}}-(1/3)\underset{\sim}{\sigma}^{\mathcal{N}}-(a/3)\underset{\sim}{\sigma}^{\lambda}\right]\equiv\mu\sum\left[(2/3)\underset{\sim}{\sigma}^{\mathcal{P}}-(1/3)\underset{\sim}{\sigma}^{\mathcal{N}}-(1/3)\underset{\sim}{\sigma}^{\lambda}\right]+\frac{A}{3}\sum\underset{\sim}{\sigma}^{\lambda} \quad (1)$$

Take the model wave functions for the lowest octet baryons:

$$\varphi_B(1,2,3)=X_{L=0}(\underset{\sim}{r}_1,\underset{\sim}{r}_2,\underset{\sim}{r}_3)\cdot W_B(1,2,3) \quad (2)$$

where $W_B$ are the spin-flavor wave functions; e.g., for the proton P:

$$W_P=\mathcal{N}\mathcal{S}[\alpha_1(\alpha_2\beta_3-\alpha_3\beta_2)\mathcal{P}_1\mathcal{P}_2\mathcal{N}_3] \quad (3)$$

where $\mathcal{N}$ is a normalization constant and $\mathcal{S}$ means symmetrization over 1,2,3; the spin-flavor wave functions $W_B$ for all octet baryons are listed in Ref.1b. On performing the expectation value:

$$\underset{\sim}{M}_{magn.B}=\langle\varphi_B|M_z|\varphi_B\rangle \quad (4)$$

and choosing the two parameters

$$\mu=2.79 \quad , \quad a=0.65 \quad (\text{or } A=0.98)$$

one gets for all the magnetic moments the fit shown in fig.1.

Now one might ask: Suppose that you were able to perform an exact calculation of the magnetic moments in QCD; and suppose that you were able to display the final result of that calculation in the form:

$$\underset{\sim}{M}_{magn.B}=\langle\varphi_B|\text{"Something"}|\varphi_B\rangle \quad (5)$$

What is the most general expression that one can obtain from QCD for the above "Something"? Here I anticipate the answer to this question (correct to the lowest order in flavor breaking); later I will show how to derive it. The answer:

$$\text{"Something" (in Eq.(5))}=\hat{\alpha}(2\underset{\sim}{J})+(\mu+KS)\sum\left[(2/3)\underset{\sim}{\sigma}^{\mathcal{P}}-(1/3)\underset{\sim}{\sigma}^{\mathcal{N}}-(1/3)\underset{\sim}{\sigma}^{\lambda}\right]+\frac{A}{3}\sum\underset{\sim}{\sigma}^{\lambda}+$$
$$+F(2\underset{\sim}{J})Q+H(2\underset{\sim}{J})S+\frac{L}{3}Q\sum\underset{\sim}{\sigma}^{\lambda}+G(2\underset{\sim}{J})QS \quad (6)$$

Here S=Strangeness, Q=Charge, $\underset{\sim}{J}$=angular momentum; $\hat{\alpha},\mu$,A,K,F,H,L,G are 8 parameters. Because, as will be shown later, $\hat{\alpha}$ is totally negligible, we remain (neglecting it) with 7 parameters; from the 7 known magnetic moments we get:

$$\mu=2.869 \qquad A=1.005$$
$$K=0.289 \;;\; F=-0.076 \;;\; H=0.086 \;;\; G=-0.14 \;;\; L=-0.17$$

Except for A, all the above parameters are less than 15% of $\mu$. As stated, the NRQM parametrization keeps only $\mu$ and A, of all the above parameters.

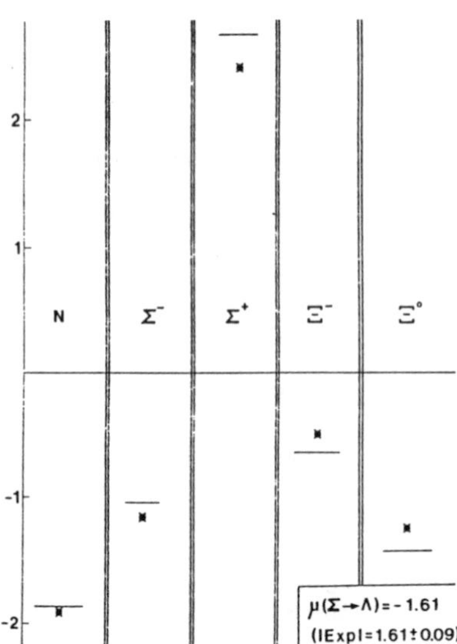


Fig.1.The magnetic moments of the 8 baryons calculated with the NRQM (Eq.1) (lines) and the exp.values (dots).It is (in Bohr proton magnetons):N=-1.86 (-1.91);P=input (+2.793) Λ=input(-0.613±0.004);Σ(-)=-1.04(-1.16±0.025) Σ(+)=2.68(+2.42±0.05);Ξ(-)=-0.50 (-0.65±0.03) Ξ(o)=-1.43 (-1.25±0.014).(values in brackets are NRQM calculated;not in brackets are exp.)

It now appears that indeed this is a fair approximation,not a chance.As to the reason why all the coefficients from K to L are small,we shall discuss this later.

<u>The Baryon Masses</u>

We display another example of parametrization,that of the baryon masses.Calculating the baryon levels with a typical NRQM Hamiltonian, a standard expression for the mass $M_B$ of baryon B (again for the L=0 states for which the model functions are $\varphi_B(1,2,3)=X_{L=0}(\underline{r}_1,\underline{r}_2,\underline{r}_3)\cdot W_B(1,2,3)$),turns out to be:

$$M_B=\langle\varphi_B(1,2,3)|M_o+B\sum_i P_i^\lambda+C\sum_{i>k}(\underline{\sigma}_i\cdot\underline{\sigma}_k)+D\sum_{i>k}(\underline{\sigma}_i\cdot\underline{\sigma}_k)(P_i^\lambda+P_k^\lambda)|\varphi_B(1,2,3)\rangle \quad (7)$$

Here I introduced the projection operator $P^\lambda$ on the strange quark:

$$P^\lambda=(1/3)(1-\lambda_8) \quad (8)$$

Once more the question is:Suppose that we perform an exact QCD calculation starting from the exact QCD Hamiltonian and suppose that,as a result of that calculation,we succeed in the rest frame of the baryon to write the exact expression of the mass as:

$$M_B^{QCD}=\langle\varphi_B|\text{"Something"}|\varphi_B\rangle \quad (9)$$

What is the most general form of "Something"?

De Rujula, Georgi and Glashow[6] calculated explicitely the masses from QCD in a semirelativistic (Darwin) approximation. As a matter of fact, to 1st-order in $P^{\lambda}$, they obtained an expression typical of the NRQM, of the form given in Eq.7. Here we are interested in a much more general answer, that is to know the most general expression for the "Something" above that results from an exact QCD calculation; I repeat, exactly, <u>with no approximation.</u>

The answer is, for the octet and decuplet L=0 baryons (I now call this "Something"=$\tilde{H}$); again $P^{\lambda}=(1-\lambda_8)/3$, the projector on the strange $\lambda$ quark:

$$\tilde{H}=M_o + B\sum_i P_i^{\lambda} + C\sum_{i>k}(\boldsymbol{\sigma}_i\cdot\boldsymbol{\sigma}_k) + D\sum_{i>k}(\boldsymbol{\sigma}_i\cdot\boldsymbol{\sigma}_k)(P_i^{\lambda}+P_k^{\lambda}) + E\sum_{\substack{i\neq k\neq j\\(i>k)}}(\boldsymbol{\sigma}_i\cdot\boldsymbol{\sigma}_k)P_j^{\lambda} + a\sum_{i>k}P_i^{\lambda}P_k^{\lambda} + b\sum_{i>k}(\boldsymbol{\sigma}_i\cdot\boldsymbol{\sigma}_k)P_i^{\lambda}P_k^{\lambda} + c\sum_{\substack{i\neq k\neq j\\(i>k)}}(\boldsymbol{\sigma}_i\cdot\boldsymbol{\sigma}_k)(P_i^{\lambda}+P_k^{\lambda})P_j^{\lambda} + d\,P_1^{\lambda}P_2^{\lambda}P_3^{\lambda} \qquad (10)$$

Note: there are 9 real parameters $M_o$, B, C, D, E, a, b, c, d, but (a+b) appear together in all the masses, so that in fact we have only 8 quantities that can be **exactly** determined from the 8 available masses (the 4 masses of the isomultiplets in the octet and the 4 in the decuplet). Their values turn out to be (in MeV):

$$M_o=1085;\ B=187;\ C=48.7;\ D=-16;\ (a+b)=-11.6;\ E=3.4;\ c=1.17;\ d\approx 1 \qquad (11)$$

One sees once more that the first four coefficients (those that a naive NRQM calculation using 2-body potentials would give) are the largest ones. Again this is the reason why the NRQM works fairly well quantitatively.

A simple consequence of the parametrization just given is the following: Neglect c and d in Eq.10 with respect to the other coefficients. Then one obtains the following mass relationship (Ref.9e):

$$(1/2)(N + \Xi) + T = (1/4)(3\Lambda + \Sigma) \qquad (12)$$

where T stays for:

$$T=\Xi^{*} - (1/2)(\Omega + \Sigma^{*}) \qquad (13)$$

This relationship is satisfied to such an accuracy that, before comparing it with the data, it is necessary to rewrite it so that it refers just to the strong part of the masses or, equivalently, that it is unaffected by the electromagnetic contributions. This can be easily done (it is sufficient to do it at zero order in flavor breaking) and the relationship becomes (now of course the charge labels appear explicitely; N is Neutron-not Nucleon):

$$\underset{(1132.4\pm0.8)}{(1/2)(N + \Xi^{o}) + T'} = \underset{(1133.9\pm0.1)}{(1/4)(3\Lambda + \hat{\Sigma}^{+})} \qquad (14)$$

where:

$$T'=\Xi^{*-} - (1/2)(\Omega + \Sigma^{*-}) \qquad (15)$$

and $\hat{\Sigma}^{+}$ stays for:

$$\hat{\Sigma}^{+}= 2\Sigma^{+} - \Sigma^{o} + 2(N - P) \qquad (16)$$

The agreement (shown by the numbers (in MeV) below the left and right sides of Eq.(14)) is rather striking.For comparison,the standard Gell-Mann Okubo mass formula (also made free from electromagnetic effects) is the same as Eq.(14) without the T' term;its two sides are (left)=1127.2±0.3 and (right)=1133.9±0.1.Note that Eq.14 was obtained,except for electromagnetic effects,long time ago,from the NRQM,neglecting three quark potentials.

Note also that:a) the agreement to 1 per 1000 is unchanged[10] if,to check the above relationship,one uses the pole values of the decuplet masses instead of the conventional ones (those listed in the Particle data tables,that were used above);b)The relationship (14) does not contain the Δ masses that still have a comparatively large error.

## The Meson Masses

As my last example of general parametrization in QCD,I consider the masses of the lowest L=0 mesons (P and V).Here,I will consider only the mesons with I≠0.A discussion of all L=0 mesons is given in Ref.9c.

In the NRQM one starts writing a wave function for meson i as:

$$\varphi^{\text{Model}}_{\text{meson } i} = f(|\mathbf{r}_1-\mathbf{r}_2|)W_i(1,2) \qquad (17)$$

where 1=quark,2=antiquark.$W_i(1,2)$ is the spin-flavor factor (Color,as before,is left understood).To calculate the mass of meson i in the NRQM one selects some simple model Hamiltonian and one takes its expectation value on the above wave function.After integrating on the space part of the model wave function one gets,typically:

$$M^{\text{NRQM}}_{\text{meson } i} = \langle W_i(1,2)|A+B\,\boldsymbol{\sigma}_1\cdot\boldsymbol{\sigma}_2+C(P^\lambda_1+P^\lambda_2)\;+D\,\boldsymbol{\sigma}_1\cdot\boldsymbol{\sigma}_2(P^\lambda_1+P^\lambda_2)|W_i(1,2)\rangle \qquad (18)$$

Once more suppose that we calculate exactly in QCD the mass of meson i and that we are able to display the exact end result in the form:

$$M^{\text{QCD}}_{\text{meson } i} = \langle W_i(1,2)|\text{"Something"}|W_i(1,2)\rangle \qquad (19)$$

We will show that in this case the "Something" turns out to be precisely equal to the above expression (in Eq.18);that is

$$\text{"Something"} = A+B\,\boldsymbol{\sigma}_1\cdot\boldsymbol{\sigma}_2+C(P^\lambda_1+P^\lambda_2)\;+D\,\boldsymbol{\sigma}_1\cdot\boldsymbol{\sigma}_2(P^\lambda_1+P^\lambda_2) \qquad (20)$$

That is in this case the NRQM naive expression coincides exactly with the most general QCD parametrization.Thus we can extract exactly from the masses of $\pi,K,\rho,K^*$ the values of the 4 parameters A,B,C,D.We get (in MeV):

$$\pi=A-3B\ (=138) \qquad \rho=A+B\ (=770) \qquad (21)$$

$$K=A-3B+C-3D\ (=495) \qquad K^*=A+B+C+D\ (=894) \qquad (22)$$

from which:

A=612 , B=158 , C=182 , D= −58

These are the values that an exact QCD calculation −knowing the quark-gluon coupling,would produce.For later use I insert here the following remark: Assume that the quark-gluon coupling were slightly different from what it is and that we had −say− B=110 and A=660 instead of B=158 and A=612.Then the pion mass would become $\pi$=330 (not so small),instead of 138,due to the factor 3 in $\pi$=A−3B.

## 3. THE PARAMETRIZATION PROCEDURE (HOW TO DERIVE, IN GENERAL, THE "SOMETHING")

### The Transformation V From The Model States To The Exact States

Because the derivation (Ref.9a) of the general parametrization from field theory is similar in all cases, I illustrate it here for the magnetic moments (for the masses it would be simpler).

Call H the exact QCD Hamiltonian, $|\Psi_B\rangle$ its eigenstates for the octet baryons at rest (I insist on "at rest" because all what I will do is relativistic but non covariant). In the rest system the magnetic moment operator can be written:

$$\underset{\sim}{\mathcal{M}} = (1/2)\int d^3\underset{\sim}{r}\ \underset{\sim}{j}(\underset{\sim}{r})\times\underset{\sim}{r} \tag{23}$$

with

$$j_\mu(x) = ie\left[\ \frac{2}{3}\,\bar{u}_R(x)\gamma_\mu u_R(x) - \frac{1}{3}\,\bar{d}_R(x)\gamma_\mu d_R(x) - \frac{1}{3}\,\bar{s}_R(x)\gamma_\mu s_R(x)\right] \equiv$$
$$\equiv \frac{ie}{2}\,[\bar{\Psi}(x)(\lambda_3+\frac{1}{3}\,\lambda_8)\gamma_\mu\Psi(x)] \tag{24}$$

The suffix R in the quark fields will be illustrated later.

The magnetic moment is

$$\underset{\sim}{M} = \langle\Psi_B|\underset{\sim}{\mathcal{M}}|\Psi_B\rangle \tag{25}$$

Of course the state $|\Psi_B\rangle$, written in the Fock space, has an extremely complicated structure. Schematically:

$$|\Psi_B\rangle = |qqq\rangle + |qqq\bar{q}q\rangle + |qqq,\text{Gluons}\rangle + \ldots\ldots$$

Now introduce an auxiliary "model" Hamiltonian $\mathcal{H}$ defined in the subspace of the Fock states of just 3 quarks (no gluon, no $\bar{q}$). As I insist, $\mathcal{H}$ operates only in the 3-quark sector. Call the eigenstates of $\mathcal{H}$ the (3-quark) "model states" $|\phi_B\rangle$:

$$\mathcal{H}|\phi_B\rangle = M_o|\phi_B\rangle \tag{26}$$

We have chosen $\mathcal{H}$ to be flavor independent so that the 8 eigenvalues (model masses) in (26) are all equal. We assume $\mathcal{H}$ to have an exceedingly simple non relativistic structure, so that writing:

$$|\phi_B\rangle = \sum\nolimits_{1,2,3}\varphi_{B,L=0}(1,2,3)\ a_1^+a_2^+a_3^+|0\rangle \tag{27}$$

the wave function $\varphi_B$ has a structure typical of the most naive NRQM. We choose $\mathcal{H}$ so that $\varphi_B$ has no configuration mixing (factorizability):

$$\varphi_B(1,2,3) = X_{L=0}(\underset{\sim}{r}_1,\underset{\sim}{r}_2,\underset{\sim}{r}_3)\cdot W_B(1,2,3) \tag{28}$$

Here $W_B$ is the spin flavor factor already discussed.

Now we assume that there exists a unitary transformation V from the naive state $|\phi_B\rangle$ (in the 3q subspace) to the exact state $|\Psi_B\rangle$:

$$|\Psi_B\rangle = V\ |\phi_B\rangle \tag{29}$$

Therefore:

$$\underset{\sim}{M}=\langle\Psi_B|\underset{\sim}{\mathcal{M}}|\Psi_B\rangle = \langle\phi_B|V^\dagger\underset{\sim}{\mathcal{M}}\ V|\phi_B\rangle \tag{30}$$

Because $\langle\phi_B|$ is a 3q Fock state,the only part of $V^\dagger\underset{\sim}{\mathcal{M}}\ V$ that intervenes in calculating $\langle\phi_B|V^\dagger\underset{\sim}{\mathcal{M}}\ V|\phi_B\rangle$ is the projection in the 3q subspace:

$$\tilde{\underset{\sim}{\mathcal{M}}} =(V^\dagger\underset{\sim}{\mathcal{M}}\ V)_{3q}= \sum_{3q,3q'}\sum |3q\rangle\langle 3q|V^\dagger\underset{\sim}{\mathcal{M}}\ V|3q'\rangle\langle 3q'| \tag{31}$$

We must therefore calculate:

$$\underset{\sim}{M} = \langle\phi_B|V^\dagger\underset{\sim}{\mathcal{M}}\ V|\phi_B\rangle = \langle\phi_B|\ \tilde{\underset{\sim}{\mathcal{M}}}\ |\phi_B\rangle \tag{32}$$

Now $\tilde{\underset{\sim}{\mathcal{M}}} \equiv V^\dagger\underset{\sim}{\mathcal{M}}\ V$ is certainly an extremely complicated field operator.But, since it has to act only on the coordinates (space,spin,flavor,color) of the three quarks present in the state $|\phi_B\rangle$ it must be,<u>after contraction of all field operators,</u>necessarily a function of three coordinates only.*We must then ask which is the most general possible expression of* $(V^\dagger\underset{\sim}{\mathcal{M}}\ V)_{3q}$ *in terms of the space,spin,flavor,color coordinates of the 3 quarks, after contraction of all the creation and destruction operators.*Clearly after contraction of all field operators the magnetic moment $\tilde{\underset{\sim}{\mathcal{M}}}$ takes the form of an ordinary operator in ordinary 3-body quantum mechanics:

$$\underset{\sim}{M} = \langle\varphi_B|\ \tilde{\underset{\sim}{\mathcal{M}}}\ |\varphi_B\rangle \tag{33}$$

where $\varphi_B$ is the 3-body function (recall, $|\phi_B\rangle=\sum_{1,2,3}\varphi_B(1,2,3)a_1^+a_2^+a_3^+|0\rangle$) and $\tilde{\underset{\sim}{\mathcal{M}}}$ (for which I keep for simplicity the same notation as before) is now - in Eq.(33)- an operator in just the ordinary 3-body space (not in Fock 3-body space).Which is the most general form of this $\tilde{\underset{\sim}{\mathcal{M}}}(1,2,3)$?

<u>The Most General Form Of $\tilde{\underset{\sim}{\mathcal{M}}}(1,2,3)$</u>.

Let us rewrite:$\underset{\sim}{M} = \langle\varphi_{B,L=0}|\ \tilde{\underset{\sim}{\mathcal{M}}}(1,2,3)\ |\varphi_{B,L=0}\rangle$.We wish to determine the most general expression of the space-spin-flavor-color operator $\tilde{\underset{\sim}{\mathcal{M}}}(1,2,3)$ of the 3 quarks that can emerge after a full QCD calculation.

The number of spin-flavor operators that can be constructed in the spin-flavor space of 3 quarks is finite.Call them $Y_\nu(\underset{\sim}{\sigma},f)$ where $\underset{\sim}{\sigma}$ are all the spin operators of the 3 quarks and f are all the flavor operators.The index $\nu$ specifies to which operator we refer.Call now: $\underset{\sim}{r} \equiv \underset{\sim}{r}_1,\underset{\sim}{r}_2,\underset{\sim}{r}_3$ ,the space coordinates of the 3 quarks.Then the most general operator of the space and spin-flavor variables of the 3 quarks has necessarily the form:

$$\sum_{\nu} R_{\nu}(\mathbf{r},\mathbf{r}')Y_{\nu}(\boldsymbol{\sigma},f) \tag{34}$$

where $R_{\nu}(\mathbf{r},\mathbf{r}')$ is an operator (not necessarily local) acting in the coordinate space $\mathbf{r} \equiv (\mathbf{r}_1, \mathbf{r}_2, \mathbf{r}_3)$ of the 3 quarks. Because to calculate a physical quantity, such as a magnetic moment, one has to take the expectation of (34) on $\varphi_B = X_{L=0}(\mathbf{r}_1,\mathbf{r}_2,\mathbf{r}_3)\cdot W_B(1,2,3)$ (Eq.23), the factorizability of $\varphi_B$ implies for the expectation value the following structure:

$$\mathbf{M} = \sum_{\nu} \langle X_{L=0}(\mathbf{r})|\ R_{\nu}(\mathbf{r},\mathbf{r}')\,|X_{L=0}(\mathbf{r}')\rangle\langle W_B|Y_{\nu}(\boldsymbol{\sigma},f)\,|W_B\rangle \tag{35}$$

Note now that $X_{L=0}$ has L=0. Thus the space part $R_{\nu}(\mathbf{r},\mathbf{r}')$ must be necessarily a space scalar. Setting:

$$g_{\nu} = \langle X_{L=0}(\mathbf{r})|\ R_{\nu}(\mathbf{r},\mathbf{r}')\,|X_{L-0}(\mathbf{r}')\rangle \tag{36}$$

we get:

$$\mathbf{M} = \sum_{\nu} g_{\nu}\langle W_B|Y_{\nu}(\boldsymbol{\sigma},f)\,|W_B\rangle \tag{37}$$

Because $\mathbf{M}$ is an axial vector the $Y_{\nu}$'s must be axial vectors formed only with the spins $\boldsymbol{\sigma}_i$ of the 3 quarks. Thus we are led to classify all the possible $Y_{\nu}(\boldsymbol{\sigma},f)$'s. We will call $\mathbf{G}_{\nu}(\boldsymbol{\sigma},f)$ the axial vectors $Y_{\nu}(\boldsymbol{\sigma},f)$.

As to the spins, note the following: we use a model state $|\phi_B\rangle$ with a non relativistic structure, thus with two component spinors and Pauli matrices. The task to generate from this state the exact state $|\Psi_B\rangle$, which is constructed in terms of Dirac quark fields with 4-components bispinors, is left to the operator V, which, in addition to transforming the 3q state into the correct state in Fock space, also performs such operation. Thus V also contains essentially a Foldy Wouthuysen transformation:

$$\exp\ [(1/2)\int\bar{\Psi}(\mathbf{x})\gamma_5\chi\ \Psi(\mathbf{x})d^3\mathbf{x}] \quad \text{with } tg^{-1}\chi = -(i/m)\boldsymbol{\sigma}\cdot\boldsymbol{\nabla}$$

for spinors of momentum $\mathbf{p}=-i\boldsymbol{\nabla}$ and mass m.

<u>The spin dependence of the</u> $\mathbf{G}_{\nu}(\boldsymbol{\sigma},f)$. The list of all possible axial vectors constructed with the spins of 3 quarks is much shorter than one might have expected. I just state the result (compare Ref.9a), noting that, for instance, for the diagonal matrix elements over real wave functions, expressions such as:

$$\boldsymbol{\sigma}_i \times \boldsymbol{\sigma}_k \qquad (i \neq k) \tag{38}$$

cannot appear (they give zero).

In conclusion the most general axial vector that can intervene in $\mathbf{G}_{\nu}(\boldsymbol{\sigma},f)$ is just a combination of:

$$\underset{\sim}{g}_1 \Gamma^J_a(f) \quad , \quad \underset{\sim}{g}_2 \Gamma^J_b(f) \quad , \quad \underset{\sim}{g}_3 \Gamma^J_c(f) \tag{39}$$

where the $\Gamma$'s are flavor operators and the index J is there to recall that they can also depend on the J of the state under consideration (Ref.9a).

<u>The flavor dependence</u>.The QCD Lagrangian for the strong interactions contains the flavor matrices only as $\lambda_8$ ,which appears in the mass breaking term of the Hamiltonian:

$$m\int d^3\underset{\sim}{r}\ \overline{\Psi}(x)\Psi(x) + \frac{1}{3}\Delta m \int d^3\underset{\sim}{r}\ \overline{\Psi}(x)(1-\lambda_8)\Psi(x) \tag{40}$$

In the strong interaction there is no other dependence from the flavor $\lambda$'s. Above I have set:

$$\Psi(x)=\begin{vmatrix} u_R(x) \\ d_R(x) \\ s_R(x) \end{vmatrix} \equiv \begin{vmatrix} \mathcal{P}(x) \\ \mathcal{N}(x) \\ \lambda(x) \end{vmatrix} \tag{41}$$

The mass m above is the so called realistic mass,the same appearing in the De Rujula,Georgi and Glashow[6] approximate one gluon exchange calculations with QCD.The use of $\mathcal{P}(x)$ for $u_R(x)$ etc. has been adopted precisely to make it clear that the QCD Lagrangian we are using has these realistic mass parameters in it.We may interprete these realistic (constituent) masses m as those that correspond to a convenient choice of the renormalization point R of the quark mass in the Lagrangian.Hence the suffix R.

As a matter of fact the only parameter that will appear in what follows is the ratio $(\Delta m/m)$ which will turn out to be $\cong(1/3)$.(Recall that in chiral QCD $((m_s-m_u)/m_u)\cong 25$).Anyway I wish to stress that the only flavor matrix appearing in the QCD Lagrangian is $\lambda_8$.Instead of $\lambda_8$ ,I will always use below the projector $P^\lambda=(1-\lambda_8)/3$.It is,of course:

$$(P^\lambda)^n=P^\lambda \tag{42}$$

If,in addition to the strong interactions,the e.m. ones are present (as when we calculate the magnetic moments) the only other flavor operator that appears in the QCD Lagrangian is the charge Q.In terms of the projectors:

$$Q=\frac{1}{2}\left[\lambda_3+\frac{1}{3}\lambda_8\right]=\frac{2}{3}P^{\mathcal{P}}-\frac{1}{3}P^{\mathcal{N}}-\frac{1}{3}P^{\lambda} \tag{43}$$

Because Q and $P^\lambda$ commute,the most general strong and e.m. QCD calculation can produce no flavor operator except $Q,P^\lambda$ and their products.

Because the magnetic moments are linear in Q,only the first power of Q can appear in the final result.Because $(P^\lambda)^n=P^\lambda$,any $P^\lambda$ can appear at most to power 1.Thus the complete list of flavor operators $\Gamma(f)$ that can appear at the end of a full exact QCD calculation–in calculating the magnetic moment of an octet baryon– is:

$$\Gamma(f) = \underset{\uparrow}{Q_i} \quad ,\underset{\uparrow}{Q_i P^\lambda_k} \quad ,Q_i P^\lambda_k P^\lambda_j \quad ,\underset{\uparrow}{\mathrm{Tr}[QP^\lambda]} \quad ,\mathrm{Tr}[QP^\lambda]P^\lambda_i \quad ,\mathrm{Tr}[QP^\lambda]P^\lambda_i P^\lambda_k \tag{44}$$

where, of course $Tr[QP^\lambda] = -(1/3)$. If we perform a calculation that neglects terms of order higher than the first in flavor breaking, only the first, second and fourth term (those with the arrow) in the above list (44) can intervene.

Now the $\Gamma(f)$'s have to be multiplied, as we saw, by $\sigma_1$ or $\sigma_2$ or $\sigma_3$ to obtain the most general $G_\nu(\sigma, f)$. Therefore, to first order in $P^\lambda$, the three flavor operators with the arrow have to be multiplied each by $\sigma_1$ or $\sigma_2$ or $\sigma_3$. But the spin flavor functions $W_B$ that intervene in calculating the expectation value of the $G_\nu(\sigma, f)$'s are spin-flavor symmetric. Thus the $G_\nu$'s must be symmetric and, to 1st order in $P^\lambda$, we get the following 8 $G_\nu$'s:

$$G_0 = Tr[QP^\lambda]\cdot\sum_i \sigma_i \qquad G_1 = \sum_i Q_i \sigma_i \qquad G_2 = \sum_i Q_i P_i^\lambda \sigma_i \tag{45}$$

$$G_3 = \sum_{i\neq k} Q_i \sigma_k \quad , \quad G_4 = \sum_{i\neq k} Q_i P_i^\lambda \sigma_k \quad , \quad G_5 = \sum_{i\neq k} Q_i P_k^\lambda \sigma_k \ ,$$

$$G_6 = \sum_{i\neq k} Q_i \sigma_i P_k^\lambda \quad , \quad G_7 = \sum_{i\neq k\neq j} Q_i P_k^\lambda \sigma_j$$

Neglecting $G_0$ (we shall see that it contributes negligibly) and reordering the other terms one obtains the expression for the magnetic moments given and discussed at the beginning:

$$\text{"Something"} = \sum_{0}^{7}{}_\nu\, g_\nu G_\nu = (\text{neglecting } g_0) = (\mu + KS)\sum\left[\frac{2}{3}\sigma^{\mathscr{P}} - \frac{1}{3}\sigma^{\mathscr{N}} - \frac{1}{3}\sigma^{\lambda}\right] + \frac{A}{3}\sum \sigma^\lambda +$$

$$+F(2J)Q + H(2J)S + \frac{L}{3}Q\sum\sigma^\lambda + G(2J)QS \tag{46}$$

## 4. THE GENERAL PARAMETRIZATION OF THE MASSES

For the Masses $=\langle\Psi_B|H|\Psi_B\rangle$ (where H is the exact QCD Hamiltonian) one has to parametrize

$$(V^\dagger H V)_{3q} \qquad \text{(a scalar)} \tag{47}$$

instead of $(V^\dagger \mathbf{M}\, V)_{3q}$ -an axial vector- that occurred for the magnetic moments. Thus the only spin operators are:

$$1 \quad , (\sigma_i \cdot \sigma_k); \qquad [(\sigma_1 \times \sigma_2)\cdot\sigma_3 \text{ gives zero between real functions}] \tag{48}$$

Moreover the (flavor) charge Q plays no role in the strong part of the mass and the only flavor operators that intervene in the parametrization (both for the octet and decuplet baryons) are:

$$1 \quad , P_i^\lambda \quad , P_i^\lambda P_k^\lambda \quad , P_1^\lambda P_2^\lambda P_3^\lambda \tag{49}$$

(There are no Tr terms now; the $Tr[P^\lambda]$ terms are included in "1" in Eq.49). From the symmetry requirement in 1,2,3, one gets from the above quantities the expression (10) already displayed, the general parametrization for the masses exact to all orders in $P^\lambda$:

$$\tilde{H}=(V^\dagger HV)_{3q}[\text{after integr. of the space coord.}]= \qquad (50)$$

$$M_o +B\sum_i P_i^\lambda +C\sum_{i>k} (\underset{\sim}{\sigma}_i\cdot\underset{\sim}{\sigma}_k) +D\sum_{i>k} (\underset{\sim}{\sigma}_i\cdot\underset{\sim}{\sigma}_k)(P_i^\lambda+P_k^\lambda) +E\sum_{\substack{i\neq k\neq j\\(i>k)}} (\underset{\sim}{\sigma}_i\cdot\underset{\sim}{\sigma}_k)P_j^\lambda +$$

$$+a\sum_{i>k} P_i^\lambda P_k^\lambda +b\sum_{i>k} (\underset{\sim}{\sigma}_i\cdot\underset{\sim}{\sigma}_k)P_i^\lambda P_k^\lambda + c\sum_{\substack{i\neq k\neq j\\(i>k)}} (\underset{\sim}{\sigma}_i\cdot\underset{\sim}{\sigma}_k)(P_i^\lambda+P_k^\lambda)P_j^\lambda + d\, P_1^\lambda P_2^\lambda P_3^\lambda$$

## 5. THE CONSTRUCTION OF V

All what has been said depends on the existence of a V such that, e.g. for baryons,

$$|\Psi_{exact}\rangle = V|\phi_{model}\rangle \qquad (51)$$

3q Fock state

I wish to give at least one construction method of V starting from the basic Hamiltonian of QCD (or from that of any relativistic QCD-like field theory). Note: all the results derived so far and to be derived depend only on two properties of QCD: a) The current is constructed only in terms of the quark fields (no additional field, e.g. no pion local field is introduced explicitely); b) Flavor breaking is due only to the $\lambda_8$ mass terms. We call QCD-like any relativistic field theory that satisfies these two properties.

A method to construct V is that of Gell Mann and Low (Phys.Rev.84,181 (1951)). To develop this application of the method (for the baryons) call $\eta$ a projection operator on the sector of the Fock 3-quark, no gluon states:

$$\eta|qqq\rangle = |qqq\rangle \qquad (52)$$
$$\eta|\neq qqq\rangle = 0 \qquad (53)$$

We rewrite H identically as:

$$H = \eta H\eta + (1-\eta)H(1-\eta) +\eta H(1-\eta) + (1-\eta)H\eta \qquad (54)$$

Introduce now the model Hamiltonian $\mathcal{H}$ written as a field Hamiltonian (having a non relativistic form) acting only in the 3q-no gluon sector of the Fock space:

$$\mathcal{H}=\eta\mathcal{H}\eta \qquad (55)$$

Decompose the exact Hamiltonian H as:

$$H= K_o+ K_1 \qquad (56)$$

with:

$$K_o = \eta\mathcal{H}\eta + (1-\eta)H(1-\eta) \qquad (57)$$
$$K_1 = \eta H(1-\eta) + (1-\eta)H\eta + \eta H\eta -\eta\mathcal{H}\eta$$

having added, to $K_o$, and subtracted, from $K_1$, the model Hamiltonian $\eta\mathcal{H}\eta$.

We assume that $\eta\mathcal{H}\eta$ has degenerate eigenvalues $M_o^o$ for all the octet and decuplet baryon states:

$$\eta\mathcal{H}\eta\ \phi_B = M_o^o\ \phi_B \qquad (B=N,\Lambda,\Sigma,\Xi,\Delta,\Sigma^*,\Xi^*,\Omega) \tag{58}$$

where $|\phi_B\rangle|0$ gluons$\rangle$ are the L=0 model states, called simply $|\phi\rangle$ in the text. Because in the three quark sector $K_o$ and $\mathcal{H}$ coincide, $|\phi_B\rangle$ are the degenerate eigenstates of $K_o$:

$$K_o|\phi_B\rangle = M_o^o|\phi_B\rangle \tag{59}$$

In the part $\eta\mathcal{H}\eta$ of $K_o$ the masses of $\mathcal{P},\mathcal{N}$ and $\lambda$ quarks are taken as equal (as implied by (58)); the flavour breaking mass term appears in $(1-\eta)H(1-\eta)$ of $K_o$ and in $\eta H\eta$ of $K_1$. The term $(1-\eta)H(1-\eta)$ of $K_o$ includes, in particular, the Hamiltonian of the non interacting gluons; $K_1$ contains the interaction terms $\eta H(1-\eta)$ and $(1-\eta)H\eta$ of the quark-gluon Hamiltonian.

We now regard $K_o$ as an unperturbed Hamiltonian, $K_1$ as a perturbation, imagine to insert $K_1$ adiabatically, and construct the true states $|\Psi\rangle$ with the procedure of Gell Mann and Low [this construction procedure is of course not compulsory, but we describe it to exemplify at least one method of construction]. With $K_1(t)=\exp(+iK_ot)K_1\exp(-iK_ot)$ the adiabatic $U(t,t_o)$ satisfies $i\dot{U}_\alpha(t,t_o) = \exp(-\alpha|t|)\cdot K_1(t)U_\alpha(t,t_o)$ [with $\alpha>0$, $U(t_o,t_o)=1$] and the $|\Psi\rangle$'s for the exact bound states corresponding to the lowest $|\phi\rangle$'s are:

$$|\Psi_B\rangle = \lim_{\alpha\to 0} \exp(-w_B/\alpha)\cdot U_\alpha(0,-\infty)|\phi_B\rangle \tag{60}$$

where $w_B$ is purely imaginary $(w_B+w_B^*=0)$ [so that the factor $\exp(-w_B/\alpha)$ in front of (60) is a pure phase factor that eliminates the singularity coming from the $\lim_{\alpha\to 0}U_\alpha(0,-\infty)$; $w_B$ in (60) is related to the $S=U(+\infty,-\infty)$ matrix element of the $\phi_B\to\phi_B$ transition by $\lim_{\alpha\to 0}\langle\phi_B|S|\phi_B\rangle = \exp(2w_B/\alpha)$]. The operator V introduced in the text can be therefore written explicitely as:

$$V = \lim_{\alpha\to 0} \exp(-w_B/\alpha)\cdot U_\alpha(0,-\infty) \tag{61}$$

Referring to magnetic moments, the formula for the magnetic moment has the form used in the text, namely:

$$\underset{\sim}{M} = \langle\phi|V^+\underset{\sim}{\mathcal{M}}\,V|\phi\rangle \tag{62}$$

As it can be shown in a few passages, this is the same as the formula used frequently for practical calculations:

$$\underset{\sim}{M} = \frac{\langle\phi|T(\underset{\sim}{\mathcal{M}}(0)S)|\phi\rangle}{\langle\phi|S|\phi\rangle} \equiv \langle\phi|T(\underset{\sim}{\mathcal{M}}(0)S)|\phi\rangle_C \tag{63}$$

where the index C means "connected".

This shows that one can reduce the calculation of some quantity to a calculation in terms of Feynman diagrams-where the "perturbation" is $K_1$. This possibility of relating the parametrized expression of "Something" to a Feynman diagram calculation -even if non covariant- is useful,because it allows to connect the orders of magnitude of the various terms in the parametrization (by various terms I mean terms with different number of indices) to the number of gluons exchanged (Ref.9g).

## 6. THE NUMBER OF QUARK INDICES AND THE MINIMUM NUMBER OF GLUONS EXCHANGED

The description in terms of Feynman diagrams allows to relate the complexity of the terms in spin-flavor space to the number of gluons exchanged and to extract in this way -from the mass parametrization and from that of the magnetic moments- an estimate of how the coefficients in the parametrization decrease in magnitude with increasing number of gluons exchanged (Ref.9g).

Schematically the expectation value of the Hamiltonian in the baryon mass parametrization receives contributions from the following classes of diagrams:

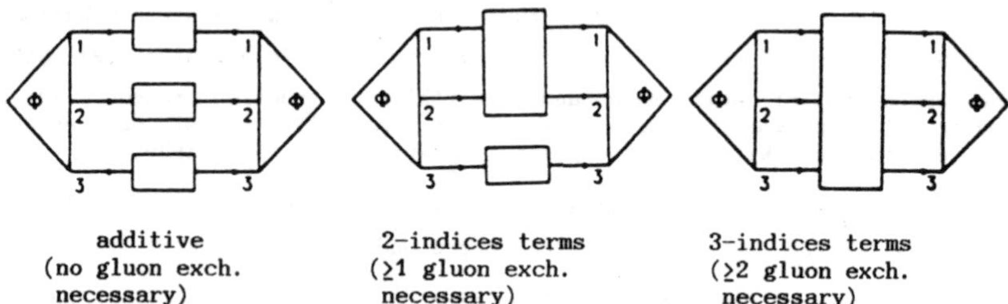


Analyzing the orders of magnitude of the coefficients of the terms in the mass spin-flavor parametrization one reaches the following conclusions:

a)Each additional exchange of a gluon between two quarks introduces a reduction factor $\cong(1/5)$

b)Each $P^\lambda$ (implying a multiplicative factor $(\Delta m/m)$),introduces a reduction of $\cong$ (1/3).

Using these factors (1/5) and (1/3) one can show that Trace terms in the magnetic moments (proportional to $Tr[QP^\lambda]$) are negligible,as already stated. Indeed due to color and the Furry theorem these terms,originated by closed loops,must have a structure of the type:

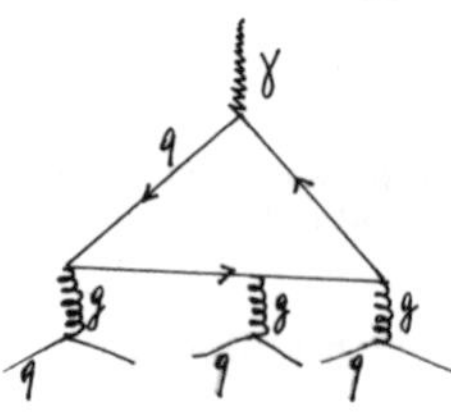


with at least 3 gluons and 1 $P^\lambda$;hence they are depressed by $\approx(1/5)^3(1/3)=$ $=1/375$. [I am much indebted to dr.D.E.Soper (Univ.of Oregon) who noted that I unduly omitted these Trace terms in my original paper (Ref.9a).Though their effect is negligible,not only on the magnetic moments but on all the other e.m. parametrizations,discarding them a priori was certainly wrong].

7.A SHORT DIGRESSION: WHY $(\mu_P/\mu_N) = -3/2$? -(ANSWER: CHANCE)

As we just showed, terms in the spin-flavor parametrization that imply the exchange of 1 or 2 gluons are reduced by ≅ 5 or ≅ 25 with respect to the "additive" (one index) terms. Each additional $P^{\lambda}$ implies a further reduction ≅ 3. Thus we understand the smallness of many coefficients and the dominance of the simple "additive" terms typical of the NRQM. However a few coefficients that are small due to other reasons or to "chance".

One amusing case is that of the magnetic moments of P and N. For them the parametrization, to all orders in flavor breaking, gives:

$$\mu(P) = \hat{\alpha} + \mu + F \cong \mu + F \text{ (neglecting } \hat{\alpha}) \qquad (64)$$
$$\mu(N) = -\frac{2}{3}\mu$$

In Eq.64, $\hat{\alpha}$ is the negligible coefficient of the Tr term of Eq.8, left out in Eq.46; the other symbols are as in Eq.46. Now $\hat{\alpha}$ is negligible, as stated; but why is F so small compared to $\mu$ ($F = -0.076$, $\mu = 2.869$) that we have:

$$\frac{\mu(P)}{\mu(N)} \cong -\frac{3}{2} \quad ?$$

Recall: $\mu$ is the coefficient of the additive expression:

$$\sum\left[\frac{2}{3}\,\underset{\sim}{\sigma}^{\mathcal{P}} - \frac{1}{3}\,\underset{\sim}{\sigma}^{\mathcal{N}} - \frac{1}{3}\,\underset{\sim}{\sigma}^{\lambda}\right] \equiv \sum_i Q_i \underset{\sim}{\sigma}_i \qquad (65)$$

while F is the coefficient of:

$$\sum_{i \neq k} Q_i \underset{\sim}{\sigma}_k \qquad (66)$$

This expression (66) needs, apparently, the exchange of just one gluon, so that F should be ≅ $(1/5)\mu$. It happens to be $\approx (1/40)\mu$. This additional factor 8 is not predicted from the above estimates of orders of magnitude. It is nevertheless very important because, if it had not been there, the famous ratio: $(\mu(P)/\mu(N)) \cong -(3/2)$ would not have been true and the reluctance to accept quarks with fractional charge would have been much stronger. In other words if F had its "natural" value $\approx(\mu/5)$ instead of the anomalously small value $(\mu/40)$, the whole story of this field might been different.

[An analogy: The "chance" coincidence of the Rutherford classical and quantum formulas, certainly played a role in the construction of Rutherford atomic model and then of the whole quantum mechanics].

A situation similar to the above one occurs for the quantity:

$$b = F - (2/3)D \qquad (67)$$

where b is the coefficient of $\sum_{i \neq k} \tau_i^{+} \underset{\sim}{\sigma}_k$ in the semileptonic decays of the baryons; b=0 corresponds to the well known result (D/F)=(3/2). It so happens that b is not zero but is appreciably smaller than one might expect on the basis of the fact that it needs just the exchange of one gluon.

## 8. SOME OTHER APPLICATIONS OF THE GENERAL PARAMETRIZATION METHOD

1) $\Delta\to P\gamma$ *decays.* The general parametrization clarifies why the ratio $[(\Delta\to P\gamma)_{NRQM}/(\Delta\to P\gamma)_{exact}](M1)$ is $\approx$1.45 times smaller than the experimental value, a comparatively large discrepancy for the NRQM where the typical deviations are less than 15% . Also we can understand the small value of the E2 ($\Delta\to P\gamma$) transition; its deviation from zero (Becchi-Morpurgo rule) is due to terms that require at least one gluon (Ref.9a).

2) *Semileptonic decays of baryons.* To zero order in flavor breaking the general parametrization leads to a 2-parameter (a,b) expression equivalent to the Cabibbo theory: D=a-b , F=(2/3)a + (1/3)b. The NRQM approximation leads to

$$(g_A/g_V)_{N\to Pe\nu}=(5/3)a \tag{68}$$

[not 5/3, as stated frequently, identifying incorrectly abstract $SU_6$ with the NRQM-compare my Erice lectures 1968, Ref.1b]. The general parametrization can be useful to include flavor breaking to 1st order. If terms with no gluon exchange (additive) prevail over the others, only one additional parameter enters; otherwise the number of new terms is rather large (Ref.9b).

3) *Meson masses and mixing angles.* Neglecting the $P_1^\lambda P_2^\lambda$ terms, a definite prediction can be made on the $\eta,\eta'$ mixing angle. It is (Ref.9c):

$$\theta_P \cong -23^\circ$$

in agreement with the data. I also recall that for the I$\neq$0 mesons the exact parametrization coincides with the form given by the NRQM, as already shown.

4) *Vector Meson$\to$PS meson $+\gamma$ decays.* The exact QCD parametrization produces also in this case several results of interest. Barring again the Trace terms, we summarize the results as follows (Ref.9d):

$$\frac{\Gamma(\omega\to\pi\gamma)}{\Gamma(\rho\to\pi\gamma)} = \left[3 + \frac{2\mu_5 f_5(P)}{\mu_1 f_1(P)}\right]^2 \cdot 1.06 \tag{69}$$

The NRQM gives 3 for the quantity in square brackets. Because the $\mu_5/\mu_1$ terms are depressed by 3-gluon exchange, once more the NRQM prediction is well fulfilled. As a matter of fact experimentally it is:

$$\frac{\Gamma(\omega\to\pi\gamma)}{\Gamma(\rho\to\pi\gamma)} = 9.9 \pm 1.6 \qquad (9\cdot 1.06\cong 9.5) \tag{70}$$

The smallness of 2-gluon exchange diagrams is also apparent in:

$$\left.\frac{\Gamma(\eta'\to\rho\gamma)}{\Gamma(\eta'\to\omega\gamma)}\right|_{\substack{\text{Gen.Par.negl.}\\ \text{2-gluon terms}}} = 11 \qquad (\text{Exp.}=11.1\pm 2) \tag{71}$$

Finally the rate of $\phi\to\pi^\circ\gamma$ is in agreement, within the errors, with the NRQM result that attributes all the decay to the small deviation ($37^\circ\pm 1.2^\circ$) of $\theta_V$ from the ideal value of $35.3^\circ$, which NRQM result again neglects 3-gluon diagrams.

To summarize, we see once more that the NRQM works because it is a parametrization that neglects only very small terms.

## 9. THE ELECTROMAGNETIC MASS DIFFERENCES OF BARYONS

### The Coleman-Glashow Formula

One can also parametrize the electromagnetic masses of the baryons. Here, in flavor space, one has to list (Ref.9f) all the operators quadratic in $Q=\frac{2}{3}P^{\mathcal{P}}-\frac{1}{3}P^{\mathcal{N}}-\frac{1}{3}P^{\lambda}$, and, if flavor breaking is taken into account, with up to three $P^{\lambda}$ s. [The Tr terms are irrelevant in all results to be displayed].

Neglecting flavor breaking, Coleman and Glashow derived-long ago

$$\delta P-\delta N = \delta\Sigma^{+}-\delta\Sigma^{-}+\delta\Xi^{-}-\delta\Xi^{0} \qquad (72)$$
$$1.29 \qquad\qquad 1.67\pm0.6\ \mathrm{MeV}$$

It has always been a mistery, at least to me, why this formula works so well since it neglects flavor breaking. The general parametrization method answers this question as follows: The Coleman-Glashow formula is correct to all orders in flavor breaking, provided that one neglects terms with three different quark indices. According to the estimates of orders of magnitude given previously, and noting that now two indices can be produced by the photon exchange (not gluon exchange), it therefore neglects terms of order $(1/5)\cdot(1/3)$ with respect to those kept; thus it is correct to $\approx(1/15)$. I add that the main exp. error ($\pm0.6$ MeV) is due presently to the $\Xi^{0}$ mass.

### The Gal-Scheck Relationships

Using the non relativistic quark model, Gal and Scheck derived long ago a set of relationships between the electromagnetic masses both for mesons and for baryons. Whereas for mesons the assumptions of Gal and Scheck are very restrictive, for baryons these assumptions amounted mostly to the neglect of three indices terms. With that neglect the NRQM leads to the following relationships that involve both the octet and the decuplet electromagnetic masses. The same relationships (with the same neglect) are derived by the general parametrization method. Of the 5 relationships that can be derived (Ref.9f), we do not write those 3 where the $\Delta$ masses appear, because the experimental errors on these are still too large. The two remaining ones are:

$$(\delta\Sigma^{*+}-\delta\Sigma^{*-})+(\delta\Xi^{*-}-\delta\Xi^{*0}) = \delta P-\delta N \qquad (73)$$
$$-1.2\pm0.9 \qquad\qquad -1.29\ \mathrm{MeV}$$

$$\frac{1}{2}(\delta\Sigma^{*+}+\delta\Sigma^{*-})-\delta\Sigma^{*0} = \frac{1}{2}(\delta\Sigma^{+}+\delta\Sigma^{-})-\delta\Sigma^{0} \qquad (74)$$
$$1.3\pm1.2 \qquad\qquad 0.85\pm0.12\ \mathrm{MeV}$$

They are fulfilled inside the errors. (These relationships are expected to be correct to 1/5, not to 1/15 as the Coleman-Glashow equation).

## 10. SOME QUESTIONS ON THE FLAVOR DEPENDENCE OF A CLASS OF QCD RESULTS

The QCD (strong) Lagrangian contains the flavor only through $\lambda_8$. Therefore no Feynman diagram calculation of the eigenstates of the QCD strong Hamiltonian can produce a flavor structure more complicated than $\lambda_8$,or,if one prefers:$P^\lambda=(1-\lambda_8)/3$.This is apparent in the parametrizations given so far.

The following important question then arises:Is QCD equivalent to effective Lagrangian theories or models in which pion local fields are introduced explicitely in addition to quarks and gluons? This class of theories includes:

a)The cloudy bag model[11],

c)QCD chiral Lagrangians[12,13] where pions and kaons are introduced (together with quarks and gluons) as quasi Goldstone bosons arising after an (assumed) spontaneous symmetry breaking.

In these cases the pion field appears as an explicit additional degree of freedom in the Lagrangian.Note that in the theories listed above the radial excitations of the $0^-$ mesons continue to be considered,at least implicitely,as bound $q\bar{q}$ S states,a curious situation.

A Lagrangian with explicit pion and kaon fields treated as local fields coupled to quarks necessarily contains the flavor matrices:

$$\lambda_1 , \lambda_2 \text{ and } \lambda_4 , \lambda_5$$

which were absent in the original QCD Lagrangian.Thus,barring cancellations that I am unable to see,the result of a calculation of the eigenstates of the effective Hamiltonian will depend in such theories not just on $P^\lambda$ but also on $\lambda_1$ ,$\lambda_2$ ,$\lambda_4$ ,$\lambda_5$ etc.

Thus the question arises of how all the models mentioned above (that are usually invoked to clarify the low energy region of QCD) can be equivalent to QCD,since they seem to produce different results,at least in the flavor dependence.(This difference is evident from what I already said but I will exemplify it again explicitely later for the cloudy bag model in the case of the magnetic moments of the $\Delta$'s).

The above argument on the difference in flavor dependence in QCD and in theories with explicit pions applies,with no modification,also in the presence of electromagnetic interactions.In fact in QCD (or in any QCD -like theory) the charge is carried only by the quarks and the flavor structure of the results can depend only on:

$$\lambda_3 \text{ and } \lambda_8$$

They commute and,no matter how complicated is the algebra,they cannot produce $\lambda_1$ and $\lambda_2$ in the final result.

This simplicity of the flavor structure of the QCD Lagrangian has been indeed one reason for the simplicity of the general parametrizations derived in the previous sections.

## 11.EXTENDING THE ABOVE ARGUMENT TO NON PERTURBATIVE TREATMENTS

That no flavor dependence different from $\lambda_3$ and $\lambda_8$ can appear in the construction of a QCD eigenstate is certainly true perturbatively.Is it also true non perturbatively? One or two remarks suggest that the answer to this question is positive.

1) If the exact QCD eigenstates cannot be obtained perturbatively as an expansion in terms of $\alpha_s(q^2)$, still a formal expansion of the result in terms of $\alpha_s$ may be possible. To each term of such a formal expansion the above argument should apply.

2) More generally, the very definition of an eigenstate of any Hamiltonian implies that it should be possible to calculate such an eigenstate to any desired approximation starting from the Hamiltonian (and a condition of normalizability) and performing a finite -though very large- number of operations. This is the basis of the lattice calculations, and applies in particular to the lowest energy state (or states), the vacuum. It should be true also if spontaneous symmetry breaking occurs.

Because with a finite number of operations (and also with an infinite number of them) we cannot generate $\lambda_1$ and $\lambda_2$ starting from $\lambda_3$ and $\lambda_8$, it is hard to see how a dependence on $\lambda_1$ and $\lambda_2$ can appear in the course of the calculation of an eigenstate of the QCD Hamiltonian and its eigenvalues.

## 12. SOME COMMENTS ON CHIRAL PERTURBATION THEORY

Chiral perturbation theory and related models are nowadays probably the source of ≈ 60% of the papers in low energy theoretical hadron physics. In spite of the assumptions that are implied in chiral perturbation theory J. Gasser, for instance, states[14]: "Chiral dynamics is a direct consequence of the symmetry properties of the standard model and chiral perturbation theory is its appropriate calculational tool. If it should turn out that experiments do not agree with the predictions of chiral dynamics, one would have to conclude that these experiments do also not agree with the predictions of the standard model".

I do not share Gasser's point of view. The basis of chiral dynamics in QCD seems to me founded on a "dogma":

QCD → Spontaneously broken QCD → Effective Lagrangians with $\pi$,K local fields as quasi-Goldstone bosons

For the believers, the arrows above stay for "unavoidable logical steps".

The basis of this "dogma" is that chiral dynamics is somehow regarded as compulsory to explain the great classical successes of PCAC+current algebra (Goldberger-Treiman, Adler-Weisberger, Weinberg's $\pi$ scattering lengths, etc.).

I think that the good results of PCAC+ current algebra <u>depend only on the empirical small value of the pion mass</u>. They were in fact derived much earlier than QCD (compare for an extraordinarily clear and detailed account the lectures by S. Coleman[15] at this school in 1967).

The small value of $m_\pi$ arises (compare the Eqs. 20,21 of the present lectures) from the fact that in the general parametrization it is:

$$m_\pi = A - 3B$$

with cancellations (by chance) between A and 3B.

The results of chiral perturbation theory (as I see them) are not so impressive as often asserted. Take, for instance, a paper by Kambor et al.[16] that presents results stated to be "impressive" and that, at first sight, (table I of ref.16) are such. But the good results (beyond old PCAC) are 7 and there are 7 arbitrary parameters in Kambor's et al. calculations. Although these parameters are not all independent and thus some predictions are possible, (table II of Ref.16), two of these are fulfilled-inside the errors- one is not and two are uncertain.

Chiral perturbation theory-in the form adopted usually (but compare for a critical analysis Stern et al.[17], and the refs. quoted there) implies the quadratic Gell Mann-Okubo meson mass formula and leads -in the natural way-to a $\eta$-$\eta'$ mixing angle of $-11^{\circ}$ in contrast with an experimental value of $\cong -22^{\circ}$ (to which the general parametrization, where the linear mass formula is natural, leads (Ref.9c)).

A final point should be mentioned on the identification of pions and kaons with quasi-Goldstone bosons and on the equivalence (at least for a subclass of phenomena) of Lagrangians written only in terms of such degrees of freedom to the original QCD Lagrangian. When dealing with such effective Lagrangians where quarks do not explicitely appear, it is not so clear if one should forbid to ever write a $\pi$-$q\bar{q}$ coupling; if not, it should be then clarified how the flavor structure arising from such coupling can be reconciled with that coming from the simple circumstance that $\lambda_3$ and $\lambda_8$ (the only flavor matrices present in QCD) cannot produce $\lambda_1$ and $\lambda_2$. This remark applies also to models a la Skyrme[18].

I will now illustrate this point on the flavor structure with another simple example: I compare the explicit calculations of the magnetic moments of the $\Delta$'s, performed with the cloudy bag model, with the exact QCD expression obtained with the general parametrization. Again their flavor dependences will be seen to be different.

## 13. THE MAGNETIC MOMENTS OF THE $\Delta$'S

*(This section has been written in collaboration with G.Dillon)*

The general parametrization method becomes extremely simple when applied to the calculation of the magnetic moments of the $\Delta$'s. (The reason is that the model wave functions of the $\Delta$'s are flavor symmetric). As a matter of fact it can be shown easily-using the technique explained previously- that writing the magnetic moments of the decuplet states as:

$$\underset{\sim}{M}_{10} = \langle \varphi_{10} | \tilde{\underset{\sim}{M}} | \varphi_{10} \rangle \tag{75}$$

where $\varphi_{10}$ are the 3-quark model states of the decuplet, the most general flavor dependence of $\tilde{\underset{\sim}{M}}$ is:

$$\tilde{\underset{\sim}{M}}(1,2,3) = \underset{\sim}{T}_{\alpha} Q_{\alpha} + \underset{\sim}{T}_{\alpha\beta} Q_{\alpha} P^{\lambda}_{\beta} + \underset{\sim}{T}_{\alpha\beta\gamma} Q_{\alpha} P^{\lambda}_{\beta} P^{\lambda}_{\gamma} + \underset{\sim}{T}_{\alpha\beta\gamma\delta} Q_{\alpha} P^{\lambda}_{\beta} P^{\lambda}_{\gamma} P^{\lambda}_{\delta} + \mathrm{tr}[QP^{\lambda}] \left[ \underset{\sim}{R} + \underset{\sim}{R}_{\alpha} P^{\lambda}_{\alpha} + \underset{\sim}{R}_{\alpha\beta} P^{\lambda}_{\alpha} P^{\lambda}_{\beta} + \underset{\sim}{R}_{\alpha\beta\gamma} P^{\lambda}_{\alpha} P^{\lambda}_{\beta} P^{\lambda}_{\gamma} \right] \tag{76}$$

where repeated indices are summed from 1 to 3. In (10) $\underset{\sim}{T}_\alpha, \underset{\sim}{T}_{\alpha\beta}, \underset{\sim}{R}, \underset{\sim}{R}_\alpha$'s etc. are each a function of all variables of the three quarks (momenta, spins, color) *except the flavor variables*, now displayed explicitely in (76).

If we limit to consider a zero strangeness baryon (such as the Δ's), $\tilde{\underset{\sim}{M}}(1,2,3)$ reduces to:

$$\tilde{\underset{\sim}{M}}(1,2,3)[\text{for the } \Delta\text{'s}] = Q_\alpha \underset{\sim}{T}_\alpha + \mathrm{tr}[QP^\lambda]\underset{\sim}{R} = Q_\alpha \underset{\sim}{T}_\alpha - (1/3)\underset{\sim}{R} \tag{77}$$

Because the wave functions $\varphi_{10}(1,2,3)$ are flavor symmetric, the only part of $Q_\alpha \underset{\sim}{T}_\alpha = Q_1\underset{\sim}{T}_1 + Q_2\underset{\sim}{T}_2 + Q_3\underset{\sim}{T}_3$ that contributes to the Δ's must have the form:

$$\underset{\sim}{A}\,\Sigma_\alpha Q_\alpha \tag{78}$$

where $\underset{\sim}{A}$ is an operator depending on all the variables, except flavor, of the three quarks. Because $\Sigma_\alpha Q_\alpha$ is the charge $Q_\Delta$ of the Δ under consideration, it results that the magnetic moments of the Δ's have the form:

$$M_\Delta = kQ_\Delta + \beta \qquad (\Delta \equiv \Delta^{++}, \Delta^{+}, \Delta^{o}, \Delta^{-}) \tag{79}$$

with $k = \langle\varphi_{10}|A_z|\varphi_{10}\rangle$ and $\beta = -(1/3)\langle\varphi_{10}|R_z|\varphi_{10}\rangle$. The Tr term R is negligible as already discussed, $\beta/k \approx (1/300)$, so that in practice the magnetic moments of the Δ's are proportional to their charges. [Note: Δm appears only in β, thus "almost" irrelevantly. The exact $M_\Delta$ is thus "almost" equal to that derived with no $SU_3$(flavor) violation, in spite of the mass difference of K and π. This is so because the diagrams in QCD have only quark and gluon lines (not pion and kaons lines). Except for the β term, all terms containing Δm are proportional to $P^\lambda$; because $f(\Delta m\, P^\lambda) = f(\Delta m)P^\lambda$ they do not contribute to states with zero strangeness].

In the cloudy bag model, where pion fields appear explicitely in addition to quark fields, the baryon magnetic moments receive, as discussed by Brown, Rho and Vento[11] a contribution of the form:

$$\Sigma_{i\neq j}(\underset{\sim}{\sigma}_i \times \underset{\sim}{\sigma}_j)(\underset{\sim}{\tau}_i \times \underset{\sim}{\tau}_j)_3 \tag{80}$$

But, as we discussed at length, such term cannot be present in QCD because a flavor structure such as $(\underset{\sim}{\tau}_i \times \underset{\sim}{\tau}_j)_3$ cannot emerge from QCD. This is confirmed by the fact that the term (80), as used in Ref. 11, leads to Δ moments of the form $M_\Delta = A(1+O(\Delta m))Q_\Delta + b(\Delta m)$ ($Q_\Delta$ is the Δ charge, A a constant independent of Δm, O(Δm) is a quantity of the order of the strange-non strange quark mass difference Δm, and b(Δm) is at least of 1st-order in Δm. As shown above, our exact result (79), even keeping the (negligible) Tr term β, is: $M_\Delta = kQ_\Delta + \beta(\Delta m)$, where β depends on Δm but k <u>*does not*</u>. Thus, in general, the Brown, Rho, Vento result differs from the exact result of QCD. Recall, moreover, that the coefficient b(Δm) in Ref. 11 is of order 0.1·A while our β is ≈(1/300)k.

Needless to say, the model of Ref. 11 does not certainly provide a particularly good fit to the octet baryon moments. Our motivation here in considering that model for the experimentally hard case of the Δ moments has been only conceptual, to compare it with our Eq. (79) which is a simple example of an exact QCD result, correct to all orders in flavor breaking.

## 14. THE PROLIFERATION OF MODELS AND THE ROLE OF THE GENERAL PARAMETRIZATION METHOD

One unpleasant feature of low energy hadron physics in the past 20 years has been the proliferation of models. This proliferation has been so huge that, as a matter of fact, it produced the impossibility of making reliable predictions in most cases. In other words one has so many models or schemes that predictive power is lost. I list some among these calculational schemes used most frequently:

1) Non relativistic quark model (NRQM)
2) Relativistic quark models (various variants: kinematical models, Bethe Salpeter, semirelativistic models $(p^2+m^2)^{1/2}$, etc.)
3) Bag model
4) Cloudy bag model
5) Skyrmions
6) QCD chiral dynamics (Lagrangians with and without quark degrees of freedom)

I have already commented upon the last three cases. I will now comment briefly on 1, 2 and 3, starting with 2 (and 3).

On the relativistic quark models my conclusion is very simple: In the rest frame of the hadron under consideration, the exact state is obtained from the model state–as we have seen–by the transformation:

$$|\Psi_{exact}\rangle = V|\phi_{model}\rangle$$

Once agreed that the model state is in all cases (the NRQM, the Relativistic Quark Model or the simple –as distinct from cloudy– bag model) very far from the exact state, simply because it is just one term of the complete Fock expansion, there does not seem to be much advantage in taking as model state a relativistic state, instead of the naive NRQM state (I am sorry to say this because I spent myself much time trying to construct relativistic wave functions[19]). Indeed choosing as $|\phi_{model}\rangle$ a relativistic state entails, in comparison to the NRQM state, an unnecessary increase in the number of parameters and therefore a decrease in predictive power. The basic advantage of the NRQM is to have very simple model states, *endowed with the maximum symmetry compatible with the quantum numbers of the system* that we describe. <u>The operator V does all the rest</u>.

It is this feature that leads to the simple type of parametrization that I described, with few coefficients of decreasing order of magnitude. This explains the quantitative success of the NRQM, as well as the fact that the NRQM is a relativistic (although non covariant) consequence of QCD. All relativistic features are contained (and hidden) in the operator V. No assumption at all is required in deriving the NRQM from QCD as to the value of v/c of the quarks inside hadrons. The non relativistic quark model is quantitatively succesfull because it selects the dominant terms in the general exact parametrization derived from QCD (or from any QCD-like field theory).

Of course, so far we have derived the general parametrization only for the lowest hadrons (those with L=0) where, due to the factorizability of the model wave function, things are simple. One should also extend the parametrization to excited hadron states, say P states, where of course the general parametrization turns out more complicated due to $\mathbf{L}\cdot\boldsymbol{\sigma}$ terms etc.

But, as far as the L=0 states are concerned, one has finally understood, I believe, the misterious quantitative success of the NRQM calculations.They describe a fully relativistic parametrization,or,to be more precise,an approximation to it that keeps the dominant terms.As a matter of fact the way to QCD was largely determined by the NRQM;a title for this paper might have been "From the NRQM to QCD and back".

## 15.SOME REFERENCES AND HISTORY


*A) Early history (1965-69) of the NRQM (also called "Constituent quark model" or "Realistic quark model" or "Potential model")*

1. G.Morpurgo:a)"Is a non relativistic approximation possible for the internal dynamics of "elementary" particles?",Physics,2,95(1965); b)"Lectures on the quark model" in "Theory and Phenomenology in particle physics,vol.A,p.83-217",A.Zichichi ed.,Academic New York 1969 (Erice Lectures 1968); c)Rapporteur talk in Proc.of the XIV Int. Conf.on High Energy Physics,Vienna 1968,p.225,ed.by J.Prentki and J.Steinberger,CERN publ.1968;d)"A short guide to the quark model", Ann.Rev.Nucl.Sci.20 (1970)p.105;
2. R.H.Dalitz,Rapporteur talk in Proc.XIII Int.Conf.on High Energy Physics Berkeley 1966,p.215,U.of Calif.Press,Berkeley 1966
3. J.Kokkedee,The Quark Model,(Benjamin,N.Y.1968)
4. D.Lichtenberg,"Unitary symmetry and Elementary particles",Academic,N.Y. 1970
5. M.Marinelli and G.Morpurgo,Physics Reports 85,161 (1982)

*B) Renaissance (1975-80)*

6. A.De Rujula,H.Georgi and S.Glashow,Phys.Rev.D 12,147(1975).For accounts of this period,that underline certain aspects compare also:
7. G.Karl and N.Isgur,Physics Today,nov.1983 p.36;also G.Karl:"Degrees of freedom in the nucleon" (Guelph preprint,lecture at Baryon 92,to be published in the Proceedings)-This paper contains a nice account of the early period of the NRQM.On one point of that account I might however make a remark,when Karl states that I was "skeptical on the evidence for negative parity [excited] baryons".Rather,for a while,I was uncertain about color,asking to myself if one had to think in terms of space antisymmetric wave functions with Majorana exchange forces of proper sign between quarks,or to accept color.For this reason I was cautious on the treatment of the baryonic excited states.
8. A.Le Yaouanc,L.Oliver,C.Pene,J.Raynal:"Hadron transitions in the quark model" (Gordon and Breach,N.Y.-1988)

*C) Modern period (1989-....)*

9. G.Morpurgo:a)Phys.Rev.D40,2997(1989);b)Phys.Rev.D40,3111(1989); c)Phys.Rev.D41,2865(1990);d)Phys.Rev.D42,1497(1990); e)Phys.Rev.Lett.68,139(1992);f)Phys.Rev.D45,1686(1992); g)Phys.Rev.D46,4068(1992);h)G.Dillon and G.Morpurgo,Univ.of Genova reports 1992,1993 (to be published);i)G.Morpurgo,Folgaria lectures, ed.T.Bressani,S.Marcello and A.Zenoni,World Sci.1992,p.3-62.
10. G.Dillon,Europhys.Lett.20,389 (1992)

*D) Other references.*

11. G.E.Brown,M.Rho and V.Vento,Phys.Lett.97B,423(1980)
12. A.Manohar and H.Georgi,Nucl.Phys.B 234,(1984)189
13. S.Peris,Phys.Lett.B 268,(1991) 415 and refs. cited there.
14. J.Gasser,"Chiral Dynamics",Proc.Workshop on DAΦNE,April 1991,G.Pancheri ed.,p.291 (publ.by Laboratori Naz. di Frascati,I-00044 Frascati (Roma))

15. S.Coleman,"Aspects of Symmetry",Lecture 2,p.36 (Cambridge U.P.1985)
16. J.Kambor,J.F.Donoghue,B.R.Holstein,J.Missimer and D.Wyler,Phys.Rev. Lett.68,1818 (1992);J.Kambor,J.Missimer and D.Wyler,Phys.Lett.B261 (1991) 496
17. J.Stern,H.Sazdijan and N.H.Fuchs,Phys.Rev.D47,3814(1993)
18. Compare e.g.I.Klebanov in "Hadrons and Hadronic matter" ed. by D. Vautherin,F.Lenz and J.Negele (Plenum,N.Y.1990),p.223
19. G.Morpurgo,in "Lepton and Hadron Structure",A.Zichichi ed. (Erice lectures 1974) Academic,N.Y.1975.

*E) Note*

I take this opportunity to correct some errors and misprints in Refs.9. In Ref.9a,Eq.36,a term $Tr[P^q P^\lambda]$ was omitted,as noted by dr.Soper,to whom I am much indebted.This produces the term $\hat{\alpha}(2\underset{\sim}{J})$ in Eq.6 (or, equivalently,the term $\underset{\sim}{G}_o$ in Eq.45) of this paper.The same term should be added to the Eqs.(62) and (63) of Ref.9a with the ensuing changes in the Eqs.64-68.As explained in the text (Sect.6),the effect of this $\hat{\alpha}(2\underset{\sim}{J})$term is fully negligible.More generally it is negligible the effect of the Tr term on the results of the papers [9a-i].In addition also the following corrections should be made in Ref.9a [with no effect,either because the correct equations were used or because the incorrect equations were written but not used]:In Eq.45 read $(1/2)\Lambda$ instead of $(1/4)\Lambda$;in the 3rd line of Eq.(63), insert a $(-2\gamma)$ in square brackets,getting $[\delta-\beta-2\gamma+(1/2)\gamma(4|\underset{\sim}{J}|^2-7)Q(2\underset{\sim}{J})]$;F in Eq.66 should be $\delta-\beta-4\gamma$ and the same combination (not $\delta-\beta+2\gamma$) also appears in Eq.64.[In footnote 1 of Ref.9f and in footnote 2 of Ref.9e, $\delta-\beta-2\gamma$ should be replaced by $\delta-\beta-4\gamma$].In Ref.9c (row after Eq.26) insert a minus sign in front of $\eta_1$.

## CHAIRMAN: G.Morpurgo

*Scientific Secretaries: M.Gibilisco, H.Petrache, O.Pisanti*

## DISCUSSION

– *Gourdin:*

During your discussion of the pseudoscalar meson masses you avoid the problem of $\eta$ and $\eta'$. Can you comment on the present status of the octet-singlet configuration mixing and more precisely on the best value from theory and experiment for the mixing angle $\theta_p$?

– *Morpurgo:*

This morning I considered only the mesons with isotopic spin $I$ different from zero, $I \neq 0$. One can consider also the $I = 0$ pseudoscalar and vector mesons. In such cases the following takes place: if you assume that second order flavor breaking terms can be neglected, then you can derive the mixing angles ( both for P and V mesons) with no further assumptions, except that you have to decide whether you wish to use a linear, a quadratic, a cubic etc. mass formula. Of course in the spirit in which I calculate the mass as the expectation value of the Hamiltonian, I naturally take a linear mass formula. In that case, the mixing angle $\eta - \eta'$ comes out $\approx (-22^o)$ whereas a quadratic mass formula gives $\approx (-11^o)$ . The experimental value is between $(-20^o)$ and $(-22^o)$ .

– *Gourdin:*

The reason for my question is that in many applications the results are very sensitive to the precise value that you choose. In fact even with a change from $(-20^o)$ to $(-22^o)$ you obtain different results, in particular for the decay to two photons.

– *Morpurgo:*

I do not believe that I can be more precise than what I said because, as I stated, in order to predict the angle I must neglect second order flavour breaking terms.

– *Giovannini:*

If I have understood correctly, your parametrisation is valid for a hadron composed of light quarks. A very predictive approach for the heavy quark sector is that of Isgur and Wise, who use an expansion in small velocities of the QCD

Lagrangian. Is it possible to extend your general construction to the heavy quark sector?

– *Morpurgo:*

Let me say that historically the difficulty in understanding the success of the NRQM had to do with the light quark sector. This has been the reason why I considered that sector. Whether the Isgur–Wise formula can be derived differently from the way they do it I do not know. I cannot answer questions on this point. Of course you may apply, probably, concepts similar to the ones that I developed to objects formed of heavy and light quarks. I think that perhaps this morning I did not emphasize one point sufficiently – my approach is such that (at least explicitly) my results do not depend at all on the ratio between the velocity of the quarks inside hadrons and the velocity of light.

– *Giovannini:*

This was the motivation for my question in fact. Because they find an expansion on the velocity it is probable that if you apply your approach to that situation, you may find something more general.

– *Morpurgo:*

Possibly, but I cannot answer on this.

– *Wegrzyn:*

Is it possible to derive from the NRQM the mass formulae for hadrons known from Regge phenomenology, like the Ademollo-Veneziano-Weinberg mass formula? If so what are the values of the parameters?

– *Morpurgo:*

Well, I am sorry this may or may not be possible but it is completely different from the question I am addressing in these two lectures. The question treated in my lectures is whether, using properties of QCD (or QCD–like relativistic field theories) you can derive the NRQM. My answer to this question is that yes, you derive parametrised expressions for several physical quantities and that you can show that the terms in these parametrised expressions that are present in the NRQM are the dominant ones. Whether you can fit some data with other formulae I do not know, but it is unrelated to the contents of my work.

– *Langacker:*

It has always been difficult to reconcile the view that the light pseudoscalars are pseudo-Goldstone bosons with the quark model. Can your new, improved quark model incorporate the chiral aspects of the pion?

– *Morpurgo:*

My theory explains why the NRQM works well starting from QCD; it is not a new, improved quark model. But aside from this, I am grateful for your question because a fraction of my next lecture will be devoted to the point you raised. I anticipate in part the answer. In the QCD Hamiltonian, only the Gell-Mann flavor matrix $\lambda_8$ is present. Now consider the transition from QCD to chiral dynamics consider for the moment just strong QCD, leaving out the e-m interaction. In the final Hamiltonian the pions, treated as independent quasi-Goldstone local fields must be coupled to the quarks via the isospin matrices $\tau_1, \tau_2$ or ( 3-dimensionally) $\lambda_1$ and $\lambda_2$. Similarly, if Kaons appear as quasi-Goldstone local fields, their coupling to the quarks must have $\lambda_4, \lambda_5$. My question is then how with any calculation that starts from a Lagrangian containing just $\lambda_8$ you can generate $\lambda_1, \lambda_2, \lambda_4, \lambda_5$? This question is the first part of my answer. I reserve the second part for tomorrow. Now I would like to add just the following: many people have been educated to think of QCD as producing necessarily quasi-Goldstone pions. In my opinion this kind of dogma is not necessarily true.

– *Beneke:*

In your derivation of the most general parametrisation of, for instance, the magnetic moment operator, you make use of the existence of a unitary transformation relating to the exact baryon state to a three constituent quark state. To me this looks like a drastic asumption. Can you give arguments in favour of the existence of this transformation?

– *Morpurgo:*

I can answer immediately. Again I must anticipate one of the subjects of tomorrow's lecture, where I will show how such a transformation can be constructed. The question is how can the transformation V connecting the exact state in Foch space to the model state be constructed. To do this, you can in principle use the Gell-Mann–Low adiabatic method. In fact tomorrow I will give a closed formula for V. If you accept that method (which is perturbative with an appropriately defined "perturbation", however), V can be constructed so as to express everything in terms of Feynmann diagrams. This turns out to be useful. I can show you the kind of formulae for V in the Gell-Mann–Low construction. The operator V is connected to "some" Dyson operator as follows: $V = \lim_{\alpha \to 0} \exp(-w_B/\alpha).U_\alpha(0, -\infty)$. I stated "some" Dyson operator because it is constructed, as I stated, in terms of a "perturbation" that I will define tomorrow. If you calculate the magnetic moments in this way, for instance, at the end you can use in this way the conventional formula:

$$\underline{M} = \frac{\langle \phi | T(\underline{M}(0) S) | \phi \rangle}{\langle \phi | S | \phi \rangle} \equiv \langle \phi | T(\underline{M}(0) S) | \phi \rangle_C$$

Here C means connected diagrams. Of course this is only one way to construct V. If you do not like it you can regard the existence of V as an assumption, but not a drastic one because it is the very basis of the definition of states in Fock space in field theory. You might of course object: how can the existence of V be reconciled with the Haag theorem that states an individual Fock state has vanishing projection on the complete solution? Here the answer is the same as I would give you in electrodynamics. As you know, for a renormalisable theory the Haag theorem is circumvented.

*–Kawall:*

Data from deep inelastic scattering and hyperon decay suggest that the total contribution of the quarks to the proton spin is $\leq (1/3)$ and that the sea quarks give a small negative contribution. In your theory can you extract values for these quantities?

*– Morpurgo:*

Your question is interesting because it touches a point that so far I have not mentioned. Namely, which are the masses of the light quarks in the Lagrangian of QCD? As a matter of fact I am not interested in the masses of the light quarks because they never appear explicitly in my parametrisation. However, I am interested in the difference between the mass of the s-quark and the masses of the non-strange quarks, more precisely in the ratio:

$$(m_s - m_{non-strange})/m_{non-strange} \equiv \Delta m/m.$$

This ratio turns out to be $\simeq (1/3)$ both from my general parametrisation and from previous calculations, in particular those of de Rujula, Georgi and Glashow. On the other hand, those people who treat the problem of the spin of the proton as seen in deep inelastic scattering experiments, have in mind a Langrangian where the quark masses are those of current quarks, with $\Delta m/m$ of the order of 25. In other words, we have here two different choices for the quark masses in the Lagrangian, which is nevertheless in both cases a relativistic Lagrangian. Now, at least in principle, it is possible (but I have not done this so far) to relate the descriptions to the two choices of masses. You may refer to the two mass choices as corresponding to two different renormalisation points or you may not. For me this is irrelevant. The point is that in any case, you may pass from one choice to the other by a certain kind of Foldy-Wouthuyson transformation which in some sense expresses, say, a constituent quark as a current quark plus a cloud of current quark-antiquark pairs plus gluons. But at the moment I cannot answer your question as to the details of the transformation that relates the spin of the proton in my description and in the current quark description.

– *Lu :*

How well does your method work for excited hadron states like $\rho$?

– *Morpurgo:*

Of course the $\rho$ has a model state with $L = 0$ and thus belongs to the objects that I have treated with my parametrisation. For mesons and baryons in states that do not have $L = 0$ model states but for instance P states, more terms intervene in the parametrisation, for instance of the type $\underline{L}.\underline{\sigma}$ and the parametrisation becomes more complicated.

– *Lu:*

Do you know anyone who has done that?

– *Morpurgo:*

No, this work is recent. Most of it has been done in the last two and a half years. It might be interesting to study how complicated the parametrisation becomes if you have to add terms of the type $\underline{L}.\underline{\sigma}$ etc. There is another point that I might mention in connection with my work that I did not talk about in my lecture: it has become apparent from this work that the only meaning of $SU_6$ is due to the fact that the model wave functions with $SU_6$ spin-flavor structure are the simplest ones having all the symmetries and good quantum numbers of the hadron under consideration. This makes them useful.

– *Lu:*

Did you try your method with heavy quarkonia states like $J/\Psi$ and Upsilon? Specifically can you calculate the decay amplitude for Upsilon(4S) to $B\bar{B}$?

– *Morpurgo:*

I answered a similar question earlier. I have dealt only with light quark phenomena for a very simple reason. People have never considered it curious that NRQM ideas could work for charmonium or bottomonium whereas it was considered mysterious that the NRQM could be successful for light quark systems. I have shown that it is not mysterious.

**CHAIRMAN: G.Morpurgo**

*Scientific Secretaries: M.Gibilisco, M.Lu, O.Pisanti*

## DISCUSSION

– *Langacker:*

You implied this morning that the classical tests of chiral symmetry such as the Adler-Weisberger relation could be obtained just from an accidentally light pion. However every derivation I have ever seen required the full apparatus of spontaneously broken chiral symmetry such as the current commutators and PCAC.

– *Morpurgo:*

Maybe I was not clear. What I meant is that you need the current algebra, of course, plus the accidental smallness of the pion mass. The point is that you can have current algebra even without spontaneously broken QCD. In fact you had it before QCD.

– *Langacker:*

I guess I still do not fully understand because you need to have PCAC which is an ingredient in addition to the current commutation rules.

– *Morpurgo:*

Yes, as I stated, you need PCAC plus current commutators, the same ingredients that Coleman used in his 1967 lectures here (before QCD). But I think that this does not make it compulsory that you have spontaneously broken QCD and all that.

– *Lu:*

What are the $u, d, s$ masses in the QCD Lagrangian?

– *Morpurgo:*

I answered this question yesterday in my answer to Dr. Kawall. For your convenience I will repeat the answer. I am interested only in the ratio $(m_s - m_{non-strange})/m_{non-strange} \equiv \Delta m/m$ which in my Lagrangian is chosen to be $\simeq (1/3)$ (corresponding, if you wish, to non-strange quarks of mass $\sim$ 380 MeV and strange quarks of mass $\sim$ 530 MeV, but as I repeat, the results from my parametrisation depend only on the above ratio being $\simeq 1/3$). You may have a perfectly defined relativistic QCD Lagrangian with quark masses having the above "high" values, the same Lagrangian that e.g. de Rujula, Georgi and Glashow used in their one gluon exchange approximation. A second description with a QCD Lagrangian – the one of the usual description related to spontaneously chiral symmetry breaking – uses in the QCD Lagrangian "low" quark mass values, say $\simeq 5$ MeV for $u$ and $d$ quarks and 15 to 25 times larger for the strange quark. You may ask what the relationship is, if any, between the two QCD's both (I repeat) relativistic field Lagrangians.

– *Lu:*

There may have been some misunderstanding. We have only one QCD Lagrangian and one set of parameters. What I want to know is what are the masses in that particular theory. Are they large or small?

– *Morpurgo:*

Well you know, Dr. Lu, I did an experiment to search for free quarks. If we had free quarks in front of us we could measure their mass precisely. We do not. So I think that you are not capable of telling me what the mass of a free quark is or the mass to be put in the QCD Lagrangian. I discussed the relationship between a QCD Lagrangian with "high" quark masses, and a QCD Lagrangian with "low" quark masses –more appropriately the latter, to describe high momentum transfer phenomena – in my answer to Dr. Kawall.

– *Lu:*

I am really asking a very simple question. I do not believe that free quarks exist. Under the assumptions that they are confined and that the QCD Lagrangian is written in quark degrees of freedom, what are the masses that you insert into that Lagrangian?

– *Morpurgo:*

I repeat that the only thing that I used in all my calculations is the ratio $(m_s - m_{non-strange}/m_{non-strange} \equiv \Delta m/m$. I already gave you its value.

– *Lu:*

Do you believe that your description is a final theory?

– *Morpurgo:*

I believe that questions of this kind cannot be answered.

– *Altarelli:*

I would like to come back to the question of the smallness of the pion mass. I would like to remember that there is strong support for the idea that the pion is the pseudo-Goldstone boson of chiral symmetry from lattice QCD. If on the lattice, you compute the pion mass and the $\rho$ mass as a function of the quark masses, you see very distinctly that while the pion mass goes to zero with vanishing quark masses

( in fact you can measure very well the slope of the line going to zero, that is $f_\pi$ ), the $\rho$ mass stays finite. So I am impressed that you called the smallness of the pion mass accidental. I think that there are good arguments for spontaneous breaking of chiral symmetry. I remember that treating the $\pi$ as the pseudo-Goldstone boson fixes all couplings of the particles in the limit of momentum zero in agreement with observations. There are some theorems that can be derived with difficulty if you do not put this hypothesis in: — for instance the Weinberg theorem on the pion scattering lengths. So, while I do agree that in many cases you can get close to the results by pole dominance or some dispersion relation, I think that, by now, it is a well established fact that the pion is the pseudo-Goldstone boson of the spontaneously broken chiral symmetry.

– *Morpurgo:*

Well let me answer your remarks in order. First of all, I believe you were not here yesterday, so that you did not follow my derivation of a formula that, in part, justifies my assertion that the small pion mass is accidental. I am rewriting that formula because it is important for my point. Using the $V$ transformation you can prove that the most general expression for the masses of the mesons with isospin $I \neq 0$ in QCD is the following ($W_i(1,2)$ is the model spin flavor function of meson $i$; $1 \equiv$ quark, $2 \equiv$ antiquark):

$$M^{QCD}_{meson\,i} = \langle W_i(1,2)|A + B\,\underline{\sigma}_1 \cdot \underline{\sigma}_2 + C(P_1^\lambda + P_2^\lambda) + D\underline{\sigma}_1 \cdot \underline{\sigma}_2(P_1^\lambda + P_2^\lambda)|W_i(1,2)\rangle$$

In this formula note two points:

*a)* A,B,C,D are four coefficients that are determined by the experimental values of the masses of $\pi, \rho, K, K^*$;

*b)* The above equation, although it is identical to the NRQM naïve formula is in fact, an <u>exact</u> result of QCD.

I insist on this point *b)* because it is essential for what follows.

Consider now only the masses $\pi$ and $\rho$. From the above equation we get $M(\pi) = A - 3B$ (equal to 140 MeV) and $M(\rho) = A + B$ (equal to 770 MeV), from which you obtain exactly A=610 MeV and B=160 MeV. Of course, if you could do exact computations in QCD, the result of such computations of A and B as a function of the quark masses and of the quark-gluon coupling should produce the above values of A and B (if QCD is correct). Now assume that you change comparatively slightly the quark-gluon coupling, $\alpha_s$. This may well produce a change of B from the above 160 to for example, 110 and a change of A ( keeping the mass of the $\rho$ at 770) from 610 to 660. Due to the factor 3 in the formula $M(\pi) = A - 3B$ you get with this change $M(\pi) = 330$ MeV, no longer so small (if I allow for a variation of $M(\rho)$ to 870 MeV , then A becomes 760 and $M(\pi)$ becomes

430). When I state that the small value of $M(\pi) = 140$ is accidental, I mean this: it is due to an accidental cancellation between A and 3B. In the reverse direction you can get $M(\pi)$ exactly equal to zero on changing the quark-gluon coupling so that B becomes 192 and A becomes 578.

To add something that I also noted yesterday on the question of the pion as a Goldstone boson, I drew attention to the following circumstance (in addition to the above quantitative remarks). It seems to me very unaesthetic that the (L=0) radial excitations of the pions or Kaons or $\eta$ are treated as $q\bar{q}$ aggregates, whereas the lowest states are treated as quasi-Goldstone bosons and this simply because the pion happens to have the small mass value of 140 MeV.

– *Altarelli:*

I am not a latticist, so I am not terribly biassed, but the lattice has the power of computing the coefficients A and B as functions of the quark masses. Now there are things on the lattice that are only roughly approximate. One of the best things that the lattice can do is really to unveil this question of the spontaneous symmetry breaking because there you can compute A and B as functions of the quark masses with rather good precision; $f_\pi$ on the lattice is computed to 15% nowadays. It is found that exactly when the masses of the quarks go to zero A equals 3B while the $\rho$ mass $A + B$ stays up. I think this is rather impressive. The latticists can also study, for instance, the breaking of chiral symmetry as a function of temperature and roughly determine the temperature for the phase transition restoring chiral symmetry. I think that their work should deserve your attention.

– *Zichichi:*

Excuse me, before you go on, I would like to understand one point. I find that Morpurgo's argument indicating how the $\rho$ can easily be massive and the pion light has a simplicity that cannot be compared to the incredible complications of the Lattice QCD calculations.

– *Altarelli:*

Morpurgo has done a parametrisation, not a calculation of A and B.

– *Morpurgo:*

This is clear of course, but the point that I have underlined is how critically the mass of the pion can possibly depend on small variations of the quark gluon coupling. This has been my point, not the dependence from the quark masses on which you have insisted so far. As to the lattice calculations I would add the following. Although I am certainly less expert than you in these calculations, I have followed their results and I must say that two things impressed me unfavourably: *a*) their instability with time *b*) the fact that the assumptions introduced in the course of the calculations are clear only to the experts in the field. As to *a*) I do not know what the number of years is, after which one can say that a Lattice result should be considered stable, but so far history has not offered many reasons for optimism in this respect.

But there is another point that I noted this morning concerning the $\pi$ as a quasi-Goldstone boson, that I would like to recall. In any calculation starting from the QCD Lagrangian you have to deal only with two flavor matrices $\lambda_8$ and, if you include electromagnetic interactions, $\lambda_3$. These form a commuting (closed) algebra and at the end of a calculation, no matter how complicated, you cannot create from them $\lambda_1$ and $\lambda_2$ or $\lambda_4$ and $\lambda_5$ that intervene in the coupling of the pions or, respectively, kaons to quarks. Then I ask you how can a Lagrangian with pions or Kaons as quasi Goldstone bosons be equivalent to the original Lagrangian if it contains $\lambda_1$ and $\lambda_2$ and the original does not.

– *Altarelli:*

Well, I do not know if I have understood the question well because $\lambda_1$ and $\lambda_2$ are matrices in the quark basis. In fact this morning I was surprised when you considered a Lagrangian with quarks and gluons appearing together with pions and kaons. In effective chiral Lagrangians they only take pions and kaons and what they are saying is that in the low energy domain you can simulate the results for pions and kaons.

– *Morpurgo:*

Well, chiral Lagrangians may contain only pions and kaons if you want to deal only with the lowest energy domain. If you want to go a little further they must also contain quarks. Do you admit this?

– *Altarelli:*

No, I would not believe it. My answer is that I would not add quarks and gluons, I would add $\rho$ mesons, for example, if you want to go to higher energies. But of course the more you go far away from low energy, the less constraining and the less motivated and the less true the effective chiral Lagrangian method becomes and so I would not believe in it. However, in the low energy region the chiral Lagrangian is proved to be equivalent to QCD in the sense that it reproduces the low energy theorems.

– *Morpurgo:*

Well, then I can reply to your statement. First of all, I feel that if you write a Lagrangian containing pions and kaons and $\rho$, excluding anything else, (for instance, the radial excitations of the above objects) one is many years back. I would also like to add that effective chiral Lagrangians with quarks are introduced, in a way, as a basis of the cloudy bag model with pions. If you calculate, in that model, the magnetic moments of the baryons you get (Brown, Rho and Vento) a term of the form $\Sigma_{i,j}(\underline{\sigma}_i \times \underline{\sigma}_j)(\underline{\tau}_i \times \underline{\tau}_j)_3$ that contains $\lambda_1$ and $\lambda_2$ ( that is the $\tau$'s) which cannot originate from QCD. If you do not wish to consider the cloudy bag model as an effective Lagrangian, take the effective Lagrangian with quarks and pions below chiral symmetry breaking introduced by Manohar and Georgi as equivalent to QCD and, in fact, used by them to "explain" the NRQM. I believe that it is inescapable that if you deal with such effective Lagrangians with quarks you also get from them flavor structures that you can never get from the original QCD Lagrangian.

– *Skenderis:*

You obtain the coefficients A,B,C,D by fitting the meson masses, What is the miracle of reproducing the masses afterwards?

– *Morpurgo:*

I never stated it as a miracle. What I stressed is that it is very conceivable, from my expression, that by just changing the quark gluon coupling by a small amount you might have a nature in which the pion instead of weighing 140 MeV weighed 350 MeV. That is all I said ( and also that my formula, on which this is based, is an exact QCD formula, not a NRQM one, although the two are identical in this case of meson masses).

– *Kawall:*

Your work was motivated by a desire to explain the successes of the NRQM. Can you explain why this model failed in the prediction of the decay $\Delta \to P\gamma$? Is this an accidental conspiracy of the parameters or is it a failure at a more fundamental level?

– *Morpurgo:*

This morning I mentioned this point. Your question allows me to explain it. If you remember the ratio of the matrix element of the $\Delta \to P\gamma$ in the NRQM to the one given by nature is $\sim 1.45$ which means a deviation of 45% to be compared with a maximum deviation typical of the NRQM of the order 10 to 15%. Now if one applies the general parametrisation to this case you find a factor of 3 multiplying the 15% typical error; this explains the deviation.

– *Kawall:*

Chiral $SU(3)_L \times SU(3)_R$ models in addition to predicting the magnetic moments also predict the magnitude of the violation of the Gottfried sum rule. In your opinion is this an artificial success of the theory?

– *Morpurgo:*

As to the magnetic moments, there are now of course, due to the proliferation of models, many different models (at least as many models as magnetic moments), so I do not think that this point is very relevant. For the Gottfried sum rule it is a completely different question, because it has to do with deep inelastic scattering and my parametrisation does not cover that case. So I cannot answer the second part of your question although the answer is possibly related to the remarks I made yesterday when precisely answering your questions.

– *Trocsanyi:*

You claimed that you had found a perturbative description of strong interactions at low energy. Is this correct?

– *Morpurgo:*

I believe that you misinterpreted, perhaps largely, what I was saying. I had the problem of showing that a way of constructing the operator V existed. Are you referring to this point?

– *Trocsanyi:*

Yes, in fact, I had the suspicion that this V does not exist. I am not convinced that this V has to exist.

– *Morpurgo:*

So you have the suspicion that an operator that transforms a 3–quark state in Fock space into the full exact baryon state does not exist. Is this your question?

– *Trocsanyi:*

Yes, I have the suspicion that such an operator does not exist.

– *Morpurgo:*

Well on what is your suspicion based, Haag's theorem or something else?

– *Trocsanyi:*

You may refer to Haag's theorem but this somehow contradicts the common sense that we have in strong interactions.

– *Morpurgo:*

No, this does nor contradict our common sense in strong interactions. Common sense has little to do with this, but if anything, it would suggest that there should exist a correspondence between model Fock states and exact states.

– *Trocsanyi:*

But this means that, for example, you can calculate magnetic moments in a perturbative way as you described this morning.

– *Morpurgo:*

No, I wished to show that if you, for a moment, accept a perturbative procedure (where the "perturbation" is defined as I did), I can give you a constructive procedure for V. If you look back at the old literature ( recall that the paper of Low and Gell- Mann is from 1951) you will find many examples of procedures, non-perturbative, that connect simple Fock states with the exact states. I just want to remind you of the Tamm-Dancoff method or similar ones. All these models are more or less related, they are non-perturbative, they have their defects (e.g. in connection with renormalisation) but I mention them here to indicate constructive procedures for V.

– *Beneke:*

One of your main arguments against the chiral Lagrangian has been the appearance of Gell- Mann flavor matrices that are not present in the QCD Lagrangian whereas your parametrisation contains only $\lambda_8$. I see no reason why new flavor structures should not appear in an effective low energy Lagrangian.

– *Morpurgo:*

Here I would like to make two remarks. The first about your suspicion and the second about your first comment. About the "suspicion", you may have seen my formula for the baryon masses which is correct to one part in a thousand. This accuracy is not common in strong interaction physics. If you allow me, I would like to say that it constitutes a check together with other checks like the Coleman-Glashow relation etc. As to your other point (that perturbation theory is not valid and therefore we cannot use it to conclude that flavor matrices different from $\lambda_8$ cannot intervene in the parametrisation), I mentioned that point this morning. I stated that no matter which method is used to solve your exact Schroedinger of QCD, $H|\Psi>= E|\Psi>$ you should be able to extract to any desired precision the eigenstate and eigenvalue of H with a finite (though as large as you like) number of operations. In not one of these operations can any flavor matrix different from $\lambda_8$ intervene and therefore flavor matrices different from $\lambda_8$ cannot appear in the final result. I invite you to generate $\lambda_1$ and $\lambda_2$.

– *Beneke:*

I have a question concerning this accuracy of one part per thousand. If I remember correctly your parametrisation formula contains eight parameters to fit 8 data. So I am not surprised that it is coincides to one part per thousand; I am not surprised that it is not exact.

–*Morpurgo:*

No, your remark is incorrect, but because what you have said may produce some confusion let me write again the parametrisation of the octet and decuplet baryon masses. The formula is :

$$\begin{aligned}\tilde{H} = M_0 &+ B\sum_i P_i^\lambda + C\sum_{i>k}(\underline{\sigma}_i \cdot \underline{\sigma}_k) + D\sum_{i>k}(\underline{\sigma}_i \cdot \underline{\sigma}_k)(P_i^\lambda + P_k^\lambda) \\ &+ E\sum_{\substack{i\neq k\neq j\\(i>k)}} \underline{\sigma}_i \cdot \underline{\sigma}_k)P_j^\lambda + a\sum_{i>k} P_i^\lambda P_k^\lambda + b\sum_{i>k}(\underline{\sigma}_i \cdot \underline{\sigma}_k)P_i^\lambda P_k^\lambda \\ &+ c\sum_{\substack{i\neq k\neq j\\(i>k)}} (\underline{\sigma}_i \cdot \underline{\sigma}_k)(P_i^\lambda + P_k^\lambda)P_j^\lambda + dP_1^\lambda P_2^\lambda P_3^\lambda\end{aligned}$$

You have in fact 8 parameters, as you say, because in the masses, only the combination $(a + b)$ appears. But in deriving my mass relation, I omitted entirely, for reasons explained in the lecture, the terms with 3-quark indices, 2nd order or more flavor breaking with coefficients $c$ and $d$. In other words I put $c$ and $d$ equal to zero. Therefore the agreement to one part in a thousand is a result. It is neither a tautology nor an accident.

– *Beneke:*

Concerning your second remark about the infinite number of operations; for example an infinite number of operations could generate the perturbative series to any order. I do not understand how else you could give an operational sense to the "infinite number of operations".

– *Morpurgo:*

No, I said something very different from that. I said that in general if you want to solve a complicated Schroedinger equation $H|\Psi >= E|\Psi >$ you necessarily have to do a certain number of operations on the objects that you have in H, no matter which method is used to solve the equation. From the fact that only $\lambda_8$ appears in each operation that you are performing, and that you have neither in the beginning nor in the course of this sequence of operations any other flavor matrices but $\lambda_8$, you will never be able to produce $\lambda_1$ and $\lambda_2$ in the end result.

– *Beneke:*

If I naïvely construct an effective Lagrangian then all operators that can appear will appear if they are not protected by some symmetry– so why not $\lambda_1$ and $\lambda_2$?

– *Morpurgo:*

I do not forbid you to construct effective Lagrangians. My question to you is if the flavor structure of such a Lagrangian reproduces that arising from the original QCD Lagrangian, that is if such Lagrangians are exactly equivalent, even in a tiny sector, to the original QCD Lagrangian.

– *Landacker:*

I disagree with your assertion that Goldstone bosons with couplings cannot occur in QCD. The pion is a $\bar{q}\lambda_i\gamma_5 q$ bound state. It is not obtained from perturbation theory.

– *Morpurgo:*

I do not obtain the pion, or any other bound state, mesonic or baryonic from perturbation theory (for mesons compare my paper in Phys. Rev D41,2865 (1990)). My assertion is that if you introduce the pion as a quasi-Goldstone local field $\varphi_i(x)$ (in addition to quarks and gluons) and you introduce then an effective interaction of type $\varphi_i(x)(\bar{q}\lambda_i\gamma_5 q)_x$, an effective Lagrangian with such a quark-gluon coupling cannot be equivalent to QCD.

# SUSY GUTs: A Practical Introduction

JORGE L. LOPEZ

Department of Physics, Texas A&M University, College Station, TX 77843–4242, USA

## 1 What are SUSY GUTs?

I will define a supersymmetric unified theory as one that incorporates the following elements in one guise or another:

(i) *GUT Symmetry*: which manifests itself in the unification of the *non-abelian* gauge couplings of the standard model above certain "unification scale" ($M_U$). Gauge groups with this property include: $SU(5)$, $SO(10)$, and $SU(5)\times U(1)$.

(ii) *GUT Fields*: which only exist because the larger GUT symmetry is present and which decouple below the unification scale. These fields play important roles in the traditional GUT processes such as gauge symmetry breaking, doublet-triplet (2/3) splitting, proton decay, baryogenesis, neutrino masses, etc.

(iii) *Yukawa Unification*: the larger gauge symmetries force the SM fermions to "share" larger representations, and therefore the gauge invariant Yukawa couplings in the GUT phase usually encompass more than one "low-energy" Yukawa coupling. Specifically one usually gets the following relations valid at $M_U$:

$$\begin{array}{ll} \lambda_b=\lambda_\tau & SU(5), \\ \lambda_b=\lambda_\tau=\lambda_t & SO(10), \\ \lambda_t=\lambda_{\nu_\tau} & SU(5)\times U(1). \end{array} \tag{1}$$

(iv) *(Universal) Soft-Supersymmetry Breaking*: spontaneous breaking of supergravity yields a global supersymmetric theory supplemented by a set of soft-supersymmetry-breaking parameters. If these parameters are $\lesssim \mathcal{O}(1\,\mathrm{TeV})$, supersymmetry effectively solves the gauge hierarchy problem.

(v) *Dynamical Evolution*: renormalization group equations (RGEs) for the gauge, Yukawa, and scalar couplings relate the values of these parameters at the high- and low-energy scales.

(v$\frac{1}{2}$) *Intermediate Scale Fields*: are apparently needed for string unification.

(vi) *Light Fields*: should include the standard model fields with two Higgs doublets plus their superpartners, and maybe other light fields such as Higgs singlets.

(vii) *Radiative Electroweak Symmetry Breaking*: allows the generation of the electroweak scale dynamically. The top-quark mass plays an important role.

The theories outlined above have the most appealing property of requiring at most *five* parameters to describe all new phenomena (excluding new sources of CP violation). These parameters are:

- The top-quark mass ($m_t \gtrsim 110\,\mathrm{GeV}$), to be measured soon at the Tevatron;
- The ratio of Higgs vacuum expectaction values (VEVs) ($1 < \tan\beta \lesssim 50$);
- Three soft-supersymmetry breaking parameters: the universal gaugino mass ($m_{1/2} \propto m_{\tilde{g}}$), the universal scalar mass ($m_0$), and the universal cubic scalar coupling ($A$).

This relative scarcity of parameters is in sharp contrast with the more than twenty parameters needed to achieve a compararable description in the minimal supersymmetric standard model (MSSM). An immediate consequence is that *all* sparticle and Higgs boson masses and couplings can be determined in terms of these five parameters, and therefore lots of non-trivial and unsuspected correlations arise. As such, these theories provide a very predictive scenario for *e.g.*, collider processes and rare decays.

An important remark to keep in mind is that it is essential to consider **all** aspects of the models under consideration, *e.g.*:

- SU(5): Gauge and Yukawa unification have been emphasized vigorously in the recent literature [1, 2, 3, 4, 5, 6, 7, 8, 9, 10, 11, 12, 13, 14, 15, 16, 17]. *However*, proton decay is a very important constraint [18, 19, 20, 21, 9, 22, 23] and so is cosmology [24, 22, 25, 26], but these aspects of the model have not received the same degree of attention (perhaps because of the perception that they entail "model-dependent" assumptions, although in practice no new unknowns are introduced). The doublet-triplet (2/3) splitting problem also needs to be addressed in a consistent way. For example, how does the missing-partner-mechanism (MPM) [27] solution affect the GUT threshold corrections [28]?
- SO(10): Predicts large $m_t$ and $\tan\beta$. Can one have radiative electroweak breaking? what about proton decay? or the 2/3 splitting problem?

# 2 Gauge Coupling Unification

## 2.1 Generalities

- Traditional GUTs: (grand) unified theories (*e.g.*, $SU(5), SO(10), E_6$, and $SU(5)\times U(1)$) generally contain:
  - Larger/new structure, revealed above some "unification" scale;
  - Some sort of gauge and Yukawa coupling unification;

- Observable proton decay, baryogenesis, neutrino masses, etc.

Examples of *non* GUTs include $SU(3)\times SU(2)_L\times U(1)_Y$ (SM) and $SU(4)\times SU(2)_L\times SU(2)_R$ (Pati-Salam)

- GUSTs: in grand unified *superstring* theories all the above generic properties are realized in sometimes novel ways:
  - Larger structure is provided by string massive states;
  - Gauge unification is automatic in this top→down approach (in simple models), and it is understandable in terms of a "primordial $SO(44)$" gauge symmetry;
  - Yukawa unification happens in a disorderly way as a consequence of remnants of higher symmetries [29].
- Intermediate Scales?

  Symmetry breaking patterns can include intermediate scales or not:

  $SU(3)\times SU(2)\times U(1)\to SU(5)$ (minimal GUT unification)

  $SU(3)\times SU(2)\times U(1)\to SO(10)$ (one-step unification)
  $SU(3)\times SU(2)\times U(1)\to SU(5)\to SO(10)$ (two-step unification)

  $SU(3)\times SU(2)\times U(1)\to SU(5)\times U(1)\to String$ (intermediate unification)
  $SU(3)\times SU(2)\times U(1)\to String\,(SU(5)\times U(1))$ (one-step string unification)[1]

- Non-minimal Matter: for example, $Q,\bar{Q},\ D^c,\bar{D}^c$ allow $SU(3)\times SU(2)\times U(1)$ to unify at $\sim 10^{18}$ GeV if their masses are chosen appropriately [30]. These fields fit snugly in the $SU(5)\times U(1)$ representations [31]:

$$\mathbf{10}=\{Q,D^c,\nu^c\},\qquad \overline{\mathbf{10}}=\{\bar{Q},\bar{D}^c,\bar{\nu}^c\}. \tag{2}$$

## 2.2 Renormalization Group Equations (RGEs)

If unification is assumed to occur in a single step and there are no intermediate scale particles with poorly determined masses, then one can study this problem in great detail (*e.g.*, in $SU(5)$)

*Problem:* solve the coupled set of two-loop gauge and Yukawa coupling RGEs, taking into account low- and high-energy threshold effects.

[1]It appears that $SU(5)\times U(1)$ only makes sense as a unified theory in the context of strings, since otherwise beyond the string scale the $SU(5)$ and $U(1)$ couplings would diverge again.

*Objective:* To obtain $\alpha_U$, $M_U$, and $\sin^2\theta_W$ in terms of $\alpha_e$ and $\alpha_3$ (both measured at $M_Z$), the light supersymmetric spectrum, and the heavy GUT spectrum.

*Philosophical Note*:

- Our approach assumes that gauge coupling unification occurs, as is the case in a unified theory. The model is tested by comparing its prediction for $\sin^2\theta_W$ against the experimental value. This is the top→down approach.
- Compare this with the ("experimental") bottom→up approach where one "tests" for unification by running up the gauge couplings. What can one conclude if it is found that the couplings do not meet?

The gauge coupling RGEs are given by

$$\frac{dg_i}{dt} = \frac{g_i}{16\pi^2}\left[b_i g_i^2 + \frac{1}{16\pi^2}\left(\sum_{j=1}^{3} b_{ij} g_i^2 g_j^2 - \sum_{j=t,b,\tau} a_{ij} g_i^2 \lambda_j^2\right)\right], \tag{3}$$

where $t = \ln(Q/M_U)$ with $Q$ the running scale and $M_U$ the unification mass. Also, $\alpha_1 = \frac{5}{3}(\alpha_e/\cos^2\theta_W)$, $\alpha_2 = (\alpha_e/\sin^2\theta_W)$, and

$$b_i = \left(\tfrac{33}{5}, 1, -3\right), \tag{4}$$

$$b_{ij} = \begin{pmatrix} \frac{199}{25} & \frac{27}{5} & \frac{88}{5} \\ \frac{9}{5} & 25 & 24 \\ \frac{11}{5} & 9 & 14 \end{pmatrix}, \tag{5}$$

$$a_{ij} = \begin{pmatrix} \frac{26}{5} & \frac{14}{5} & \frac{18}{5} \\ 6 & 6 & 2 \\ 4 & 4 & 0 \end{pmatrix}. \tag{6}$$

## 2.3 Analytic solutions

Neglecting the very small effect of the Yukawa couplings in the gauge coupling evolution [15], one can write down analytic solutions to the RGEs to the desired accuracy (see *e.g.*, [10]). First we neglect all heavy thresholds, *i.e.*, we assume unification occurs at one point. The solutions are:

$$\begin{aligned} \ln\frac{M_U}{M_Z} &= \frac{\pi}{10}\left(\frac{1}{\alpha_e} - \frac{8}{3\alpha_3}\right) && \text{(one − loop)} \\ &\quad -\frac{1}{40}\left(C_2 + \tfrac{5}{3}C_1 - \tfrac{8}{3}C_3\right) && \text{(two − loop)} \\ &\quad +\sum_i p_i \ln\frac{\widetilde{m}_i}{M_Z} && \text{(light thresholds)} \end{aligned} \tag{7}$$

$$
\begin{aligned}
\frac{\alpha_e}{\alpha_U} &= \frac{3}{20}\left(1+\frac{4\alpha_e}{\alpha_3}\right) \\
&\quad +\frac{\alpha_e}{80\pi}\left[b_3C_2+\tfrac{5}{3}b_3C_1-\left(b_2+\tfrac{5}{3}b_1\right)C_3\right] \\
&\quad +\frac{\alpha_e}{2\pi}\sum_i q_i \ln\frac{\widetilde{m}_i}{M_Z} \qquad (8)\\
\sin^2\theta_W &= 0.2+\frac{7\alpha_e}{15\alpha_3} \\
&\quad -\frac{5}{3}\frac{\alpha_e}{80\pi}\left[(b_1-b_2)C_3+(b_3-b_1)C_2+(b_2-b_3)C_1\right] \\
&\quad +\frac{\alpha_e}{20\pi}\sum_i r_i \ln\frac{\widetilde{m}_i}{M_Z} \\
&\quad +\Delta T_H \qquad \text{(heavy thresholds)} \qquad (9)
\end{aligned}
$$

The $p_i, q_i, r_i$ coefficients are given in Table 1. Also,

$$
C_i=\sum_j \frac{b_{ij}}{b_j}\ln\frac{\alpha_j^{-1}(M_Z)}{\alpha_U^{-1}}+\left(\sum_j\frac{b'_{ij}}{b'_j}-\sum_j\frac{b_{ij}}{b_j}\right)\ln\frac{\alpha_j^{-1}(M_Z)}{\alpha_j^{-1}(\widetilde{m})} \tag{10}
$$

with the $b'_i, b'_{ij}$ the one- and two-loop non-supersymmetric RGE coefficients,

$$
b'_i = \left(\tfrac{41}{10}, -\tfrac{19}{6}, -7\right), \tag{11}
$$

$$
b'_{ij} = \begin{pmatrix} \frac{199}{50} & \frac{27}{10} & \frac{44}{5} \\ \frac{9}{10} & \frac{35}{6} & 12 \\ \frac{11}{10} & \frac{9}{2} & -26 \end{pmatrix}. \tag{12}
$$

*Note:* In these equations the non-supersymmetric regime includes only the lighter Higgs doublet. The symbol $\widetilde{m}$ in the definition of the $C_i$ coefficients is an average sparticle mass.

• **Two comments about threshold effects:**

– The sparticles are decoupled in a single-step approximation at a mass scale equal to their physical mass in both the $\overline{MS}$ and $\overline{DR}$ schemes used to treat the light and heavy sectors of the theory respectively [6, 13]. (Exception: in the $\overline{MS}$ scheme the spin-1 particles are decoupled at $e^{-1/21} \approx 0.95$ of their mass.)

– Note in the above equations (7,8,9) that the threshold effects are comparable in size to the two-loop effects.

• **Heavy Thresholds:** If symmetry breaking occurs because of the VEV of the **24** of Higgs, then only three GUT masses are needed to parametrize the relevant effects: (i) the masses of the $X, Y$ gauge bosons $M_V$, (ii) the mass of the adjoint Higgs multiplet $M_\Sigma$, and (iii) the

Table 1: The coefficients which weigh the light threshold corrections to the unification mass ($p_i$), the unified coupling ($q_i$), and $\sin^2\theta_W$ ($r_i$).

| $i$ | $p_i$ | $q_i$ | $r_i$ |
|---|---|---|---|
| $t$ | $\frac{1}{120}$ | $\frac{83}{120}$ | $-3$ |
| $\tilde{w}$ | $\frac{1}{15}$ | $\frac{1}{5}$ | $-\frac{32}{3}$ |
| $\tilde{g}$ | $-\frac{4}{15}$ | $\frac{6}{5}$ | $-\frac{28}{3}$ |
| $\tilde{h}$ | $\frac{1}{15}$ | $\frac{1}{5}$ | $-4$ |
| $H$ | $\frac{1}{60}$ | $\frac{1}{20}$ | $-1$ |
| $\tilde{q}$ | $-\frac{5}{48}$ | $\frac{65}{48}$ | $\frac{5}{2}$ |
| $\tilde{t}_L$ | $\frac{1}{240}$ | $\frac{43}{240}$ | $-\frac{19}{6}$ |
| $\tilde{t}_R$ | $0$ | $\frac{1}{6}$ | $\frac{5}{3}$ |
| $\tilde{l}_L$ | $\frac{1}{20}$ | $\frac{3}{20}$ | $-3$ |
| $\tilde{l}_R$ | $\frac{1}{20}$ | $\frac{3}{20}$ | $2$ |

mass of the color triplet Higgs fields $M_H$ (which mediate proton decay). The contribution to $\sin^2\theta_W$ is given by [6, 16, 9]

$$\Delta T_H = \frac{\alpha_e}{20\pi}\left(-6\ln\frac{M_U}{M_H} + 4\ln\frac{M_U}{M_V} + 2\ln\frac{M_U}{M_\Sigma}\right), \tag{13}$$

where $M_U$ is the largest of the three masses, *i.e.*, a "unification scale" does not exist. Since proton decay requires a large $M_H$, most likely $\Delta T_H > 0$.

## 2.4 Numerical Status

Initially it was thought possible to determine the supersymmetry scale by a "best fit" to unification. Early studies even claimed that the supersymmetric spectrum had to lie in the TeV region to possibly achieve unification. However, it was eventually realized that several uncertainties in the calculations (most notably the heavy GUT thresholds) [16, 11, 7] do not allow to constrain the supersymmetric parameters more than within a few TeV, that is, there is no real constraint on the supersymmetric particle masses from these analyses. Implementing the objective described above, one finds that by varying all parameters in the calculation one obtains $\sin^2\theta_W$ within the experimental range (see *e.g.*, [6, 13]), although certain combinations of the parameters are not allowed. The most up-to-date analysis is by Langacker and Polonsky [13]:

– Experimental data:

$$M_Z = 91.187 \pm 0.007\,\text{GeV} \tag{14}$$

$$\alpha_e^{-1} = 127.9 \pm 0.1 \tag{15}$$

$$\alpha_3(M_Z) = 0.120 \pm 0.010 \tag{16}$$

– Two-parameter fit to all $W^{\pm}$,$Z$, and neutral current data:

$$\sin^2\theta_W = 0.2324 \pm 0.0006 \tag{17}$$
$$m_t = 138^{+20}_{-25} + 5\,\mathrm{GeV} \tag{18}$$

where +5 is due to the supersymmetric Higgs variation ($m_h = 50 - 150\,\mathrm{GeV}$)

– Varying *all* parameters, $SU(5)$ GUT gives[2]

$$\begin{aligned}\sin^2\theta_W = \; & 0.2334 && (m_t = 138, m_h = \widetilde{m} = M_Z)\\ & \pm 0.0025 && (\alpha_e, \alpha_3)\\ & \pm 0.0014 && (\text{light thresholds})\\ & \pm 0.0006 && (m_t, m_h)\\ & {}^{+0.0013}_{-0.0005} && (\text{heavy thresholds})\end{aligned} \tag{19}$$

The individual contributions to each of these effects are shown in Fig. 1 [13]. Since all of the parameters inducing uncertainties in the predicted value of $\sin^2\theta_W$ are independent, contrasting the prediction in Eq. (19) with the experimental determination in Eq. (17), one can rule out some combinations of the various parameters. However, the various uncertainties appear to be too large to make any definite statements. Alternatively, to cirmcumvent the large uncertainty on $\alpha_3$, one can input $\sin^2\theta_W$ and obtain the predicted value of $\alpha_3$

$$\alpha_3(M_Z) = 0.125 \pm 0.001 \pm 0.005 \pm 0.002^{+0.005}_{-0.002} \tag{20}$$

where the first error is now due to $\sin^2\theta_W$ and the others are as above. Here again, it is clear that gauge coupling unification in the minimal $SU(5)$ supergravity model is in very good agreement with low-energy data.

[2]Note: $0.2334 = \underbrace{0.2304}_{\text{one-loop}} + \underbrace{0.0030}_{\text{two-loop}}$, thus threshold effects are comparable to two-loop effects.

Figure 1: Contributions from individual correction terms to the $SU(5)$ GUT prediction for $\sin^2\theta_W$ (from Ref. [13]). Dashed line: error bar on $\sin^2\theta_W$. Dashed-dotted line: uncertainty induced by $\alpha_3$ on the prediction for $\sin^2\theta_W$ Dotted line: two-loop contribution to $\sin^2\theta_W$.

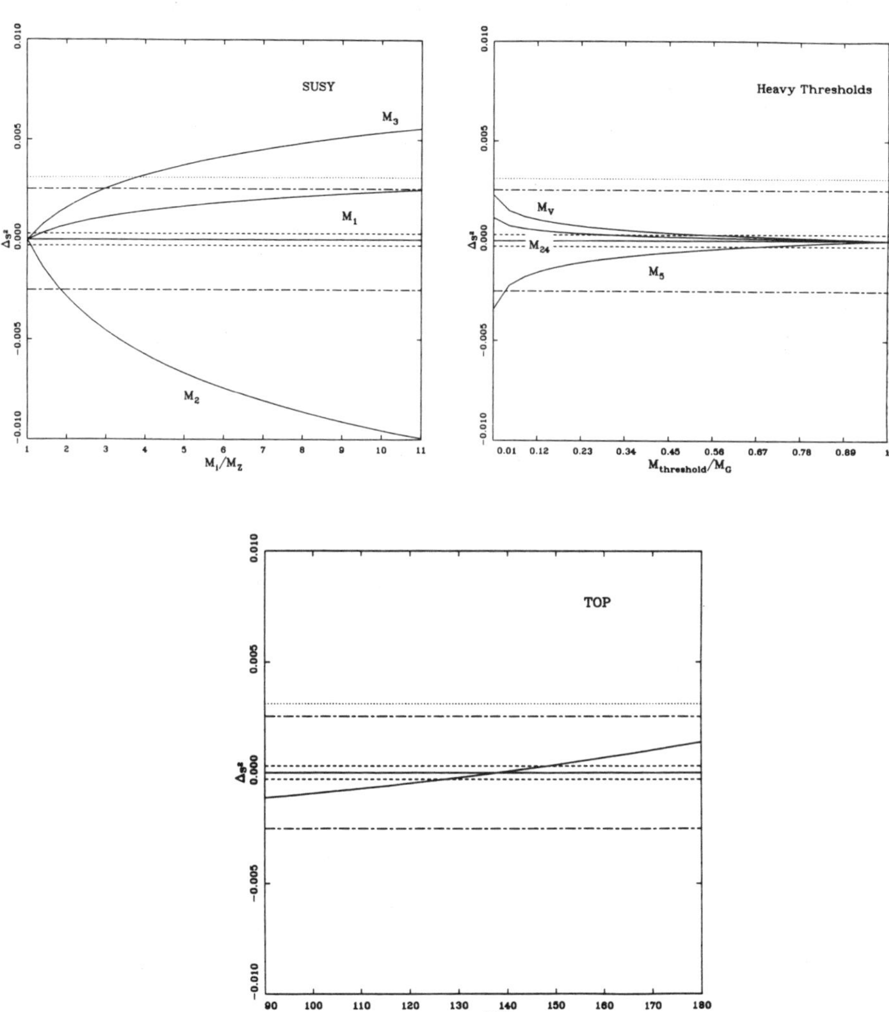

# 3 Yukawa Coupling Unification

We now consider the further constrain where two or three of the third-generation Yukawa couplings are unified at the scale $M_U$. For completeness, first we present the relevant two-loop RGEs (from [15]) which should be used in conjunction with the gauge coupling RGEs in the previous section.

## 3.1 Two-loop RGEs

$$\begin{aligned}\frac{d\lambda_t}{dt} = \frac{\lambda_t}{16\pi^2}\Bigg[\Bigg( & - \sum_i c_i g_i^2 + 6\lambda_t^2 + \lambda_b^2\Bigg) \\ & + \frac{1}{16\pi^2}\Bigg(\sum_i \left(c_i b_i + c_i^2/2\right) g_i^4 + g_1^2 g_2^2 + \frac{136}{45} g_1^2 g_3^2 + 8 g_2^2 g_3^2 \\ & \quad + \lambda_t^2\left(\frac{6}{5} g_1^2 + 6 g_2^2 + 16 g_3^2\right) + \frac{2}{5}\lambda_b^2 g_1^2 \\ & \quad - \left\{22\lambda_t^4 + 5\lambda_t^2\lambda_b^2 + 5\lambda_b^4 + \lambda_b^2\lambda_\tau^2\right\}\Bigg)\Bigg]\end{aligned} \tag{21}$$

$$\begin{aligned}\frac{d\lambda_b}{dt} = \frac{\lambda_b}{16\pi^2}\Bigg[\Bigg( & - \sum_i c_i' g_i^2 + \lambda_t^2 + 6\lambda_b^2 + \lambda_\tau^2\Bigg) \\ & + \frac{1}{16\pi^2}\Bigg(\sum_i \left(c_i' b_i + c_i'^2/2\right) g_i^4 + g_1^2 g_2^2 + \frac{8}{9} g_1^2 g_3^2 + 8 g_2^2 g_3^2 \\ & \quad + \frac{4}{5}\lambda_t^2 g_1^2 + \lambda_b^2\left(\frac{2}{5} g_1^2 + 6 g_2^2 + 16 g_3^2\right) + \frac{6}{5}\lambda_\tau^2 g_1^2 \\ & \quad - \left\{22\lambda_b^4 + 5\lambda_t^2\lambda_b^2 + 3\lambda_b^2\lambda_\tau^2 + 3\lambda_\tau^4 + 5\lambda_t^4\right\}\Bigg)\Bigg]\end{aligned} \tag{22}$$

$$\begin{aligned}\frac{d\lambda_\tau}{dt} = \frac{\lambda_\tau}{16\pi^2}\Bigg[\Bigg( & - \sum_i c_i'' g_i^2 + 3\lambda_b^2 + 4\lambda_\tau^2\Bigg) \\ & + \frac{1}{16\pi^2}\Bigg(\sum_i \left(c_i'' b_i + c_i''^2/2\right) g_i^4 + \frac{9}{5} g_1^2 g_2^2 \\ & \quad + \lambda_b^2\left(-\frac{2}{5} g_1^2 + 16 g_3^2\right) + \lambda_\tau^2\left(\frac{6}{5} g_1^2 + 6 g_2^2\right) \\ & \quad - \left\{3\lambda_t^2\lambda_b^2 + 9\lambda_b^4 + 9\lambda_b^2\lambda_\tau^2 + 10\lambda_\tau^4\right\}\Bigg)\Bigg]\end{aligned} \tag{23}$$

With

$$c_i = \left(\frac{13}{15}, 3, \frac{16}{3}\right), \quad c_i' = \left(\frac{7}{15}, 3, \frac{16}{3}\right), \quad c_i'' = \left(\frac{9}{5}, 3, 0\right). \tag{24}$$

## 3.2 Fixed points

Independently of the GUT relations among the Yukawa couplings, $\lambda_{b,t,\tau}$ must be bounded above at low energies (*i.e.*, $\lambda_{b,t,\tau} \lesssim 1$), otherwise they would blow up before reaching $M_U$ (*i.e.*, a Landau pole is encountered, see *e.g.*, [32]). Using this fact one can obtain an upper bound on the top-quark mass,

$$\begin{aligned} m_t &= \lambda_t v_2 = \lambda_t \frac{v_0}{\sqrt{2}} \sin\beta \\ &\lesssim (174\,\mathrm{GeV})\lambda_t^{max} \frac{1}{\sqrt{1+1/\tan^2\beta}} \\ &\approx 135, 170, 180, 190\,\mathrm{GeV} \qquad (25) \\ \text{for } \tan\beta &= 1, 2, 3, \infty \qquad (26) \end{aligned}$$

The numerical upper bound ($\lambda_t^{max} \approx 1.09$) depends on $\alpha_3$, the light thresholds, etc [15]. These $\tan\beta$-dependent upper bounds are quite relevant nowadays, and could rule out a whole class of supersymmetric unified theories, or more likely, provide (somewhat mild) lower bounds on $\tan\beta$. Analogously, the bottom-quark Yukawa coupling upper bound entails an upper bound on $\tan\beta$,

$$\begin{aligned} m_b(M_Z) &= \frac{1}{\eta_b} m_b(m_b) = \lambda_b \frac{v_0}{\sqrt{2}} \cos\beta \\ &\lesssim (174\,\mathrm{GeV})\lambda_b^{max} \frac{1}{\sqrt{1+\tan^2\beta}} \\ \Rightarrow \tan\beta &\lesssim \frac{190\eta_b}{m_b(m_b)} \approx 50, 55, 58 \qquad (27) \\ \text{for } m_b(m_b) &= 5.0, 4.5, 4.25\,\mathrm{GeV} \qquad (28) \end{aligned}$$

with $\eta_b = m_b(m_b)/m_b(M_Z) \approx 1.3$.

## 3.3 Unification conditions

(A) **SU(5)**: the relation $\lambda_b(M_U) = \lambda_\tau(M_U)$, entails a constraint on the $(m_t, \tan\beta)$ plane for given $m_b, \alpha_3$ [6, 33, 34, 15, 35, 36]. The procedure to determine this constraint is somewhat complicated: (a) for a given $\tan\beta$ and $m_b(m_b)$ (and $m_\tau$) one determines the low-energy values of $\lambda_{b,\tau}$; (b) one runs these Yukawa coupling up to the given value of $m_t$, and determines $\lambda_t(m_t)$; (c) then all three $\lambda_{b,t,\tau}$ are run up to the unification scale and the GUT relation is tested; (d) the given value of $m_t$ is adjusted until the GUT relation is satisfied. The result of the calculations is a set of curves in the $(m_t, \tan\beta)$ plane for fixed values of $m_b(m_b)$ and $\alpha_3$. A sample set of these curves is shown in Fig. 2 (taken from Ref. [15]) and show:

- If $m_b = 4.25 \pm 0.15\,\mathrm{GeV}$ (the shaded areas), then either: (i) $\tan\beta \sim 1$ or $\gtrsim 40$, for a wide range of $m_t$ values, or (ii) $m_t \gtrsim 180\,\mathrm{GeV}$ for a wide range of $\tan\beta$ values.

Figure 2: The constraint on the $(m_t, \tan\beta)$ plane from the SU(5) Yukawa unification condition (from Ref. [15]).

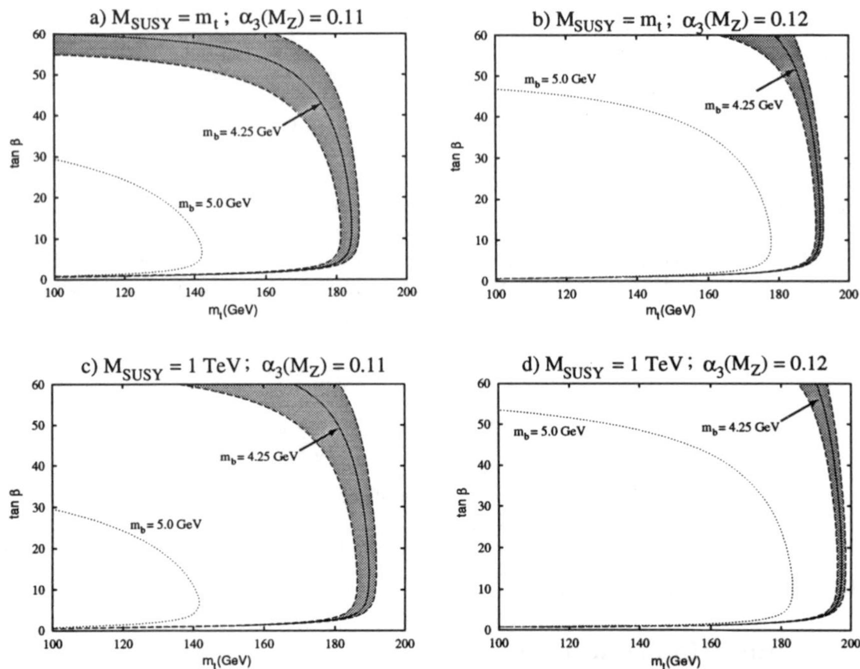

Figure 3: Effect of "threshold corrections" on the SU(5) Yukawa unification condition for $m_b = 4.25\,\mathrm{GeV}$ (from Ref. [35]).

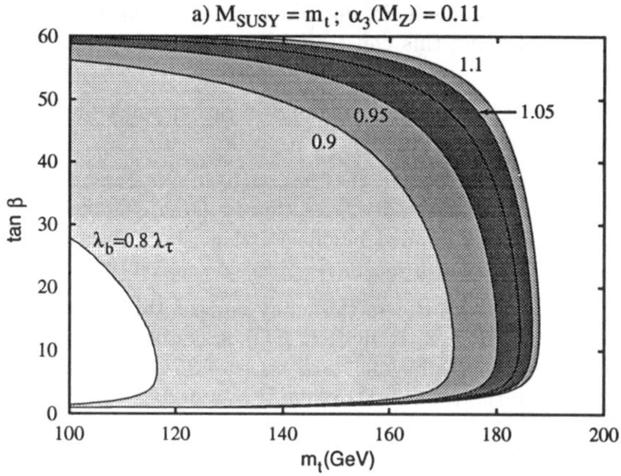


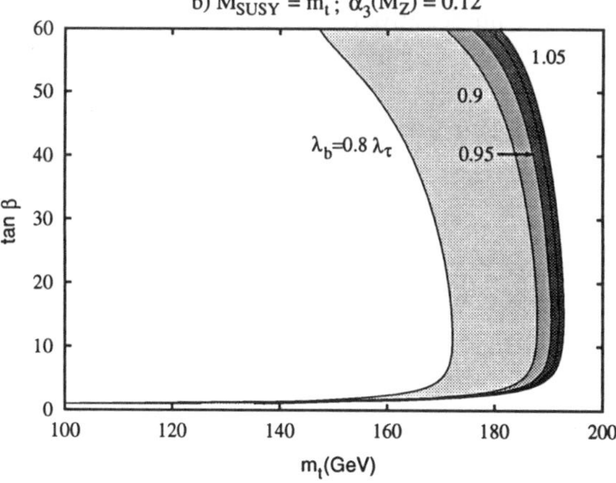

The value of $\alpha_3$ has a moderate effect in this case. If $\alpha_3 \gtrsim 0.12$ then the second possibility (*i.e.*, $\tan\beta \gtrsim 40$) is eliminated. It is important to note that these predictions have been obtained without enforcing the gauge coupling unification constraint to a high degree of precision, that is, the points in the shaded areas do not necessarily give a value of $\sin^2\theta_W$ which is consistent with the experimentally allowed range. Enforcing this constraint more precisely leads to much narrower allowed bands [36].

- If $m_b \approx 5\,\mathrm{GeV}$ then the constraint on $\tan\beta$ is rather weak and the value of $\alpha_3$ is quite relevant to the results [6, 33].
- With what confidence can we say that $m_b \approx 5\,\mathrm{GeV}$ is excluded? The early discussions on this matter did not settle this issue satisfactorily; subsequent discussions were effectively quelled by the Particle Data Group published value of $m_b = 4.25 \pm 0.15\,\mathrm{GeV}$. For a recent reappraisal see Ref. [37].
- What about GUT threshold effects that may correct this relation somewhat? Allowing $\lambda_b < \lambda_\tau$ (see Fig. 3, from Ref. [35]) is equivalent to increasing $m_b$ and viceversa. For example, solutions with $m_b \approx 4.25\,\mathrm{GeV}$ would lead to much relaxed constraints in the $(m_t, \tan\beta)$ plane if $\lambda_b \approx 0.8\lambda_\tau$.

(B) **SO(10)**: the relation $\lambda_b(M_U) = \lambda_\tau(M_U) = \lambda_t(M_U)$ is obtained in the simplest $SO(10)$ GUT models and determines $m_t$ and $\tan\beta$ for given $m_b, \alpha_3$ (*i.e.*, one point on the $SU(5)$ curves in Fig. 2). The procedure is similar to that described above for $SU(5)$ but the additional relation allows $\tan\beta$ to be adjusted also. Because of this delicate tuning, the results are quite sensitive to the various correction factors, but generally give $m_t \gtrsim 160\,\mathrm{GeV}$ and $\tan\beta \gtrsim 40$ (see *e.g.*, [38, 3, 33, 15, 37]).

# 4 GUT Properties

## 4.1 SU(5)

- GUT superpotential [39]:

$$W_G = \lambda_1(\tfrac{1}{3}\Sigma^3 + \tfrac{1}{2}M\Sigma^2) + \lambda_2 H(\Sigma + 3M')\bar{H} \tag{29}$$

where $\Sigma$ is the **24** of Higgs whose VEV $\langle\Sigma\rangle = M\,\mathrm{diag}(2,2,2,-3,-3)$ breaks $SU(5)$ down to $SU(3)\times SU(2)\times U(1)$, and $H = \{H_2, H_3\}$ is the Higgs pentaplet.

- Doublet-triplet (2/3) splitting: The choice $M = M'$ makes the triplet $H_3$ heavy, while keeping the doublet $H_2$ light. This fine-tuning is avoided by the "missing partner mechanism" where the **75** breaks the GUT symmetry (instead of the **24**) and the couplings $\mathbf{50}\cdot\mathbf{75}\cdot h$, $\overline{\mathbf{50}}\cdot\mathbf{75}\cdot\bar{h}$ effect the 2/3 splitting. These representations ($\mathbf{75}$,$\mathbf{50}$,$\overline{\mathbf{50}}$) have seldom been considered in heavy threshold analyses [28]. For other methods to solve this problem see Ref. [40].

- GUT fields: The heavy GUT fields, their transformation properties under $SU(3) \times SU(2)$, their mass, and their usual notation are given below

| Field | $SU(3) \times SU(2)$ | Mass | Name |
|---|---|---|---|
| $H_3, \bar{H}_3$ | $(3,1), (\bar{3},1)$ | $5\lambda_2 M$ | $M_H$ |
| $\Sigma^8$ | $(8,1)$ | $\frac{5}{2}\lambda_1 M$ | $M_\Sigma$ |
| $\Sigma^3$ | $(1,3)$ | $\frac{5}{2}\lambda_1 M$ | $M_\Sigma$ |
| $\Sigma^0$ | $(1,1)$ | $\frac{1}{2}\lambda_1 M$ | (SM singlet) |
| $X, Y$ | $(3,2), (\bar{3},2)$ | $5gM$ | $M_V$ |

One can see that only three mass parameters $(M_H, M_\Sigma, M_V)$ are needed to describe the heavy GUT fields, and thus the heavy threshold effects.

- Proton Decay:

*Dimension-six*: is mediated by the $X, Y$ gauge bosons. The largest mode is $p \to e^+\pi^0$, which if dominant would give $\tau_p \sim 3.3 \times 10^{35}(M_U/10^{16})^4\,\text{y}$, thus it is basically unobservable. However, the experimental bound on this mode implies $M_U \gtrsim 10^{15}\,\text{GeV}$.

*Dimension-five*: is mediated by the Higgs triplet fields $H_3, \bar{H}_3$ through the couplings $\lambda_u 10_f \cdot 10_f \cdot 5_h \supset Q \cdot Q \cdot H_3$ and $\lambda_d 10_f \cdot \bar{5}_f \cdot \bar{5}_h \supset Q \cdot L \cdot \bar{H}_3$ and the mixing term $\sim M_U H_3 \bar{H}_3$. This operator needs to be "dressed" by a chargino loop, and the largest contribution comes from CKM mixing with the second generation. The largest mode is $p \to \bar{\nu}_{\mu,\tau} K^+$ and a schematic expression for this "partial lifetime" is given by [19]

$$\tau(p \to \bar{\nu}_{\mu,\tau} K^+) \sim \left| M_H \sin 2\beta \frac{1}{f} \frac{1}{1+y^{tK}} \right|^2 . \tag{30}$$

In this expression: (i) $M_H$ is the Higgs triplet mass: large $M_H$ makes the lifetime longer; (ii) $\sin 2\beta = 2\tan\beta/(1+\tan\beta)$: small $\tan\beta$ needed to keep lifetime long enough; (iii) $f$: one-loop dressing function which goes as $f \sim m_{\chi_1^\pm}/m_{\tilde{q}}^2$: heavy squarks and light charginos are preferred; (iv) $1 + y^{tK}$: ratio of third-to-second generation contribution to dressing. Strong constraints on the parameter space of the minimal $SU(5)$ supergravity model follow [21, 22, 23]. The experimental constraint on the proton lifetime can be evaded by either: (a) increasing the Higgs triplet mass, which is taken to be $M_H < (3-10)M_U$ in these analyses (higher values produce too large heavy threshold corrections to $\sin^2\theta_W$); (b) relaxing the naturalness constraint of $m_{\tilde{q},\tilde{g}} < 1\,\text{TeV}$. It must be emphasized that the values calculated this way are *upper* bounds (since one uses an upper bound on $M_H$) and could well be much larger if $M_H \approx M_U$ ($M_H \gtrsim M_U$ is required). In any event, the next generation of proton decay experiments (SuperKamiokande and Icarus) should carve out a large fraction of the remaining parameter space in this model.

## 4.2 SU(5)xU(1)

- GUT superpotential [41]:

$$W_G = \lambda_4 HHh + \lambda_5 \bar{H}\bar{H}\bar{h} + \lambda_6 F\bar{H}\phi + \mu h\bar{h} \tag{31}$$

where $H = \{Q_H, d^c_H, \nu^c_H\}$, $\bar{H} = \{Q_{\bar{H}}, d^c_{\bar{H}}, \nu^c_{\bar{H}}\}$ are $SU(5)$ decaplets, $h = \{H_2, H_3\}, \bar{h}$ are $SU(5)$ pentaplets, $\phi$ in an $SU(5)$ singlet, and the matter fields are in $F = \{Q, d^c, \nu^c\}$, $\bar{f} = \{L, u^c\}$, and $l = e^c$. The vevs $\langle \nu^c_{H,\bar{H}} \rangle$ break $SU(5) \times U(1)$ down to $SU(3) \times SU(2) \times U(1)$. This property, *i.e.*, no need for adjoint representations to break the GUT symmetry, is central to the appeal of flipped $SU(5)$ as a string-derived model.

- Doublet-triplet (2/3) splitting: no need for additional representations

$$\left.\begin{array}{ll} H \cdot H \cdot h & \supset \langle \nu^c_H \rangle d^c_H H_3 \\ \bar{H} \cdot \bar{H} \cdot \bar{h} & \supset \langle \nu^c_{\bar{H}} \rangle d^c_{\bar{H}} \bar{H}_3 \end{array}\right\} \begin{array}{l} H_3, \bar{H}_3 \text{ heavy} \\ H_2, \bar{H}_2 \text{ light} \end{array}$$

- Proton decay:

*Dimension-six*: as in the minimal $SU(5)$ case.

*Dimension-five*: are suppressed relative to the $SU(5)$ case by $\sim (M_Z/M_U)^2 \lesssim 10^{-28}$, *i.e.*, negligible. Reason: $H_3, \bar{H}_3$ mixing is $\sim \mu \sim M_Z$, as opposed to $\sim M_U$ in $SU(5)$, while the Higgs triplet mass is also $\sim M_U$.

- Neutrino masses: the singlet field $\phi$ enlarges the usual $2 \times 2$ see-saw matrix to $3 \times 3$. This mechanism is essential for the consistency of the model, since otherwise the neutrinos would acquire large masses

$$\left.\begin{array}{lll} & F \cdot \bar{H} \cdot \phi & \to \langle \nu^c_{\bar{H}} \rangle \nu^c \phi \\ \lambda_u & F \cdot \bar{f} \cdot \bar{h} & \to m_u \nu \nu^c \end{array}\right\} M_\nu = \begin{array}{c} \\ \nu \\ \nu^c \\ \phi \end{array} \begin{array}{c} \begin{array}{ccc} \nu & \nu^c & \phi \end{array} \\ \begin{pmatrix} 0 & m_u & 0 \\ m_u & 0 & M_U \\ 0 & M_U & - \end{pmatrix} \end{array}$$

The see-saw mechanism then gives $m_{\nu_{e,\mu,\tau}} \sim m^2_{u,c,t}/M^2_U$. Incorporating all the appropriate renormalization group factors, it has been shown that the $\nu_e, \nu_\mu$ sector could reproduce the needed MSW effect in the solar neutrino flux [42], $\nu_\tau$ could be a hot dark matter candidate [43], and $\nu_\mu - \nu_\tau$ oscillations could be observed at forthcoming experiments [42]. Moreover, the out-of-equilibrium decays of the "flipped neutrinos" ($\nu^c$) could generate a lepton asymmetry, which would later be processed into a baryon asymmetry by electroweak non-perturbative interactions [44].

# 5 Soft Supersymmetry Breaking

Spontaneous breaking of supergravity (*e.g.*, induced dynamically by gaugino condensation in the hidden sector) results in a global supersymmetric theory *plus* a set of calculable soft supersymmetry breaking terms. Here "soft" mean operators of dimension $\leq 3$ which do not regenerate the quadratic divergences which supersymmetry avoids de facto. There are three classes of such soft-supersymmetry-breaking terms:

• **Gaugino masses** (parameters = 3)

– $M_3, M_2, M_1$ for the three $SU(3)_C, SU(2)_L, U(1)_Y$ gauginos, respectively.

– $SU(5)$ symmetry implies $M_3 = M_2 = M_1$.

– Universal soft-supersymmetry breaking decrees: $M_3 = M_2 = M_1 = m_{1/2}$ at $M_U$

– This relation is almost universally adopted in low-energy supersymmetric phenomenological studies (for exceptions see Ref. [45]). For recent studies of two-loop effects on the running of the $M_i$ see Ref. [46].

• **Scalar masses** (parameters = $5 \times 3 + 2 = 17$)

– $(\tilde{Q}, \tilde{U}^c, \tilde{D}^c, \tilde{L}, \tilde{E}^c)_i$, $i = 1, 2, 3$ squarks and sleptons; $H_1, H_2$ Higgs bosons.

– Universality decrees: $m_{(\tilde{Q},\tilde{U}^c,\tilde{D}^c,\tilde{L},\tilde{E}^c)_{1,2,3}} = m_{H_{1,2}} = m_0$ at $M_U$.

– Departures from universality are strongly constrained by flavor-changing-neutral-current (FCNC) processes in the $K - \bar{K}$ system (most notably the CP-violating $\epsilon$ parameter) [47].

– Such departures are generic in string-inspired supergravities [48, 49].

• **Scalar couplings** (parameters = $3 + 1 = 4$)

– To each superpotential coupling there corresponds one scalar coupling:

$$\begin{array}{ll} \lambda_t Q t^c H_2 & \to \lambda_t A_t \tilde{Q} \tilde{t}^c H_2 \\ \lambda_b Q b^c H_1 & \to \lambda_b A_b \tilde{Q} \tilde{b}^c H_1 \\ \lambda_\tau L \tau^c H_1 & \to \lambda_\tau A_\tau \tilde{L} \tilde{\tau}^c H_1 \\ \mu H_1 H_2 & \to \mu B H_1 H_2 \end{array}$$

– Universality decrees: $A_t = A_b = A_\tau = A$ at $M_U$.

- **Some particular (string-inspired) soft-supersymmetry-breaking scenaria**

  (a) No-Scale: $m_0 = A = 0$ [50];

  (b) Strict No-scale: $m_0 = A = 0$ and $B(M_U) = 0$;

  (c) Dilaton: $m_0 = \frac{1}{\sqrt{3}} m_{1/2}$, $A = -m_{1/2}$ [51];

  (d) Special Dilaton: $m_0 = \frac{1}{\sqrt{3}} m_{1/2}$, $A = -m_{1/2}$ and $B(M_U) = \frac{2}{\sqrt{3}} m_{1/2}$;

  (e) Moduli: $m_0 \sim m_{3/2}$, $m_{1/2} \sim (\alpha/4\pi) m_{3/2}$ (*i.e.*, $m_{1/2}/m_0 \ll 1$) [48, 49].

All the soft-supersymmetry-breaking parameters, and the gauge and Yukawa couplings evolve to low energies as prescribed by the appropriate set of coupled RGEs. It is important to note that the values of $\mu$ and $B$ do not feed into the other RGEs, and therefore they need not be specified at high energies.

- **Parameter count at low energies:**

| Parameter | MSSM | SUGRA | |
|---|---|---|---|
| $M_1, M_2, M_3$ | 3 | 1 | $(m_{1/2})$ |
| $(\tilde{Q}, \tilde{U}^c, \tilde{D}^c, \tilde{L}, \tilde{E}^c)_i$ | 15 | 1 | $(m_0)$ |
| $\tilde{H}_1, \tilde{H}_2$ | 2 | 0 | $(m_0)$ |
| $A_t, A_b, A_\tau$ | 3 | 1 | $(A)$ |
| $B$ | 1 | 1 | (determined by radiative |
| $\mu$ | 1 | 1 | electroweak breaking) |
| $\lambda_{b,t,\tau}, \tan\beta$ | 2 | 2 | |
| | — | — | |
| Total : | 27 | 7 | |

The two minimization conditions of the electroweak scalar potential (to be discussed in the next section) impose two additional constraints which can be used to determine $\mu, B$ (at low energies) and thus reduce the parameter count down to **5** (versus 25 in the MSSM): $m_t, \tan\beta, m_{1/2}, m_0, A$. In the particular scenarios mentioned above, the parameters are just three $(m_t, \tan\beta, m_{1/2})$, or even two $(m_t, m_{1/2})$ in scenarios (b) and (d).

# 6 Radiative Electroweak Breaking

We now discuss the mechanism by which the electroweak symmetry is broken. In supergravity theories this occurs via radiative corrections in the presence of soft-supersymmetry-breaking masses [52]. This mechanism connects in a nontrivial way various aspects of these theories, such as the physics at the high and low scales, the breaking of supersymmetry, and the value of the top-quark mass. In the MSSM electroweak symmetry breaking is put in by hand and the top-quark mass plays no special role.

## 6.1 Tree-level minimization

The tree-level Higgs potential is given by

$$\begin{aligned} V_0 &= (m_{H_1}^2+\mu^2)|H_1|^2+(m_{H_2}^2+\mu^2)|H_2|^2+B\mu(H_1H_2+\text{h.c.}) \\ &\quad +\tfrac{1}{8}g_2^2(H_2^\dagger\sigma H_2+H_1^\dagger\sigma H_1)^2+\tfrac{1}{8}g'^2\left(|H_2|^2-|H_1|^2\right)^2, \end{aligned} \tag{32}$$

where $H_1\equiv\binom{H_1^0}{H_1^-}$ and $H_2\equiv\binom{H_2^+}{H_2^0}$ are the two complex Higgs doublet fields. Assuming that only the neutral components get vevs, the expression for $V_0$ simplifies to

$$V_0=(m_{H_1}^2+\mu^2)h_1^2+(m_{H_2}^2+\mu^2)h_2^2+2B\mu h_1h_2+\tfrac{1}{8}(g_2^2+g'^2)(h_2^2-h_1^2)^2, \tag{33}$$

where $h_i=\text{Re}\,H_{1,2}^0$. One can then write down the minimization conditions $\partial V_0/\partial h_i=0$ and obtain

$$\mu^2 = \frac{m_{H_1}^2-m_{H_2}^2\tan^2\beta}{\tan^2\beta-1}-\tfrac{1}{2}M_Z^2, \tag{34}$$

$$B\mu = -\tfrac{1}{2}\sin 2\beta(m_{H_1}^2+m_{H_2}^2+2\mu^2). \tag{35}$$

The solutions to these equations will be physically sensible only if they reflect a minimum away from the origin

$$\mathcal{S}=(m_{H_1}^2+\mu^2)(m_{H_2}^2+\mu^2)-B^2\mu^2<0 \tag{36}$$

of a potential bounded from below

$$\mathcal{B}=m_{H_1}^2+m_{H_2}^2+2\mu^2+2B\mu>0. \tag{37}$$

Taking the second derivatives $\partial^2V_0/\partial h_i\partial h_j$ one can determine the tree-level masses of the five physical Higgs bosons: the CP-even states $h,H$, the CP-odd state $A$, and the charged Higgs boson $H^\pm$. An important result is that the lightest Higgs boson mass is bounded above $m_h\le|\cos 2\beta|M_Z$, although one-loop corrections relax this constraint considerably. Also, $m_h<m_A$, $m_H>M_Z$, and $m_{H^\pm}>M_W$, which may also be affected by radiative corrections.

## 6.2 One-loop minimization

The minimization of the RGE-improved tree-level Higgs potential described above suffers from a scale-dependence problem. That is, the physical output obtained depends considerably on the scale one chooses to perform the minimization, *i.e.*, the scale at which the RGEs are stopped. This problem is most simply stated as

$$\frac{dV_0}{d\ln Q} \neq 0, \tag{38}$$

that is, the tree-level Higgs potential does not satisfy the renormalization group equation. In practice, as the scale is lowered one typically has the sequence of events pictured below

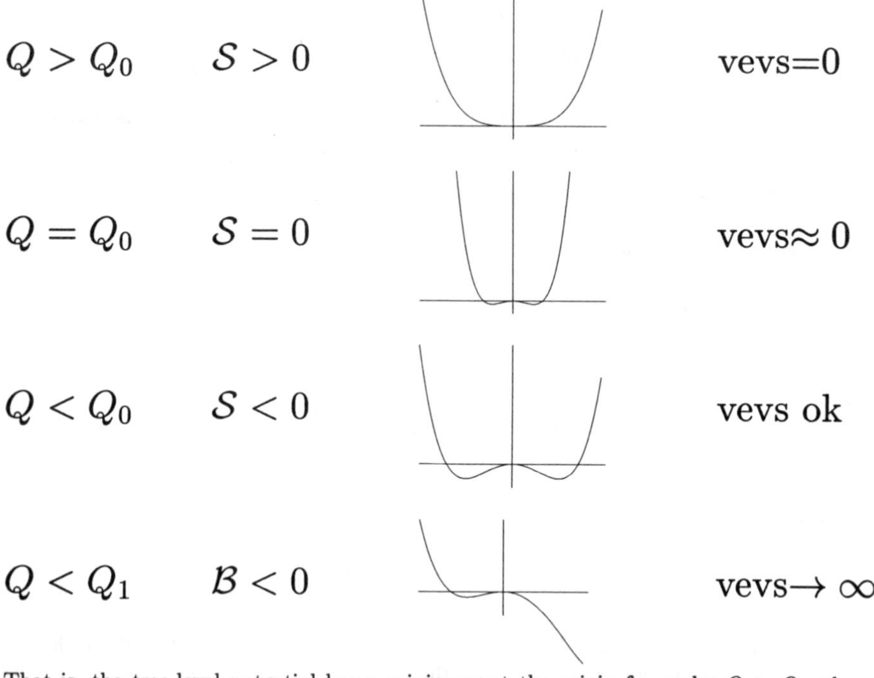


That is, the tree-level potential has a minimum at the origin for scales $Q > Q_0$, then it develops at minimum away from the origin ($\mathcal{S} < 0$) with vevs which grow to be such that $M_Z^2 = \frac{1}{2}(g^2 + g'^2)(v_1^2 + v_2^2)$. However, for scales $Q < Q_1$ this minimum becomes unbounded from below ($\mathcal{B} < 0$) and the vevs run away to infinity. Thus, the vevs vary a lot for scales $Q \lesssim 1\,\text{TeV}$ [53], as shown schematically in the following figure:

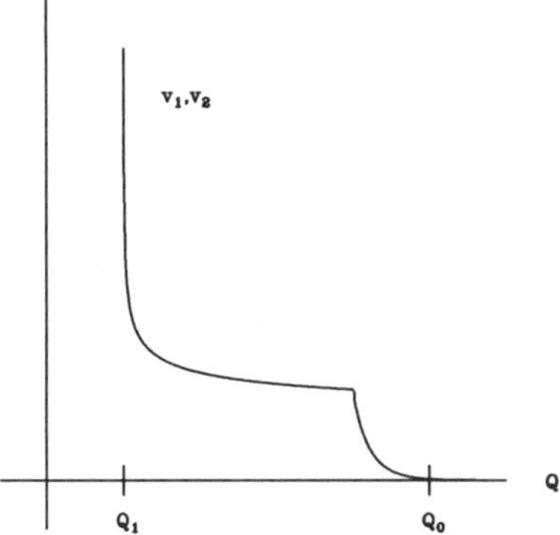


The solution to this problem is to use the one-loop effective potential $V_1 = V_0 + \Delta V$, which satisfies $\frac{dV_1}{d\ln Q} = 0$ (to one-loop order), where

$$\Delta V = \tfrac{1}{64\pi^2} \mathrm{Str}\, \mathcal{M}^4 \left( \ln \frac{\mathcal{M}^2}{Q^2} - \tfrac{3}{2} \right) \tag{39}$$

with $\mathrm{Str}\mathcal{M}^2 = \sum_j (-1)^{2j}(2j+1)\mathrm{Tr}\mathcal{M}_j^2$. Following this procedure the vevs are $Q$-independent (up to two-loop effects) in the range of interest ($\lesssim 1\,\mathrm{TeV}$) [53]. However, one must perform the minimization numerically (a non-trivial task) and *all* the spectrum enters into $\Delta V$ (although $\tilde{t}, \tilde{b}$ are the dominant contributions) (see *e.g.*, [54]). This procedure also gives automatically the *one-loop corrected* Higgs boson masses (taking second derivatives of $V_1$).

## 6.3 Radiative Symmetry Breaking

To grasp the concept most easily, let us consider the simple (although unrealistic) case of $\mu = 0$. Let us also not worry about the one-loop correction to the Higgs potential. Neither of these simplifications will affect the physical mechanism which we want to illustrate. In this case the "stability" condition becomes

$$\mathcal{S} \to m_{H_1}^2 \cdot m_{H_2}^2 < 0. \tag{40}$$

This means that one must arrange that one $m_{H_i}^2 < 0$ somehow. No help is available from low-energy physics inputs alone (*i.e.*, this is put in by hand in the MSSM). To proceed, consider RGEs for the (first- and second-generation) scalar masses schematically (setting $\lambda_b = \lambda_\tau = 0$)

$$\frac{d\widetilde{m}^2}{dt} = \frac{1}{(4\pi)^2} \left\{ -\sum_i c_i g_i^2 M_i^2 + c_t \lambda_t^2 \left( \sum_i \widetilde{m}_i^2 \right) \right\}, \tag{41}$$

where the various coefficients are given below

| | $c_t$ | $c_3$ | $c_2$ |
|---|---|---|---|
| $H_1$ | 0 | 0 | 6 |
| $H_2$ | 6 | 0 | 6 |
| $\widetilde{Q}$ | 0 | $\frac{32}{3}$ | 6 |
| $\widetilde{U}^c$ | 0 | $\frac{32}{3}$ | 0 |
| $\widetilde{D}^c$ | 0 | $\frac{32}{3}$ | 0 |
| $\widetilde{L}$ | 0 | 0 | 6 |
| $\widetilde{E}^c$ | 0 | 0 | 0 |

The result of running these RGEs is illustrated below for the indicated values of the parameters. Note that $m_{H_2}^2 < 0$ while $m_{H_1}^2 > 0$ for $Q < Q_0$. Since one can show that $\mu$ is small in

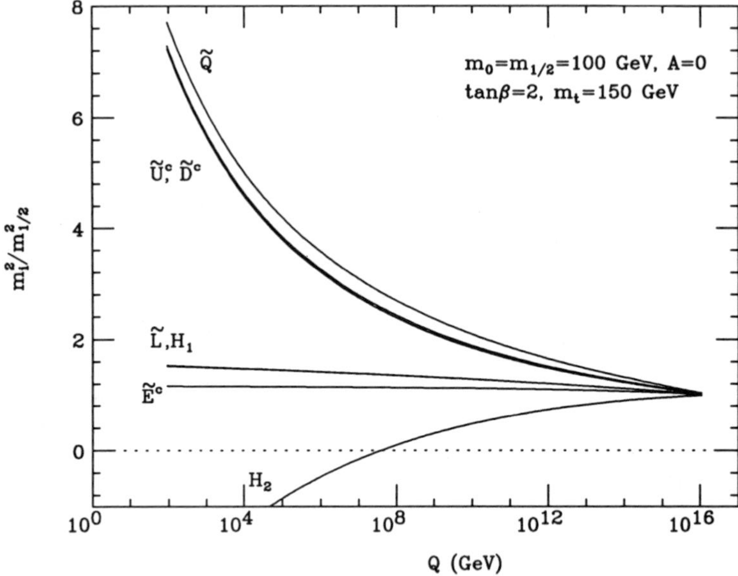


this case ($\mu \approx 30\,\text{GeV}$), this implies that $\mathcal{S} < 0$, *i.e.*, that a minimum away from the origin has developed. If $\mu$ were not small, radiative breaking will require a large enough negative value for $m_{H_2}^2$ (in order to possibly get $\mathcal{S} < 0$, see Eqs. (34,36)). This will generally be the case in models where $m_0 \ll m_{1/2}$. The top-quark Yukawa coupling ($\lambda_t$) plays a *fundamental* role in driving $m_{H_2}^2$ to negative values. However, this is only possible if it is large enough to counteract the effect of the gauge couplings. This is why early on people said that this mechanism required a "heavy top quark". Nowadays, any allowed top-quark mass is heavy enough. Note also that $m^2_{\tilde{Q},\tilde{U}^c,\tilde{D}^c} > 0$ because of the large $\alpha_3$ contribution to their running. For the same reason the sleptons $(\tilde{L}, \tilde{E}^c)$ renormalize much less.

## 6.4 Constraints on the $(m_t, \tan\beta)$ plane

As we have seen above, in supersymmetric unified models with radiative electroweak symmetry breaking, there are only five parameters: three soft-supersymmetry-breaking parameters $(m_{1/2}, m_0, A)$ and $m_t, \tan\beta$. It turns out that the two-dimensional area spanned by the last two parameters is completely bounded [55, 54] using the tree-level or one-loop minimization procedures discussed above. This is not the case for any other pair of variables. The shape of the resulting boundaries depends on the values of the soft-supersymmetry-breaking parameters, the sign of $\mu$, and the phenomenomenological constraints which are imposed on the spectrum.

First let us study the case of *no phenomenological constraints.* In the figures below, solid (dashed) lines denote the allowed boundaries obtained by enforcing radiative breaking using the tree-level (one-loop) Higgs potential. We have taken $m_0 = A = 0$ (the no-scale scenario) and $m_{1/2} = 150, 250\,\mathrm{GeV}$.

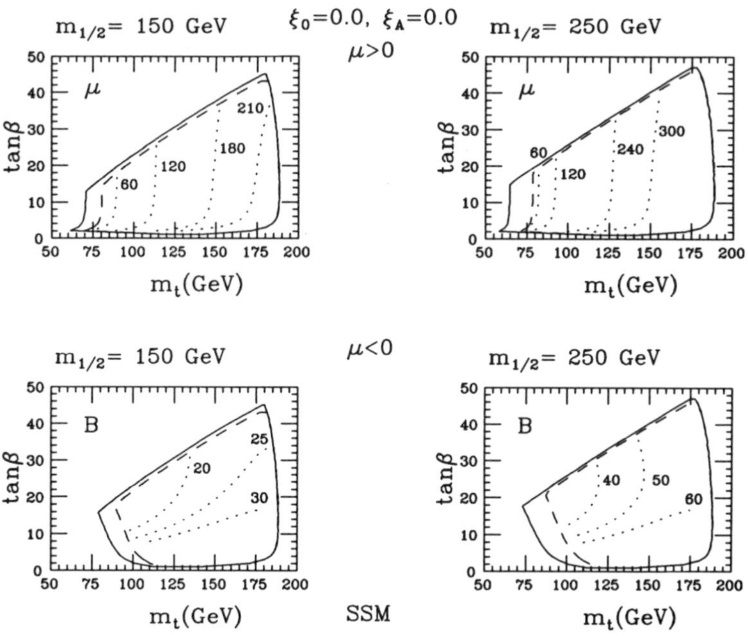

The bounded region is seen to have five boundaries:

(i) Top boundary: this is the slanted line which limits the magnitude of $\tan\beta$. Along this line the CP-odd Higgs boson mass vanishes ($m_A = 0$) and $\mathcal{B} = 0$, *i.e.*, the potential becomes unbounded for points above this line. Also, on this boundary the relation $\lambda_b \approx \lambda_t$ holds, which implies that $\tan\beta \approx m_t/m_b(M_Z) \approx m_t/3.77$.

(ii) Upper corner: the largest allowed value of $\tan\beta$ is determined by the fixed point in $\lambda_b$ (as discussed in Sec. 3.2); this value is $m_b(m_b)$ dependent. The rounded portion at the top and towards the larger values of $m_t$ results from the strengthening of the fixed point bound because of the non-zero top-quark Yukawa coupling.

(iii) Right boundary: this is determined by the fixed point in $\lambda_t$ and entails a $\tan\beta$-dependent upper bound on $m_t$, which is quite restrictive for low $\tan\beta$ (see Sec. 3.2).

(iv) Bottom boundary: here $\tan\beta > 1$ which is a direct and important consequence of the radiative breaking mechanism. This lower bound is routinely assumed in phenomenological analyses and has no other known explanation.

(v) Left boundary: on this line $\mu$ vanishes, *i.e.*, to the left of the line $\mu^2 < 0$. Moreover, the one-loop procedure yields the largest correction to the tree-level result precisely on this boundary. For example, for $\mu > 0$ and $m_{1/2} = 250\,\mathrm{GeV}$, on the one-loop left boundary $\mu_{loop} = 0$, whereas $\mu_{tree} \approx 60\,\mathrm{GeV}$.

For comparison, we now consider the case of $m_0 = m_{1/2}$ and $A = 0$, to study the effect of the soft-supersymmetry-breaking parameters on the shape of the boundary. (Still no phenomenological constraints have been applied.) The thing to note are the larger $\mu$ values which are generated by the larger values of $m^2_{H_{1,2}}$ in the expression for $\mu^2$. The position of the left boundary has also shifted (to the right).

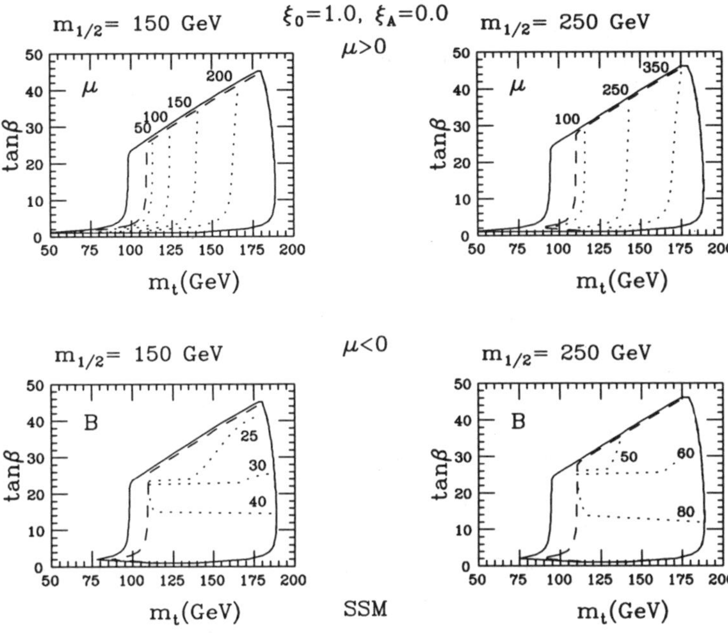

Now let us include the various *phenomenological cuts*: (i) the lightest supersymmetric particle (LSP) should be neutral, in fact, one demands that it be the lightest neutralino; (ii) the chargino mass is bounded below by LEP, $m_{\chi_1^\pm} > 45\,\text{GeV}$; (iii) and so are the slepton masses, $m_{\tilde{l}} > 45\,\text{GeV}$; (iv) the neutralino contributions to $\Gamma_Z$ should be small enough; (v) the lightest Higgs boson mass is also bounded below by LEP, $m_h > 43\,\text{GeV}$.

The resulting allowed region in the $(m_t, \tan\beta)$ plane for the case $m_0 = A = 0$ is shown below. The most notable change is the more restricted range of $\tan\beta$ values which are allowed: the maximum allowed $\tan\beta$ is decreased and for $m_t \gtrsim 100\,\text{GeV}$ this value decreases with $m_t$. It is interesting to note how large a constraint it is to have a lower bound on the top-quark mass which is ever increasing ($m_t \gtrsim 110\,\text{GeV}$ at present). In this figure one can also see the effect of $A$ (dotted line: $A = m_{1/2}$; dashed line: $A = -m_{1/2}$).

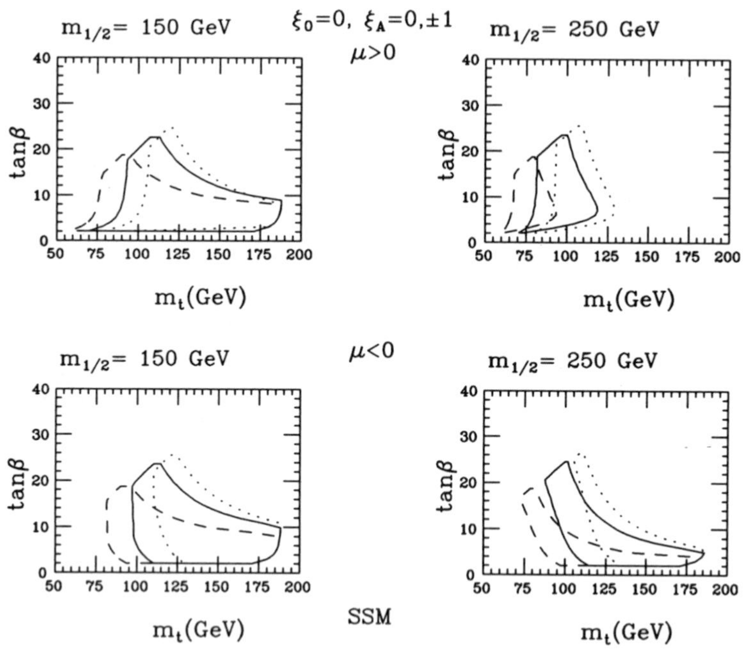

To conclude we show below the case of $m_0 = m_{1/2}$ when the phenomenological constraints are imposed. Note how much weaker these cuts become. This is simply because the spectrum is heavier than in the previous (no-scale) case.

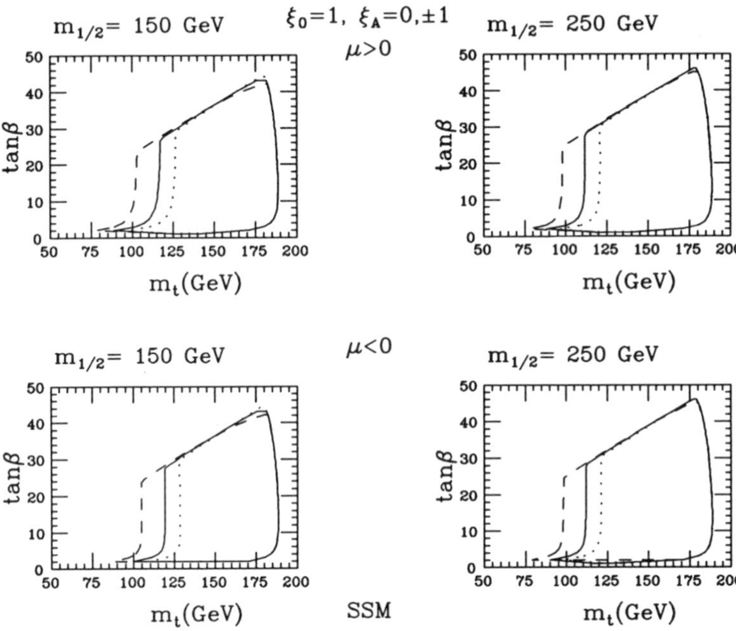


Without further assumptions about the soft-supersymmetry-breaking parameters (such as those mentioned in Sec. 5) or more phenomenological constraints (*e.g.*, proton decay), the five-dimensional parameter space is still rather vast. However, this situation changes drastically in specific models, such as the no-scale [56] and dilaton [57] flipped $SU(5)$ supergravity models, where $m_0$ and $A$ are functions of $m_{1/2} \propto m_{\tilde{g}}$. Furthermore, more restricted versions of each of these models allow to determine $\tan\beta$ as a function of $m_t$ and $m_{1/2}$ and a two-dimensional parameter space results. In this case, the allowed areas in $(m_t, \tan\beta)$ space shown above degenerate into a line on those figures. Such two- or three-parameter models have been shown to be quite predictive, and the expectations for collider processes at the Tevatron [58], LEP [59, 60], and HERA [61] have been explicitly calculated, as well as indirect probes through one-loop precision electroweak corrections at LEP [62], the FCNC $b \to s\gamma$ rare decay at CLEO [63], the anomalous magnetic moment of the muon [64], and also the prospects for cosmological dark matter [65, 56, 57], and indirect neutralino detection at neutrino telescopes [66]. Some of these calculations have also been performed in the minimal $SU(5)$ supergravity model, where the constraints from proton decay are so powerful that the general three-dimensional soft-supersymmetry-breaking parameter space becomes quite manageable [21, 22, 23, 25, 26]. For recent reviews of this line of research see *e.g.*, Ref. [67, 68].

# 7 Conclusions

In this lecture I have shown that supersymmetric unified theories with radiative electroweak symmetry breaking are highly predictive models of low-energy supersymmetry, once all necessary elements are incorporated. As such they stand in sharp contrast with the minimal supersymmetric standard model, whose "minimality" in the field content is all but washed out by the large number of unknown parameters needed to describe it. I would also like to emphasize that one must consider all aspects of any such unified models before being able to judge their experimental viability. It is not unusual for various sectors of the theory to complement each other as far as constraints on the parameter space are concerned. Finally, building on the structure I have described, it is possible to construct well motivated (and more constrained) models which can be used to calculate all sparticle and Higgs boson masses, as well as all conceivable observables of interest at present and future experimental facilities. These computations and their intricate correlations should play a very important role in accumulating direct and indirect evidence for the supersymmetric model which is actually realized in nature.

# Acknowledgements

I would like to thank the Ettore Majorana Center for Scientific Culture and Professor Zichichi for their kind invitation to deliver this Lecture. This work has been supported by an SSC Fellowship.

# References

[1] J. Ellis, S. Kelley, and D. V. Nanopoulos, Phys. Lett. B **249** (1990) 441.

[2] P. Langacker and M.-X. Luo, Phys. Rev. D **44** (1991) 817; U. Amaldi, W. de Boer, and H. Fürstenau, Phys. Lett. B **260** (1991) 447; L. Clavelli, Phys. Rev. D **45** (1992) 3276.

[3] H. Arason, *et. al.*, Phys. Rev. Lett. **67** (1991) 2933, Phys. Rev. D **46** (1992) 3945, Phys. Rev. D **47** (1992) 232.

[4] F. Anselmo, L. Cifarelli, A. Peterman, and A. Zichichi, Nuovo Cim. **104A** (1991) 1817 and **105A** (1992) 1025.

[5] J. Ellis, S. Kelley, and D. V. Nanopoulos, Phys. Lett. B **260** (1991) 131.

[6] J. Ellis, S. Kelley and D. V. Nanopoulos, Nucl. Phys. B **373** (1992) 55.

[7] J. Ellis, S. Kelley and D. V. Nanopoulos, Phys. Lett. B **287** (1992) 95.

[8] G. Ross and R. Roberts, Nucl. Phys. B **377** (1992) 571.

[9] J. Hisano, H. Murayama, and T. Yanagida, Phys. Rev. Lett. **69** (1992) 1014 and Nucl. Phys. B **402** (1993) 46.

[10] F. Anselmo, L. Cifarelli, A. Peterman, and A. Zichichi, Nuovo Cim. **105A** (1992) 581; F. Anselmo, L. Cifarelli, and A. Zichichi, Nuovo Cim. **105A** (1992) 1357.

[11] F. Anselmo, L. Cifarelli, A. Peterman, and A. Zichichi, Nuovo Cim. **105A** (1992) 1201.

[12] F. Anselmo, L. Cifarelli, A. Peterman, and A. Zichichi, Nuovo Cimento **105A** (1992) 1179; F. Anselmo, L. Cifarelli, and A. Zichichi, Nuovo Cimento **105A** (1992) 1335.

[13] P. Langacker and N. Polonsky, Phys. Rev. D **47** (1993) 4028.

[14] M. Carena, S. Pokorski, and C. Wagner, MPI-Ph/93-10.

[15] V. Barger, M. Berger, and P. Ohman, Phys. Rev. D **47** (1993) 1093.

[16] R. Barbieri and L. Hall, Phys. Rev. Lett. **68** (1992) 752; A. Faraggi, B. Grinstein, and S. Meshkov, Phys. Rev. D **47** (1993) 5018.

[17] L. Hall and U. Sarid, Phys. Rev. Lett. **70** (1993) 2673.

[18] J. Ellis, D. V. Nanopoulos, and S. Rudaz, Nucl. Phys. B **202** (1982) 43; B. Campbell, J. Ellis, and D. V. Nanopoulos, Phys. Lett. B **141** (1984) 229; K. Enqvist, A. Masiero, and D. V. Nanopoulos, Phys. Lett. B **156** (1985) 209.

[19] P. Nath, A. Chamseddine, and R. Arnowitt, Phys. Rev. D **32** (1985) 2348; P. Nath and R. Arnowitt, Phys. Rev. D **38** (1988) 1479.

[20] M. Matsumoto, J. Arafune, H. Tanaka, and K. Shiraishi, Phys. Rev. D **46** (1992) 3966.

[21] R. Arnowitt and P. Nath, Phys. Rev. Lett. **69** (1992) 725; P. Nath and R. Arnowitt, Phys. Lett. B **287** (1992) 89.

[22] J. L. Lopez, D. V. Nanopoulos, and H. Pois, Phys. Rev. D **47** (1993) 2468.

[23] J. L. Lopez, D. V. Nanopoulos, H. Pois, and A. Zichichi, Phys. Lett. B **299** (1993) 262.

[24] J. L. Lopez, D. V. Nanopoulos, and A. Zichichi, Phys. Lett. B **291** (1992) 255.

[25] R. Arnowitt and P. Nath, Phys. Lett. B **299** (1993) 58 and **307** (1993) 403(E); P. Nath and R. Arnowitt, Phys. Rev. Lett. **70** (1993) 3696.

[26] J. L. Lopez, D. V. Nanopoulos, and K. Yuan, Texas A & M University preprint CTP-TAMU-14/93 (to appear in Phys. Rev. D, Sep. 15).

[27] A. Masiero, D. V. Nanopoulos, K. Tamvakis, and T. Yanagida, Phys. Lett. B **115** (1982) 380; B. Grinstein, Nucl. Phys. B **206** (1982) 387.

[28] K. Hagiwara and Y. Yamada, Phys. Rev. Lett. **70** (1993) 709; B. Brahmachani, P. Patra, U. Sarkar, and K. Sridhar, Mod. Phys. Lett. A **8** (1993) 1487.

[29] See *e.g.*, J. L. Lopez and D. V. Nanopoulos, Phys. Lett. B **251** (1990) 73 and Phys. Lett. B **268** (1991) 359; A. Faraggi, Nucl. Phys. B **387** (1992) 239 and Nucl. Phys. B **403** (1993) 101.

[30] I. Antoniadis, J. Ellis, S. Kelley, and D. V. Nanopoulos, Phys. Lett. B **272** (1991) 31; D. Bailin and A. Love, Phys. Lett. B **280** (1992) 26.

[31] S. Kelley, J. L. Lopez, and D. V. Nanopoulos, Phys. Lett. B **278** (1992) 140.

[32] L. Durand and J. L. Lopez, Phys. Lett. B **217** (1989) 463, Phys. Rev. D **40** (1989) 207.

[33] S. Kelley, J. L. Lopez, and D. V. Nanopoulos, Phys. Lett. B **274** (1992) 387.

[34] S. Dimopoulos, L. Hall, and S. Raby, Phys. Rev. Lett. **68** (1992) 1984, Phys. Rev. D **45** (1992) 4192; G. Anderson, S. Dimopoulos, L. Hall, and S. Raby, Phys. Rev. D **47** (1992) R3702.

[35] V. Barger, M. Berger, and P. Ohmann, MAD/PH/758 (1993).

[36] P. Langacker and N. Polonsky, UPR-0556T (May 1993).

[37] L. Hall, R. Ratazzi, and U. Sarid, LBL-33997 (June 1993).

[38] B. Ananthanarayan, G. Lazarides, and Q. Shafi, Phys. Rev. D **44** (1991) 1613.

[39] For reviews see *e.g.*, R. Arnowitt and P. Nath, *Applied N=1 Supergravity* (World Scientific, Singapore 1983); H. P. Nilles, Phys. Rep. **110** (1984) 1.

[40] A. Anselm and A. Johansen, Phys. Lett. B **200** (1988) 331; G. Dvali, Phys. Lett. B **287** (1992) 101; R. Barbieri, G. Dvali, and A. Strumia, Nucl. Phys. B **391** (1993) 487; R. Barbieri, G. Dvali, and M. Moretti, Phys. Lett. B **312** (1993) 137.

[41] For a recent review see J. L. Lopez, D. V. Nanopoulos, and A. Zichichi, CERN-TH.6926/93 and Texas A & M University preprint CTP-TAMU-33/93.

[42] J. Ellis, J. L. Lopez, and D. V. Nanopoulos, Phys. Lett. B **292** (1992) 189.

[43] J. Ellis, J. L. Lopez, D. V. Nanopoulos, and K. Olive, Phys. Lett. B **308** (1993) 70.

[44] J. Ellis, D. V. Nanopoulos, and K. Olive, Phys. Lett. B **300** (1993) 121.

[45] A. Bartl, H. Fraas, W. Majerotto, and N. Oshimo, Phys. Rev. D **40** (1989) 1594; M. Drees and X. Tata, Phys. Rev. D **43** (1991) 2971; K. Griest and L. Roszkowski, Phys. Rev. D **46** (1992) 3309.

[46] Y. Yamada, KEK-TH-363,371 (1993); S. Martin and M. Vaughn, NUB-3072-93TH.

[47] J. Ellis and D. V. Nanopoulos, Phys. Lett. B **110** (1982) 44. For a recent reappraisal see J. Hagelin, S. Kelley, and T. Tanaka, MIU-THP-92-59 (Sep. 1992).

[48] L. Ibáñez and D. Lüst, Nucl. Phys. B **382** (1992) 305.

[49] B. de Carlos, J. Casas, and C. Muñoz, Nucl. Phys. B **399** (1993) 623 and Phys. Lett. B **299** (1993) 234; A. Brignole, L. Ibáñez, and C. Muñoz, FTUAM-26/93.

[50] For a review see A. B. Lahanas and D. V. Nanopoulos, Phys. Rep. **145** (1987) 1.

[51] V. Kaplunovsky and J. Louis, Phys. Lett. B **306** (1993) 269.

[52] L. Ibáñez and G. Ross, Phys. Lett. B **110** (1982) 215; K. Inoue, *et. al.*, Prog. Theor. Phys. 68 (1982) 927; L. Ibáñez, Nucl. Phys. B **218** (1983) 514 and Phys. Lett. B **118** (1982) 73; H. P. Nilles, Nucl. Phys. B **217** (1983) 366; J. Ellis, D. V. Nanopoulos, and K. Tamvakis, Phys. Lett. B **121** (1983) 123; J. Ellis, J. Hagelin, D. V. Nanopoulos, and K. Tamvakis, Phys. Lett. B **125** (1983) 275; L. Alvarez-Gaumé, J. Polchinski, and M. Wise, Nucl. Phys. B **221** (1983) 495; L. Ibañéz and C. López, Phys. Lett. B **126** (1983) 54 and Nucl. Phys. B **233** (1984) 545; C. Kounnas, A. Lahanas, D. V. Nanopoulos, and M. Quirós, Phys. Lett. B **132** (1983) 95 and C. Kounnas, A. Lahanas, D. V. Nanopoulos, and M. Quirós, Nucl. Phys. B **236** (1984) 438.

[53] G. Gamberini, G. Ridolfi, and F. Zwirner, Nucl. Phys. B **331** (1990) 331.

[54] S. Kelley, J. L. Lopez, D. V. Nanopoulos, H. Pois, and K. Yuan, Nucl. Phys. B **398** (1993) 3.

[55] S. Kelley, J. L. Lopez, D. V. Nanopoulos, H. Pois, and K. Yuan, Phys. Lett. B **273** (1991) 423.

[56] J. L. Lopez, D. V. Nanopoulos, and A. Zichichi, CERN-TH.6667/92, Texas A & M University preprint CTP-TAMU-68/92.

[57] J. L. Lopez, D. V. Nanopoulos, and A. Zichichi, CERN-TH.6903/93, Texas A & M University preprint CTP-TAMU-31/93.

[58] J. L. Lopez, D. V. Nanopoulos, X. Wang, and A. Zichichi, Phys. Rev. D **48** (1993) 2062.

[59] J. L. Lopez, D. V. Nanopoulos, H. Pois, X. Wang, and A. Zichichi, Phys. Lett. B **306** (1993) 73.

[60] J. L. Lopez, D. V. Nanopoulos, H. Pois, X. Wang, and A. Zichichi, Texas A & M University preprint CTP-TAMU-89/92 and CERN-PPE/93-16 (to appear in Phys. Rev. D).

[61] J. L. Lopez, D. V. Nanopoulos, X. Wang, and A. Zichichi, Texas A & M University preprint CTP-TAMU-15/93 and CERN-PPE/93-64 (to appear in Phys. Rev. D).

[62] J. L. Lopez, D. V. Nanopoulos, G. Park, H. Pois, and K. Yuan, Texas A & M University preprint CTP-TAMU-19/93 (to appear in Phys. Rev. D, Oct. 1).

[63] J. L. Lopez, D. V. Nanopoulos, and G. Park, Phys. Rev. D **48** (1993) R974; J. L. Lopez, D. V. Nanopoulos, G. Park, and A. Zichichi, Texas A & M University preprint CTP-TAMU-40/93.

[64] J. L. Lopez, D. V. Nanopoulos, and X. Wang, Texas A & M University preprint CTP-TAMU-44/93.

[65] S. Kelley, J. L. Lopez, D. V. Nanopoulos, H. Pois, and K. Yuan, Phys. Rev. D **47** (1993) 2461.

[66] R. Gandhi, J. L. Lopez, D. V. Nanopoulos, K. Yuan, and A. Zichichi, Texas A & M University preprint CTP-TAMU-48/93.

[67] A. Zichichi, these proceedings; J. L. Lopez, D. V. Nanopoulos, and A. Zichichi, CERN-TH.6934/93, Texas A & M University preprint CTP-TAMU-34/93.

[68] R. Arnowitt, these proceedings; R. Arnowitt and P. Nath, Texas A & M University preprint CTP-TAMU-23/93, NUB-TH-3062/92.

## CHAIRMAN: J.LOPEZ

*Scientific Secretaries: H. Petrache, Z. Trocsanyi*

## DISCUSSION

– *Giannakis:*

a) During your presentation this morning you compared the MSSM based on global SUSY with a GUT model based on local SUSY. Are there models with gauge group $SU(3) \times SU(2) \times U(1)$ and local SUSY? Is the number of parameters also constrained in this case?

b) Also in the case of $SU(5) \times U(1)$ you seem to lose the explanation of charge quantisation. Do you need to embed it in a larger group in order to recover this feature?

– *Lopez:*

a) There are of course models based on $SU(3) \times SU(2) \times U(1)$ which are supergravity models. The choice of a gauge group did not play much of a role in the counting of parameters that I have. There are three supersymmetry breaking parameters that are always there no matter what you choose for a gauge group. If you choose SU(5) then $m_t$ and $\tan\beta$ will be related so you reduce one parameter. If you choose $SO(10)$ then you know what $m_t$ and $\tan\beta$ are, and we have even one parameter less. What I want to stress is that in the counting of parameters I did not use the GUT symmetry so the five parameters also apply to $SU(3) \times SU(2) \times U(1)$

Why do I like to examine a GUT model as opposed to a standard-like model? There are some properties of GUT models that were invented for a reason e.g. for proton decay (of course the $SU(3) \times SU(2) \times U(1)$ can also have proton decay unless you put additional symmetry). GUTs are also used to explain the origin of neutrino masses or to explain the origin of baryon number in the universe. So there are several motivations for considering GUT models. I do not like non-GUT models particularly but people have looked at $SU(3) \times SU(2) \times U(1)$ or $SU(3)^3$ models and things like that.

b) As to the second part of the question, indeed it is true that in $SU(5) \times U(1)$ the charge quantisation relation that you have in $SU(5)$ is in principle lost. If the model is derivable from string theory then we would not care if it could or could not explain charge quantisation. But that particular feature of traditional Grand Unified models is not there. But there are other things, e.g. the $\lambda_B = \lambda_\tau$ relation that you have in $SU(5)$ which are also, in principle, not present in $SU(5) \times U(1)$ .

However, in string models for no apparent reason, you find many times that those Yukawa couplings are the same. There are all kinds of things like that which have no apparent explanation but they occur.

– *Beneke:*

You repeatedly emphasized that you can calculate everything in terms of the five parameters. Can you predict the spectrum of the light fermion masses?

– *Lopez:*

Perhaps I didn't tell the whole truth. I have been talking all along about the third generation of quarks and leptons. If you input the value of the $b$ quark mass and the value of the $\tau$ mass, then these five parameters follow. So they are not a prediction in this sense. As to the lighter quark masses, more complicated GUT models based on, for example, $SO(10)$ have been recently investigated by e.g. Dinopoulos et al. and they have found some ways of predicting new things as opposed to postdictions. But it's true you have to do something to account for quark masses and KM mixing angles beyond what I have described.

– *Langacker:*

I have two comments. First, the relation between the top Yukawa to the $b$ and $\tau$ Yukawas is very specific to only the simplest $SO(10)$ model which has only a simple 10-plet. Secondly, you showed a plot showing that only limited values of $\tan\beta$ versus $m_t$ are allowed if 2-Yukawa unification ($\lambda_b = \lambda_\tau$) is assumed. However, those calculations ignored the correlation between $\alpha, \alpha_s, \sin^2\theta_w$, and $m_t$ in models with gauge unification (which is likely to hold if there is Yukawa unification). If one includes that correlation the allowed $\tan\beta - m_t$ region is even smaller.

– *Lopez:*

That is true. In fact, I think that the way it was presented you can put in $\alpha_3$ and get out $\sin^2\theta_w$.

– *Langacker:*

In that case if you turn it around that way most of the points in those bands would get the incorrect value of $\sin^2\theta_w$. In your language that you require that $\sin^2\theta_w$ comes out correctly you get a much smaller band.

– *Lopez:*

That is correct.

– *Vassilevskaya:*

It is known that Standard Theory isn't able to predict the existence of several fermion generations. How can this problem be solved in the framework of Supersymmetry theory?

– *Lopez:*

I don't think that Supersymmetry sheds any light on the problem of why there are three generations and not more. This question is at the moment unsolved. There are some unification models in which things are so constrained that you can fit three generations inside a big representation. There are also some models in which the number of generations is restricted because of gauge symmetry e.g. because of anomalies, because of the way you can place particle representations etc. I do not think that any of these explanations is particularly compelling. The answer is that I do not know how Supersymmetry or Unification can help in showing that there are a certain number of generations.

– *Vassilevskaya:*

At present the problem of the 17 KeV neutrinos is being widely discussed. What is your opinion about the possibility of the existence of such a heavy neutrino in the framework of Supersymmetry?

– *Lopez:*

It is true that the 17 KeV neutrino has received a lot of attention. The last I heard at the Dallas Conference last Summer was that the signals they were seeing were probably not true or correct somehow. Therefore, I would say that there is no 17 KeV neutrino. But if there were a 17 KeV neutrino, there would be a lot of problems arising in cosmology and astrophysics. People spent a lot of time building models to somehow get around these problems. There are all kinds of complications because, for example, heavy particles have to decay very quickly or very late in the history of the universe. The simple answer would be that I do not think there is a 17 KeV neutrino but people did worry a lot about what if there were one. I remember that half-way through that epidemic I did a search and there were some 52 papers written on the subject. I think that some conceivable models were found but they were very complicated

– *Lahanas:*

A particular class of models mentioned in your lecture you termed No–Scale. Is there a particular reason for that or is it just a definition?

– *Lopez:*

To some extent the No–Scale dogma has two parts. One of them is an assumption that the hidden sector has a particular structure, like $SU(1,1)$ or $SU(N,1)$. If that is the case then there is a $SU(N,1)$ scenario. When you have a Unified theory with heavy particles then it follows that $m_0$ and $A$ will be zero. So when I said No–Scale I really meant $SU(N,1)$. Now, the second part of the No–Scale problem is that this particular structure of the hidden sector does not determine the value of the Supersymmetry breaking parameters. So back in 1984, you co-authored a paper in which there was a way in which the light supersymmetry breaking scale could be determined through radiative corrections introducing one–loop effective potential. That part of the problem we have not pursued. So it is No–Scale as it was meant originally. So, if you go back in the literature before this second part of the program was implemented it was already called No–Scale.

*–Lahanas:*

I am worried particularly because the gaugino mass for instance will hopefully be determined dynamically sooner or later. In addition you have the new parameter which mixes the two Higgs so in your models, especially $SU(5) \times U(1)$ you have additional scales below the GUT scale.

*–Lopez:*

Yes, in the $SU(5) \times U(1)$ model there are additional particles which introduce intermediate scales. The idea is that if you succeed in constructing string models of that kind, those mass scales will be calculable, not determined dynamically, but perhaps also determined indirectly. We have a string inspired version of that model in which we have written down mass matrices for all these particles and there are specific formulae that you get for the masses of these particles. For example, it could be the vacuum expectation value of some particular Higgs field. So in principle all these things are calculable. I would definitely make a distinction between the determination of those masses and the determination of the Supersymmetry scale.

*– Anselm:*

There is a problem that you have mentioned in your lecture: how to get triplet–doublet splitting of the Higgs. There is one possibility of solving this problem without fine tuning. The idea is to make the doublet be a Goldstone at the GUT scale. To achieve this one can enlarge the gauge $SU(5)$ symmetry to be $SU(6)$ global symmetry of the superpotential (but of the whole theory). The 24–plet and 5–plet of $SU(5)$ Higgs enter then into adjoint representation of $SU(6)$. It is easy to see that if this $SU(6)$ symmetry is broken down to $SU(4) \times SU(2) \times U(1)$ additional Goldstone bosons, not eaten up by Higgs mechanism, appear and they are exactly what is necessary: the normal Salaam–Weinberg doublet. A very economical $SU(5)$ theory would eventually appear with many parameters fixed, as compared to the general case.

*– Lopez:*

I am aware of your paper. You have an additional $SU(6)$ global symmetry in your theory. All I have to say about global symmetries is that in this string model normally global symmetries do not occur.

– *Zichichi:*

I have a remark to make in connection with what Professor Lahanas has said. If I understood his question he said that he is worried about whether the introduction of the extra masses breaks the No–Scale theorem. Is this your question?

– *Lahanas:*

No, I just want to understand what the definition of a No–Scale model is. A No–Scale Model should have no arbitrary scale in it. All scales have to be generated dynamically.

– *Lopez:*

That is right. But the definition in the original paper was that it was just a hidden sector structure $SU(1,1)$ or $SU(N,1)$. That was the definition of No-Scale. But this is a matter of semantics. I think we have an explanation from where the other scales can come.

– *Zichichi:*

But there is a nice feature of $SU(5) \times U(1)$ in that it is derivable from strings. No–Scale supergravity is the infrared solution of Superstring. This is Witten's theorem, Therefore, you should not worry.

–*Lahanas:*

Yes, but in Witten's approach there was no Supersymmetry breaking in the effective Lagrangian. Actually, you have to assume a non–perturbative mechanism to break it.

– *Lopez:*

But these are two different things. One is that you can have No–Scale with or without strings. I was talking about No–Scale in the context of GUTs. My lecture had very little contact with string unification. As Professor Kaplanovsky explained, the whole subject is very tricky. You have to start from the top and go down. No scale existed before strings and what we have done here, I think, follows the spirit of the original No–Scale.

–*Arnowitt:*

Is it possible to construct a flipped $O(10)$ model which would allow grand unification at $M_G$ instead of $M_p$ as in flipped $SU(5)$?

– *Lopez:*

I remember there was a paper by Barr in 1989 where he looked at the generalisations of flipped $SU(5)$ applied to bigger things, possibly $SO(10)$. I do not recall how the $SO(10)$ symmetry would be broken. If it is possible to break it with a non-adjoint represation, then these models would be of some interest. But what was your interest in such a model?

– *Arnowitt:*

You made the interesting remark that if you had $SU(5) \times U(1)$ and you let the coupling constants unify at the GUT scale they would begin to move apart if you go towards the Planck scale. Therefore, you do not have true unification because it is a product group. If, however, you had a $SO(10)$ you could have an alternative possibility which in fact would be like ordinary $SO(5)$ or $O(10)$.

–*Lopez:*

There was a remark in small print that I actually did not mention. What I really meant is that real string unification was necessary in order to make sense of $SU(5) \times U(1)$ as a unified group. You are saying that if you had $SO(10) \times U(1)$ that would work better. But what happens to the other $U(1)$? By flipped $SO(10)$ do you mean $SO(10) \times U(1)$?

–*Arnowitt:*

I meant $SO(10)$ with a flipped particle assignment.

– *Lopez:*

Then I would be doubtful. The breaking occurs because of simple gauge group properties. If you had $SO(10)$ that would require the standard representations to break it, not something that would be compatible with strings I would think. I do not know really.

– *Wegrzyn:*

a) You mentioned bounds on dark matter as a positive feature of flipped $SU(5)$ model. Do these bounds differ much from usual $SU(5)$ model predictions?

b) If you want to solve RGE for all relevant couplings and masses you need some boundary or initial conditions. Could you briefly explain what is assumed?

– *Lopez:*

a) The predictions for relic density of the neutralino are different in $SU(5)$ versus $SU(5) \times U(1)$. In $SU(5)$ the proton decay constraint is very stringent and requires particular values of the masses e.g. the squark and slepton masses should be heavy relative to the chargino masses. Because of these correlations it happens that the dark matter density is most of the time very large, when proton decay is satisfied, so that is the prediction for $SU(5)$. In the case of $SU(5) \times U(1)$ with the No–Scale scenario, the predictions are numbers which are always below 1 or 1/4. You do not want to get numbers bigger than 1, so the predictions are quite different. The real reason why they are different is not because of $SU(5)$ or $SU(5) \times U(1)$ but because of the supersymmetry breaking parameters that you have. Maybe you are referring to the $\nu_\tau$ mass, because this is about the cold dark matter. Hot dark matter would be provided by neutrinos which are in the range of a few eV. There again $SU(5)$ does not accommodate neutrino masses whereas in $SU(5) \times U(1)$ you can have neutrino masses and in fact you can have $\nu_\tau$ in that range (a few up to 10–20 eV) which could give a significant contribution to cold dark matter in the universe which as of today seems to be necessary to fit cosmological observations.

b) All these equations start running at the unification scale. At that scale you have to put in initial values of the parameters. What I have described is that when you have universal soft supersymmetry breaking all the scalar masses get a value of $m_0$, all the gaugino masses get a value of $m_{1/2}$ and all the scalar couplings get a value of $\lambda$. So those are the initial conditions at that high scale.

There are gauge coupling RGEs which start at the same value of the unified coupling and then they split. You also need initial conditions for the Yukawa couplings. In that case if you have $SU(5)$ for instance, you pick two to be the same $b$ and $\tau$ and to know the third one, the top, you have to do something indirect. Sometimes you have to start from low energies and run up and find what the value is, then start again. It is a tricky game, but not difficult to understand.

– *Arnowitt:*

I wanted to make a comment concerning the relic density in regular $SU(5)$. As Jorge commented, the parameter space is very large (5 dimensional) and in the regular $SU(5)$ one can accommodate a relic density that is $\Omega h^2 < 1$ over still a fairly sizeable amount of the parameter space. One is more predictive when one includes the relic density, one lowers the value of the Higgs mass and the top mass, for example, a little bit. Then it is still possible to accommodate both proton decay and relic densities.

– *Lopez:*

This is true.

– *Hoang:*

a) You mentioned that at the RG evolution of the gauge couplings in SUSY GUTS the point where additional particles come into the game are generally not exactly at the mass value of these particles. Can you explain this more precisely?

b) You told us that 1–loop corrections to the SUSY Higgs potential have to be taken into account. Now, it's well known that the one–loop corrected Higgs mass relations are extremely important for phenomenology because the light Higgs can be much heavier than the $Z$. Can the two–loop corrections be of similar importance?

– *Lopez:*

a) How to incorporate threshold effects in the running of the RGEs is a technical issue. It depends on the renormalisation scheme that you are using. It turns out low energy physics is best described using the $\overline{MS}$ scheme whereas very high energy physics is usually described in terms of the $\overline{DR}$ scheme. The real threshold is a smooth threshold while in calculations you introduce a discontinuity. The question is : what is the proper point to introduce the discontinuity in order to reproduce the same end points that the smooth threshold gives? This point will be exactly equal to the mass of the particle in the $\overline{DR}$ scheme. On the $\overline{MS}$ scheme the point will be the mass of the particle for scalars and fermions but for vector bosons it is $e^{-1/21} \approx 0.95$. So it is a matching condition that you have to work out.

b) Presumably not. The lightest Higgs mass at tree level can take values from 0 up to 90 GeV, and the really large corrections occur when you start with a tree level value very close to zero. When $\tan\beta = 1$ the tree level Higgs mass is 0. The one–loop correction, let us say 20 GeV, is a huge correction relatively speaking. As you asymptote to larger values of $\tan\beta$, then the tree level value does not change that much when you put in the one–loop correction. If you, in addition, put in a two–loop correction, I think the effect will be there, probably most importantly for close to $\tan\beta = 1$ when you have zero initially. But since you have already moved away from zero so much it probably never matters too much. These calculations have been done and the conclusion was that two–loop corrections are not nearly as important as one–loop. Whether or not two–loop corrections to the effective potential should be included is also a question. Maybe or maybe not. The fact is that there is no such formula. Nobody has ever calculated the two–loop correction to the effective potential. So even if they were important we would not know.

– *Beneke:*

You mentioned the Landau pole constraints on $m_t$ and $m_b$, in particular in relation to the possibility of ruling out the model if the top is found to be heavier than $\approx 180$ GeV. Now, the existence of the Landau pole in the RG flow is a non–perturbative issue. Does the closeness of the pole in the two–loop approximation to the realistically expected mass for $m_t$ imply that one can no longer trust the two–loop RG equation?

– *Lopez:*

You can do the analyses that I described. You start increasing the value of the top Yukawa at low energies and then you see what happens at high energies. Then you can say that the maximum value of the top Yukawa coupling is the one that will give infinity right at the unification scale and you get some bound say 1.11. If you say instead, I will be more conservative and won't let it blow up, it cannot be as big as 5. In fact, in 1988 I did a calculation of what would be the tree level unitarity constraint that you have to impose at the unification scale on this Yukawa coupling in order not to let it get too big, not even non–perturbative. The result is that the low energy upper bound goes down from 1.11 to 1.09, which is a very small difference, basically negligible. The upper bound on the top quark mass, 190 GeV that I talked about, changes very slightly. In fact it is a lot more dependent on what you put in for the strong coupling. So this kind of question is probably not important—the result is very robust in that sense.

– *Kawall:*

't Hooft and Polyakov showed that magnetic monopoles are a rigorous prediction of a larger gauge group breaking into $SU(2) \times U(1)$. Presumably the density of magnetic moments is dependent on one's cosmological model of the early universe. So far no monopole candidates have been corroborated. Do you take this failure very seriously and do you see monopole searches as a sensitive way of deciding between models?

– *Lopez:*

I do not know. One has to distinguish between cosmological and laboratory constraints. So far we have not studied those sectors of the theory. I do not even know that any of these gauge groups actually contain magnetic monopoles. So I do not know what to tell you.

– *Langacker:*

Could I just comment on that? The monopole problem first surfaced many years ago and that was what motivated Alan Guth to invent inflation in the first place and then it solved many other problems. If there was any period of inflation subsequent to any GUT period then there would be no monopoles left. So there is a sort of easy way out, you just would not expect them.

– *J. Lopez:*

Thanks.

# WHERE CAN SUSY BE?

Antonino Zichichi

CERN

Geneva, Switzerland

## 1) Introduction

The purpose of this lecture is to propose a first attempt to discriminate between Field Theory and String Theory. To do this I would like to choose two examples of Supergravity models, $SU(5)$ and $SU(5) \times U(1)$. Why are they interesting? Because $SU(5)$ cannot easily be derived from String Theory whereas $SU(5) \times U(1)$ has already been derived from it. I would like to first point out that String Theory is a mathematical structure and not a model and that it attempts to describe all possible phenomena existing in Subnuclear Physics. It has many appealing features, the most important being that it is a non point-like theory. However, we need more basic theoretical studies and detailed computations to come up with predictions in order to make experimental tests. So far we do not have enough studies to bring these interesting speculations to real life. I well remember Rabi, one of the greatest supporters of this School, continually asking the theorists what experiments should be done to determine whether a model was right or wrong. We need experimental conclusions from theoretical speculations.

This lecture consists of five components: First, we will look at the shortcomings of the Standard Model and the features of Superstring theory that make it so appealing. Then, the Phenomenology of $SU(5)$ and $SU(5) \times U(1)$ Supergravity will be considered. Next, a striking result in $SU(5) \times U(1)$ Supergravity will be shown and we will consider the detailed calculations necessary for making predictions at LEP-I and II, FNAL, HERA, Gran Sasso and Superkamiokande. Finally, we will present some Conclusions and point out some Notes for Pessimists.

## 2) From Standard Model to Superstrings

The Standard Model has 14 masses, 6 angles, 3 couplings and also many free choices. For example, the attributes of electroweak isospin are completely arbitrary.

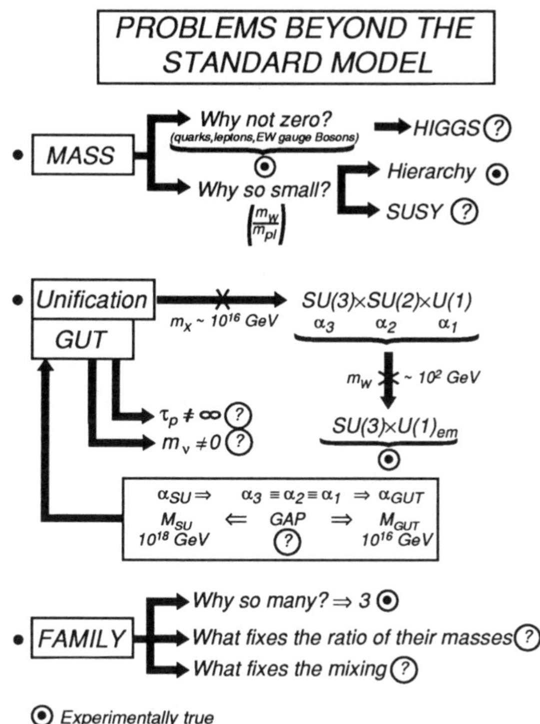


Fig. 1. Problems beyond the Standard Model.

In addition we know there are problems beyond the Standard Model (Fig. 1). These crucial problems are those of mass, unification and family. Why are quarks, leptons and electroweak gauge bosons not zero in mass? We know experimentally that they have mass. Is the answer the Higgs? Also if the masses are there why are they so small? Is there a hierarchy there? Is the answer in SUSY? Unification can be reached from above and from below. If you run the couplings from below you get a Unification Mass of $10^{16}$ GeV but from String Theory the String Unification Mass is $10^{18}$ GeV. Is there a gap? This was discussed at the School last year. Finally, why are there so many families? Why are there three? What fixes the ratio of their masses? What determines their mixing?

| | | | SPIN |
|---|---|---|---|
| Gauge forces | ⇒ | Gauge Bosons | 1 |
| SUSY | ⇒ | Gauginos | $^1/_2$ |
| 3 generations | ⇒ | quarks & leptons | $^1/_2$ |
| SUSY | ⇒ | squarks & sleptons | 0 |
| 2 doublets $H_1$, $H_2$ | ⇒ | Higgs Bosons | 0 |
| SUSY | ⇒ | Higgsinos | $^1/_2$ |

Multiplicative exact R-parity

$$R = (-1)^{2s+L+3B}$$

s ≡ spin
L ≡ leptonic number
B ≡ baryonic number

Two complex Higgs doublets needed to cancel the ABJ (triangle) anomaly. In fact, higgsinos, being fermions, contribute to ABJ anomaly.

$$H_1=(H_1^0,H_1^-)$$
$$H_2=(H_2^+,H_2^0)$$

Table 1. The main ingredients of the MSSM.

Table 1 shows the particle contents of the Minimal Supersymmetric extension of the Standard Model (MSSM): gauge bosons, gauginos, quarks and leptons, squarks and sleptons and two Higgs doublets because of ABJ cancellation. In the Standard Model there are 23 parameters and also many free options, in Global Supersymmetry there are 131 parameters plus the Standard Model ones and if you introduce soft Supersymmetry breaking there are 21 more parameters. However, with Superstring Theory we go from hundreds of parameters down to only one.

3) Phenomenology

Let me say a few words on phenomenology. First, let us consider Unification. What do we want to unify? We want to unify gauge couplings and we want to unify Yukawa couplings. What about breaking? We have SUSY breaking and electroweak breaking. A great step forward in this field is the radiative induced electroweak breaking. When we speak about SUSY do we speak of global SUSY or local SUSY? What about threshold

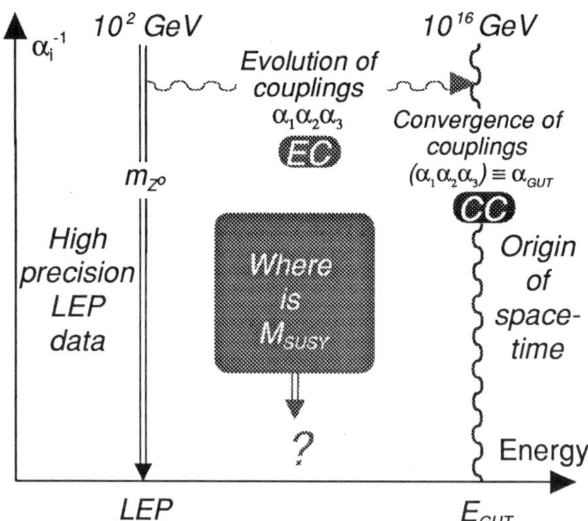


Fig. 2. Pictorial representation of the problem.

effects? There are two thresholds, one is light at $10^2$ to $10^4$ GeV energy range and the other heavy near the Planck mass at the Unification Scale.

Next we consider the Renormalisation Group Equations (RGEs). When you extrapolate over 18 orders of magnitude you have to consider the evolution of gauge couplings, the evolution of masses and the evolution of Yukawa couplings. All these need to be combined. Now let me emphasize the following: $M_{SUSY}$ is a totally meaningless quantity. Let me remind you what $M_{SUSY}$ is. The quantity $M_{SUSY}$ was introduced early on when one-loop RGEs for the gauge couplings were believed to be a good enough approximation and the threshold effects were altogether neglected. The running of the inverse of the gauge couplings $\alpha_1^{-1}, \alpha_2^{-1}, \alpha_3^{-1}$ was described by straight lines, the slopes of which had to change due to the changing $\beta$-functions from the Supersymmetric regime to the non-Supersymmetric one. The change of slope was, for simplicity, described by a single parameter $M_{SUSY}$, but in no case was $M_{SUSY}$ supposed to represent a physical quantity. It is, in fact, out of the question that the real spectrum of the Supersymmetric particles can be degenerate and therefore be represented by a single mass value. But this is how $M_{SUSY}$ is identified in the paper by Amaldi et al. [1]. The important thing is to know what is the lightest Supersymmetric particle observable and its mass. To determine this you have to do difficult detailed calculations.

The starting point is high precision LEP data at 100 GeV energy range for $\sin^2\theta_W$, $\alpha_3$ and $\alpha_{em}$ (Fig. 2). Evolution of couplings with all conditions specified has to be calculated. Then you have the convergence of couplings. Let us start with this high precision LEP data. The largest uncertainty on $\sin^2\theta_W(M_Z)$ comes from the unknown top mass, $m_t$, which induces both logarithmic and quadratic corrections. Langacker and Polonsky have done a best fit to $\sin^2\theta_W(M_Z)$ [2] which comes to $0.2324 \pm 0.0003$ using $m_t = 138$ GeV. For other values of $m_t$ the best fit is

$$\sin^2\theta_W(M_Z) \simeq 0.2324 - 1.03 \cdot 10^{-7}\mathrm{GeV}^{-2}(m_t^2 - m_{t_0}^2),$$

with $m_{t_0} = 138$ GeV. The higher the top mass the lower is $\sin^2\theta_W(M_Z)$. When we consider the value of the electromagnetic gauge coupling we see there is no quadratic divergence, only a logarithmic dependence on $m_t$. There is a general tendency for the values of $\alpha_3$ obtained experimentally at low energies to be slightly lower than those obtained from high energy data. A light gluino could improve the agreement between high and low energy data [3]. Many theoretical uncertainties exist in the determination of $\alpha_3$ at both high and low energies and these may account for the discrepancies in values of $\alpha_3$. I would like to point out that last year, we [4] worked out the two-loop Renormalisation Group Equations including evolution of masses and have shown that $\alpha_3(M_Z)$ can be as high as 0.130. At this value of $\alpha_3(M_Z)$ nothing tragic occurs so that it is not true that high values of $\alpha_3(M_Z)$ give a value for $m_{1/2}$ below $M_Z$ and that they therefore should be excluded; $m_{1/2}$ can be as high as 300 GeV with $\alpha_3(M_Z) = 0.130$.

Two years ago the publication of the already mentioned paper by Amaldi et al. [1] created a great deal of confusion. In it only one value of $\alpha_3(M_Z)$ is used, the smallest value published, the convergence of couplings is analysed, a "geometrical" $\chi^2$ is worked out and the result is that $M_{GUT}$ is at $10^{16}$ GeV and the SUSY-breaking scale at 1 TeV (Fig. 3).

This caused a lot of discussions among the people looking for SUSY at LEP. We did a detailed analysis in answer to this paper and came up ([4]–[10]) with the following conclusions.

You can have a smooth convergence of the couplings at an $E_{GUT}$ value which is in agreement with the experimental lower limit for the proton lifetime and the couplings do

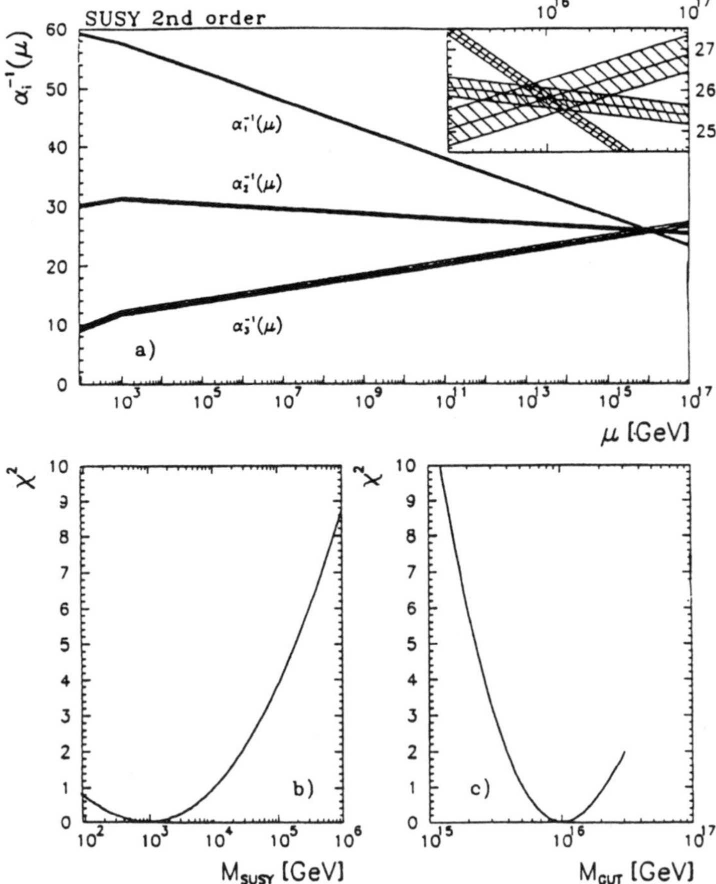


Fig. 3. In this well publicised example of unification of couplings (Fig. 2 of Ref. [1]), the divergence of the couplings for scales above $E_{GUT}$ ($\mu$ and $M_{GUT}$ in the figure) is clear, as is the sharp change in slope of the lines at low energies. Note that neither threshold effects ("light" or "heavy") nor the evolution of the gaugino masses (EGM) are included, and that these are crucial in the determination of where the Superworld could start showing evidence for its existence. Notice that the best fit for the "geometric" convergence of the couplings predicts $M_{SUSY}$ at $10^3$ GeV, and that a low value of $\alpha_3(M_Z) = 0.108$ has been used in this analysis.

| | Input data | Errors | EC | $M_{SUSY}$ | CC | UC | $\Delta T_L$ | $M_X$ | $\Delta T_H$ | EGM |
|---|---|---|---|---|---|---|---|---|---|---|
| *ACPZ [5-11]* | *WA* | $\pm 2\sigma$ | *all possible solutions (4)* | *Yes* | *physical* | *Yes* | *Yes* | *Yes* | *Yes* | *Yes* |

| | Input data | Errors | EC | $M_{SUSY}$ | CC | UC | $\Delta T_L$ | $M_X$ | $\Delta T_H$ | EGM |
|---|---|---|---|---|---|---|---|---|---|---|
| *AdBF [1]* | *only one experiment* | $\pm 1\sigma$ | *only one solution* | *Yes* | *geo-metrical* | *No* | *No* | *No* | *No* | *No* |

*WA* $\equiv$ *World Average*
*EC* $\equiv$ *Evolution of Couplings*
$M_{SUSY}$ $\equiv$ *Parameter used by AdBF and others*
*CC* $\equiv$ *Convergence of Couplings*
*UC* $\equiv$ *Unification of Couplings above* $E_{GUT}$
$\Delta T_L$ $\equiv$ *Light Threshold (sparticle spectrum)*
$M_X \equiv (M_X \equiv M_V \equiv M_H \equiv M_\Sigma)$
$\Delta T_H$ $\equiv$ *Heavy Threshold (* $M_V$, $M_H$, $M_\Sigma$ *)*
*EGM* $\equiv$ *Evolution of Gaugino Masses*

Table 2. Comparison between the analyses in Refs. [4-10] and Ref. [1].

not diverge again, which is very important. Our $\chi^2$ was not "geometrical" but had a physical meaning. It was based on the comparison of the experimental values of $\sin^2\theta_W(M_Z)$ and $\alpha_{em}(M_Z)$ to the respective predicted values.

We showed that the evolution from the GUT scale down to the $M_Z$ scale does not need any SUSY breaking. Therefore, we can forget about the other precise predictions on the SUSY-breaking scale [1]. I would like to show you a summary of the input for the two analyses and make a comparison. From Table 2 you can see that whereas we take the world average for $\alpha_3(M_Z)$ they take one experimental value. We use $\pm 2\sigma$ as error limits while they use $\pm\sigma$ and we consider four possible solutions of the evolution of the couplings whereas they only use one. We have a "physical" $\chi^2$ rather than a "geometrical" $\chi^2$. Then we have unification of couplings above $E_{GUT}$, we consider light threshold, $M_X$, heavy threshold and evolution of the gaugino masses. Figure 4 summarises our results.

Thus, we see that the analysis from Ref. [1] cannot be used to claim that the mass of the Supersymmetric world will only be in the energy range of colliders of the next generation.

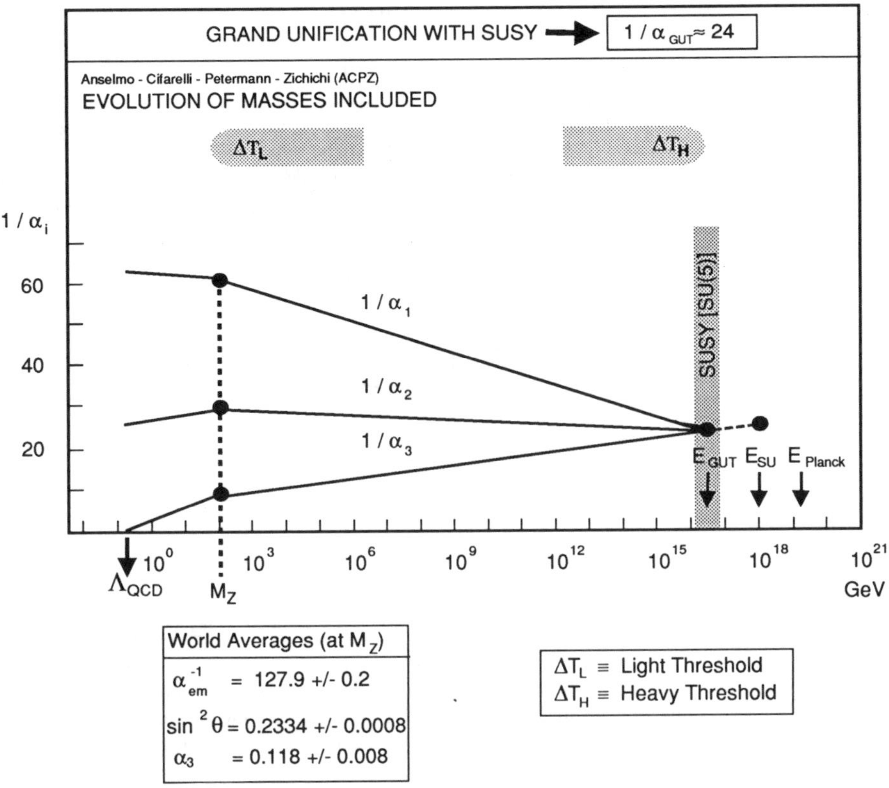


Fig. 4. This is the best proof that the convergence of the gauge couplings can be obtained with $M_{SUSY}$ at an energy level as low as $M_Z$. Notice that the effects of "light" and "heavy" thresholds have been accounted for, as well as the Evolution of Gaugino Masses. $E_{SU}$ is the String Unification scale.

The important conclusion is that SUSY breaking can be degenerate with electroweak breaking. This means that Supersymmetric particles could be seen at existing (or soon to be operating) facilities. Therefore we should indeed make detailed calculations on SUSY processes at LEP-I, LEP-II, Tevatron, HERA, Gran Sasso and Superkamiokande. We can anticipate some of these results. At Fermilab [11] tri-lepton events due to SUSY particle production could be seen. We have shown [12] that the LEP-I lower limit on the Standard Model Higgs can be applied to some Supergravity models. At LEP-I and II acoplanar multi-lepton and "mixed" events could also be observed [13].

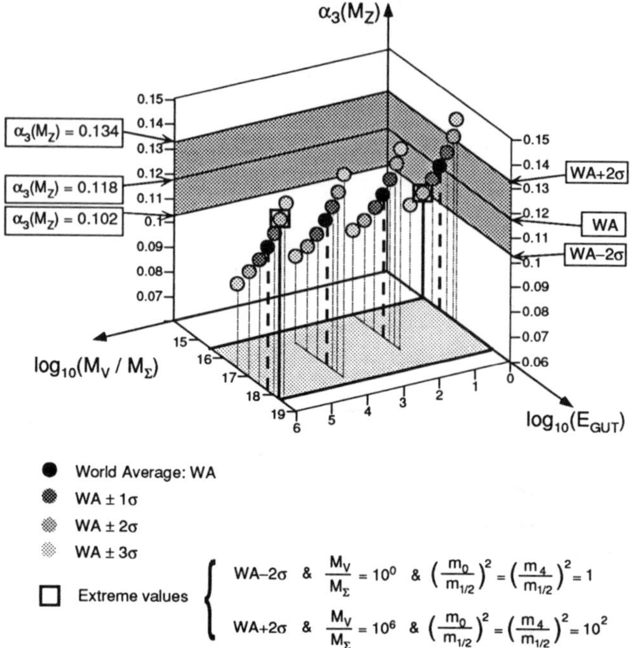


Fig. 5. The dependence of $E_{GUT}$ on $\alpha_3(M_Z)$ and on the ratio of the two crucial heavy-threshold masses, $M_V/M_\Sigma$. Here $m_0$ parametrizes the squark and slepton masses, $m_{1/2}$ the gaugino masses, $m_0$ and $m_4$ the Higgs-boson masses, and $m_4$ the Higgsino mass ($m_4$ is more commonly denoted by $\mu$). Note that the extreme value for $E_{GUT}$ is above $10^{18}$ GeV.

The statement that at HERA there could be no observation of SUSY particles is not true. If you take $SU(5) \times U(1)$ Supergravity predictions, then there is a novel result. Contrary to other model predictions, the right selectron could be light, in fact light enough to produce an effect observable in the leading proton spectrometer of the ZEUS detector.

Finally, according to certain predictions of the minimal SU(5) Supergravity model there could be evidence for proton decay at Gran Sasso and Superkamiokande [14].

We have done an extensive analysis of where the Unification Energy could be. For example we take the evolution of $\alpha_1, \alpha_2, \alpha_3$ at two loops and the evolution of the gaugino masses. When you improve the accuracy of RGEs to two loops, $E_{GUT}$ goes down. If you introduce the heavy threshold effect, $E_{GUT}$ goes up. All this is synthesized in Fig. 5. The purpose of this was to see what reasonable assumptions could be made for the gap between $E_{SU}$ and $E_{GUT}$ to exist. Superstring Theory tells you the Unification Energy is about $10^{18}$ GeV. Is this correct? Now if you develop the couplings from below to up again, can you reach $10^{18}$ GeV? The answer is yes, with $\alpha_3(M_Z) = \alpha_3^{WA}(M_Z)+2\sigma$, a very high ratio of the heavy threshold masses ($10^6$) and the Supersymmetry breaking parameters: $m_0/m_{1/2}$, also called $\xi_0$, and $m_4/m_{1/2}$, also called $\xi_A$, equal to 10. There is another possibility: you use $\alpha_3(M_Z) = \alpha_3^{WA}(M_Z)-2\sigma$, degenerate heavy thresholds, equal Supersymmetry-breaking parameters and then you get $E_{GUT} \lesssim 10^{16}$ GeV. There is nothing magical about $E_{GUT}$ at $10^{16}$ GeV. It could be that the gap is there.

4) A Striking Result in $SU(5) \times U(1)$ Supergravity

Using $SU(5) \times U(1)$ we have worked out a possible spectrum for positive $\mu$ and for negative $\mu$ (Figs. 6,7). We have taken a range of $\tan\beta$ from 1 to 30. We now do more detailed calculations with precise prescriptions and we find that a very striking result is obtained in the no-scale $SU(5) \times U(1)$ case when $m_0 = 0$, $A$ (the coefficient of the trilinear coupling) is zero and $B$ (the coefficient of bilinear coupling) is zero. If what we do is correct the solutions obtained indicate that for positive $\mu$ the top mass must be below 135 GeV and for negative $\mu$, $m_t$ must be above 140 GeV. If Fermilab discovers the top, the sign of $\mu$ will be established. If $m_t \leq 135$ GeV, $\mu$ is positive and the Higgs must be below 100 GeV. If $m_t > 140$ GeV then $\mu$ is negative and Higgs must be above 100 GeV. A further consequence of a negative $\mu$ is that $\tan\beta$ is determined only in terms of $m_t$ and $m_{\tilde{g}}$ and is in a certain range (Fig. 8). For $\mu$ positive this is also true, and for a range of $m_t$ $\tan\beta$ can be double valued. Why is this interesting? Because it implies that $\mu$ cannot become as large as in other cases. $\mu_{MAX}$ can be 440 GeV and therefore $\chi_3^0$ and $\chi_4^0$ and $\chi_2^{\pm}$ are going to be limited and the heavy Higgs is not going to be above 500 GeV.

So this striking result limits the predictions to reasonable values. So far we are still not talking about exact predictions. What are we going to do at the existing facilities? Contrary to the claim that the predictions from RGEs and high precision LEP data excluded using present existing facilities, we have proven with detailed cases that the reverse is true. In fact if we take $\alpha_3(M_Z) = 0.126$ and $\mu$ positive, then you see (Fig. 9) that many points in parameter space for chargino masses in the range 60 to 120 GeV predict a value of the proton lifetime which could allow existing facilities to observe proton decay. For $\mu$ negative it is even better.

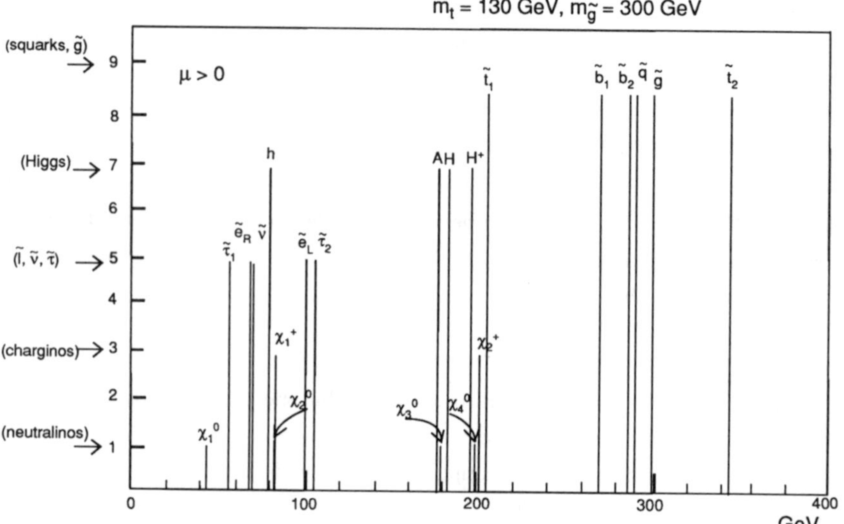


Fig. 6. Spectra of sparticles predicted by the $SU(5) \times U(1)$ Supergravity model for $\mu > 0$.

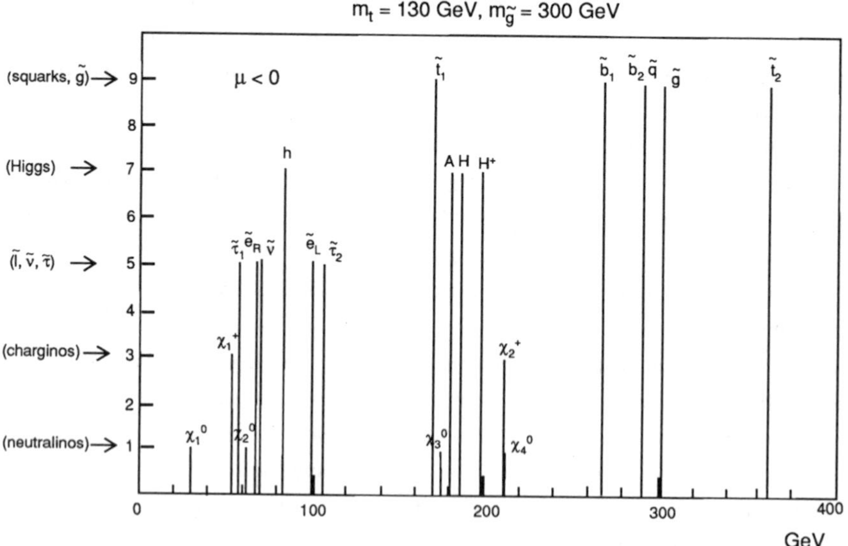


Fig. 7. Same as Fig. 6 but for $\mu < 0$.

Now I would like to make a note about minimal $SU(5)$ Supergravity. Before the evolution of gaugino masses at two-loops was calculated, the proton lifetime was four times higher, $\tan\beta$ had to be less than 6, and $\xi_0$ had to be greater than 3. If you now take into account the two-loop evolution of gaugino masses for the Renormalisation Group Equations, then the proton lifetime is as shown in Fig. 9. But the important point is that $\tan\beta$ is limited more from above and $\xi_0$ is limited more from below and the points in the parameter space which are shown in Fig. 9 are within these boundary conditions.

5) FNAL, LEP I and II, HERA, Gran Sasso and Superkamiokande: Detailed Calculations

5.1) Fermilab

At Fermilab there are 3 possibilities: observation of a tri-lepton signal, top and stop. At 100 $pb^{-1}$ luminosity nearly all points in parameter space of $SU(5)$ Supergravity will be explored via the tri-lepton signal (due to the annihilation of antiproton and proton into $\chi_2^0$ and $\chi_1^{\pm}$ which decay respectively into two leptons and one lepton giving a tri-lepton signal). However, some regions of the parameter space remain out of reach for $SU(5)\times U(1)$. In the dilaton scenario, which gives you a special relation between $m_0$ and $m_{1/2}$ and between $A$ and $m_{1/2}$, weaker signals are expected since the sparticle masses are heavier in this model.

What about the top? Here there are two possibilities for the strict no-scale case: $\mu > 0$ and $\mu < 0$. The discovery of the top would allow you to know the sign of $\mu$ and therefore whether or not the Higgs could be seen at LEP. In fact, if $\mu$ is negative, the Higgs will not be seen at LEP-II and if the top is above 150 GeV the dilaton scenario is practically out. Finally if $m_t$ is below 100 GeV, the dilaton scenario presents an interesting way out: namely that the stop could be light, as light as 100 GeV.

Now let me mention some possible conclusions. Suppose that it is found that the slepton masses are light and that the branching ratio of the second neutralino is large, then this is typical of $SU(5) \times U(1)$ Supergravity. However, if the slepton masses are large and the branching ratio of the second neutralino is small, this is typical of $SU(5)$ Supergravity. So you see these two Supergravity models do produce different results. Let me mention another difference. If we compute the tri-lepton event rate for an integrated luminosity of 100 $pb^{-1}$, summing on all possible combinations, minimal $SU(5)$ Supergravity would produce at least one event for all allowed points in parameter space. But what is important is that for low chargino masses, with positive $\mu$, we would get 140 events while, with negative $\mu$, we would get 84 events. So this difference in rate could be a clear way of

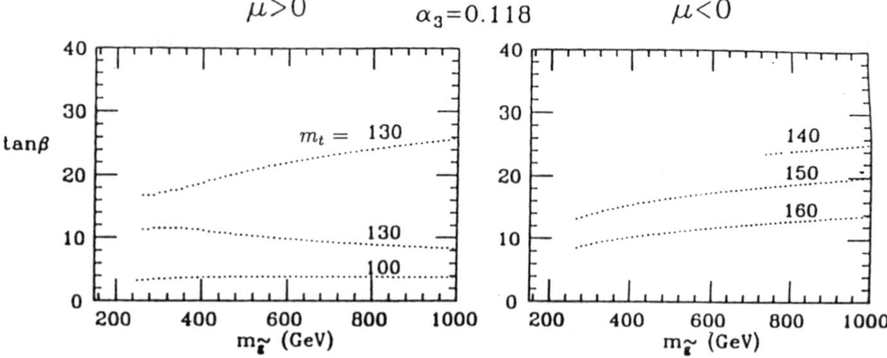


Fig. 8. Scatter plots of $\tan\beta$ versus $m_{\tilde{g}}$ for the strict no-scale case (where $B(M_U) = 0$) for the indicated values of $m_t$. Note that the sign of $\mu$ is determined by $m_t$ and that $\tan\beta$ can be double valued for $\mu > 0$.

distinguishing the sign of $\mu$. However, the interesting point is that, for $\mu < 0$, $SU(5)\times U(1)$ could produce under some conditions 420 events, instead of 84 for $SU(5)$ Supergravity.

5.2) LEP

The LEP lower limit for the Standard Model Higgs, from the process $e^+e^- \to Z^*H$, where the Higgs decays into two jets, is about 60 GeV. This limit can be extended to Supergravity models but not to the MSSM. The reason is that the ratio of the cross section for Supersymmetric Higgs production divided by the Standard Model one is equal to $\sin^2(\alpha-\beta)$, where $\alpha$ is the SUSY Higgs mixing angle and $\tan\beta$ is the ratio of the Higgs vacuum expectation values. If this quantity is one and if the ratio, $f$, of the branching ratio of the SUSY Higgs into two jets divided by the branching ratio of the Standard Model Higgs into two jets is one, the lower limit on the Standard Model Higgs mass is going to be same for the Supergravity models. Now the MSSM cannot limit either $\sin^2(\alpha-\beta)$ or $f$ to be one.

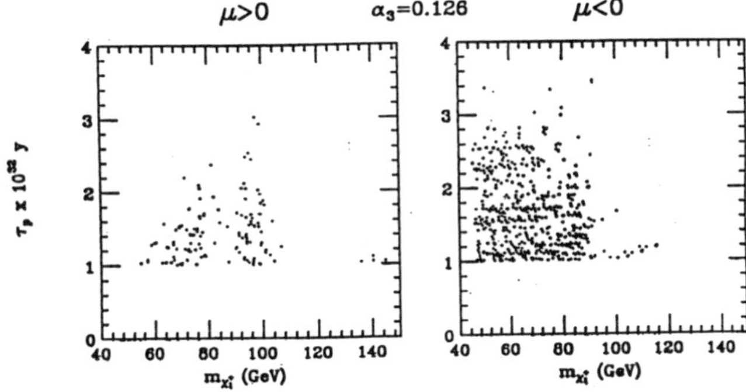


Fig. 9. The calculated values of the proton lifetime into $p \to \bar{\nu}_\mu K^+$ versus the lightest chargino (or second-to-lightest neutralino) mass for both signs of $\mu$, the more accurate value of the unification mass (which includes two-loop and low-energy supersymmetric threshold effects). Note that we have taken $\alpha_3(M_Z) = \alpha_3^{WA}(M_Z) + 1\sigma = 0.126$ in order to maximise $\tau_p$. Note also that future proton decay experiments should be sensitive up to $\tau_p \approx 20 \times 10^{32}\ y$.

This is the reason why the Higgs lower bound due to MSSM is as low as 43 GeV. But if you give up this model and choose a Supergravity model instead, this quantity is very near to one and therefore the limit on the Higgs mass typical of a Supergravity model is going to be the same as the limit of the Standard Model Higgs mass. Suppose by working hard at LEP, having found the Higgs, we discovered that the SUSY Higgs sector is Standard Model-like, we claim that this would be evidence for a radiative electroweak breaking mechanism. The fact that $f$ is equal to one can be understood as a decoupling phenomenon in the Higgs sector that takes place as the Supersymmetry breaking scale rises. This is communicated to the Higgs sector through the radiative electroweak symmetry breaking mechanism.

Now let us consider LEP-II. For LEP-II we propose that "mixed" jets-lepton and pure leptonic states should be searched for and we have computed (see Fig. 10) the number of events for 100 $pb^{-1}$ at LEP in the minimal $SU(5)$ case versus the mass of $\chi_1^\pm$. The numbers are very encouraging.

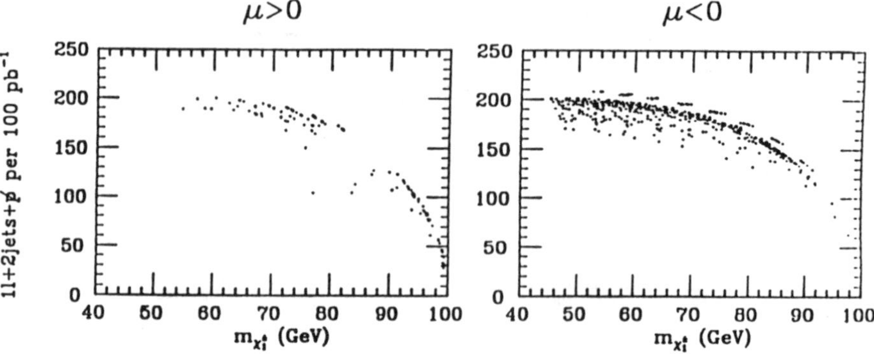


Fig. 10. The number of "mixed" (1-lepton + 2-jets + $\not{p}$ ) events per $\mathcal{L} = 100\ pb^{-1}$ for minimal $SU(5)$ model.

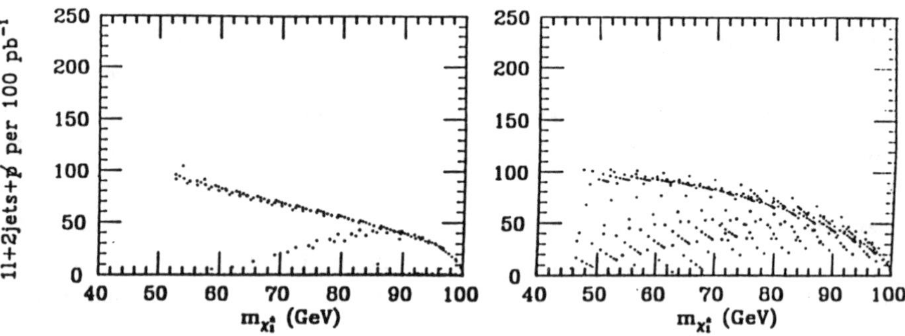


Fig. 11. The number of "mixed" (1-lepton + 2-jets + $\not{p}$ ) events per $\mathcal{L} = 100\ pb^{-1}$ for $SU(5) \times U(1)$ model.

For $SU(5) \times U(1)$ (see Fig. 11) the numbers are different but this again allows us to distinguish between a string-derivable model and a non-string-derivable one.

5.3) HERA

Due to the fact that in $SU(5) \times U(1)$ (not in $SU(5)$) Supergravity, right-handed selectrons are light, an interesting process to look for at HERA would be $ep \to \tilde{e}_R \chi_1^0 p$. Since the right-handed selectron decays into $e^- \chi_1^0$, the signature would be a high $p_t$ electron plus missing momentum.

Another measurable signal is the slowed down outgoing proton. Since the transverse momentum of the outgoing proton is very small, the relative energy loss of the proton energy $z = (E_p^{in} - E_p^{out})/E_p^{in}$ is given by $z = 1 - x_L$, where $x_L$ is the longitudinal momentum of the leading proton. The $z$-distribution is peaked at a value not much larger than its minimum value,

$$z_{min} = \frac{1}{s}(m_{\tilde{e}_R} + m_{\chi_1^0})^2.$$

Therefore, the smallest measured value in the $z$-distribution should be a good approximation to $z_{min}$. Since the Leading Proton Spectrometer (LPS) of the ZEUS detector at HERA can measure this distribution accurately, one may have a new way of probing the Supersymmetric spectrum, as follows. If elastic cross sections could be measured down to $\approx 10^{-3} pb$ then $\tilde{z}_{min}$ could be fully probed up to $\approx 0.17$.

5.4) Gran Sasso and Superkamiokande

Within the minimal $SU(5)$ Supergravity model, using the two-loop RGEs, with the inclusion of Supersymmetric threshold corrections, the proton decay rate has been calculated [14]. It comes out that $\tau_p < 3.1(3.4) \times 10^{32}\ y$ for $\mu > 0$ ($\mu < 0$). The $p \to \bar{\nu} K^+$ mode should then be readily observable at Superkamiokande and Gran Sasso since those experiments should be sensitive up to $\tau_p \approx 2 \times 10^{33}\ y$.

6) Conclusions and Note for Pessimists

It could be that at Fermilab the top will not be discovered but tri-lepton events will be there. Possibly at LEP-I and II this big emphasis on the Higgs will disappear but "mixed" jets-lepton and multi-lepton events will show up. It is also possible that at HERA the right-handed selectron will be discovered.

Now a remark for pessimistic fellows! I was nearly 25 years younger when at Frascati we measured and published the $\sigma(e^+e^- \to \text{hadrons})$ cross section stopping at $\sqrt{s} = 3$ GeV. These were the best measurements of hadronic production. I had a proposal to run the

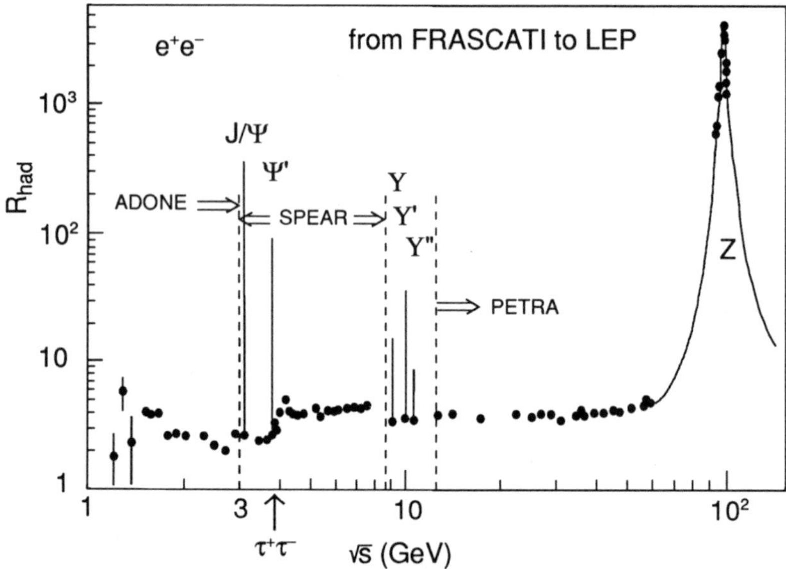


Fig. 12. $R_{\text{had}} = \sigma(e^+e^- \to \text{hadrons})/\sigma(e^+e^- \to \mu^+\mu^-)$ versus $e^+e^-$ collider energy $\sqrt{s}$, from ADONE (Frascati) to LEP-I.

Frascati machine (ADONE) at a higher energy in order to look for narrow resonances and heavy leptons. The reason why this was rejected was the standard statement at Frascati that "Zichichi was looking for butterflies" and the machine would break down. Now after 25 years it is interesting to see how many butterflies were present and where they were (Fig. 12).

After the $J/\Psi$ had been discovered by Ting and Richter, it was seen in one night at Frascati so it was not true that the machine would have broken. This was the first butterfly. The second butterfly, the Upsilon, was also just around the corner. An incredible thing was this Lederman's butterfly which was above SPEAR energy and below PETRA, where there was a hole. This is a warning for when we speak about parameter space: "Don't leave a hole in parameter space because it could be that there are new butterflies there!"

Finally, to summarise, a comparison between the two Supergravity models is shown in Table 3.

| Minimal $SU(5)$ Supergravity model | No-scale $SU(5)\times U(1)$ Supergravity model |
|---|---|
| Not easily string-derivable, no known examples | Easily string-derivable, several known examples |
| Symmetry breaking to Standard Model due to vev of **24** and independent of Supersymmetry breaking | Symmetry breaking to Standard Model due to vevs of $\mathbf{10},\overline{\mathbf{10}}$ and tied to onset of Supersymmetry breaking |
| No simple mechanism for doublet-triplet splitting | Natural doublet-triplet splitting mechanism |
| No-scale Supergravity excluded | No-scale Supergravity by construction |
| $m_{\tilde{q}}, m_{\tilde{g}} < 1\,\mathrm{TeV}$ by ad-hoc choice: naturalness | $m_{\tilde{q}}, m_{\tilde{g}} < 1\,\mathrm{TeV}$ by no-scale mechanism |
| Parameters 5: $m_{1/2}, m_0, A, \tan\beta, m_t$ | Parameters 3: $m_{1/2}, \tan\beta, m_t$ |
| Proton decay: $d=5$ large, strong constraints needed | Proton decay: $d=5$ very small |
| Dark matter: $\Omega_\chi h_0^2 \gg 1$ for most of the parameter space, strong constraints needed | Dark matter: $\Omega_\chi h_0^2 \lesssim 0.25$, ok with cosmology and big enough for dark matter problem |
| $1 \lesssim \tan\beta \lesssim 3.5$, $m_t < 180\,\mathrm{GeV}$, $\xi_0 \gtrsim 6$ | $2 \lesssim \tan\beta \lesssim 32$, $m_t < 190\,\mathrm{GeV}$, $\xi_0 = 0$ |
| $m_{\tilde{g}} \lesssim 500\,\mathrm{GeV}$ | $m_{\tilde{g}} \lesssim 1\,\mathrm{TeV}$, $m_{\tilde{q}} \approx m_{\tilde{g}}$ |
| $m_{\tilde{q}} > m_{\tilde{l}} > 2m_{\tilde{g}}$ | $m_{\tilde{l}_L} \approx m_{\tilde{\nu}} \approx 0.3 m_{\tilde{g}} \lesssim 300\,\mathrm{GeV}$ |
| | $m_{\tilde{l}_R} \approx 0.18 m_{\tilde{g}} \lesssim 200\,\mathrm{GeV}$ |
| $2m_{\chi_1^0} \sim m_{\chi_2^0} \sim m_{\chi_1^\pm} \sim 0.3 m_{\tilde{g}} \lesssim 150\,\mathrm{GeV}$ | $2m_{\chi_1^0} \sim m_{\chi_2^0} \approx m_{\chi_1^\pm} \sim 0.3 m_{\tilde{g}} \lesssim 285\,\mathrm{GeV}$ |
| $m_{\chi_3^0} \sim m_{\chi_4^0} \sim m_{\chi_2^\pm} \sim \lvert\mu\rvert$ | $m_{\chi_3^0} \sim m_{\chi_4^0} \sim m_{\chi_2^\pm} \sim \lvert\mu\rvert$ |
| $m_h \lesssim 100\,\mathrm{GeV}$ | $m_h \lesssim 135\,\mathrm{GeV}$ |
| No analogue | Strict no-scale: $\tan\beta = \tan\beta(m_{\tilde{g}}, m_t)$<br>$m_t \lesssim 135\,\mathrm{GeV} \Rightarrow \mu > 0, m_h \lesssim 100\,\mathrm{GeV}$<br>$m_t \gtrsim 140\,\mathrm{GeV} \Rightarrow \mu < 0, m_h \gtrsim 100\,\mathrm{GeV}$ |

Table 3. Comparison of the most important features describing the minimal $SU(5)$ Supergravity model and the no-scale flipped $SU(5)$ Supergravity model.

## References


[1] U. Amaldi, W. de Boer, and H. Fürstenau, *Phys. Lett.* **260B** (1991) 447.

[2] P. Langacker and N. Polonsky, *Phys. Rev.* **D47** (1993) 4028.

[3] I. Antoniadis, J. Ellis and D.V. Nanopoulos, *Phys. Lett.* **B262** (1991) 109.

[4] F. Anselmo, L. Cifarelli, A. Peterman and A. Zichichi, *Il Nuovo Cimento* **105A** (1992) 1201.

[5] F. Anselmo, L. Cifarelli, A. Peterman and A. Zichichi, *Il Nuovo Cimento* **104A** (1991) 1817.

[6] F. Anselmo, L. Cifarelli, A. Peterman and A. Zichichi, *Il Nuovo Cimento* **105A** (1992) 1025.

[7] F. Anselmo, L. Cifarelli, A. Peterman and A. Zichichi, *Il Nuovo Cimento* **105A** (1992) 581.

[8] F. Anselmo, L. Cifarelli, A. Peterman and A. Zichichi, *Il Nuovo Cimento* **105A** (1992) 1179.

[9] F. Anselmo, L. Cifarelli and A. Zichichi, *Il Nuovo Cimento* **105A** (1992) 1335.

[10] F. Anselmo, L. Cifarelli and A. Zichichi, *Il Nuovo Cimento* **105A** (1992) 1357.

[11] J. Lopez, D.V. Nanopoulos, X. Wang and A. Zichichi, *Phys. Rev.* **D48** (1993) 2062.

[12] J. Lopez, D.V. Nanopoulos, H. Pois, X. Wang and A. Zichichi, *Phys. Lett.* **B306** (1993) 73.

[13] J. Lopez, D.V. Nanopoulos, H. Pois, X. Wang and A. Zichichi, *Phys. Rev.* **D48** (1993) 4062.

[14] J. Lopez, D. Nanopoulos, H. Pois and A. Zichichi, *Phys. Lett.* **B299** (1993) 262.

## CHAIRMAN: A. Zichichi

*Scientific Secretaries: A. Hoang, P. Wegrzyn*

## DISCUSSION

– *Morpurgo:*

In your talk you pointed out that there are two completely different approaches for analysing the convergence of couplings towards $E_{GUT}$ and getting information about the SUSY thresholds. Can you summarize the main differences between these two studies?

– *Zichichi:*

As I said during my talk the main question we have to answer is where SUSY could be. There is an attempt to discuss this topic in the paper by Amaldi and collaborators (*Phys. Lett.* **B260** (1991) 447): this is what I call the Geometrical Approach. The method used in their studies is the following:

- It is assumed that the couplings must converge at a point where we have the unification of all interactions.

- An energy scale $M_{SUSY}$, the SUSY breaking scale, is introduced and it is defined as the transition point where, in the running of the Renormalization Group Equations (RGEs), the slopes of the three couplings change going from the Standard Model (SM) $\beta$-functions to the $\beta$-functions of the Minimal Supersymmetric extension of the Standard Model (MSSM). Only one solution of the RGEs has been used to analyse the convergence of $\alpha_1$, $\alpha_2$, $\alpha_3$ towards $E_{GUT}$.

- Only one experimental result for $\alpha_3(M_Z)$ is used and the uncertainties have been estimated using only $\pm 1\sigma$.

- $M_{SUSY}$ separates the Supersymmetric region from the non-Supersymmetric one. This is the energy scale at which SUSY must appear.

- To obtain an estimation of $M_{SUSY}$ and $E_{GUT}$ (the Unification Scale) a fit ($\chi^2$) is used with which the convergence of the three couplings w.r.t. the change of the slope is studied.

- The best fit (minimization of the $\chi^2$) indicates the energy scale at which SUSY must be produced (around 1 TeV in the Geometrical Approach) and the energy at which the Unification Scale must appear (around $10^{16}$ GeV in this study).

This is the method used by the authors of the previous paper. In this way they follow only a geometrical idea of physical quantities and of their relations without a clear analysis and understanding of the underlying dynamics. This is the reason why I call their method the Geometrical Approach. But Physics is not Euclidean Geometry. It is straightforward to show the limits of this approach and the erroneous assumptions.

I will now clarify the main points I have studied with my collaborators: they are the basic ingredients of what I call the Physical Approach. Some of them have been considered also in the Geometrical Approach but in a non complete and/or erroneous way:

- We have used the world average (WA) measurement of $\alpha_3(M_Z)$. The confidence level used to quote the uncertainties has been defined as $\pm 2\sigma$ and not $\pm 1\sigma$ as in the other approach.

- We have analysed many solutions of the coupled evolution equations, studying in several cases also the effects we get going from the one-loop approximation to the two-loop formula. For the last one we have also considered the influence on the final result of different choices.

- The light thresholds ($\Delta T_L$) have been extensively studied either assuming mass degeneracy (the only case analysed in the Geometrical Approach) or working out a more realistic scenario with resulting particle spectra.

There are also other points that have been completely neglected in the Geometrical Approach and have been extensively analysed in the Physical Approach because they are crucial problems for studying SUSY thresholds and the convergence of couplings:

- We have shown that the heavy thresholds ($\Delta T_H$) are an important topic in order to get a smooth convergence of the couplings at $E_{GUT}$. We have in particular analysed the effect of assuming a degenerate or a non-degenerate $\Delta T_H$ spectrum.

- We have worked out the evolution of all masses and shown that in particular the evolution of gaugino masses can strongly influence the SUSY thresholds.

- In our method the three couplings do not diverge after their smooth convergence towards $E_{GUT}$ therefore solving problems associated with non-realistic final configurations after the Unification Scale.

- We have studied different approaches to $M_{GUT}$ and $\alpha_{GUT}$ and the resulting scenarios w.r.t. mass evolutions, threshold effects and different levels of theoretical accuracy.

Keeping in mind the previous points, the $\chi^2$-test we have used to study the convergence of the couplings at $E_{GUT}$ in the differently performed numerical analyses is inspired by physical considerations and not only by pure geometrical constraints (i.e. crossing of couplings at a point). As I said, the underlying dynamics of this convergence has been carefully evaluated: $\Delta T_L$, $\Delta T_H$, evolution of all masses and complete treatment of the coupled RGEs. The full analysis of the previous problems is a very complicated subject and all people interested in having a detailed description of each point can go back to our papers. The main conclusion of these studies is that any claim for SUSY at the TeV scale relies on analysis following prejudices and does not have any support from our present experimental and theoretical knowledge. On the other hand any statement about $M_{SUSY}$ and $M_{GUT}$ and related topics needs detailed and complicated studies of a large number of parameters and effects. If we skip one and/or part of them we can only contribute to increasing the amount of prejudice.

– *Shabelski:*

You are saying that light and heavy thresholds can play an important role in studying the SUSY spectrum. Can you clarify the reasons and give some examples?

– *Zichichi:*

First of all I want to point out that the study of $\Delta T_L$ and $\Delta T_H$ can contribute to removing the wrong idea that we need the TeV scale in order to obtain the best convergence of the gauge couplings towards $E_{GUT}$. We have to introduce $\Delta T_H$ in the RGEs analysis because if we want to understand the convergence at $E_{GUT}$ we have to take care of what happens at this energy scale. In fact what happens at $E_{GUT}$ is supposed to have consequences in the energy range many orders of magnitude below. On the other hand if the main purpose of ours as in other studies is to get information about the lightest observable SUSY particle, we have shown that $\Delta T_L$ spans nearly an order of magnitude in mass. For this reason it is clearly misleading to use only $M_{SUSY}$, i.e. a degenerate mass spectrum, as parameter. The problems associated with $\Delta T_L$ and $\Delta T_H$ have been extensively analysed in our work; now I will give only two examples and mention some of the resulting conclusions. For what concerns $\Delta T_L$, in our paper: "The Simultaneous Evolution of Masses and Couplings: Consequences on Supersymmetry Spectra and Thresholds" (*Il Nuovo Cimento* **105A** (1992) 1179), we have worked out all the fundamental quantities $\alpha_{GUT}$, $M_{GUT}$, $\alpha_1$, $\alpha_2$, $\alpha_3$, including a detailed spectrum of all particles and sparticles thresholds. This analysis has been done using the RGEs at one-loop level and considering the simultaneous evolution of all masses and couplings in the framework of the MSSM. We

SUSY PARTICLE MASS SPECTRUM

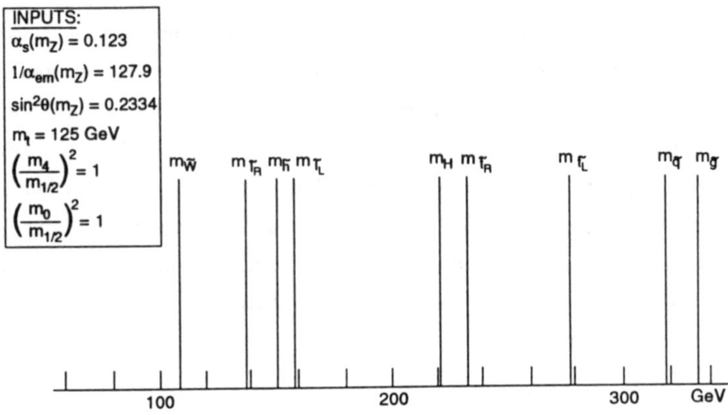


Fig. 13. An example of a SUSY particle mass spectra.

have also worked out in a detailed way the light SUSY spectrum using only experimentally measured quantities ($\alpha_1(M_Z)$, $\alpha_2(M_Z)$, $\alpha_3(M_Z)$ and consequently $\sin^2\theta(M_Z)$) and nothing else, like naturalness conditions. As the experimental input for the three gauge couplings we have used their respective WA and considered their $2\sigma$ deviations from the WA. Some of the conclusions that define a trend for the topics we are discussing are:

- The dominant effect correlated with the definition of the spectrum is the one due to $\alpha_3(M_Z)$.

- Higher values of $\alpha_3(M_Z)$ push $M_{GUT}$ towards higher values.

- There is an anticorrelation between $M_{GUT}$ and $M_{SUSY}$: increasing $M_{GUT}$, $M_{SUSY}$ is lowered.

- Higher values of $\alpha_3(M_Z)$ lower the $M_{SUSY}$ threshold.

- When the theoretical accuracy improves (i.e. we use the two-loop instead of the one-loop formula for the RGEs), the values of $\alpha_3(M_Z)$ are preferred to be on the high side (above WA) and the values of $M_{GUT}$ on the low side.

As a final remark I want to point out that we have also shown that the splitting among the different masses of the threshold is in the same order of magnitude for all experimental

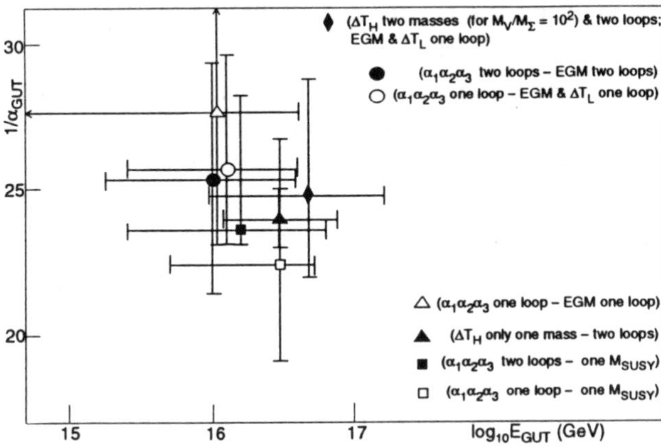


Fig. 14. Results for $\alpha_{GUT}$ and $E_{GUT}$ from different approaches.

values of $\alpha_3(M_Z)$, $\sin^2\theta(M_Z)$ and $\alpha_1(M_Z)$ between their WA$\pm 2\sigma$, but depends very much on the choice of the primordial parameters $m_{1/2}$, $m_0$, $m_4$ of SUSY breaking (these quantities parametrize at the GUT scale respectively the gauginos, the squarks and sleptons, and the higgsino; $m_0$ and $m_4$ parametrize the Higgs). There are also some subtle effects associated with the light thresholds. In Fig. 13 an example for $\Delta T_L$ is shown, where we assume that we know the model and mixing angles exactly, so that we can use the wino instead of the chargino and the higgsino instead of the neutralino.

One of the main conclusions of this study is that for all the various spectra considered we can get the convergence of couplings towards $E_{GUT}$ and therefore the unification for some $\alpha_{GUT}$. As an example this is illustrated in Fig. 14. The corresponding value of $E_{GUT}$ is plotted versus a given value of $1/\alpha_{GUT}$. The graph illustrates how many possibilities there are of getting convergence choosing a light threshold.

For the contribution given by the heavy thresholds to define the framework of the Physical Approach, I want to mention only two interesting results. Considering $\Delta T_H$ we have a smooth convergence of all couplings up to $E_{GUT}$ without any necessity of introducing a SUSY breaking above $M_Z$ and a change of the slope for the couplings. This confirms the result obtained by working out $\Delta T_L$: the SUSY threshold can be as low as the $Z^0$ mass. There is another challenging question strongly correlated with the heavy spectrum, which has been studied by our group, and is still awaiting a clear understanding: the gap between the Grand Unification scale $E_{GUT}$ and the String Unification scale $E_{SU}$. The question is, whether there are mechanisms, without introducing extra hypotheses, to push the convergence energy $E_{GUT}$ as high as possible in order to eliminate the gap. In the SU(5)×U(1) Supergravity model this gap vanishes in a natural way without any fine tuning. In the SU(5) Supergravity model the gap can only be removed by taking extreme parameter values as is shown in Fig. 5. On the other hand in our paper: "A Study of the Various Approaches to $E_{GUT}$ and $\alpha_{GUT}$" (*Il Nuovo Cimento* **105A** (1992) 1335), we have shown that the highest possible value for the energy level where the low energy measurements allow the couplings to be merged is $M_{GUT} \sim 3 \times 10^{18}$ GeV: around an order of magnitude below the Planck scale. The existence of this gap and its consequences, if we consider the range of accessible energies, are a very interesting puzzle and should be extensively studied.

– *Vassilevskaya:*

You have explained that in your studies you have included the evolution of all masses. Why is the evolution of the gaugino masses so important to understand the convergence of all couplings towards $E_{GUT}$?

– *Zichichi:*

The evolution of gaugino masses is a crucial point not only to understand the convergence of couplings towards $E_{GUT}$ but also to work out possible lower bounds for the SUSY thresholds. We have studied this problem in the paper "The Evolution of the Gaugino Masses and the SUSY threshold" (*Il Nuovo Cimento* **105A** (1992) 581). By the RGEs and using an iterative procedure we calculate the set of SUSY masses and we are able to reach the convergence and get a stable solution for the previous masses. Our expression for $\Delta T_L$ agrees with that one worked out by J. Ellis et al. (*Phys. Lett.* **B260** (1991) 131 and *Nucl. Phys.* **B373** (1992) 55). We can use the result for $\Delta T_L$ to extract $\log(m_{1/2}/m_Z)$ versus $m_{\tilde{g}}$ and $m_{\tilde{W}}$. The basic equation can be written as:

$$\log(\frac{m_{1/2}}{m_Z}) = [(\mathrm{F(X)} \pm \Delta) + 7\log(\frac{\alpha_3(m_{\tilde{g}})}{\alpha_{GUT}}) - 8\log(\frac{\alpha_2(m_{\tilde{W}})}{\alpha_{GUT}})]$$

in which we assume $\alpha_{GUT}$ equal to $\alpha_{GUT}(m_{1/2})$. The function F(X), with its error $\pm\Delta$, synthesizes all our knowledge (experimental and theoretical) except the effect due to the

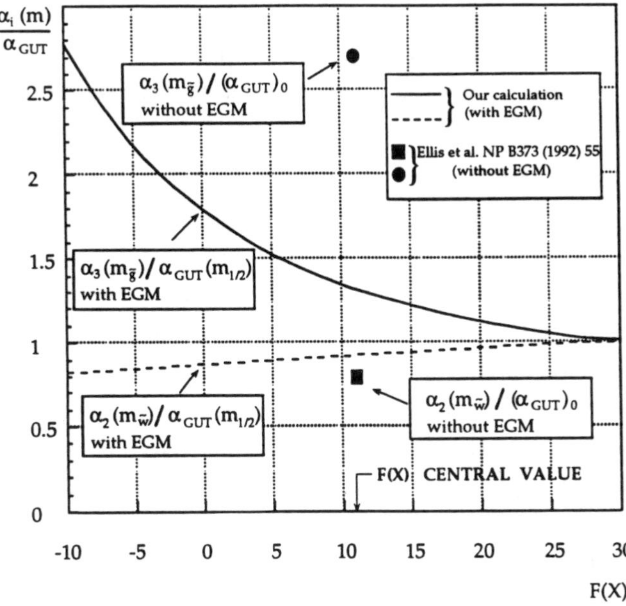


Fig. 15. The EGM effect.

evolution of the gaugino masses (EGM) (see *Il Nuovo Cimento* **105A** (1992) 581 for details). This effect is accounted for by the last two terms in the previous equation. We have shown that if we consider this evolution of masses the lower bound for SUSY threshold found by the previous authors to be 21 TeV may be lowered down to 45 GeV even if we consider only the result at $1\sigma$ level. This gives a clear idea of how the EGM effect is important in order to make predictions about the SUSY spectrum. The reason for this dramatic change by two orders of magnitude can be explained if we consider the correlation (see Fig. 15) between the central value of F(X) used in our numerical analysis and the two couplings $\alpha_2$ and $\alpha_3$ that enter into the definition of $m_{1/2}$ according to the above equation: $\alpha_2$ is not very much influenced by the EGM effect but $\alpha_3$ is much lower with EGM effect than without EGM. Since $\alpha_2$ and $\alpha_3$ are used to solve the RGEs, it is obvious that the EGM plays a crucial role in the determination of the SUSY thresholds.

– *Habig:*

Do you really believe that at the existing facilities we may have a chance of discovering a SUSY signal? And in particular would this be within the predictions of SU(5)×U(1) and SU(5) Supergravity models?

– *Zichichi:*

We do not know if we can see a signal. What I have tried to explain during my lecture and during this discussion is that there is no reason from a phenomenological and theoretical point of view to exclude this possibility: this is what we have studied and proven. On the other hand we do not have to become discouraged and in particular for two reasons. First of all because all the mass limits we are able to establish for the existence of SUSY are very important for phenomenological and theoretical knowledge: for example we know that non-Supersymmetric SU(5) is ruled out due to data on proton decay and $\sin^2\theta(M_Z)$, while SU(5) Supergravity is still alive. On the other hand at the existing facilities there are a lot of possible exciting discoveries we can make as I have explained during my lecture. People at LEP should seriously study the events with lepton-jets and multi-leptons, originating from charginos and sleptons. At HERA it could be that the number of new messages is small, but instead of looking at the expected Physics, we should notice if we find a bump in the electron spectrum coming from the Superpartners of the right-handed electrons. As a cross-check in case we find a bump, we can switch off the machine, turn to a machine with left-handed polarized beams and the effect should disappear. At Gran Sasso and Superkamiokande it could be that the K-channel proton decay is just round the corner and SU(5) Supergravity is going to be the source of this decay. At Fermilab the tri-lepton signal could be within the possibility of the existing facilities. My conclusion, as I have explained in my lecture showing the results of detailed calculations done by our group, is that at Fermilab and LEP-II both Supergravity models SU(5) and SU(5)×U(1) can be tested, at LEP-I all Supergravity models, if the light Higgs is heavier than about 60 GeV, at HERA only SU(5)×U(1) and at Gran Sasso and Superkamiokande only SU(5).

– *Lopez:*

We know the decision of the USA parliament to cut the financial support for SSC. Is it a clear signature that the future of High-Energy Physics could be problematic?

– *Zichichi:*

Our field needs unity, collaboration and removal of logical inconsistencies. When these do not exist and this becomes known to decision making leaders, the consequences cannot be positive. I have been asked on several occasions how the members of "Scientific Committees" are chosen and appointed. These committees do have consequences on billion dollar facilities. Are they elected? If so, how and by whom? Do they represent the "Scientific Community"? For example, how are ECFA members appointed? How are other decision making or advisory committees structured? We need to have a lot of good and logically consistent answers if we want to have the support of smart decision making leaders. Furthermore everyone of us should do whatever he can to let people know the value of our scientific work. In a democratic society we need public opinion to support our projects.

**CHAIRMAN: D.V.Nanopoulos**

*Scientific Secretaries: N.Sarlis, K.Skenderis*

## DISCUSSION

– *Lopez:*

What do you need to do in order to get an $SO(10)$ GUT model from strings? Similarly what do you need to do to the Dimopoulos et al programme, for example, for fermion masses?

– *Nanopoulos:*

You need to go to a level greater than one. To get adjoints you need k=2 and to get the 2/3 splitting you need $k \geq 5$ and with such a high level, representations of dimension > 100 will occur. This appears to be an improbable thing to happen in nature.

– *Peccei:*

Is it true that for all kinds of string models with Kak-Moody level 1 you never get adjoints? If so, then you never really have a GUT group or you have to have a flipped representation. Is this correct?

– *Nanopoulos:*

What you are saying is correct. This is a theorem. In $k = 1$ string theory you don't get adjoints. You can prove it at the level of Conformal Field Theory.

– *Peccei:*

Which means that effectively you always have to flip something.

– *Nanopoulos:*

Yes, you either go directly from String Theory to $SU(3) \times SU(2)$ or you flip something.

– *Peccei:*

That's very peculiar. People are saying that you unify but basically you disunify. I understand what you want to do but it is peculiar. Do you bypass Grand Unification completely?

– *Nanopoulos:*

This is a beautiful thing in String Theory. In the string you can give me for example 35 groups and if they are coming from String Theory you know they are going to unify.

– *Wegrzyn:*

Your gauge group chosen for the 4–D Heterotic String Model contains a $U(1)^4$ subgroup. Could you repeat the arguments for such a choice and say something about the interpretation of this sector of the theory?

– *Nanopoulos:*

In this free fermionic formulation we have a set of rules you have to satisfy. It basically boils down to finding "consistent" boundary conditions for these two dimensional world sheet fermions. We have a set of fermions that we call a "vector" and then we have to have several of these vectors that get some phases when you take them around the world sheet. What is happening is the following: we want to build an $SU(5) \times U(1)$ so basically we try to give the minimal basis to do this. We start with $SO(28) \times E_8$ and we finally want to get an $SU(5) \times U(1)$. Going through this procedure we get:

$$[(SU(5) \times U(1)) \times U(1)^4] \otimes [SO(10) \times SU(4)]$$

where $SO(10) \times SU(4)$ is the hidden sector. The existence of $U(1)^4$ cannot be avoided and it plays a very fundamental role. We use it to distinguish between different Yukawa couplings and different gauge symmetries. For instance we have these generations which have standard $SU(5) \times U(1)$ numbers but there is a distinction in the other extra $U(1)$ charges and because of this we are getting this disparity of masses. This is dynamical (the existence of $U(1)^4$) and it is not something you put in by hand. What I am claiming is that this is not the only $SU(5) \times U(1)$ but this is one of those which makes sense phenomenologically.

– *Arnowitt:*

Does the theorem in which Kac–Moody level k=1 algebra forbids adjoint representations also forbid them forming composite adjoints as one evolves down from the Plank scale to some GUT scale e.g. a 5 and $\bar{5}$ forming a composite 24?

– *Nanopoulos:*

No, it does not. One has to be careful though with the mechanism of confinement of the 5 and $\bar{5}$. One would need at least an $SU(4)$ confining, included in the <u>observable</u> <u>sector</u> and then one needs to prove why all other 5 and $\bar{5}$, 10 and $\bar{10}$ do not confine! In conclusion, while it is not excluded by the "No–adjoints Theorem", it looks highly non-trivial to implement.

– *Stoilov:*

What is the status of the higher mass string states when you build up effective field theory? Does the Virasoro symmetry break down when you just throw massive states out?

– *Nanopoulos:*

There are many effects from the towers of massive string states, from which the most significant is the string threshold in the Renormalisation Group Equations. As regards Virasoro symmetry, everything is O.K. We don't break conformal invariance.

– *Vassilevskaya:*

How can you explain the existence and Dirac nature of superheavy neutrinos in the framework of this model?

– *Nanopoulos:*

The point is that when we make the "flip" we can't have a charged lepton in the decouplet $F_i$, we have to have a neutral lepton. This introduces three (i.e. 3 families) right-handed neutrinos. This is not a kind of Gell-Mann et al. formulism where you introduce Majorano neutrinos. We don't have any Majorana neutrinos. What we have is the right-handed neutrinos and the singlet $\phi_m$. There is a very nice coupling between the decouplet $F_i$, the $\bar{H}(\bar{10})$ and $\phi_m$. The singlet immediately absorbs this right handed neutrino and automatically gives a super heavy Dirac mass. In general, it is going to be some kind of mixing and actually we have used some kind of mechanism in a realistic model. We found some non-trivial mixing between $\nu_\mu$ and $\nu_\tau$ and things like that. These numbers can be tested directly in experiments such as CHORUS and NOMAD at CERN.

– *Giovannini:*

At which scale does inflation take place in your model?

– *Nanopoulos:*

If I wanted to be inside traditional physics I could use almost any of these singlets that are getting vacuum expectation values close but not too close to the string scale and I somehow could write for you potentials that could give inflation. This is a kind of fake and that is why we haven't done it. A much more interesting approach is through what Dr. Mavromatos was talking about this morning. In this type of approach you have a very natural way of creating entropy and solving the horizon problem without really using any kind of inflatons, potentials etc.

– *Kaplunovsky:*

String has so many candidate vacua that chaotic inflation is much more likely than any other type of inflation.

– *Mavromatos:*

Commenting on Kaplunovsky's remark on inflation of strings I would like to say the following: In the framework of the identification of time with a Liouville mode in non-critical string theory, there seems to be an initial cosmological era, when the "gravitational function" is very large and leads to a very rapid inflationary scenario for the universe, characterised by a large increase in entropy. This does not necessitate the existence of an inflaton but the conditions of inflation as derived in a recent article by Turner et al., i.e. large initial entropy and super luminal expansion, seem to be satisfied. This seems similar to traditional inflation and is probably related to Kaplunovsky's remarks, for conventional string theory. But there is no necessity for inflatons, I stress again. As I stressed in my talk such Cosmological applications are very important in establishing any idea about String Theory. Of course, it is understood that in this novel framework of time $\equiv$ Liouville mode, such scenaria on inflation are still to be considered very preliminary.

– *Giovannini:*

Do you have any problems with monopoles?

– *Nanopoulos:*

Actually, no. One of the reasons we wrote this $SU(5) \times U(1)$ paper in 1984 (with J.Kim, and J.D. Derendinger) was that because the U(1) is not inside a simple group, the Polyakov–t'Hooft condition doesn't have to be satisfied. So we don't have any monopoles at all. We don't need inflation for resolving the monopole problem.

## CHAIRMAN: V. Kaplunovsky

*Scientific Secretaries: H.Petrache, P.Wegrzyn*

## DISCUSSION

– *Lopez:*

What did you mean when you said that having a vacuum energy $V_0 \simeq m_{3/2}^2 M_{Pl}^2$ was O.K. regarding the cosmological constant?

– *Kaplunovsky:*

Let me clarify this point. Nobody ever solved the cosmological constant problem completely. The cosmological constant should be zero at the millivolt level. Of course, it is beyond the capacity of the theory we understand now. But suppose we know some mechanism to get rid of the cosmological constant. The question is how much adjustment I can tolerate before I say the entire theory is violated by this mechanism. In non-supersymmetric cases you can say whatever you wish, but in supersymmetric theories if you want to adjust it, the number of things you can do is quite limited. To end up with a vacuum and its parameters predicted by a given potential, the mechanism of supersymmetry breaking cannot be modified too much. If the naïve cosmological constant is less than the geometric mean of the gravitino mass and the Planck scale taken to appropriate powers then it may be O.K. If it is bigger than that, whatever mechanism you have used to eliminate the cosmological constant will likely change the theory so much as to invalidate the whole procedure completely. On the other hand, it does not mean there are no problems having the cosmological constant just below $m_{3/2}^2 \mu_p^2$, only that there is a theoretical possibility of solving the problem without trashing everything else.

– *Langacker:*

You emphasized that there are "zillions" of possible string vacua.

a) Is there any hope of finding a theoretical principle to pick out the correct vacuum?

b) If that fails is there any hope of classifying the vacua in such a way that any effective search (perhaps by computer) could be made to find a phenomenologically acceptable one?

– *Kaplunovsky:*

Yes. My hope is that the inherent mechanism of String Theory might tell you what class of vacua you are in. But there are still some theoretical problems left to contend with. The aim is to find out criteria if you need to solve theoretical problems in terms which make sense in string theoretical language. Otherwise you need to make a choice of the vacuum, work out the SUSY breaking, and go back. If the choice is not good you have to try again. This seems to be hopeless. The procedure presented up to now hardly resembles classification of vacua, but a lot of people are working and maybe one day a large subset of vacua will be classified and we will start looking at the most gross features of the theory. So in addition to looking at the suitable observable sector of three generations, we can also look for the hidden sector and after this interplay, a game of how to stabilise dilatons starts. I repeat again that this project will not be done tomorrow. The chances are that the next accelerator will be built first.

– *Stoilov:*

a) Which is the string theorem you are starting from?

b) At the beginning of your lecture you mentioned that the dilaton does not enter the superpotential on the three level. Is this an assumption or result?

– *Kaplunovsky:*

a) My implicit assumption is that I am starting from heterotic strings, but my definition of the heterotic string is strictly in world-sheet terms. In principle, I can also start from type-II String Theory, but classically this theory is not good, there is no room for both space-time and the observable sector. My assumption is that if we have 4D String Theory with some internal sector and at the Planck scale perturbation theory works, at least qualitatively, everything else then follows in the framework of an effective Field Theory. There is no inherent string theoretical reason why I should start there. But it simply happens to be the only thing we understand with the present technology.

b) As the superpotential is not renormalisable at any order of the perturbation theory, both in String Theory and Field Theory, the three level dilaton does not enter and it would not enter into all orders of perturbation theory. Non-perturbatively, if the effect is field theoretical then you end up with an extra term in the superpotential which is a function of condensates only.

– *Giannakis:*

a) This morning you listed a set of requirements which an effective theory should satisfy. Could you translate these requirements into a conformal Field Theory language in order to reduce the number of possible vacua of String Theory?

b) Could you clarify a point. Is the dilaton independent of the graviton or is it intertwined with it?

– *Kaplunovsky:*

a) That was precisely my point that unless we do something like that there is no hope of finding the needle in the haystack, more precisely in the field of haystacks. That is precisely what my biggest wish was to do, but I could not.

b) In String Field Theory the dilaton comes with gravity as an inseparable package. Even if there is no supersymmetry, the dilaton is still there. But from the effective Field Theory point of view a connection is no longer there and a dilaton is just a field.

– *Lopez:*

Are there any general predictions of String Theory or should one always work out precise models which satisfy all consistency constraints?

– *Kaplunovsky:*

At the moment String Theory has absolutely zero power of experimentally predictable predictions. It is a framework for model building which has some theoretical consistencies. You have to assume something about the vacuum and then you can proceed. For any particular model you make predictions. Anybody who claims that these are predictions of the String Theory as such is a charleton pure and simple! I don't want to give names but almost from the beginning of String Theory, that has been a tendency. There have been a lot of papers which have analysed some particular model and said this is a prediction of String Theory, this is the ratio of the super-partners. There seems to be a high correlation between those kind of results and a total lack of understanding of String Theory. I would not claim that the correlation is 100% but it is there.

– *Zichichi:*

But also Field Theory cannot be tested unless there is a model. So probably the answer is that if you have a model you could say this model is within the context of String Theory. Are you saying that this isn't possible?

– *Kaplunovsky:*

There are some very weak general predictions of Field Theory, for example CPT theorem. As for direct predictions of String Theory which were to hold in a vacuum independent manner there is nothing I could state. The best you can do as a compromise between specific model building and full understanding of String Theory is to adopt a scenario and in some particular case of the scenario you are able to parametrise your ignorance in a few parameters and make some general prediction of that scenario. But this is not a prediction of String Theory and this was precisely the point I had to argue with Mike Dine at the Berkeley Conference where he misinterpreted the results of myself, Jan Louis, Barbieri and Masiero. This is not a prediction of a String Theory but a prediction of one particular scenario.

– *Zichichi:*

Is it interesting to contrast a particular scenario in QFT with a particular scenario in String Theory? Do you also deny this possibilty?

– *Kaplunovsky:*

Some scenarios have limited predictive powers, e.g. the biggest predictive power is in a scenario where somehow the connection is between the hidden sector and the observable sector, where the supersymmetry is broken via the dilaton. In this case, if you also assume minimality, you have quite a lot of predictive power. But no-one can promise you minimality either so that can substantially weaken the prediction. So if someone tomorrow finds out the super-partners and says this is not in accordance with this model, you maybe have to modify the model or maybe you have to look for a very different model. The paradoxical result of this theoretical mess is that we desperately need experimental data. We do not have confidence in any model that will tell experimentalists what to look for. They have better brains than that. They know that nobody's prediction about specific mass of super-partners should be taken seriously. They should just look for super-partners wherever they can find them. And when they do, all the theorists will be eternally grateful because that will put at least some physical foundation into the game. Let us consider the question: "Are the squarks and sleptons with equal charges more or less degenerate in mass?" There are some arguments for degeneracy, namely the flavor changing neutral currents. There are some theoretical arguments against, namely in String Theory. In most of the models you do not get degeneracy unless you arrange for some miracle, and that would be helpful in finding out if that is true or not. Theorists should not be so arrogant as to tell experimentalists exactly what is going to happen.

– *Zichichi:*

But it could nevertheless be very interesting to make an effort to see if a special model could be built in the framework of a pointlike field theory and be confronted with any prediction you like, with other models built in the Field Theory. Do you deny this possibility because if you are very negative ...

– *Kaplunovsky:*

Of course, the more assumptions you make the more predictions you can make.

– *Zichichi:*

Could you tell me what experimental data you would like to have tomorrow morning?

– *Kaplunovsky:*

Super-partners' masses.

– *Zichichi:*

But this could still be in the framework of Field Theory.

– *Kaplunovsky:*

Everything String Theory predicts has to make sense in terms of Field Theory. This is simply inevitable. The question is what kind of unification you get. And if you get all super-partners' masses, you get a much better idea of what happens to the couplings. And then if the couplings say something has to happen that should be a GUT at $10^{16}$ GeV, that would be very bad for String Theory. On the other hand Yukawa couplings may suggest a unification that makes no sense in terms of conventional GUT. That would be a good inclination for String Theory. String Theory very often predicts some numerical ratio between the Yukawa couplings without having anything resembling the multiplets of a bigger gauge group. From this point of view, string unification is very different from conventional Grand Unification.

– *Zichichi:*

Give me three examples of experimental data you would like to have.

– *Kaplunovsky:*

Unfortunately my appetite is bigger than that. Three is not enough, I want all the super-partners!

# SUPERGRAVITY MODELS

R. Arnowitt*[a] and Pran Nath[b]
[a]Center for Theoretical Physics, Department of Physics
Texas A&M University, College Station, TX 77843-4242
[b]Department of Physics, Northeastern University, Boston, MA 02115

ABSTRACT

Theoretical and experimental motivations behind supergravity grand unified models are described. The basic ideas of supergravity, and the origin of the soft breaking terms are reviewed. Effects of GUT thresholds and predictions arising from models possessing proton decay are discussed. Speculations as to which aspects of the Standard Model might be explained by supergravity models and which may require Planck scale physics to understand are mentioned.

## 1. INTRODUCTION

Supergravity has become the main vehicle for efforts to construct grand unified models. There are now both experimental and theoretical reasons for examining the consequences of such models. On the experimental side, there is the well known fact that measurements of $\alpha_1 \equiv (5/3)\ \alpha_Y$, $\alpha_2$ and $\alpha_3$ at mass scale Q $= M_Z$ (where $\alpha_Y$ is the hypercharge coupling constant) allows a test of whether these three couplings of the Standard Model unify at some high scale Q $= M_G$. What is found [1] is that unification does not occur with the Standard Model (SM) mass spectrum but unification does appear to occur with the supersymmetrized Standard Model with one pair of Higgs doublets. Thus using the two loop renormalization group equations and making the approximation of neglecting both mass splitting of the supersymmetry (SUSY) spectrum and mass splitting of the GUT mass spectrum, one finds that

$$M_G = 10^{16.19\pm0.34}\ \text{GeV}; \quad M_S = 10^{2.37\pm1.0}\ \text{GeV}$$
$$\alpha_G^{-1} = 25.4 \pm 1.7 \tag{1}$$

where $M_S$ is the common SUSY mass, and $\alpha_G$ is the gauge coupling constant at the unification scale $M_G$. (The errors in Eq. (1) are due to the errors in $\alpha_3(M_Z)$ and we use $\alpha_3(M_Z) = 0.118 \pm 0.007$ [2].)

There are several points worth noting about the above result:

* Speaker

(i) Unification occcurs only for the choice $\alpha_1 \equiv (5/3)\alpha_Y$, which states the way in which the hypercharge is embedded into the GUT group G. Thus unification is not completely a property of the low energy particle spectrum, but depends also on the nature of the high energy group G.

(ii) Unification is indeed obtained by adjusting the parameter $M_S$. However, the significant point is that $M_G$ and $M_S$ come out at values that are physically acceptable, i.e. $M_G$ is sufficiently large to inhibit proton decay, and $M_S$ is in the correct mass region for the SUSY particles to solve the gauge hierarchy problem discussed below. (Thus $M_S \cong 10^{2.5} \cong 300$ GeV.)

(iii) Acceptable unification occurs only with one pair of Higgs doublets. With more Higgs doublets, $M_G$ is so small that proton decay would already have been observed, and $M_S$ is so large that the hierarchy problem remains.

Of course, we have no real knowledge of what the particle spectrum is above the electroweak scale. There may be additional particles at higher energies which delay or prevent grand unification from occurring. However, the simplest and most natural implication of the above result is that grand unification occurs at scale $M_G$, and the particle spectrum between $M_Z$ and $M_G$ is the supersymmetrized Standard Model with one pair of Higgs doublets.

There are also several theoretical arguments supporting the building of supersymmetric particle models. From the high energy side, string theory implies the validity of N = 1 supergravity as an effective field theory below the Planck scale, $M_{P\ell} = (1/8\pi G_N)^{1/2}$ where $G_N$ is the Newtonian constant ($M_{P\ell} = 2.4 \times 10^{18}$ GeV). Note, however, that $M_G/\mathrm{M}_{P\ell} \approx 10^{-2}$ and so the GUT theory is moderately isolated from Planck scale physics. However, we do not expect it to be a precisely accurate theory as it may possess (1-10) % corrections from "Planck slop" terms (non-renormalizable terms scaled by powers of $1/\mathrm{M}_{P\ell}$).

From the low energy electroweak scale, supersymmetry offers a solution to the well-known gauge hierarchy problem. Thus in the Standard Model, the loop corrections to the Higgs mass $m_H$ (Fig.1) is quadratically divergent:

$$m_H^2 = m_0^2 + c(\tilde{\alpha}/4\pi)\Lambda^2 \tag{2}$$

where $m_0$ is the bare mass, $\tilde{\alpha}$ is a coupling constant, $c$ is a numeric and $\Lambda$ is the cut-off. If one takes the bare Lagrangian as fundamental, then the existence of the divergence implies that the theory is valid at energies below $\Lambda$, and $\Lambda$ is the scale of new physics which intervenes to converge the integral. How large can $\Lambda$ be, i.e. at what scale does new physics enter? Now $\mathrm{m}_H$ sets the electroweak scale. However, as $\Lambda$ gets large, $\mathrm{m}_H$ and eventually other particle masses all get close to the large scale $\Lambda$. This is the "gauge hierarchy" problem which states that it is not possible to maintain a hierarchy of masses, some small at the electroweak scale and some large (e.g. at $M_G$ or $\mathrm{M}_{P\ell}$ scale). An alternate way of thinking of this problem is to try to choose $\mathrm{m}_0^2$ to cancel the large $\Lambda^2$ term. However, for $\Lambda \approx M_G \approx 10^{15}$ GeV, this requires fine tuning $m_0^2$ to 24 decimal places (!) and trouble begins already for $\Lambda \gtrsim 1$ TeV. This alternate view of the problem is known as the "fine tuning" problem. Of course, the same difficulties enter with the other divergences of relativisitic quantum field theory. However, these only grow logarithmically with $\Lambda$, and so hierarchy difficulties only set in at the Planck scale where we already know new physics must occur.

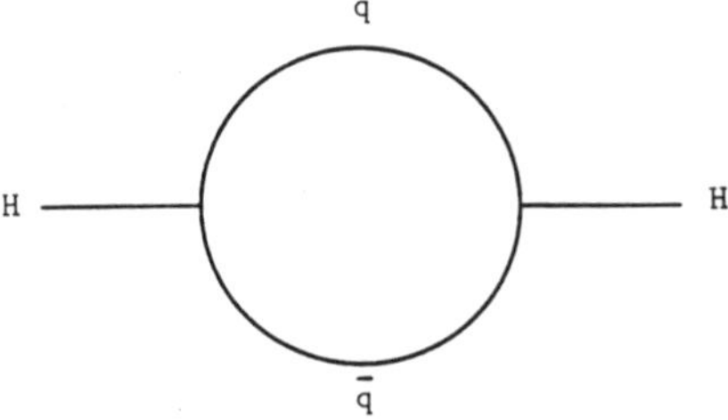


Fig. 1 One loop correction to Higgs self mass from Higgs coupling to quarks.

Solutions to the gauge hierarchy problem fall into two categories: either one assumes the Higgs is composite (e.g. as in technicolor or $t\bar{t}$ condensate models) and hence dissociates at scale $\Lambda$, or one assumes a symmetry exists to cancel the quadratic divergences. The latter possibility is supersymmetry where the Bose-Fermi symmetry causes this cancellation. For perfect supersymmetry, the two diagrams of Fig. 2 precisely cancel. If supersymmetry is broken by lifting the squark-quark degeneracy then the quadratic divergence still cancels leaving an underlying logarithmic divergence:

$$\Lambda^2 \to (m_{\tilde{q}}^2 - m_q^2)\ell n(\Lambda^2/m_{\tilde{q}}^2) \tag{3}$$

Thus to avoid fine tuning we need $m_{\tilde{q}} \gtrsim 1$ TeV, i.e. $M_S \lesssim 1$ TeV and the new SUSY particles lie within the range for detection by current and planned accelerators. In fact, for a wide class of models it has been shown that $m_h \lesssim 146$ GeV [3] (and usually $m_h \lesssim 120$ GeV) which would make the light Higgs accessible to LEP200 or its upgrades.

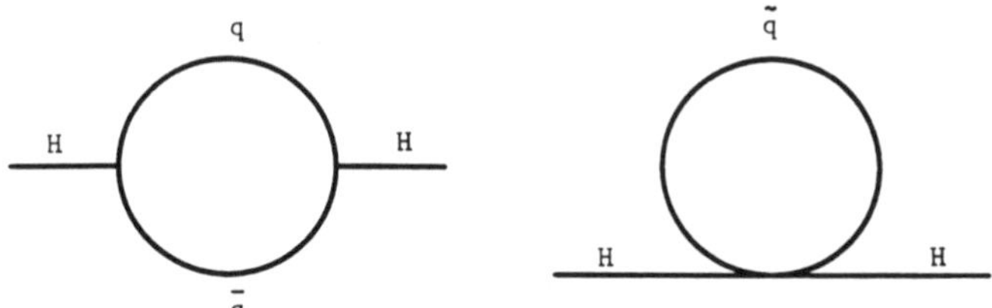


Fig. 2. Higgs one loop corrections in supersymmetric models. $\tilde{q}$ are spin zero squarks.

## 2. TRIVIALITY BOUND: AN ALTERNATE VIEW

The analysis given above takes the viewpoint that the bare Lagrangian is the fundamental quantity. However, the Standard Model is a renormalizable field theory. One can therefore pre-renormalize it (by introducing counter terms) and deal only with finite renormalized Green's functions. Masses and coupling constants can then be defined by these Green's functions at fixed momenta e.g. for the renormalized Higgs propagator $\Delta_H^{(R)}(q^2)$ one may define the Higgs mass parameter $m_H$ by $m_H^2 = [\Delta_H^{(R)}(0)]^{-1}$. The $Z_2$ rescaling of $\Delta_H^{(R)}$ can be defined by the condition $[\partial(\Delta_H^{(R)})^{-1}/\partial q^2]_{q^2=0} = 1$. Similarly, the $\lambda\phi^4$ coupling constant may be defined from the renormalized 4-point vertex $\Gamma_4^{(R)}(p_1, p_2, p_3)$ by $\lambda = \Gamma_4^{(R)}(0,0,0)$.

In the tree approximation, one has $V_H = -m^2\phi^+\phi + \lambda(\phi^+\phi)^2$ with $m^2$, $\lambda > 0$, and defining $\langle\phi\rangle \equiv v/\sqrt{2}$ one finds $\langle\phi\rangle^2 = m^2/2\lambda$ and the Higgs mass to be $m_H^2 = 2m^2$. Since $M_W = g_2 v/2$ (and hence $v \cong 247$ GeV) one may write

$$M_W = \frac{g_2}{2\sqrt{2\lambda}} m_H \tag{4}$$

showing that the Higgs mass scales electroweak physics, and also that

$$\lambda = \frac{g_2^2}{8} \frac{m_H^2}{M_W^2} \tag{5}$$

If one takes now the alternate viewpoint that the renormalized field theory is the fundamental theory, one never sees a quadratic divergence (or any other divergence). Thus the theory has no problems unless it is internally inconsistent (under which circumstances it would self-destruct). This actually happens, as the theory develops a Landau pole. Letting $\lambda(Q)$ be the running coupling constant, one finds, in the approximation of keeping only the Higgs self-couplings of $V_H$, the result

$$\lambda(Q) = \frac{\lambda(M_W)}{1 - \frac{3\lambda(M_W)}{4\pi^2} \ell n(Q^2/M_W^2)} \tag{6}$$

where $\lambda(M_W)$ is the low energy value given approximately by Eq. (5). A pole occurs in Eq. (6) at scale $Q_0$ where the denominator vanishes. The theory breaks down at $Q \approx Q_0$ and so $Q_0$ must be a scale where new physics sets in. Using Eq. (5) one finds for this scale

$$\frac{3}{4\pi} \alpha_2 \frac{m_H^2}{M_W^2} \ell n(Q_0/M_W) = 1 \tag{7}$$

In this viewpoint, the scale of new physics is determined by the *experimental* value of the Higgs mass, and the lighter the Higgs mass the larger $Q_0$ is. For example, if $m_H = 146$ GeV one finds $Q_0 \cong M_{P\ell}$ while if $m_H = 500$ GeV then $Q_0 \cong 2$ TeV. Thus, if the Higgs is light, the Standard Model could hold all the way up to the Planck scale. If the Higgs is heavy, the Standard Model must break down in the TeV range implying an upper limit on $m_H$. (Of course, the argument does not exclude new physics from arising before $Q_0$ from some other cause, but only that $Q_0$ is an upper bound on the validity of the SM.)

The analysis given here can be extended to include gauge and Yukawa couplings, and has been performed using lattice gauge theory (as the theory becomes non-perturbative near the Landau pole). The above results remain qualitatively correct. (See, e.g. Ref [4].) Which viewpoint, the previous discussion of the gauge hierarchy problem or the Landau pole problem, determines the scale where new physics must arise depends on whether one believes the bare or renormalized theory is fundamental. In this discussion we take the gauge hierarchy problem as fundamental, and discuss the consequences of the supersymmetric solution to this difficulty.

## 3. SUSY BASICS

In supersymmetry, multiplets must have an equal number of Fermi and Bose helicity states. To build a supersymmetrized Standard Model, one needs two types of massless multiplets, chiral multiplets and vector multiplets.

Chiral Multiplets: $(z(x),\ \chi(x))$

Here $z(x)$ is a complex scalar field (s=0) and $\chi(x)$ is a left-handed (L) Weyl spinor (s = 1/2). Thus $\chi(x)$ can be used to represent quarks and leptons and also the spin 1/2 Higgsino partners of the Higgs boson, while the $z(x)$ can be used to represent the Higgs boson and the spin 0 squarks and slepton partners of the quarks and leptons

Vector Multiplets: $(V^{\mu}(x), \lambda(x))$

Here the $V^{\mu}(x)$ are real vector fields (s = 1) representing the gauge bosons, and $\lambda(x)$ are Majorana spinors (s = 1/2) representing the gaugino partners.

The Higgs doublets must come in pairs in supersymmetry to cancel anomalies. The minimal number is just two:

$$H_1 = (H_1^0, H_1^-); H_2 = (H_2^+, H_2^0) \tag{8}$$

The dynamics of global supersymmetry consists of gauge interactions (supersymmetrized) and Yukawa interactions governed by the superpotential $W$. In general, $W(z_a)$ is a holomorphic function of the scalar fields $z_a$ and hence independent of the $z_a^{\dagger}$. For renormalizable interactions, $W$ is at most cubic in the fields. Thus the most general renormalizable $SU(3)_C \times SU(2)_L \times U(1)_Y$ and $R$ parity invariant form for $W$ is

$$\begin{aligned} W = \mu H_1^{\alpha} H_{2\alpha} + [\lambda_{ij}^{(u)} q_i^{\alpha} H_{2\alpha} u_j^C + \lambda_{ij}^{(d)} q_i^{\alpha} H_{1\alpha} d_j^C \\ + \lambda_{ij}^{(\ell)} \ell_i^{\alpha} H_{1\alpha} e_j^C] \end{aligned} \tag{9}$$

Here $i, j = 1, 2, 3$ are generation indices, $\alpha = 1, 2$ is the $SU(2)_L$ index ($H_{\alpha} = \varepsilon_{\alpha\beta} H^{\beta}$, $\varepsilon_{\alpha\beta} = -\varepsilon_{\beta\alpha}$, $\varepsilon_{12} = +1$), $C$ = charge conjugate, $\lambda_{ij}^{(u,d,\ell)}$ are Yukawa coupling constants and $\mu$ is a mass scaling the Higgs mixing term. Note that the gauge invariant $u$-quark interaction requires the $H_{2\alpha}$ Higgs doublet to appear, since $H_{1\alpha}^{\dagger}$ cannot enter as $W$ is holomorphic. Thus the existance of two Higgs doublets is also necessary to obtain mass growth of both the up and down quarks.

The supersymmetry invariant dynamics can be described by an effective potential

$$V = \sum_a \mid \frac{\partial W}{\partial Z_a} \mid^2 + V_D; \;\; V_D = \frac{1}{2}\, g_i^2 D_{ir} D_{ir} \tag{10}$$

(where $g_i$ are the $SU(3) \times SU(2) \times U(1)$ coupling constants, $D_{ir} = z_a^{\dagger} (T^{ir})_{ab} z_b$, $T_{ab}^{ir}$ =group generators), and fermionic interactions

$$\mathcal{L}_Y = -\frac{1}{2} \sum_{a,b} (\bar{\chi}^{aC} \frac{\partial^2 W}{\partial z_a \partial z_b} \chi^b + h.c.) \tag{11}$$

and

$$\mathcal{L}_{\lambda} = -i\sqrt{2}\, \sum g_i \bar{\lambda}^{ir} z_a^{\dagger} (T^{ir})_b^a \chi_b + h.c. \tag{12}$$

Note that $V_D$ plays the role of the $\lambda(\phi^+\phi)^2$ term in the SM, but with $\lambda$ replaced by the gauge coupling constants $g_i$. It is this that allows SUSY predictions of Higgs mass bounds since the $g_i$ are known.

After $SU(2)\times U(1)$ breaking, the Higgsinos and $SU(2)\times U(1)$ gauginos mix. There result 32 new SUSY particles: (i) 12 squarks (s=0, complex): $\tilde{q}_i = (\tilde{u}_{iL}, \tilde{d}_{iL})$; $\tilde{u}_{iR}$, $\tilde{d}_{iR}$; (ii) 9 sleptons (s=0, complex) $\tilde{\ell}_i = \tilde{\nu}_{iL}, \tilde{e}_{iL}); \tilde{e}_{iR}$; (iii) 1 gluino ($s=\frac{1}{2}$, Majorana) $\lambda^a$, $a = 1\ldots 8 = SU(3)_C$ index; (iv) 2 Winos (Charginos) ($s=\frac{1}{2}$, Dirac). $\tilde{W}_i$, $i=1,2$, $m_i < m_j$ for $i<j$; (v) 4 Zinos (Neutralinos) (s = 1/2, Marjorana) $\tilde{Z}_i$, $i = 1\ldots 4$, $m_i < m_j$ for $i<j$; and (vi) 4 Higgs (s = 0) $h^0$, $H^0$ real CP even; $A^0$ real CP odd; $H^{\pm}$ charged.

The $h^0$ is the particle which most resembles the SM Higgs.

## 4. SUPERGRAVITY BASICS

The global SUSY models discussed in the previous section possesses one serious drawback: it is not possible to achieve a phenomenologically satisfactory spontaneous breaking of supersymmetry. There are a number of reasons for this. Most obvious is that the breaking of a global symmetry implies the existence of a massless Goldstone particle, in this case a spin 1/2 particle (the Goldstino), and no candidate exists experimentally. (The neutrino interactions do not obey the correct threshold theorems.[9a]) An obvious solution to this difficulty is to promote supersymmetry to a local symmetry. The gauge particle is then spin 3/2 (the gravitino) and upon breaking of supersymmetry it absorbs the spin 1/2 Goldstino to become massive. However, supersymmetry requires that the gravitino be embedded in a massless multiplet $(g_{\mu\nu}(x);\ \psi_\mu(x))$. Here $g_{\mu\nu}(x)$ is a massless spin 2 field i.e. one is led to supergravity theory [5] where gravity is automatically included.

The coupling of supergravity to chiral and vector matter multiplets depends upon the following functions [6-9]: the superpotential $W(z_a)$, the Kähler potential $d(z_z, z_a^\dagger)$, and the gauge kinetic function $f_{\alpha\beta}(z_a, z_a^\dagger)$ where $\alpha, \beta$ are gauge indices. Actually, $W$ and $d$ enter only in the combination $\mathcal{G} = -\kappa^2 d - \ell n\ [\kappa^6 W W^\dagger]$ where $\kappa \equiv 1/M_{P\ell}$. We will assume in the following that $d$ and $f_{\alpha\beta}$ can be expanded in powers of fields with the higher non-linear terms scaled by $\kappa$:

$$d(z_a, z_a^\dagger) = c_b^a z_a z_b^\dagger + \ (a^{ab} z_a z_b + h.c.) + \kappa c_c^{ab} z_a z_b z_c^\dagger + \cdots \tag{13}$$

$$f_{\alpha\beta}(z_a z_a^\dagger) = c_{\alpha\beta} + \kappa(c_{\alpha\beta}^a z_a + h.c.) + \cdots \tag{14}$$

Supersymmetry breaking can occur at the tree level [10] or via condensates [11] due to supergravity interactions. The simplest example is to choose $W = m^2(z+B)$, $d = z_a z_a^\dagger$ and minimizing the effective potential one finds $\langle z\rangle = \pm\kappa^{-1}(\sqrt{2}-\sqrt{6}) = O(M_{P\ell})$. (One may further chose B to fine tune the cosmological constant to zero.) The quantity $M_S = O(\langle\kappa^2 W\rangle) \sim \kappa m^2$ will turn out to scale the SUSY masses.

The full supergravity dynamics is quite complicated. (For a discussion see Refs. [8,12].) We list here some of the important terms. The effective potential is given by

$$V = e^{\kappa d}[(g^{-1})_b^a(\frac{\partial W}{\partial z_a} + \kappa^2 d_a W)(\frac{\partial W}{\partial z_b} + \kappa^2 d_b W)^\dagger - 3\kappa^2 \mid W \mid^2] + V_D \tag{15}$$

where

$$V_D = \frac{1}{2}g^2 Re(f^{-1})_{\alpha\beta} D_\alpha D_\beta;\ D_\alpha = d^a (T^\alpha)_{ab} z_b \tag{16}$$

$d_a = \partial d/\partial z_a^\dagger$, $d^a = \partial d/\partial z_a$ and $g_b^a = \partial^2 d/\partial z_b \partial z_a^\dagger$, and $\alpha, \beta$ are gauge indices. Thus there are $\kappa = 1/M_{P\ell}$ corrections to Eq. (10). The scalar field kinetic energy is $-\frac{1}{2}g_b^a(D_\mu z_b)(D_\mu z_a)^\dagger$ (where $D_\mu$ is the covariant derivative). From Eq.(13), $g_b^a = c_b^a + O(\kappa)$ and so diagonalizing $c_b^a$ brings the scalar kinetic energy into canonical form. The $O(\kappa)$ correction are non-renormalizable corrections scaled by $1/M_{P\ell}$, and presumably small below the GUT scale. The gauge field kinetic energy is $-\frac{1}{4}(\mathrm{Re}\ f_{\alpha\beta})F^\alpha_{\mu\nu}F^{\mu\nu\beta}$ and from Eq.(14) one sees that one obtains the canonical kinetic energy plus possible $1/M_{P\ell}$ corrections. Finally, there is a gaugino term

$$[e^{\kappa^2 d}\kappa^2 \mid W \mid (g^{-1})_b^a d^b f^\dagger_{\alpha\beta a}]\bar{\lambda}^\alpha \lambda^\beta \tag{17}$$

where $f_{\alpha\beta a} \equiv \partial f_{\alpha\beta}/\partial z_a$.

To see the effects of supersymmetry breaking we consider the simple tree model discussed above where $W = m^2(z+B)$ and $d = z_a z_a^\dagger$. From Eq. (15) one has the term

$$(\kappa^2 d_a W)(\kappa^2 d_a W)^\dagger \to (\kappa^2\ \langle W\ \rangle)^2 z_a z_a^\dagger \tag{18}$$

Thus each scalar field grows a universal mass $m_0^2 = (\kappa^2\langle W\rangle)^2 = O(M_S^2)$. From Eqs. (17) and (14) for the case $z_a = z$, a universal gaugino mass, $m_{1/2}$, forms of size

$$\langle \kappa^2 \mid W \mid (\partial d/\partial z) f^\dagger_{\alpha\beta z}\rangle = \langle \kappa^2 \mid W \mid z^\dagger \kappa c^z_{\alpha\beta}\rangle \tag{19}$$

and so $m_{1/2} = O(M_S)$. Further, by transforming the second term of Eq. (13) from the Kähler potential to the superpotential by a Kähler transformation, a Higgs mixing parameter $\mu_0$ forms where

$$\mu_0 = \langle \kappa^2 W\rangle a^{H_1 H_2} = O(M_S) \tag{20}$$

Finally, two additional supersymmetry breaking structures arise from Eq. (15) when the matter parts of the superpotential are included:

$$A_0 W^{(3)} + B_0 W^{(2)} \tag{21}$$

where $W^{(2,3)}$ are the (quadratic, cubic) parts of the matter superpotential. One finds here also that

$$A_0, B_0 = O(M_S) \tag{22}$$

so that supersymmetry breaking gives rise to four soft breaking terms scaled by $m_0$, $m_{1/2}$, $A_0$, $B_0$ and a supersymmetric Higgs mixing parameter $\mu_0$. All these parameters are $O(M_S)$.

## 5. RADIATIVE BREAKING

A remarkable feature of supergravity GUT models is that they offer a natural explanation of $SU(2)\times U(1)$ beaking via radiative corrections [13]. In the Standard Model, $SU(2) \times U(1)$ breaking is *accomodated* by the device of having a negative $(mass)^2$ for the Higgs. However, no explanation is given as to why this choice should be made. We saw in Eq. (18), that supersymmetry breaking gives rise to a universal positive $(mass)^2$, $m_0^2 > 0$, at the scale $Q = M_G$. One may now run the renormalization group equations (RGE) down to the electroweak scale. As shown schematically in Fig. 3, $m_{H_2}^2$ bends downward and eventually turns negative (due to the t-quark Yukawa couplings) signaling the breaking of $SU(2) \times U(1)$.

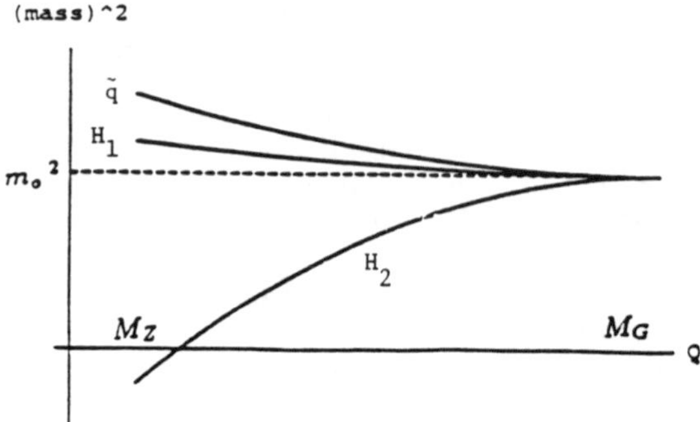


Fig. 3. Schematic diagram of running masses showing that $m_H^2$ turns negative at the electroweak scale due to the heavy top interactions.

To see the above more quantitatively, the renormalizable Higgs interactions from Eq. (15), have the form

$$\begin{aligned} V_H = m_1^2(t) \mid H_1 \mid^2 + m_2^2(t) \mid H_2 \mid^2 - m_3^2(t)(H_1H_2 + h.c.) \\ + \frac{1}{8}[g_2^2(t) + g_Y^2(t)][\mid H_1 \mid^2 - \mid H_2 \mid^2]^2 + \Delta V_1 \end{aligned} \tag{23}$$

where $\Delta V_1$ is the one loop addition, and all parameters are running with respect to the variable $t = \ell n[M_G^2/Q^2]$. In Eq. (23), the masses are defined by $m_i^2(t) = m_{Hi}^2(t) + \mu^2(t)$, $i = 1, 2$ and $m_3^2(t) = -B(t)\mu(t)$ subject to the boundary conditions at $Q = M_G$ of $m_i^2(0) = m_0^2 + \mu_0^2, m_3^2(t) = -B_0\mu_0$. Minimizing the effective potential, $\partial V_H/\partial v_i = 0$, $v_i \equiv \langle H_i \rangle$ one finds

$$\frac{1}{2}M_Z^2 = \frac{\mu_1^2 - \mu_2^2\tan^2\beta}{\tan^2\beta - 1}; \ \sin2\beta = \frac{2m_3^2}{\mu_1^2 + \mu_2^2} \tag{24}$$

where $\mu_i^2 = m_i^2 + \Sigma_i$ and $\tan\beta \equiv v_2/v_1$. ($\Sigma_i$ are the one loop corrections.) The RGE allows one to express all the parameters in Eq. (24) in terms of the GUT scale constants $m_0, m_{1/2}, A_0, B_0$ and $\mu_0$. One may use Eq. (24) to eliminate $B_0$ and $\mu_0^2$ in terms of the remaining constants and tan $\beta$. Thus one can express all 32 SUSY masses in terms of the four parameters $m_0, m_{1/2}, A_0$ and tan $\beta$ and the as yet undetermined top quark mass $m_t$. Since the sign of $\mu_0$ is not determined there are two branches: $\mu_0 < 0$ and $\mu_0 > 0$. It is interesting to ask under what conditions will a satisfactory electroweak breaking occur. Three necessary conditions are: (i) Not all the soft breaking parameters, $m_0, m_{1/2}$, $A_0$, and $B_0$ can be zero; (ii) $\mu_0$ must be non-zero; (iii) $m_t$ must be large ($m_t > \gtrsim 90$ GeV). Thus in a real sense, item (i) implies that supersymmetry breaking triggers electroweak breaking, and from (iii) the existence of electroweak breaking predicts that the top must be heavy.

## 6. SIMPLE GUT MODEL

In Sec. 1, grand unification was discussed neglecting, however, the existence of possible GUT states which would produce threshold corrections in the vicinity of $M_G$. In order to see the size of these effects, we examine here a simple $SU(5)$ model first proposed within the framework of global supersymmetry [14]. GUT physics here is characterized by the superpotential

$$W_G = \lambda_1[\frac{1}{3}Tr\Sigma^3 + \frac{1}{2}MTr\Sigma^2] + \lambda_2 H^Y[\Sigma_Y^X + 3M'\delta_Y^X]\bar{H}_X) \tag{25}$$

Here $\Sigma_Y^X(X, Y = 1\ldots5)$ is a 24 of $SU(5)$, while $H^Y$ and $\bar{H}_X$ are a 5 and $\bar{5}$ of $SU(5)$. The $SU(2)$ doublets of $H^X$ and $\bar{H}_X$ are just the $H_1$ and $H_2$ doublets of low energy theory. They are kept light by the choice $M = M'$ (which we will make here) though more natural ways of keeping the Higgs doublets light exist [15]. Upon minimizing the effective potential $\Sigma_Y^X$ grows the VEV

$$\mathrm{diag}\langle\Sigma_Y^X\rangle = M(2, 2, 2, -3, -3) \tag{26}$$

breaking $SU(5)$ to the SM. We have then that $M = O(M_G)$. The states that become superheavy are the color triplets of $H^X$ and $\bar{H}_X$ transforming like (3,1) and $(\bar{3}, 1)$ under $SU(3)_C \times SU(2)_L$ with mass $M_{H_3} = 5\lambda_2 M$, massive vector multiplets, transforming as (3,2) and $(\bar{3}, 2)$ with mass $M_V = 5\sqrt{2}gM$ ($\alpha_G \equiv g^2/4\pi$) and the superheavy components of $\Sigma_Y^X$ transforming as (8,1), (1,3) and (1,1) with masses $M_\Sigma^8 = 5\lambda_1 M/2 = M_\Sigma^3$ and $M_\Sigma^0 = \lambda_1 M/2$. This model has been considered previously [16] (though with inaccurate arguments).

We limit here $\lambda_{1,2} < 2$ (so that one stays within the perturbative domain) and also require $\lambda_{1,2} > 0.01$ (so that the superheavy spectra stay in the GUT range). When thresholds are ignored, the RGE can be used to predict a value for $\alpha_3(M_Z)$. With thresholds, one gets instead a correlation between $\alpha_3(M_Z)$ and $M_{H_3}$. As seen in Fig. 4 [17], one obtains an upperbound of $\alpha_3(M_Z) < 0.135$. Since current proton decay data requires $M_{H_3} \gtrsim 1 \times 10^{16}$ GeV, one also gets a lower bound of $\alpha_3(M_Z) > 0.114$. These are consistent with the current experimental bounds of $\alpha_3(M_Z) = 0.118 \pm 0.007$. For the $1\sigma$ upper limit of $\alpha_3(M_Z) = 0.125$, one finds $M_{H_3} < 2 \times 10^{17}$GeV, and so $M_{H_3}$ is always below the Planck scale [18]. Thus the model gives generally reasonable results. Measurements of the SUSY particle masses would determine $M_S$ which corresponds in Fig. 4 to a line in between the $M_S = 1$ TeV and $M_S = 30$ GeV bounding lines. That, plus an accurate measurement of $\alpha_3(M_Z)$, would determine a point within the quadrilateral and hence fix $M_{H_3}$. Thus accurate low energy measurements would allow a prediction of the proton decay rate for $p \to \bar{\nu}K^+$, i.e. the model can also be experimentally tested!

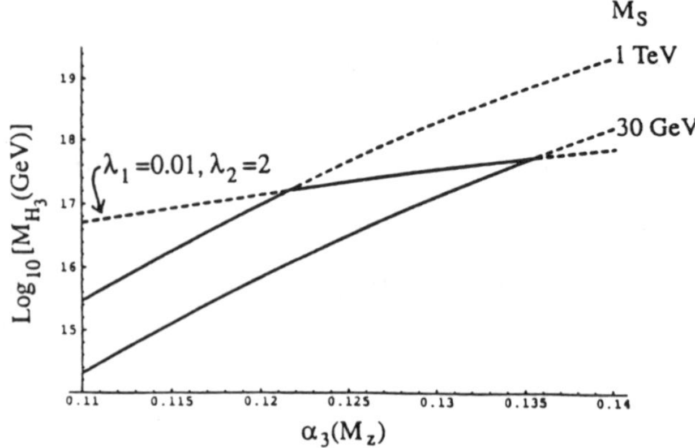


Fig. 4. Grand unification constraints for the GUT model of Eq. (25). Grand unification correlates $\alpha_3(M_Z)$ with $M_{H_3}$. The allowed region is within the solid quadrilateral.

## 7. GUT PHYSICS OR PLANCK PHYSICS?

Supergravity GUT models do not represent a fundamental theory, but rather an effective theory valid at energies below $M_G$. One may ask what aspects of the theory can be understood at the GUT level, and what requires higher scale physics, presumably unknown Planck scale physics, to understand. We list here a few speculations.

(i) Unification of gauge couplings. This is presumably GUT physics since it depends on the particle spectrum below $M_G$ and on the grand unification group $G$ which holds above $M_G$.

(ii) Quark/lepton masses, KM matrix elements, Yukawa coupling constants are presumably Planck scale physics (e.g. as in string theory) except for possible symmetry constraints that the GUT group G may impose.

(iii) Nature of supersymmetry breaking. The structure of the hidden sector where supersymmetry breaking takes place is presumably Planck physics. However, it can be parameterized at the GUT scale in terms of five parameters $m_0$, $m_{1/2}$, $A_0$, $B_0$ and $\mu_0$.

(iv) Squark/slepton masses and widths. This is GUT physics, once the five hidden sector parameters are chosen.

(v) Electroweak breaking. This is GUT physics, once the hidden sector parameters are chosen.

(vi) Proton stability. GUT physics depending on the interactions which break $G$ to the SM group.

We see from the above, that while supergravity grand unified models add significantly to our understanding of low energy physics, there are a number of areas, notably in the Yukawa couplings and in the structure of the hidden sector, where it offers no new insights. For these one must make a phenomenological treatment.

## 8. PROTON DECAY

There are two main modes of proton decay in GUT models: $p \to e^+ + \pi^0$ and $p \to \bar{\nu} + K^+$. The former can occur in both SUSY and non-SUSY grand unification, and generally will occur for any model whose grand unification group $G$ possesses $SU(5)$ as a subgroup. The latter is a specifically supersymmetric mode. Thus the observation of $p \to \bar{\nu}K^+$ would be a strong indication of the validity of supergravity grand unification. This decay can also occur when $G$ possesses an $SU(5)$ subgroup and if the light matter below $M_G$ is embedded in the usual way in 10 and $\bar{5}$ representations of the $SU(5)$ subgroup. However, it is possible to construct a complicated Higgs sector where one fine tunes the $p \to \bar{\nu}K$ amplitude to zero and still maintains only too light Higgs doublets below $M_G$. However, such models appear somewhat artificial, and the $p \to \bar{\nu}K$ decay mode is generally expected to arise, though it can be evaded.

(i) $p \to e^+\pi^0$. This mode proceeds as in non-SUSY GUTs through the superheavy vector bosons of mass $M_V = O(M_G)$. For SUSY models one has [19]

$$\tau(p \to e^+\pi^0) = 10^{31\pm1}(\frac{M_V}{6 \times 10^{14}\ \text{GeV}})^4\text{yr} \tag{27}$$

The current experimental bound is [20] $\tau(p \to e^+\pi^0) > 5.5\times10^{32}$ yr (90% CL). Super Kamiokande expects to be sensitive up to a lifetime $\tau(p \to e^+\pi^0) < 1 \times 10^{34}$ yr [21]. From Eq. (27) this would require $M_V \lesssim 6 \times 10^{15}$ GeV for the decay mode to be observable.

(ii) $p \to \bar{\nu}K^+$. For the models discussed above, this mode proceeds through the exchange of the superheavy Higgsino color triplet as can be seen in Fig. 5 [22,23]. Current data [20] gives the bound $\tau(p \to \bar{\nu}K^+) > 1 \times 10^{32}$ yr (90% CL). From Fig. 5, one sees that the amplitude for decay depends on $1/M_{H_3}$. The current data then puts a bound of $M_{H_3} \gtrsim 1 \times 10^{16}$ GeV [24]. Future experiments expect an increased sensitivity for Super Kamiokande of up to $\tau(p \to \bar{\nu}K^+) < 2\times10^{33}$ yr [21], and for ICARUS of up to $\tau(p \to \bar{\nu}K^+) < 5 \times 10^{33}$ yr [25]. Thus the GUT model of Sec. 6, where $M_{H_3} < 2M_V$, would predict that if the $p \to e^+\pi^0$ mode at future experiments were observed, the $p \to \bar{\nu}K^+$ should be seen very copiously as then $M_{H_3}$ would be less than $1.2 \times 10^{16}$ GeV.

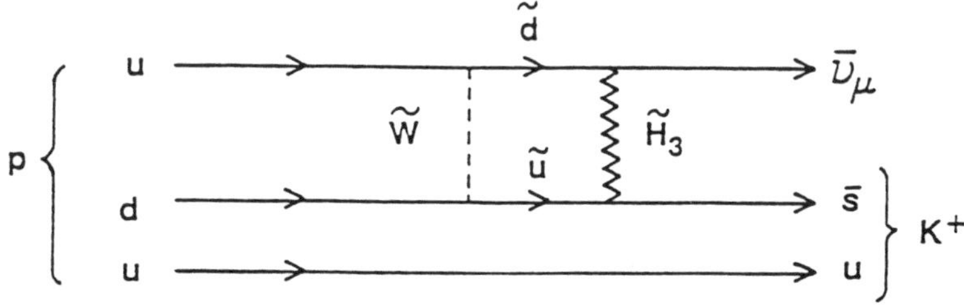


Fig. 5. Example of diagram contributing to the decay $p \to \bar{\nu}_\mu K^+$. There are additional diagrams with $\bar{\nu}_\tau$ and $\bar{\nu}_e$ final state. CKM matrix elements appear at the $\tilde{W}$ vertices.

The p$\to$ $\bar{\nu}K^+$ amplitude depends not only on $M_{H_3}$ but also in a detailed way, on the SUSY masses of the particles in the loop of Fig. 5 [23]. Since as discussed in Sec. 5, these masses are functions of the basic parameters, which we may choose to be $m_0$, $m_{\tilde{g}} = [\alpha_3(m_{\tilde{g}})/\alpha_G]m_{1/2}, A_t$ (The t-quark A parameter at the electroweak scale) and tan $\beta$, the current bounds on p-decay give rise to bounds in this parameter space. If we restrict $M_{H_3} < 2 \times 10^{17}$GeV (which keeps

$M_{H_3}/M_{P\ell} < 1/10$ and is what is implied by the GUT model of Sec. 6) one finds the restrictions $\tan\beta l8, | A_t/m_0 | l2$ and in most of the parameter space $m_0 > m_{\tilde{g}}$. Fig. 6 [26] shows what can be expected from future proton decay experiments. Thus if we require $m_0 \leq 1$ TeV (to prevent excessive fine tuning), we see that ICARUS should detect $p \to \bar{\nu}K^+$ proton decay for even the largest value of $M_{H_3}$ considered here (and Super Kamiokande should similarly detect this mode for $m_0 \leq 950$ GeV) if $m_{\tilde{W}_1} > 100$ GeV. Thus if these experiments do not see proton decay, then $m_{\tilde{W}_1} < 100$ GeV, and hence the light Wino should be observable at LEP 200. In either case, $m_{\tilde{W}_1} < 100$ GeV or $m_{\tilde{W}_1} > 100$ GeV these models with $SU(5)$ type proton decay imply that a signal of supersymmetry should be observed, and this could occur prior to the turning on of the LHC or SSC.

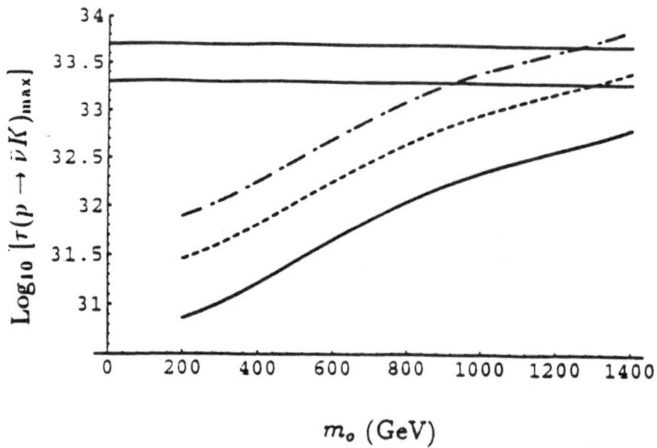


Fig. 6. Maximum value of $\tau(p \to \bar{\nu}K^+)$ for $m_t = 150$ GeV, $\mu < 0$ subject to the constraint $m_{\tilde{W}_1} > 100$ GeV. The dash-dot curve is for $M_{H_3} = 2 \times 10^{17}$ GeV. The dashed curve for $M_{H_3} = 1.2 \times 10^{17}$ GeV, and the solid curve for $M_{H_3} = 6 \times 10^{16}$ GeV. The horizontal upper and lower lines are the bounds of ICARUS and Super Kamiokande.

## 9. CONCLUSIONS

Supersymmetry represents a natural way of solving the gauge hierarchy problem. Local supersymmetry, i.e. supergravity, supplies a formal structure for treating supersymmetric grand unified models which allow for a consistent treatment of spontaneous breaking of supersymmetry. The supergravity GUT models have a large amount of predictive ability in that the 32 SUSY particle masses are determined from only five parameters. One set of mass relations which holds in several models over most of the parameter space is the following scaling relations [24, 27]:

$$2m_{\tilde{Z}_1} \cong m_{\tilde{W}_1} \cong m_{\tilde{Z}_2} \tag{28}$$

$$m_{\tilde{W}_2} \cong m_{\tilde{Z}_3} \cong m_{\tilde{Z}_4} >> m_{\tilde{Z}_1} \tag{29}$$

$$m_{\tilde{W}_1} \simeq \frac{1}{3} m_{\tilde{g}} \text{ for } \mu < 0; m_{\tilde{W}_1} \simeq \frac{1}{4}\, m_{\tilde{g}} \text{ for } \mu > 0 \tag{30}$$

and

$$m_{H^0} \cong m_A \cong m_{H^\pm} >> m_h \tag{31}$$

These relations are actually the remnants of the gauge hierarchy problem. Thus in most of the allowed parameter space one has $m_0^2$, $m_{\tilde{g}}^2 >> M_Z^2$ (which occurs already when $m_0$, $m_{\tilde{g}} \gtrsim (2-3) M_Z$). In the radiative breaking equations, this usually means then that $\mu^2 >> M_Z^2$ to guarantee enough cancellation so that the r.h.s. of the first equation in Eq. (24) correctly adds up to only $\frac{1}{2}M_Z^2$. One can then check that Eqs. (28) - (31) are a consequence of $\mu^2 >> M_Z^2$ etc. A verificaiton of Eqs. (28) - (31) would be strong support of supergravity GUT models as they depend strongly on how the structure of the theory at the GUT scale accomplishes $SU(2) \times U(1)$ breaking at the electroweak scale.

Finally, we should like to stress that in spite of the ability of supergravity GUT models to make testable predictions such as the ones discussed above, even if it is a valid idea, it must still be viewed as an approximate effective theory holding at scales below $M_G$. The closeness of $M_G$ to $M_{P\ell}$, i.e. $M_G/M_{P\ell} \simeq 1/10 - 1/100$, implies then that the theory may possess $\approx (1 - 10)\%$ errors in its predictions, and precision experiments on the validity of these models could conceivably yield information on the nature of Planck scale physics.

ACKNOWLEDGEMENTS

This work was supported in part by the National Science Foundation Grants Nos. PHY-916593 and PHY-93-06906. One of us (R.A.) would like to thank the Department of Theoretical Physics, Oxford University for its kind hospitality during the writing of this report.

References and Footnotes

[1] P. Langacker, Proc. PASCOS 90-Symposium, Eds. P. Nath and S. Reucroft (World Scientific, Singapore 1990); J. Ellis, S. Kelley and D. V. Nanopoulos, Phys. Lett. 249B, 441 (1990); B260, 131, (1991); V. Amaldi, W. De Boer and H. Fürstenau, Phys. Lett. 260B, 447 (1991); F. Anselmo, L. Cifarelli, A. Peterman and A. Zichichi, Nuov. Cim. 104A, 1817 (1991); 115A, 581 (1992).

[2] H. Bethke, XXVI Conference on High Energy Physics, Dallas, 1992, Ed. J. Sanford, AIP Conf. Proc. No. 272 (1993); G. Altarelli, talk at Europhysics Conference on High Energy Physics, Marseille, 1993.

[3] G. L. Kane et al, Phys. Rev. Lett. 70, 2686 (1993).

[4] J. F. Gunion, H. E. Haber, G. Kane and S. Dawson, "The Higgs Hunter's Guide" (Addison-Wesley, Reading, MA, 1990) [Erratum: SCIPP-92/58 (1992)].

[5] D. Freedman, S. Ferrara and P. van Nieuwenhuizen, Phys. Rev. D13, 3214 (1976); S. Deser and B. Zumino, Phys. Lett. B62, 335 (1976).

[6] E. Cremmer, S. Ferrara, L. Girardello and A. van Proeyen, Phys. Lett. 116B, 231 (1982); Nucl. Phys.B212, 413 (1983).

[7] A. H. Chamseddine, R. Arnowitt and P. Nath, Phys. Rev. Lett. 29, 970 (1982).

[8] P. Nath, R. Arnowitt and A. H. Chamseddine, "Applied N=1 Supergravity" (World Scientific, Singapore, 1984).

[9] E. Witten and J. Bagger, Nucl. Phys. B222, 125 (1983).

[9a] B. de Wit and D.Z. Freedman, Phys. Rev. Lett. 35, 827 (1975).

[10] J. Polonyi, Univ. of Budapest Rep. No. KFKI-1977-93 (1977).

[11] H. P. Nilles, Phys. Lett. B115, 193 (1981); S. Ferrara, L. Girardello and H. P. Nilles, Phys. Lett. B125, 457 (1983).

[12] H. P. Nilles, Phys. Rep. 110, 1 (1984).

[13] K. Inoue et. al., Prog. Theor. Phys. 68, 927 (1982); L. Ibañez and G. G. Ross, Phys. Lett. B110, 227 (1982); L. Alvarez-Gaumé, J. Polchinski and M. B. Wise, Nucl. Phys. B250, 495 (1983); J. Ellis, J. Hagelin, D. V. Nanopoulos and K. Tamvakis, Phys. Lett. B125, 2275(1983); L. E. Ibañez and C. Lopez, Phys. Lett. B128, 54 (1983); Nucl. Phys. B233, 545 (1984); L. E. Ibañez, C. Lopez and C. Muños, Nucl. Phys. B256, 218 (1985); J. Ellis and F. Zwirner, Nucl. Phys. B388, 317 (1990).

[14] E. Witten, Nucl. Phys. B177, 477 (1981); B185 513 (1981), S. Dimopoulos and H. Georgi, Nucl. Phys. B193, 150 (1981); N. Sakai, Zeit. f. Phys. C11, 153 (1981).

[15] K. Inoue, A. Kakuto and T. Tankano, Prog. Theor. Phys. 75, 664 (1986); A. Anselm and A. Johansen, Phys. Lett. B200, 331 (1988); A. Anselm, Sov. Phys. JETP. 67, 663 (1988);R. Barbieri, G. Dvali and A. Strumia, Nucl. Phys. B391, 487 (1993).

[16] R. Barbieri and L. J. Hall, Phys. Rev. Lett. 68, 752 (1992); A. Faraggi, B. Grinstein and S. Meshkov, Phys. Rev. D47, 5018 (1993); L. Hall and U. Sarid, Phys. Rev. Lett. 70, 2673 (1993).

[17] D. Ring and S. Urano, unpublished (1992).

[18] While the world average for $\alpha_3(M_Z)$ is $0.118 \pm 0.007$ [2], the $1\sigma$ value on the high side, $\alpha_3(M_Z) = 0.125$, is already $2.6\sigma$ above the low energy deep inelastic scattering result of $\alpha_3(M_Z) = 0.112 \pm 0.005$. Thus one can not really go much above the $\alpha_3(M_Z) = 0.125$ without assuming some systematic error exists in the low energy measurements of $\alpha_3$.

[19] P. Langacker and N. Polonsky, Phys. Rev. D47, 4028 (1993).

[20] Particle Data Group, Phys. Rev. D45, Part 2 (June, 1992).

[21] Y. Totsuka, XXIV Conf. on High Energy Physics, Munich, 1988, Eds. R. Kotthaus and J. H. Kuhn (Springer Verlag, Berlin, Heidelberg, 1989).

[22] S. Weinberg, Phys. Rev. D26, 287 (1982); N. Sakai and T. Yanagida, Nucl. Phys. B197, 533 (1982); S. Dimopoulos, S. Raby and F. Wilczek, Phys. Lett. B112, 133 (1982); S. Chadha and M. Daniels, Nucl. Phys. B229, 105 (1983); B. A. Campbell, J. Ellis, and D. V. Nanopoulos, Phys. Lett. B141, 224 (1984).

[23] R. Arnowitt, A. H. Chamseddine and P. Nath, Phys. Lett. B156, 215 (1985); P. Nath, R. Arnowitt and A. H. Chamseddine, Phys. Rev. D32, 2348 (1985).

[24] R. Arnowitt and P. Nath, Phys. Rev. Lett. 69, 725 (1992).

[25] ICARUS Detector Group, Int. Symposium on Neutrino Astrophysics, Takayama, 1992.

[26] R. Arnowitt and P. Nath, CTP-TAMU-32/93-NUB-TH-3066/93-SSCL Preprint-440 (1993).

[27] J. L. Lopez, D. V.Nanopoulos and A. Zichichi, CERN-TH 6667 (1993).

**CHAIRMAN: R.Arnowitt**

*Scientific Secretaries: R.McPherson, M.Stoilov*

## DISCUSSION

– *Lopez:*

The mass relations among neutralinos and charginos that hold in SU(5) do not seem to depend on the basis of proton decay but occur more generally in supergravity models. I think it is a consequence of the radiative breaking mechanism.

– *Arnowitt:*

Yes, the real reason for these relations is because $\mu$ is large, which occurs because of proton decay; however, other models which give $\mu$ large will give the same relations. This will not necessarily happen in an arbitrary theory such as the MSSM.

– *Nanopoulos:*

1) What are you doing with the 2-3 splitting of Higgs particles in the minimal SU(5) case?

2) If you employ the "missing partner" mechanism or "pseudo Goldstone" bosons effect, don't you change the particle spectrum?

– *Arnowitt:*

1) First of all, everything I said was not restricted to SU(5); that is only one possible model for which the hypotheses that I laid down can hold. There are a number of ways for dealing with the doublet-triplet problem, even within SU(5). One is the method that Professor Lopez described, the missing partner models, and another is the one that Professor Anselm discussed in which you have a global symmetry of higher size which, when the local symmetry of SU(5) breaks, generates a Goldstone boson from the breaking of the global symmetry and the Higgs doublets become the Goldstone bosons.

2) I don't personally like the non-renormalizable interactions which have some problems. In the renormalizable interactions, the 75s of SU(5) lead to a complicated analysis but is, as far as I know, an acceptable model. It has not been looked at in any detail. The global SU(6) example has some additional fields which become heavy and might affect GUT thresholds, but those are probably much more minor than what would happen with 75s. I agree with you that such models need to be looked at.

– *Skenderis:*

The Standard Model suffers from the so-called triviality problem; namely, lattice calculations have shown that the theory is self-consistent only for vanishing $\phi^4$ coupling constants. Do the models you described solve that problem or not?

– *Arnowitt:*

First let me say that the triviality problem is precisely the one I called the alternate viewpoint to the fine-tuning problem that SUSY is supposed to resolve. The triviality problem is serious if the Higgs mass is high, and much less of a problem if the Higgs mass is low. The SUSY models don't have that problem because the $\phi^4$ coupling constant is replaced by the gauge coupling constant. There is a different Landau pole for high top masses, at around 190 GeV. In all of the models I presented one has an upper bound on top mass of around 175 GeV, so we don't run into that pole.

– *Habig:*

Some of the mentioned models had a "hidden sector" that interacts with the "physical sector" via gravity scaled by $1/M_P$. What are the cosmological implications of this, especially at very early times?

– *Arnowitt:*

This hasn't been studied, so I can't answer that.

– *Peccei:*

Is $M_V/M_G = O(1)$ a reasonable value or would you expect something much smaller?

– *Arnowitt:*

First of all I used as a measure what I called $M_{GUT}$, the GUT scale you would get if you ignore all thresholds and said that everything was degenerate. For the simple GUT model that I had, essentially the super-heavy vector bosons are lighter than the Higgs triplet, and the Higgs triplet can be roughly three times the super-heavy vector bosons. I said that Super Kamiokande should be able to see $p \to e^+\pi^0$ if the vector bosons have mass $\lesssim 6 \times 10^{15}$ GeV, and if you multiply that by three you get $1.8 \times 10^{16}$ GeV, and if that's the case Super Kamiokande should see the $\nu K$ mode in huge numbers.

– *Trocsanyi:*

You presented many constraints on proton decay in the $p \to \bar{\nu} + K^+$ decay mode which you said were independent of the unification group. Are these results independent of the pattern of SUSY breaking?

– *Arnowitt:*

The SUSY breaking is in the hidden sector, therefore the answer is that the results do not depend on SUSY breaking (other than the dependence on the soft breaking parameters).

– *Gibilisco:*

Can you speak about the most relevant cosmological consequences of the presence in the universe of SUSY particles like axinos?

– *Arnowitt:*

The $\mu$ parameter, if it were zero, would produce an axino. The $\mu$ parameter in the models we've been talking about is fairly large, so there isn't any axino in this theory. The cosmological consequences that are of interest is the issue of the relic density of the Lightest Supersymmetric Particle, the LSP. This restricts the parameter space a bit more, and calculations have been made for both ordinary and flipped SU(5).

– *Giannakis:*

Why can't we associate the charged Higgs component as the superpartner of the electron, and the neutral component as the superpartner of the neutrino?

– *Arnowitt:*

I think it's an old idea to hope that the supersymmetric partners are really particles that we already have. If you try to do that, you can't build an acceptable phenomenology. This was studied in the 70's.

– *McPherson:*

The higgs scalar, $h^0$, should be very close to potential experiments, and other SUSY particles should show up by around 1 TeV. If no SUSY signal is seen in current experiments or at the SSC, is SUSY dead?

– *Arnowitt:*

I certainly wouldn't want to work on it. Some will always say "we need $SSC^2$", of course.

– *Hoang:*

At the beginning of your talk you mentioned acceptable unification is only achievable for 2 Higgs doublets which are light. I would like you to comment on this statement. So my question is: what would happen if you in one case increased the number of Higgs doublets, and in the other case used heavy Higgs doublets?

– *Arnowitt:*

If you had four light Higgs doublets, then the unification scale drops precipitously to about $10^{14}$ GeV from $10^{16}$ GeV, and $p \to e^+\pi^0$ would be seen. Also, the SUSY mass rises to around $10^9$ GeV, if unification exists, which means that you have a very serious fine-tuning problem that most people would reject. So, both on experimental and theoretical grounds having four light Higgs doublets is essentially disastrous for unification. If you have heavy Higgs doublets, you can certainly put in lots of stuff at $10^{10} - 10^{12}$ GeV, and then you can move the unification scale up or down, and this is an issue related to model-building. For example, in flipped SU(5) they want the unification scale to be at the Planck scale rather than at the GUT scale where it arises from simply one Higgs doublet and the SUSY standard model spectrum, and they add additional terms to do this because that's more reasonable within their framework.

– *Beneke:*

Is there any physical reason for requiring R-parity apart from simplicity? So far, physicists have had bad experience with all kinds of parity.

– *Arnowitt:*

I think that requiring R-parity is a sense of taste. It leads to models which are simpler than models that don't have R-parity, and you have to make sure that if you break R-parity you don't do something very bad like breaking lepton number or baryon number in an experimentally unacceptable way. But you certainly can construct viable models with broken R-parity, and as far as I'm concerned they're perfectly acceptable. I just consider them the next level of complexity.

# NO-SCALE SUPERGRAVITY - A VIABLE SCENARIO FOR UNDERSTANDING THE SUSY BREAKING SCALE?

A.B. Lahanas[1]

Theoretical Physics Division, CERN
CH - 1211 Geneva 23

## ABSTRACT

We are reviewing in a pedagogical way the no-scale mechanism for the generation of the low-energy electroweak and supersymmetry breaking scales. Such an approach was used in the past in supergravity theories and later gained support from string theories, some of which yield no-scale supergravities in their low-energy limit. This same mechanism can also explain the hierarchies $M_W/M_P, M_{SUSY}/M_P \sim 0(10^{-16})$ if employed in string perturbation theory with a large internal dimension. It therefore offers as an alternative to other non-perturbative schemes for understanding the origin of supersymmetry breaking scale.

## INTRODUCTION

In the last years, supersymmetry (SUSY) has gained ground as a reasonable extension of the Standard Model (SM) with, as yet, unconfirmed experimental evidence for its validity. It is a common belief that the SM despite all its successes cannot be the ultimate theory since it involves a large number of arbitrary parameters which are not predicted but are constrained by experiment. It is therefore natural to seek extensions of the SM and at present SUSY serves as a good candidate. Besides in a supersymmetric scheme electroweak and strong forces seem to unify as the precision LEP data suggests; this provides further support for considering SUSY as a reasonable extension encompassing all interactions, except gravity, beyond the TeV scale. The local supersymmetry, or supergravity (SUGRA) as otherwise is called, offers, as a proper mathematical framework, to incorporate gravitational forces too and what is

[1]On leave of absence from University of Athens, Nuclear and Particle Physics Section, Physics Department, Athens 157-71, Greece.

perhaps more important, is the fact that such theories come out as low-energy effective theories of the superstring. Therefore supergravity theories interpolate between low-energy physics, accessible to experiments, and string theory which describes the particle dynamics beyond the Planck scale. At present supersymmetric theories are a challenge to both theoreticians and experimentalists and the hunt for experimental signatures leading to its establishment is of utmost importance.

## 1. - FROM $N = 1$ SUSY TO $N = 1$ SUGRA

In its early days, SUSY was invoked in an attempt towards understanding the masslessness of the neutrino; later it was employed in order to resolve the gauge hierarchy problem [1]. As a mathematical structure, it is based on a graded Lie algebra, which is a natural generalization of the Poincaré algebra, with elements

$$P_m \text{ (momentum)}, M_{mn} \text{ (angular momentum)}, Q_\alpha \text{ (spinorial charge)}$$

affecting translations ($\alpha_m$), rotation/boosts ($\lambda_{mn}$) and SUSY transformations ($\epsilon$). In its simplest form ($N = 1$, SUSY) it has just one spinorial charge $Q_\alpha$ of Majorana type, the generator of supersymmetry transformations, whose commutation/anticommutation relations with the remaining elements and with itself are

$$\begin{aligned} &[P_m, Q_\alpha] = 0 \ , \quad [M_{mn}, Q_\alpha] = -(\sigma_{mn} Q)_\alpha \\ &\{Q_\alpha, \bar{Q}_\beta\} = 2\gamma^m_{\alpha\beta} P_m \end{aligned}$$

Other algebraic structures involving more spinorial charges exist and in this case we talk about extended supersymmetries.

Theories based on such a mathematical structure predict a degenerate boson/fermion mass spectrum which is not realized in Nature; therefore only broken supersymmetric theories can make physical sense. In a spontaneously broken supersymmetry there appears a massless fermion field $\psi_g$ called Goldstino, the analogous of the Goldstone boson in an ordinary spontaneously broken gauge theory, as a result of the fact that the fermionic generator $Q_\alpha$ does not annihilate the vacuum, i.e., $Q_\alpha|0\rangle \neq 0$. Its coupling "$e_g$" to a multiplet accommodating a boson $b$ and a fermion $f$ lifts their mass degeneracy by an amount

$$m_b^2 - m_f^2 = f \ e_g \neq 0 \ .$$

In it, the order parameters "$f$" is related to the vacuum energy by $V_0 = f^2$ and to the transformation properties of the Goldstino under supersymmetry

$$\delta\psi_g = -\sqrt{2} f \epsilon + \text{(field dependent terms)} \ .$$

Therefore the v.e.v. of the $\delta\psi_g$ is non-vanishing and is given by $\langle\delta\psi_g\rangle = -\sqrt{2} f \epsilon$.

One promotes rigid supersymmetry to local by making the parameters $\alpha_m, \lambda_{mn}, \epsilon$ local that is space-time dependent. The corresponding gauge fields necessary to build invariant actions are the vierbein $e^m_\mu$, which is related to the graviton spin-2 field in the usual manner, the spin connection $\omega_\mu^{mn}$ and the gravitino $\psi_\mu$, a spin-3/2 field. The massless graviton and its superpartner the Rarita-Schwinger spin-3/2 field naturally appears in local supersymmetry. Therefore theories based on such a scheme naturally involve gravitational interactions and hence the name supergravity (SUGRA).

In building up $N = 1$ SUGRA actions one needs in principle three arbitrary functions of the scalar fields $\phi_i, \phi_i^*$ involved. A real function $K(\phi_i, \phi_i^*)$ known as the *Kähler potential* and two chiral functions, the *superpotential* $g(\phi_i)$ and the *gauge chiral function* $f_{\alpha\beta}(\phi_i)$. The latter transforms as the symmetric product of the adjoint representation of the gauge group $G$. Due to its *Kähler invariance* the action depends on the particular combination of

$$\mathcal{G} = K - \ln |g|^2$$

and not separately on $K$ and $g$. The final Lagrangian is then found to depend on $\mathcal{G}$, $f_{\alpha\beta}$ and derivatives of these functions. The form of the $N = 1$ SUGRA Lagrangian is rather involved and can be traced in the literature [2]. However, for the benefit of the reader we exhibit below some of its terms which will be of interest for the rest of our discussion

$$\begin{aligned}
e^{-1}L = & -\tfrac{1}{2}R && \text{(Einstein term)} \\
& +\mathcal{G}_{\bar{i}j} D_\mu \phi_i D^\mu \phi_j^* && \text{(scalar kinetic terms)} \\
& +e^{-\mathcal{G}}(3 + \mathcal{G}_k \mathcal{G}^{-1}_{\bar{k}\ell} \mathcal{G}_{\bar{\ell}}) - \tfrac{g^2}{2} Re f^{-1}_{\alpha\beta} (\mathcal{G}^i T_i^{\alpha\ j} \phi_j)\, (\mathcal{G}^k T_k^{\beta\ \ell} \phi_\ell) && \text{(}-\text{potential)} \\
& -\tfrac{1}{4} Re f_{\alpha\beta} (F^\alpha_{\mu\nu} F^{\beta,\mu\nu}) && \text{(gauge field kinetic terms)} \\
& +\mathcal{G}_{ij} \bar{\chi}_{iR} \not{D} \chi_{jR} + (h.c.) && \text{(matter fermion kinetic terms)} \\
& -\tfrac{e^{-1}}{4} \epsilon^{\mu\nu\rho\sigma} \bar{\psi}_\mu \gamma_5 \gamma_v D_\rho \psi_\sigma + (h.c.) && \text{(spin} - \tfrac{3}{2} \text{ field kinetic term)} \\
& +e^{-\mathcal{G}/2} \bar{\psi}_{\mu_L} \sigma^{\mu\nu} \psi_{\nu_R} + (h.c.) && \text{(spin}- \tfrac{3}{2} \text{ mass term)} \\
& +\tfrac{e^{-\mathcal{G}/2}}{4} \mathcal{G}_{\bar{\ell}} \mathcal{G}^{-1}_{\ell\bar{k}} \left( \tfrac{\partial f_{\alpha\beta}}{\partial \phi_k} \right)^* \bar{\lambda}^\alpha_L \lambda^\beta_R + (h.c.) && \text{(gaugino mass term)} \\
& + \ldots
\end{aligned} \tag{1}$$

Regarding the form of the $N = 1$ supergravity action some comments are in order.

(i) It is non-renormalizable. This means that we do not know how to handle its ultra-violet infinities.

(ii) For its construction, two arbitrary functions $\mathcal{G}$ and $f_{\alpha\beta}$ are needed. Besides we also have to specify the gauge symmetry and its particular physical content as in ordinary gauge theories.

Certainly such theories cannot be regarded as fundamental. However, they can make sense as effective non-renormalizable theories of some fundamental underlying theory describing the particle dynamics beyond the Planck scale (string theories); for a comparison, recall QCD and $\sigma$-model chiral Lagrangians of strong interaction. Viewed in this way, supergravities interpolate between strings and low energy physics as in the following scheme:

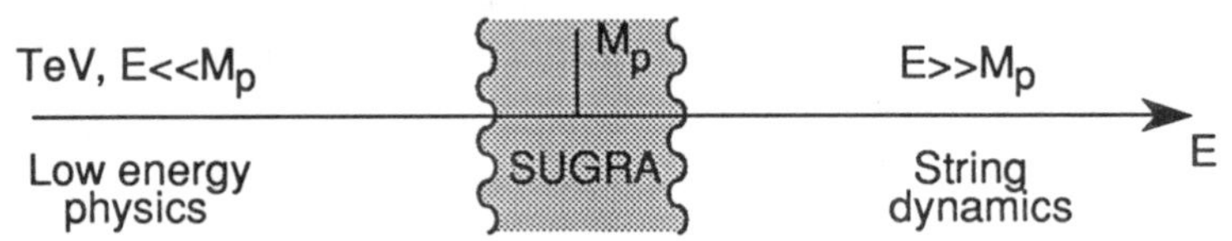


Still in a SUGRA the local supersymmetry has to be broken if we are to deal with a physical theory in which bosons and fermions are not degenerate. It has been shown [2] that if the tree-level cosmological constant vanishes, as required in any theory involving gravity,

$$V_0 \equiv \langle V \rangle = 0 \ ,$$

and $\langle \exp -\mathcal{G}/2 \rangle \neq 0$, then spontaneous symmetry breaking of local supersymmetry occurs. This breaking makes the otherwise massless spin-3/2 gravitino field (two spin states) absorb the Goldstino of the spontaneously broken SUSY (two spin states) and become massive (four spin states) with a mass given by

$$m_{3/2} = \langle \exp -\mathcal{G}/2 \rangle$$

This mechanism is called superHiggs mechanism for obvious reasons. To compare how the Higgs mechanism operates in ordinary gauge theories and in supergravities, we display the table [2].

For phenomenology we certainly need the effective theory valid at energies $E \lesssim M_{Planck}$. At such energies the non-renormalizable interactions can be neglected and the Lagrangian emerging out has the following form

$$\begin{aligned} L^{eff} = \quad & L^{SUSY} - \textstyle\sum_i m_i^2 |\phi_i|^2 - \frac{1}{2} \sum_\alpha M_\alpha \bar{\lambda}_\alpha \lambda_\alpha \\ & - \textstyle\sum \left[ A_{ijk} h_{ijk} \phi_i \phi_j \phi_k + B_{ij} \mu_{ij} \phi_i \phi_j + \text{h.c.} \right] \end{aligned} \tag{2a}$$

where $L^{SUSY}$ is a globally supersymmetric Lagrangian specified by a superpotential

$$W = \sum \left( h_{ijk} \phi_i \phi_j \phi_k + \mu_{ij} \phi_i \phi_j \right) \tag{2b}$$

[2] $m_s$ is the local supersymmetry breaking scale. The order of magnitude argument relates this to the gravitino mass as $m_{3/2} \sim m_s^2/M_P$ although the particular relation between these two scales depends actually on the so-called hidden sector of the theory.

| | Ordinary Gauge Theories | SUGRA |
|---|---|---|
| Symmetry generator | $T^a$ (spin-0) | $Q$ (spin-1/2) |
| Gauge field | $V_\mu^a$ (spin-1) | $\psi_\mu$ (spin-3/2) |
| Goldstone field | $\begin{cases} G \ \text{(spin}-0) \\ \delta_\omega G = \omega(gv) + \ldots \end{cases}$ | $\begin{cases} \eta \ \text{(spin}-1/2) \\ \delta_\epsilon \eta = \epsilon m_s^2 + \ldots \end{cases}$ |
| Gauge field mass | $M_V = g\langle H\rangle \equiv gv$ <br> ↑ <br> Higgs mechanism | $m_{3/2} = \frac{1}{\sqrt{G_N}}\langle e^{-\mathcal{G}/2}\rangle \sim m_s^2/M_P$ <br> ↑ <br> SuperHiggs mechanism |

All terms appearing in these expressions are consistent with the gauge symmetries. Note the presence of the terms $m_i^2$ (soft scalar masses) $M_\alpha$ (gaugino masses), $A_{ijk}, B_{ij}$ ($A, B$ soft terms) whose presence breaks global supersymmetry, that is without introducing quadratic divergences; this property is important for the resolution of the gauge hierarchy. These terms, usually referred to as $m_0, m_{1/2}, A, B$, emerge as a result of the spontaneous breaking of local supersymmetry[3] and are welcome since due to them the boson/fermion mass degeneracies are lifted by the following amounts

$$m_B^2 - m_F^2 = \Delta^2(m_0, m_{1/2}, A, B) \neq 0 \ . \tag{3}$$

$\Delta^2$ sets therefore the scale of the global supersymmetry breaking which, in order to protect the gauge hierarchy, must be less than $\sim$ O (1 TeV). The values of the parameters $m_0, m_{1/2}, A, B$ which are essential for phenomenology are known once the $N=1$ SUGRA is completely specified and local SUSY is broken spontaneously. If we believe that supergravities have anything to do with string theories the latter should provide us with definite answers concerning the nature of the effective supergravity (gauge group; physical content; $\mathcal{G}$ and $f_{\alpha\beta}$ functions, etc.). In this way we can have well-defined predictions that can be tested in the laboratory. Conversely, by studying phenomenologically low-energy supergravity models we may get information which will be useful in picking up the correct string theory.

## 2. - LOW ENERGY SUPERGRAVITIES AND THE RADIATIVE ELECTROWEAK SYMMETRY BREAKING

The effective supergravity theory given by Eq. (3) is meant at energies less but near the Planck scale $M_P$. For scales $\mu$ much less than $M_P$ and especially in the region of energies accessible to future experiments, $\gtrsim$ TeV, the theory is still valid but the parameters $m_0, m_{1/2}, A, B$ should be considered as running with the scale $\mu$. Their

[3]To implement local SUSY breaking one usually assumes the existence of a set of hidden fields $z^i$ singlets under the low energy gauge group which decouple from the observable sector fields at energies $E \lesssim M_{Planck}$.

evolution from $M_P$ to $\mu$ are given by their renormalization group equations (RGE) whose one-loop expression can be traced in the literature [3]. A nice feature of the low-energy supergravity models associated with the running (evolution) of the soft parameters $m_0, m_{1/2}, A, B$ is the determination of the electroweak symmetry breaking scale $\sim M_W$ through radiative corrections; the so-called radiative breaking mechanism [4]. This takes place for relatively large top Yukawa couplings $h_t \sim g_2$ and yields naturally the right order of magnitude for the electroweak breaking scale. To get an idea how this is implemented, consider for instance the case of the minimal supersymmetric Standard Model with gauge group $SU(3) \times SU(2) \times U(1)$ and allow for the appearance of soft global SUSY breaking terms $m_0, m_{1/2}, A, B$ as a result of local SUSY breaking. We assume that the superpotential has the following form[4]:

$$W = h_t Q U_c H_2 \ . \tag{4}$$

In (4), $h_t$ stands for the top Yukawa coupling and $Q, U_c, H_2$ are multiplets accommodating the left-handed top quark, right-handed top quark and the Higgs field giving mass to the top if it happens to develop a non-zero v.e.v. The RGE for the mass squared $m^2$ of the Higgs scalar $H_2$ is given in this case by[5]

$$\mu \frac{dm_2^2}{d\mu} = \frac{1}{(4\pi)^2} \Big\{ -(6g_2^2 M_2^2 + \frac{6}{5} g_1^2 M_1^2) + 6h_t^2 (m_Q^2 + m_U^2 + m_2^2 + A^2) \Big\} \tag{5}$$

where $m_{Q,U_c}^2$ are the soft masses associated with $Q, U_c$ scalars and $A$ is the soft trilinear coupling [see Eq. (3a)]. It is obvious from Eq. (5) that for relatively large values of $h_t$ the second term wins over the first and hence $\mu \frac{dm_2^2}{d\mu} > 0$. Therefore, starting from $m_0^2$ at $M_P$ $m_2^2$ decreases as we move to lower values of the scale $\mu$ and the evolution of $m_2^2$ in this case is depicted in Fig. 1

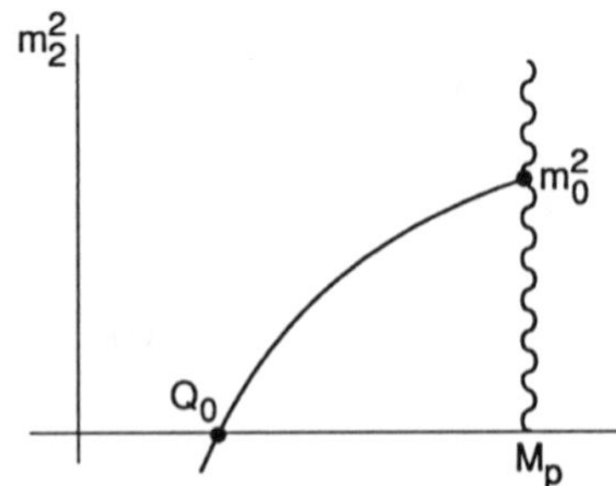


Figure 1 - Evolution of the Higgs mass $m_2^2(\mu)$ with the scale $\mu$ for the model specified by the superpotential of Eq. (4).

[4]Other terms involving the remaining quark/lepton and Higgs multiplets are possible but we consider the simplest case for reasons of clarity.

[5]$g_1 = \sqrt{5/3}\, g_Y$; $g_Y$ is the gauge coupling of the $U(1)$ group.

At the critical scale $Q_0, m_2^2$ vanishes and it becomes neative below $Q_0$, signalling therefore electroweak symmetry breaking. A rough estimate of the scale $Q_0$ yields then that

$$Q_0 = M_P \exp\{-\frac{O(1)}{\alpha_t}\} , \quad \alpha_t = \frac{h_t^2}{4\pi} \tag{6}$$

The precise value of the O(1) constant in Eq. (6) depends on dimensionless parameters, gauge coupling constants, ratios of masses at the Planck scale $M_P$, etc. $Q_0$ turns out to be naturally of the order of 100 GeV for values of $h_t$ such that $\alpha_t \gtrsim \alpha_2$. This in turn implies values for the top quark mass, of the order of the electroweak scale $\sim M_W$. Thus the heaviness of the top quark mass is closely related with the scale at which electroweak symmetry breaks down in this minimal supersymmetric extension of the Standard Model. Phenomenological models based on this scheme seem to prefer values around 150 GeV for $m_t$. The radiative breaking mechanism of the electroweak symmetry breaking is a basic ingredient of the $N = 1$ supergravity models since the weak scale $M_W$ is not put in by hand but is generated via the dynamics of the theory itself[6]. This is essential in order to understand the hierarchy of the electroweak and the Planck scale $M_W/M_P \sim 10^{-16}$.

## 3. - THE "NO-SCALE" SUPERGRAVITY

### *The standard "no-scale" model*

In the rest of our discussion we shall focus on a particular class of supergravity models the so-called no-scale models [6],[7]. These models have the virtue that both electroweak and supersymmetry breaking scales are determined through radiative corrections. Although particular models studied in the past are not acceptable, for reasons which will be explained in the sequel, however, they set up a framework within which a dynamical determination of the low-energy scales is possible. During the last years there has been a revived interest in these models since the effective low-energy Lagrangians of some string theories seem to be $N = 1$ supergravities of the no-scale type.

The early electroweak supergravity models suffered a serious fine-tuning problem associated with the vanishing of the cosmological constant. One had to fine tune some of the parameters of the theory in order to achieve vanishing vacuum energy $V_0 = 0$. Any deviation from these fine-tuned values would produce a huge cosmological constant $V_0 \sim m_0^2 M_P^2$ where $m_0$ is the global SUSY breaking scale. A classical example exhibiting clearly this pathology is the well-known Polonyi model [8]. $m_0$ is related to the sparticle masses and is therefore constrained by experiment to be larger than $\sim$ 100 GeV while $m_0 \lesssim 0$ (1 TeV) is demanded in order not to destabilize the gauge

[6] We should remark that Eq. (6) holds provided that the physical masses of the particles participating are not larger than $Q_0$ otherwise the evolution of $m_2^2$ stops before electroweak symmetry breaking takes place. Note also that small variations of $\alpha_t$ induce large variations of $Q_0$. Thus $\alpha_t$ has actually to be fine tuned. We do not discuss such subtleties associated with the radiative breaking mechanism and we refer the reader to the literature [5].

hierarchy. As a result, $V_0 \sim m_0^2 M_P^2 \sim 10^{-32} M_P^4$ certainly smaller than $m_P^4$, owing to the supersymmetric character of the theory (boson/fermion cancellations), but much larger than $10^{-120} M_P^4$ as required. How could we then avoid this fine tuning of parameters and still have vanishing vacuum energy? The answer is rather simple: "have no parameters at all!" In order to see how this can be implemented, consider a simplified toy model with just one chiral multiplet $T$ in which case the scalar potential reads

$$V = -e^{-\mathcal{G}}(3 + \mathcal{G}_T \mathcal{G}_{T^*} \mathcal{G}_{TT^*}^{-1}) \tag{7}$$

One can have $V = 0$ for any value of $T$ if $\mathcal{G} = 3\ln(T + T^*)$ [7], up to trivial field redefinitions $T \to f(T)$ [7]. Thus we have vanishing energy for any $T$ without fine tuning any parameter; in this case we actually deal with a theory which has a "flat" potential. In this particular model the gravitino mass is non-vanishing, $m_{3/2} = \langle e^{-\mathcal{G}/2} \rangle \neq 0$, that is local supersymmetry breaks but the size of the breaking is not fixed; that is the gravitino mass is sliding. This flatness of the potential is associated with the particular geometrical structure of the scalar kinetic terms. To be more precise, consider the scalar kinetic energy term [see Eq. (1)], which has a structure reminiscent of a non-sinear $\sigma$-model, given by

$$e^{-1}\mathcal{L}_{kinetic} = \mathcal{G}_{TT^*} \partial_\mu T \ \partial^\mu T^* \ . \tag{8}$$

Its form resembles the line element of a Kählerian manifold whose metric $\mathcal{G}_{TT^*}$ is derived from the Kähler potential $\mathcal{G}$. We shall then say that $T$ and $T^*$ parametrize a Kähler manifold with metric derived from $G$. However, in the case under consideration, with $\mathcal{G} = 3\ln(T+T^*)$, the Kählerian manifold is the special coset space $SU(1,1)/U(1)$, where the non-compact group $SU(1,1)$ is the group of isometries and $U(1)$ its maximal compact subgroup. Moreover, the $SU(1,1)/U(1)$ at hand is a maximally symmetric space with a curvature given by

$$R_{TT^*} = -2\mathcal{G}_{TT^*}/3$$

Hence we are actually dealing with an Einstein/Kähler manifold whose scalar curvature has the particular value [9]

$$R = 2/3$$

This value for the scalar curvature is closely associated with the vanishing of the scalar potential. Therefore there is an interplay between the particular geometric structure of the scalar kinetic terms and flatness of the scalar potential. This property of the potential was crucial in order to achieve natural vanishing of the tree-level cosmological constant. It is worth remarking that in extended $N \geq 4$ supergravities the scalars belong to irreducible representations of the graviton, multiplet and their kinetic terms always span coset spaces $G/H$ where $G$ is non-compact and $H$ is the maximal compact subgroup of $G$. Hence in such theories, scalar potentials having flat directions may exist and the vacuum energy may naturally be vanishing.

---

[7] $M_P = 1$.

The observation that there exists a particular class of supergravity theories that have naturally a vanishing cosmological constant and undetermined the scale of the spontaneously broken local supersymmetry, was crucial for the construction of the "no-scale" supergravity [6]. In order to give a very general idea of how a dynamical determination of electroweak and SUSY breaking scales, in terms of the Planck scale, is possible in such models consider a $N = 1$ supergravity involving non-single fields $\phi_i$, in addition to $T$, with a Kähler function given by

$$\mathcal{G} = 3\ln(T + T^*) - \phi_i^*\phi_i - \ln|g(\phi_i, T)|^2 \tag{9}$$

With this $\mathcal{G}$ the non-singlets under the gauge group fields $\phi_i$ have canonical kinetic terms but this is by no means mandatory and other choices may well do. We have in mind of course a low-energy $SU(3) \times SU(2) \times U(1)$ supergravity model with a minimal physical content. Later we shall discuss the case of a unified scheme. The superpotential "$g$" mixes $\phi_i$'s with $T$ and there are reasonable choices [for details, see Ref. [6]] of $g$ for which the scalar potential $V$ is positive semi-definite, $V \geq 0$, attaining its minimum at $\langle\phi\rangle = 0$ leaving undetermined the value of the v.e.v. $\langle T\rangle$;

$$V_0 \equiv V(\langle\phi\rangle, \langle T\rangle) = 0 \tag{10}$$

Therefore the cosmological constant at the classical level vanishes naturally. Also local supersymmetry breaks but owing to the flatness of the potential along the $T$-direction, the SUSY breaking scale $m_{3/2} \sim \langle T + T^*\rangle$ is not fixed; that is we have a sliding gravitino mass. The global SUSY breaking parameters $m_0, A, B$ are non-vanishing and proportional to $m_{3/2}$ in this simplified version of the no-scale model under discussion and hence undetermined too. We have thus managed to construct a model with naturally vanishing vacuum energy and undetermined, as yet, the logal/global SUSY breaking scales.

Departing from the tree-level dynamics and considering the quantum corrections the picture changes and the electroweak symmetry is broken via the radiative breaking mechanism at a scale $Q_0 \sim$ O (100 GeV) [8]. Also the tree-level potential is corrected receiving a form (in the $\overline{\text{DR}}$ scheme)

$$V(\phi, m_{3/2}) = V_{tree}(\phi, m_{3/2}) + \frac{1}{64\pi^2}\sum_J(-)^{2J}(2J+1)m_J^4(\phi)\left(\ln\frac{m_J^2(\phi)}{\mu^2} - \frac{3}{2}\right) \tag{11}$$

where $\phi$ collectively stands for all the observable fields involved. In Eq. (11), $m_{3/2}$ appears instead of the combination $T + T^*$. It follows then that the one-loop contributions to the potential in Eq. (11) deform the flatness along the $T + T^*$ direction, or equivalently $m_{3/2}$, and a minimum for $\langle T + T^*\rangle$ is developed. This is best seen by studying the vacuum energy as a function of the gravitino mass $m_{3/2}$

$$V_0(m_{3/2}) \equiv V(\langle\phi\rangle, m_{3/2}) \tag{12}$$

[8]The equation determining $Q_0$ does not depend on $m_{3/2}$!

($\langle\phi\rangle$ is non-vanishing and proportional to $m_{3/2}$ for the Higgs fields involved). If $V$ develops a minimum then $V_0(m_{3/2})$ has a minimum determined by the minimizing equation

$$\frac{dV_0}{dm_{3/2}} = 0 \tag{13}$$

In Fig. 2 we display the behaviour of $V_0$ as a function of the sliding gravitino mass $m_{3/2}$, which sets the order parameter of the local SUSY breaking scale. We observe that a minimum is developed at a point $\widetilde{m_{3/2}}$ which is close to the critical scale $Q_0$ signalling electroweak symmetry breaking. That is $\widetilde{m_{3/2}} = O(1)\ Q_0$. Hence in this model, electroweak as well as SUSY breaking scales are not put in by hand but are determined through radiative corrections to be $\sim 10^{-16} M_P$. Although we reached these conclusions based on a particular supergravity model, however, this "no-scale" mechanism sets a general framework for building phenomenological low-energy supergravity models bearing the following features:

(i) Natural vanishing of the tree-level cosmological constant, i.e., by avoiding adjustment of parameters.

(ii) Determination of the electroweak ($M_W$) and supersymmetry breaking scale ($M_S$) through radiative effects.

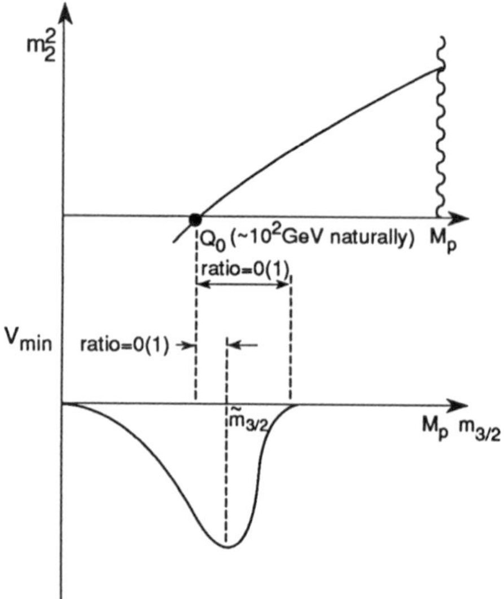


Fig. 2 - $V_{min} \equiv V(<\phi>, m_{3/2})$ as a function of the floating gravitino mass $m_{3/2}$. $V_{min}$ becomes deepest at $\tilde{m}_{3/2} = 0(1)Q_0$.

Appealing as it is, however, this scheme has its own drawbacks and criticisms have been raised about its validity:

(i) There is a great deal of arbitrariness because of the ad hoc choice of the Kähler function $\mathcal{G}$ (the same occurring in any other SUGRA model).

(ii) The tree-level cosmological constant vanishes naturally but at the one loop $V_0 \sim m_{3/2}^4 \sim 0(M_W^4)$ which is of the order of $10^{-64}M_P^4$; a huge cosmological constant indeed compared to its upper limit $10^{-120}M_P^4$ but certainly much smaller than $10^{-32}M_P^4$ of other supergravity models.

(iii) More seriously than the previous, gravitational effects may induce radiative corrections to the scalar potential given by

$$\delta V \sim M_P^2(Str\mathcal{M}^2) .$$

where $Str\mathcal{M}^2 \equiv \sum_J(-)^{2J}(2J+1)m_J^2$. If $Str\mathcal{M}^2 \sim m_{3/2}^2$, as is the case in the $SU(1,1)/U(1)$ Kählerian manifold, these corrections drive $m_{3/2}$ to values of the order of the Planck scale and the whole scheme breaks down.

Regarding these criticisms, one should not forget that we are dealing with an effective theory which is assumed to be the low-energy limit of a more fundamental theory (string?) valid at energies $E \gtrsim M_P$. Hence, answers to such questions are postponed until we have a better control of the physics beyond the Planck scale and know how to break the gravitational interactions. It is only in that case that we can have a firm answer how the gravitational effects affect the no-scale scenario.

*No scale GUTs*

The no-scale idea can also be applied to a Grand Unified Theory. The first such model to be constructed was based on the $SU(5)$ GUT embedded in a supergravity whose Kähler function was assumed to have the following form [10]

$$\mathcal{G} = 3\ln(T + T^* - \phi_i^*\phi_i) - \ln|W(\phi)|^2 . \tag{14}$$

$T$ is the previously discussed singlet field and $\phi_i, i = 1, N$ are non-singlets, in general, under the $SU(5)$ group. Moreover the superpotential $W(\phi)$ was assumed to be of the form $W = e^F$ where $F = h_{ijk}\phi_i\phi_j\phi_k$ yielding

$$W(\phi) = c + h_{ijk}\phi_i\phi_j\phi_k + \dots \quad ; \tag{15}$$

The Yukawa couplings $h_{ijk}$ are consistent with the gauge symmetries. This form for $\mathcal{G}$ is reminiscent of the structure discussed in the standard no-scale model. In the case under consideration, the scalar fields $\phi_i, T$ parametrize the Kählerian space

$SU(N,1)/U(1) \times SU(N)$ and their kinetic terms are invariant under the non-compact $SU(N,1)$ symmetry. This property makes the scalar potential be positive semi-definite with a flat direction along $T$. Without embarking on details we list in the following the salient features of the model.

(a) The scalar fields $T, \phi_i = 1,2\ldots N$ parametrize the non-compact Kählerian manifold $SU(1,N)/SU(N) \times U(1)$ with metric derived from the Kähler function $\mathcal{G} = 3\ln(T + T^* - \phi_i\phi_i^*) - \ln|W(\phi)|^2$. This entails to a positive semi-definite scalar potential $V \geq 0$ with naturally vanishing vacuum energy.

(b) All soft SUSY breaking terms, except the gaugino masses, vanish, that is

$$m_0 = A = B = 0 \ .$$

This leaves the gaugino masses $M_{1/2}$ which must therefore be non-vanishing as the only source of SUSY breaking in the observable sector. This is obtained by having a non-trivial gauge chiral function $f_{\alpha\beta} \neq \delta_{\alpha\beta}$ [see Eq. (1)].

(c) $M_{1/2}$ is undetermined at the classical level, along with the local SUSY breaking scale $m_{3/2}$, due to the flatness of the potential along the $T$ direction. Radiative effects set the magnitude of $M_{1/2} = \mathrm{O}\ (M_W)$ as in the standard "no-scale" model but $m_{3/2}$ is not fixed. Various models of no-scale GUTs can be constructed in which $m_{3/2}$ can be very heavy $\sim \mathrm{O}\ (M_P)$ or very light $\ll \mathrm{O}\ (M_P)$ depending on the choice of the gauge chiral function $f_{\alpha\beta}$ (for details, see Ref. [10]).

It has become evident by now that within the context of the no-scale supergravity models the hierarchy of scales $M_W/M_P, M_{SUSY}/M_P$ can be naturally explained. Also since the SUSY breaking parameters are not put in by hand but they rather come out, they are characterized by a major degree of predictive power. However, as in any other supergravity model, we have the arbitrariness for the choice of the Kähler function $\mathcal{G}$ and the gauge chiral function $f_{\alpha\beta}$ whose particular form was essential to implement the no-scale scenario. The question then arises: "who orders this particular supergravity model?"

## 4. - DO STRINGS INDUCE "NO-SCALE" SUPERGRAVITIES?

After the proposal that the Heterotic Superstring [11] may be the fundamental theory encompassing all existing forces of Nature, much effort has been expanded towards building up realistic models accommodating the standard electroweak theory. With some rather plausible assumptions, such a theory in its low-energy limit induces an effective $N = 1$ supergravity which astonishingly enough has striking similarities with the no-scale model we discussed in the previous sections. Skipping all the technical details, some of which are quite involved, I will try to sketch how a no-scale type supergravity arises out of the string theory.

The heterotic superstring is an extended object in $D = 10$ target space dimensions admitting as a gauge group the $E_8 \times E_{8'}$ (or $SO(32)$) so that one does not have anomalies [12]. In its zero slope limit $\alpha' \to 0$ (or same $T$(ension) $\to \infty$) only the massless modes are of relevance and an effective local field theory emerges which is a modified $N = 1$ supergravity in $D = 10$ space/time dimensions [12]-[14]:

$$\begin{aligned} e_{10}^{-1}\mathcal{L}^{(10)} &= -\frac{1}{2k_{10}^2}R^{(10)} + \frac{3k_{10}^2}{2}\varphi^{-3/2}H_{MNP}^2 + \frac{9}{16k_{10}^2}\,\frac{(\partial_M\phi)^2}{\phi^2} \\ & -\frac{1}{4}\phi^{-3/4}F_{MN}^2 + \text{(fermionic terms)} \end{aligned} \tag{16}$$

This involves the graviton multiplet $e_M^A(35)$ , $\psi_M(56)$ , $B_{MN}(28)$ , $\lambda(8)$ , $\varphi(1)$ with 64 bosonic and 64 fermionic degrees of freedom. The physical degrees of freedom of the fields involved are shown in brackets; for instance, 56 degrees of freedom for the Majorana/Weyl gravitino and 28 for the antisymmetric tensor field $B_{MN}$ whose strength in Eq. (16) is denoted by $H_{MNP}$. In addition to these we have a gauge multiplet $(A_\mu^\alpha, \lambda^\alpha)$. The subscript/superscript (10) in Eq. (16) denotes that we are dealing with a theory in space-time ten dimensions. Still in this theory we have to get rid of the surplus six dimensions which is done by the compactification procedure. The ten space $M_{10}$ is

$$M_{10} = M_4 \times K_6$$

where $K_6$ is a compact six-dimensional space with internal dimension (radius) $\sim R$. As in any compactification, Kaluza-Klein scheme heavy massive states of mass $\sim \frac{1}{R}$ appear which decouple at lower energies. By integrating out these heavy modes one has the effective four-dimensional theory of the massless spectrum. In a path integral language this is formally written as

$$e^{iS_{eff}[\phi]} = \int [\mathcal{D}_{\text{heavy modes}}]\ e^{iS[\phi,\text{heavy}]} \tag{17}$$

Obviously, such an integration of the heavy modes is a difficult task to carry out unless we make some plausible physical assumptions concerning the effective theory. The minimum set of requirements towards this goal are the following [15]:

(i) Preserve $N = 1$ SUSY (in order to resolve the gauge hierarchy problem).

(ii) Have a realistic four-dimensional group able to accommodate fermions (vector-like theories are not acceptable).

(iii) The ground state (vacuum) of the string to be $M_4 \times K_6$ with $M_4$ a four-dimensional maximally symmetric space (homogeneous and isotropic).

(iv) Have a realistic (now three!) number of generations.

All these can be met if $K_6$ is a manifold which, in mathematics, is classified as a Calabi-Yau manifold[9]. This, in conjuction with (iii) obliges $M_4$ to be a Minkowski space, so that the tree-level cosmological constant is zero. Requirement (iv) can be satisfied only if the manifold is not simply connected. Actually the number of generations $N_g$ and the Euler characteristic $\chi(K_6)$ of the manifold are related by $N_g = |\chi(K_6)|/2$ and $\chi(k_6) \gg 1$ in simply connected Calabi-Yau manifolds. This topological property of the $K_6$ Calabi-Yau manifold is also important for the symmetry breaking of the gauge group $E_6 \times E_8'$, which survives after compactification, into smaller groups by the so-called Hosotani mechanism [16],[17].

By a proper truncation scheme which mimics the compactification procedure, Witten was able to derive the effective $N = 1, D = 4$ supergravity which is specified with the following Kähler ($\mathcal{G}$) and gauge chiral function ($f_{\alpha\beta}$)

$$\begin{aligned} f_{\alpha\beta} &= \delta_{\alpha\beta} S \\ \mathcal{G} &= \ln(s + s^*) + 3\ln(T + T^* - \phi_i \phi_i^*) - \ln|W(\phi_i)|^2 \end{aligned} \tag{18a}$$

In (18a), the superpotential $W(\phi)$ involves only trilinear couplings of the non-singlet fields $\phi_i$ which belong to $\{27\}$ representations of $E_6$, i.e.,

$$W(\phi) = C_{ijk}\phi_i\phi_j\phi_k \; . \tag{18b}$$

The fields $S, T$ appearing in Eq. (18a) are defined as

$$S = e^{3\sigma}\phi^{-3/4} + 3_i\sqrt{2}D \; , \quad T = e^{\sigma}\phi^{3/4} - i\sqrt{2}a + \phi_i\phi_i{}^* \; . \tag{18c}$$

In (18a), $D$ is the axion field, $a$ is an axionic internal field while $\phi$ is the ten-dimensional dilaton field encountered in Lagrangian[10] (16). The ten metric $\mathcal{G}^{(10)}_{MN}$ is decomposed as

$$g^{(10)}_{MN} = \begin{pmatrix} e^{-3\sigma}g_{\mu\nu} & \vdots \\ \cdots\cdots\cdots & \cdots\cdots \\ & \vdots \;\; e^{\sigma}g^{(0)}_{mn} \end{pmatrix} \qquad \begin{aligned} &\mu, \nu = 0, 1, 2, 3 \; , \\ &m, n = 4, 5, \ldots, 9 \; , \end{aligned}$$

so that the field $\sigma$ scales the compactifying manifold. With these it is not actually hard to see how the real part of the field $S, e^{3\sigma}\phi^{-3/4}$ arises. The gauge boson kinetic term in the action of (16) is written as

$$-\frac{1}{4}\int dx^{10}\sqrt{-g^{(10)}}\;\phi^{-3/4}\;F^2_{MN} = -\frac{1}{4}\int dx^4\sqrt{-g^{(4)}}e^{3\sigma}\phi^{-3/4}(TrF_{\mu\nu})^2 + \ldots$$

which automatically yields [see Eq. (1)]

$$Ref_{\alpha\beta} = \delta_{\alpha\beta}(e^{3\sigma}\phi^{-3/4}) \equiv \delta_{\alpha\beta}ReS$$

[9] A six-dimensional complex Kähler manifold with metric $ds^2 = \partial^2 K/\partial z_i\partial\bar{z}_j \;\; dz^i d\bar{z}^j$ which is Ricci flat $R_{ij} \equiv \partial_i\partial_{\bar{j}} \log\det G_{i\bar{j}} = 0$. The latter implies that it has an $SU(3)$ holonomy.

[10] $D$ and $a$ are related to $B_{MN}$ as $B_{mn} \sim \epsilon_{mn}a$, $\;H_{\mu\nu\rho} \sim \epsilon_{\mu\nu\rho\sigma}\partial^\sigma D$; $\;m, n = 4, 5, \ldots, 9$, $\;\mu\nu\rho\sigma = 0, 1, 2, 3$.

with $S$ as given before. One then notices that the gauge coupling $g^2$ in this scheme is field dependent, given by

$$g^2 = 1/ReS$$

and hence the v.e.v. of $S$ fixes its magnitude in this theory.

The resemblance of the Kähler function $\mathcal{G}$ of Eq. (18a) with the no-scale GUT discussed in the previous section is at least worth noticing. Except the field $S$ which parametrizes an $SU(1,1)/U(1)$ coset space, $T$ and $\phi_i$ couple in the Kähler potential in exactly the manner prescribed by the no-scale models.

The utility of the supergravity theory specified by $f_{\alpha\beta}, \mathcal{G}$ given by Eq. (18a) is questionable since local supersymmetry is unbroken as one can verify by explicitly working out of the model. We are facing the problem of how to break the supersymmetry in this model which, as we will discuss in the following, proves to be the Achille's heel of a large class of models. In the particular model, sources of supersymmetry breaking are sought within the hidden $E_8'$ sector [19] which is assumed to be strongly coupled forming gaugino condensates $\langle\bar{\lambda}\lambda\rangle \neq 0$. The renormalization group equation yields a non-perturbatively large coupling at a condensation scale

$$\Lambda_c \simeq M_P \exp\Big(-\frac{1}{2b_Q g^2(M_P)}\Big) \tag{19}$$

where $b_Q$ is the one-loop beta function coefficient of the subgroup $Q \subset E_8'$ which interacts strongly. $b_Q$ can be large so that $\Lambda_c$ is near but smaller than $M_P$. Thus this dynamical mechanism produces, as in the case of chiral symmetry breaking of QCD, fermion (gaugino) condensates

$$\langle\bar{\lambda}\lambda\rangle = h\Lambda_c^3 \neq 0$$

resulting in breaking of supersymmetry. The constant $h$ parametrizes this particular gaugino condensate at hand but other condensates can be formed, actually needed, as, for instance, the antisymmetric tensor field condensate parametrized by a constant $c$

$$\langle H_{MNP}\rangle = c\ \epsilon_{MNP} \neq 0$$

The effect of the strongly coupled sector of the theory modifies the superpotential $W(\phi)$ to include in addition an $S$-dependent piece given by

$$\Delta W(s) = c + h\exp\Big\{-\frac{24\pi^2}{b_Q}S\Big\} \tag{20}$$

With all these taken into account one is led to a theory with a positive semi-definite scalar potential which attains its minimum $V_0 = 0$ at the points $\langle\phi_i\rangle = 0, \langle S\rangle \neq 0$. Thus the cosmological constant vanishes and in addition the gravitino mass

$$m_{3/2} = \frac{|W(\langle S\rangle)|^2}{16\mathrm{Re}\langle S\rangle\ (\mathrm{Re}\langle T\rangle)^3}$$

is non-vanishing signalling spontaneous breaking of local supersymmetry at an undetermined scale $m_{3/2}$. The determination of $m_{3/2}$, or equivalently, the v.e.v. of the "modulus" field $T$ which sets the size of the compactifying manifold, cannot proceed as in the old no-scale scenario. This is associated with the fact that all global SUSY breaking parameters $m_0, M_{1/2}, A$ and $B$ vanish at the classical level resulting in unbroken global supersymmetry, a very bad feature indeed. Phrased in a different way, the local SUSY breaking fails to be transmitted to the low-energy sector of the theory and this forbids the possibility of constructing realistic electroweak models. Various attempts to remedy the situation and produce non-vanishing soft-breaking parameters via radiative effects leads to results that were contradictory and at any rate not very convincing.

The lack of a satisfactory supersymmetry breaking mechanism is not related to the particularities of the model under discussion. It seems to be a problem associated with the underlying string theory for which we are lacking also a satisfactory SUSY breaking mechanism.

## 5. - THE "NO-SCALE" SUPERSYMMETRY BREAKING MECHANISM IN STRING-INSPIRED MODELS

The more sophisticated techniques developed in the field for the construction of string models (orbifold compactification, bosonic/fermionic constructions) opened the way to depart from the Calabi-Yau scheme and Witten's approach and enlarge the class of the phenomenological models available. Having as guidelines the basic properties of string theories conformal and modular invariance, and considering strings on four space-time dimensions we allowed for a wider option of the gauge group and evaded problems associated with the compactificaton procedure [20]. Several supergravity models have been proposed since then as low-energy effective theories which exhibit a "no-scale" structure similar to the one we discussed in the previous sections [21].

Among the various phenomenological supergravity models built, some of which claiming to have phenomenological virtues, as for instance the flipped $SU(5) \times U(1)$ [22], there is a category bearing the name "no-scale" for the following reasons:

(i) They are described by a scalar potential $V \geq 0$ which leads to spontaneous breaking of the gauge symmetry with naturally vanishing cosmological constant owing to the flatness of $V$ along some directions.

(ii) The soft global SUSY breaking parameters $m_0$, $A$ vanish $m_0 = A = 0$ while $B$ and $M_{1/2}$ are free. When, in addition, $B = 0$ too, one speaks of a strict no-scale case.

In the context of these string-derived models, neither the origin nor the magnitude of the SUSY breaking scale $M_S$ are explained. The size of $M_S$ is put in by hand which, for phenomenological purposes, suffices, but the hierarchy $M_S/M_P$ is not explained through radiative effects. This was the essence of the no-scale scenario. However, still it may not be unconceivable that the no-scale mechanism may be a viable alternative, to other proposed schemes, for the determination of the supersymmetry breaking scale as we shall see.

The supersymmetry breaking mechanism is a hard issue in string theory. In such theories, the typical scale is the slope parameter $\alpha'^{-1/2}(\sim M_P)$ and the phenomenological desired SUSY breaking scale $M_S$ must be many orders of magnitude smaller in order to protect the gauge hierarchy. In supergravities such a large hierarchy of $M_S$ and $M_P$ could be explained by use of the no-scale mechanism. The non-perturbative properties of string theory, which would in principle be able to explain such a hierarchy, are out of reach. In string perturbation theory on the other hand, starting with a spontaneously broken string solution, local and global SUSY break together and their corresponding scales are related to internal radii $m_{3/2} \sim m_S \sim 1/R \sim M_P$ which are therefore phenomenologically unacceptable [23]. For this reason, one prefers to look for non-perturbative no-stringy mechanisms originating at scales smaller than the Planck scale. The gaugino condensation [24] discussed in the previous section is such an example. In this approach one starts with a gauge group that accommodates the standard $SU(3) \times SU(2) \times U(1)$ and a hidden group $H$. The $H$ is assumed large and the corresponding gauge coupling becomes strong at $\Lambda_c$, the condensation scale, resulting in gaugino condensation and hence breaking of supersymmetry. However, still in this mechanism, more quantitative analysis at the non-perturbative level is needed in order to extract the SUSY breaking scale.

As we said before, at the perturbative level with a spontaneously broken solution already at the string construction level, the SUSY breaking scale $m_S$ is necessarily inverse proportional to some internal dimension $\sim 1/R$ and hence large. However, large internal radii $R^{-1} \sim$ TeV result to acceptable values of $m_S$ [25],[26]. At first size this looks catastrophic. With such a large compactification radius all Kaluza-Klein excitations are moved from the Planck scale to the TeV region. Above the scale $1/R$, the theory behaves like a higher-dimensional non-renormalizable theory and the decoupling theorem is not operative for the massive Kaluza-Klein (K-K) states. Thus we lose predictability of the theory even at energies below the compactification scale. However, this is cured as a consequence of the finite properties of the string. In fact although the theory of large $R$ becomes effectively higher-dimensional string calculations yield finite results in terms of the higher-dimensional coupling [25],[26]

$$g_D^2 = R^{D-4} g_4^2 \ .$$

Needless to say that above the extra dimension the effective low-energy description is not valid and the full string theory takes over. However, although predictability is maintained as a result of the string finiteness properties calculability is in general lost due to huge threshold [27] corrections of the $K-K$ excitations which make the coupling constant growing rapidly above $1/R$. In this regime the theory becomes strongly interacting. Actually the coupling constant of the $i^{th}$ gauge group factor at a scale $\mu$ is

$$\frac{1}{g_i^2(\mu)} = \frac{K_i}{g_4^2} + \frac{b_i}{4\pi^2}\log\frac{M_P^2}{\mu^2} + \frac{1}{4\pi^2}\Delta_i \tag{21}$$

where $K_i$ is the level of the Kac-Moody algebra, $b_i$ the beta function coefficients, and $\Delta_i$ the threshold corrections which have an $R$ dependence roughly given by

$$\Delta_i \sim R^2 + \ln R + \mathrm{O}(1) \tag{22}$$

However, in certain cases, the K-K excitations form $N = 4$ supergravity multiplets which are known to give vanishing contributions to the beta functions resulting therefore in threshold corrections $\Delta_i$ which do not grow as the radius becomes large. They are actually exponentially suppressed after supersymmetry breaking [25] takes place.

Therefore, perturbative supersymmetry breaking is not entirely ruled out and may offer an alternative to the non-perturbative approaches. What is more important is the fact that the no-scale mechanism for the determination of the SUSY breaking scale can be implemented in this class of models. The key ingredient in such an approach is the breaking of supersymmetry along a flat direction corresponding to arbitrary values of $R$ with vanishing tree-level vacuum energy. The scale $R^{-1}$ is arbitrary since it is given by the v.e.v. of a modulus field which remains a flat direction even after the SUSY breaking[11]. The supersymmetry breaking mechanisms employed at the string level is that of Scherk and Schwarz [28] which was used originally in extended supergravity theories later being applied also to string theories. This mechanism exploits the fact that higher-dimensional fields are periodic up to symmetry transformations. For instance, upon compactifying, let us say, the fifth co-ordinate $x_5$ on a circle of radius $R$ and for a $U(1)$ symmetry the higher-dimensional fields can be chosen so that

$$\phi(x, x_5 + 2\pi R) = e^{iQ}\phi(x, x_5)$$

where $x$ denotes collectively the rest of the co-ordinates and $e^{iQ}$ is an element of the $U(1)$ symmetry transformation of the action. The $U(1)$ charge $Q$ leads to $Q$ dependent mass shifts in the four-dimensional theory which can be different for bosons/fermions of the same multiplet if they have different charges $Q_B \neq Q_F$. This happens for instance if we have an $R$-symmetry which indeed assigns different charges to bosons and fermions of a given multiplet. Since $Q_B \neq Q_F$ the induced mass shifts for bosons/fermions are different leading to supersymmetry breaking.

In a particular example worked out using the aforementioned supersymmetry breaking mechanism with orbifold compactification, the authors of Ref. [29] reached the following conclusion. A superpotential $W$ which couples the two Higgs multiplets $\hat{H}_1, \hat{H}_2$ as

$$W = \frac{1}{2R}\hat{H}_1\hat{H}_2 + \ldots \tag{23a}$$

[11] In the early no-scale and string derived models the modulus was the field $T$.

with soft SUSY breaking terms given by

$$L_{SB} = \frac{1}{4R^2}\ |H_1|^2 - \frac{3}{4R^2}\ |H_2|^2 \tag{23b}$$

and all common gaugino masses by

$$M_{1/2} = 1/2R \tag{23c}$$

With these one gets the following values for the soft SUSY breaking parameters of the effective SUGRA theory

$$\left.\begin{array}{l} A = B = 0 \\ m^2_{\tilde{q},\tilde{\ell}} = 0\ ,\ \ m^2_{H_1} = -\frac{1}{4R^2}\ ,\ \ m^2_{H_2} = +\frac{3}{4R^2} \\ m_{1/2} = 1/2R \end{array}\right\} \tag{24}$$

where $m^2_{\tilde{q},\tilde{\ell}}$ refer to the squark/slepton soft masses. The magnitude of the internal radius $R$ is, however, undetermined owing to the fact that we have a flat direction in the space of the moduli fields $T_i$. In order to find the minima we have to minimize the potential with respect to the moduli fields and hence with respect to $R$ too. Then the one-loop effects as in the standard no-scale model distort the flatness along the direction $R$ and the minimizing equation

$$\frac{dV}{dR} = 0 \tag{25}$$

fixes the value of $R$ in terms of the critical scale $Q_0$. Thus both scales are found to be hierachically smaller than $M_P$

$$\frac{1}{R} \sim Q_0 \sim M_P \exp\left\{-\frac{4\pi}{h_t^2}\mathrm{O}(1)\right\}$$

Such a determination fixes also the value of the so-called $\mu$-parameter, the coupling of the two Higgs multiplets in the superpotential $W$ [see Eq. (23a)] which, in this model, is given by $\mu = 1/2R$, thus offering a resolution to the so-called $\mu$-problem.

As long as the particle spectrum is concerned, the mass of the largely wanted top quark is very much constrained. If one takes into account the existing experimental limits on both supersymmetric Higgs detection and top quark mass, then

$$140 \lesssim m_t \lesssim 155\ \mathrm{GeV}$$

and this is the only free parameter of the model. The full particle spectrum in terms of $m_t$ is discussed in detail in Ref. [29] and will not be related here.

A crucial point regarding the model under consideration is the fact that the cosmological constant does not have an $R^{-2}$ dependence. In fact the one-loop cosmological constant is calculated to be

$$\Lambda \sim \frac{1}{R^4} + 0(e^{-\lambda R^2})$$

implying that

$$Str\ M^2 = 0\ .$$

This is a very important result since it shows that there are no dreadful $M_P^2(Str\ M^2)$ corrections to the effective potential which would drive $1/R$, and hence the supersymmetry breaking scale, to be of the order of the Planck scale. Recall that this was assumed in the early no-scale models and it was one of the weak points often having been criticized. Within the string theory framework, we see that we have a natural resolution to this problem.

A generic feature of models with large internal radii $1/R \sim$ TeV is that new physics opens in the TeV region just beyond $1/R$. Particles of the so-called untwisted sector, the Higgses and gauge-boson multiplets, have light Kaluza-Klein excitations with masses $\sim 1/R$. The lightest of these is the first excitation of the photon $\gamma^*$ with mass $m_{\gamma^*} = 1/R$ accessible to future accelerators with a clear signal in the $\ell^-\ell^+$ channel. The exchange of these Kaluza-Klein excitations produces dimension-six effective interactions of the type

$$\kappa\ R^2\ \alpha_{em}(\bar{\psi}\gamma^\mu\psi)^2 \tag{26}$$

where $\kappa$ is a constant. We have thus a composite-like picture with the compositeness scale given by

$$\Lambda_c^{-2} = \kappa R^2 \alpha_{em} \tag{27a}$$

The experimental bounds on $\Lambda_c$ constrain $R$. The strongest bound comes from $ee\mu\mu$ which yields

$$R^{-1} \geq \frac{\sqrt{6}}{\pi}\ \frac{\sqrt{\alpha_{em}}}{2.2}\ \Lambda_{ee\mu\mu} \sim 140 \text{ GeV} \tag{27b}$$

In (27b), the value of the factor $\kappa^{1/2}$ has been taken equal to $\sqrt{6}/2.2\pi$ as calculations of exchange of the infinite tower of massive photons and excitations of $Z$ reveal. The bound (27b) on $R^{-1}$ implies that $\gamma^*$ may even be detected at LEP 200.

The model we have just talked about cannot be considered as complete but it serves as a prototype to show how the no-scale mechanism can be realised in string-derived models. From the analysis presented, it seems that a perturbative supersymmetry scheme at the string level under certain circumstances can coexist with a no-scale dynamical determination of the electroweak and SUSY breaking scales [30].

## 6. - DISCUSSION/CONCLUSION

Nowadays, supersymmetry is the best candidate for extending the standard electroweak theory. Its local version incorporates gravity and supergravity theories seem to play a key rôle both phenomenologically and theoretically. From a theoretical point of view, these theories are the link between the Planck scale physics (string) and the low-energy physics in the TeV region which will be probed in the near future. Study of physics at the TeV scale may yield important information of relevance for the underlying string theory. From a phenomenological point of view, supergravities make a lot of spectacular predictions that may be tested in the near future. For their phenomenological study, we need to know the scale of supersymmetry breaking $M_S$. Resolution of the gauge hierarchy problem as well as coupling unification seem to favour values of $M_S$ in the TeV range. However, a theoretical explanation of both its origin and magnitude is of fundamental importance.

Understanding the SUSY breaking mechanism in the context of string theories is a hard issue. The presently available non-perturbative dynamical schemes do not provide us with any quantitative information concerning $M_S$. In a certain class of supergravity theories, there is an approach towards understanding both hierarchies $M_W/M_P$ and $M_S/M_P$ through radiative corrections. This is the "no-scale" mechanism which was applied almost ten years ago to particular supergravity models. Such models were further supported since some string theories in their low-energy limit yielded no-scale-type supergravities. It seems that this scenario can successfully work for string theories as well. Perturbative supersymmetry breaking with a large internal dimension, which can be realized in a wide class of orbifold models, can explain the hierarchies $M_W/M_P, M_S/M_P \sim O(10^{-16})$. Such models are characterized by a major degree of predictability since there are no free parameters except the mass of the undiscovered as yet top which is unknown. The attempts towards constructing realistic string models with such a mechanism for the determination of low energy scales should be pursued. The no-scale picture may be a viable alternative or at least can be accommodated along with the other efforts towards understanding the origin of supersymmetry breaking.

ACKNOWLEDGEMENTS

I thank Prof. A. Zichichi for inviting me to participate and enjoy the splendid atmosphere of this School, and for the warm hospitality extended to me.

# References

[1] For reviews on Supersymmetry/Supergravity, see:
P. Fayet and S. Ferrara, *Physics Reports* **32** (1977) 249;
P. van Nieuwenhuizen, *Physics Reports* **68** (1981) 189;
J. Bagger and J. Wess, Supersymmetry and Supergravity (University Press, Princeton, NJ, 1983);
P. Nath, R. Arnowitt and A.H. Chamseddine, Applied $N = 1$ Supergravity (World Scientific, Singapore, 1984);
H.P. Nilles, *Physics Reports* **110** (1984) 1;
H.E. Haber and G.L. Kane, *Physics Reports* **117** (1985) 75;
A.B. Lahanas and D.V. Nanopoulos, *Physics Reports* **145** (1987) 1.

[2] E. Cremmer, B. Julia, J. Scherk, S. Ferrara, L. Girardello and P. van Nieuwenhuizen, *Nucl.Phys.* **B147** (1979) 105;
E. Cremmer, S. Ferrara, L. Girardello and A. Van Proyen, *Phys.Lett.* **B116** (1982) 231; *Nucl.Phys.* **B212** (1983) 413.

[3] K. Inoue, A. Kakuto, H. Komatsu and S. Takeshita, *Progr.Theor.Phys.* **68** (927; **71** (1984) 348.

[4] L.E. Ibañez and G.G. Ross, *Phys.Lett.* **B110** (1982) 219;
J. Ellis, L.E. Ibãnez and G.G. Ross, *Phys.Lett.* **B113** (1982) 283;
L. Alvarez-Gaumé, M. Claudson and M. Wise, *Nucl.Phys.* **B207** (1982) 96;
L. Alvarez-Gaumé, J. Polchinski and M. Wise, *Nucl.Phys.* **B221** (1983) 495;
J. Ellis, D.V. Nanopoulos and K. Tamvakis, *Phys.Lett.* **B121** (1983) 123;
J. Ellis, J.S. Hagelin, D.V. Nanopoulos and K. Tamvakis, *Phys.Lett.* **B125** (1983) 275;
L.E. Ibañez and C. Lopez, *Phys.Lett.* **B126** (1983) 54; *Nucl.Phys.* **B233** (1984) 511;
C. Kounnas, A.B. Lahanas, D.V. Nanopoulos and M. Quiros, *Nucl.Phys.* **B236** (1984) 438;
L.E. Ibañez, C. Lopez and C. Muñoz, *Nucl.Phys.* **B256** (1985) 218;
A. Bouquet, J. Kaplan and C. Savoy, *Nucl.Phys.* **B262** (1985) 299;
U. Ellwanger, *Nucl.Phys.* **B238** (1984) 665;
G. Gamberini, G. Ridolfi and F. Zwirner, *Nucl.Phys.* **B331** (1990) 331.

[5] G.G. Ross and R.G. Roberts, *Nucl.Phys.* **B377** (1992) 571.

[6] J. Ellis, A.B. Lahanas, D.V. Nanopoulos and K. Tamvakis, *Phys.Lett.* **134B** (1984) 429.
See also: A.B. Lahanas and D.V. Nanopoulos in Ref. [1].

[7] E. Cremmer, S. Ferrara, C. Kounnas and D.V. Nanopoulos, *Phys.Lett.* **B133** (1983) 61.

[8] J. Polonyi, Budapest preprint KFKI-1977-93 (1977).

[9] J. Ellis, C. Kounnas and D.V. Nanopoulos, *Nucl.Phys.* **B241** (1984) 406; *Phys.Lett.* **143B** (1984) 410.

[10] J. Ellis, C. Kounnas and D.V. Nanopoulos, *Nucl.Phys.* **B331** (1984) 373.

[11] D.J. Gross, J. Harvey, E. Martinec and R. Rohm, *Phys.Rev.Lett.* **54** (1985) 502; *Nucl.Phys.* **B256** (1985) 253; *Nucl.Phys.* **B267** (1986) 75.

[12] M.B. Green and J.H. Schwarz, *Phys.Lett.* **B149** (1984) 117; *Phys.Lett.* **B251** (1985) 21.

[13] A.H. Chamseddine, *Nucl.Phys.* **B185** (1981) 403;
E. Bergshoeff, M. de Roo, B. de Wit and P. van Nieuwenhuizen, *Nucl.Phys.* **B195** (1982) 97;
G.F. Chapline and N.S. Manton, *Phys.Lett.* **120B** (1983) 105.

[14] M.B. Green, J.H. Schwarz and P.C. West, *Nucl.Phys.* **255** (1985) 93.

[15] P. Candelas, G. Horowitz, A. Strominger and E. Witten, *Nucl.Phys.* **258** (1985) 46.

[16] Y. Hosotani, *Phys.Lett.* **129B** (1983) 193.

[17] E. Witten, *Nucl.Phys.* **B258** (1985) 75;
M. Dine, V. Kaplunovsky, M. Mangano, C. Nappi and N. Seiberg, *Nucl.Phys.* **B259** (1986) 519;
S. Cecotti, J.P. Derendinger, S. Ferrara, L. Girardello and M. Roncadelli, *Phys.Lett.* **156B** (1985) 318;
J.D. Breit, B.A. Ovrut and G. Segré, *Phys.Lett.* **198B** (1985) 37.

[18] E. Witten, *Phys.Lett.* **155B** (1985) 151.

[19] J.P. Derendinger, L.E. Ibañez and H.P. Nilles, *Phys.Lett.* **B155** (1985) 65;
M. Dine, R. Rohm, N. Seiberg and E. Witten, *Phys.Lett.* **B156** (1985) 55.

[20] K.S. Narain, *Phys.Lett.* **B169** (1986) 41;
W. Lerche, D. Lüst and A.N. Schellekens, *Nucl.Phys.* **B287** (1987) 477;
H. Kawai, D.C. Lewellen and S.H.-H. Tye, *Phys.Rev.* **D34** (1986) 3794; *Nucl.Phys.* **B288** (1987) 1;
I. Antoniadis, C. Bachas and C. Kounnas, *Nucl.Phys.* **B289** (1987) 87;
I. Antoniadis and C. Bachas, *Nucl.Phys.* **B298** (1988) 586.

[21] S. Ferrara, C. Kounnas and M. Porrati, *Phys.Lett.* **B181** (1986) 263;
S. Ferrara, L. Girardello, C. Kounnas and M. Porrati, *Phys.Lett.* **192B** (1987) 368; *Phys.Lett.* **194B** (1987) 96;
I. Antoniadis, J. Ellis, E. Floratos, D.V. Nanopoulos and T. Tomaras, *Phys.Lett.* **191B** (1987) 96;
S. Ferrara and M. Porrati, *Phys.Lett.* **216B** (1989) 289.

[22] I. Antoniadis, J. Ellis, J.S. Hagelin and D.V. Nanopoulos, *Phys.Lett.* **194B** (1987) 231; *Phys.Lett.* **205B** (1988) 459; *Phys.Lett.* **208B** (1988) 209; *Phys.Lett.* **231B** (1989) 69;
J.L. Lopez, D.V. Nanopoulos and A. Zichichi, Texas A & M University preprint CTP-TAMU-68/92, CERN Preprint TH. 6667/92 (1992) and CERN-PPE/92-188 (1992);
J.L. Lopez, D.V. Nanopoulos and K. Yuan, *Nucl.Phys.* **B399** (1993) 654;
I. Antoniadis, J. Ellis, S. Kelley and D.V. Nanopoulos, *Phys.Lett.* **B272** (1991)

31;
D. Bailin and A. Love, *Phys.Lett.* **B280** (1992) 26;
J.L. Lopez, D.V. Nanopoulos and A. Zichichi, CERN Preprint TH. 6926/93 (1993), CTP-TAMU-33/93, ACT-12/93, and references therein.

[23] J. Antoniadis, C. Bachas, D. Lewellen and T. Tomaras, *Phys.Lett.* **207B** (1988) 441;
T. Banks and L. Dixon, *Nucl.Phys.* **B307** (1988) 93;
M. Dine and N. Seiberg, *Nucl.Phys.* **B301** (1988) 357.

[24] For a review see, for example:
D. Amati, K. Konishi, Y. Meurice, G.C. Rossi and G. Veneziano, *Physics Reports* **162** (1988) 169;
H.P. Nilles, *Int.J.Mod.Phys.* **A5** (1990) 4199;
J. Louis, in: Proceedings 1991 DPF Meeting (Vancouver, B.C., Canada) (World Scientific, Singapore, 1992) and references therein.

[25] I. Antoniadis, *Phys.Lett.* **246B** (1990) 377.

[26] I. Antoniadis, Talk at the 23rd Workshop, "Properties of SUSY Particles", Erice, 28 Sept. - 4 Oct. 1992.

[27] V.S. Kaplunovsky, *Nucl.Phys.* **B307** (1988) 145;
L.J. Dixon, V.S. Kaplunovsky and J. Louis, *Nucl.Phys.* **B355** (1991) 649;
J.P. Derendinger, S. Ferrara, C. Kounnas and F. Zwirner, *Nucl.Phys.* **B372** (1992) 145;
I. Antoniadis, K.S. Narain and T. Taylor, *Phys.Lett.* **B276** (1991) 37.

[28] J. Scherk and J.H. Schwarz, *Phys.Lett.* **82B** (1979) 60; *Nucl.Phys.* **B153** (1979) 61;
R. Rohm, *Nucl.Phys.* **B237** (1984) 553.

[29] I. Antoniadis, C. Muñoz and M. Quirós, CPTH-A206-1192, FTUAM 92/35, IEM-FT-63/92, 1992.

[30] See also:
C. Kounnas, Talks presented at the 23rd Workshop "Properties of SUSY Particles", Erice, 28 Sept. - 4 Oct. 1992.

**CHAIRMAN: A.Lahanas**

*Scientific Secretaries: N.Sarlis, K.Skenderis*

## DISCUSSION

– *Arnowitt:*

At the very end of your talk you mentioned a model in which, you said that $Q_0$ came out to be about $m_w$ where $w_t = 150$ GeV. Does that come out naturally or is this value of $Q_0$ due to the value of $m_t$?

– *Lahanas:*

First of all, I want to say that what I mentioned was not a complete theory. It was just an attempt towards a model. Anyway in this attempt they calculate $m_t = (150 \pm 5)$ GeV which gives the right scale. As you know well in this theory regarding the radiative symmetry breaking, you have a rather severe fine tuning problem, and really in this model I don't know how they overpass it. The fine-tuning problem is the following: $Q_0$ is given by:

$$Q_0 = M_{pl} \exp\left(\frac{-O(1)}{h_t^2}\right)$$

You can arrange things in such a way as to have the correct order of magnitude for $Q_0$. However in doing this any small deviation from its original value may disturb $Q_0$ by two or three orders of magnitude. I don't know if this has been overpassed. But I repeat this is not a complete model but an attempt towards a model.

– *Cadoni:*

You said that all masses of the low energy theory are dynamically determined in terms of the Planck mass. Does this hold also for the quark and lepton masses?

– *Lahanas:*

It is premature to say that this framework can give you the mass of the top quark at this stage. I remind you that it is not yet a theory,if you consider Supergravity as an effective low energy model coming from string theory, in principle you can calculate everything.

– *Nanopoulos:*

May I comment on something. If you determine dynamically the vacuum expectation value of the Higgs particle then this is the main source of the fermion masses also. So the thing that remains is not the source of the mass but the source of the Yukawa coupling. In order to interpret Yukawa couplings, which are from O(1) to O($10^{-2}$), you have to go to a higher theory. Supergravity does not provide a solution to this problem but other extended theories like String Theory do.

– *Lahanas:*

Eventually if effective Supergravity comes out from String Theory, if String Theory is the correct theory, all these problems will be solved.

– *Giannakis:*

How can you get from Callabi–Yau manifold the No–Scale Supergravity model? In general you get a Supergravity model but not a No–Scale one. What determines the no-scale model?

– *Lahanas:*

I have to refer you to Witten's original paper. I cannot discuss right now such a technical point.

– *Giannakis:*

So not any Callabi–Yau manifold leads necessarily to a No–Scale Supergravity model.

– *Lahanas:*

No it does not necessarily lead to a No–Scale Supergravity model.

– *Lopez:*

I have a question coming from the No–Scale models from the old days. There the determination of $m_{1/2}$ did not depend on the physics of the hidden sector. Could this picture be altered by more complicated hidden sectors?

– *Lahanas:*

If you have a hidden sector which couples in a funny way with low-energy observables this picture is certainly disturbed. The guideline towards building No–Scale models is to have flatness in a certain direction which is lifted by perturbative effects giving rise to a Supersymmetry breaking scale.

# QUARK AND LEPTON SUBSTRUCTURE: ISSUES, PROMISES AND PROBLEMS *

R. D. Peccei

*Department of Physics*

*University of California, Los Angeles, Los Angeles, CA 90024-1547*

## ABSTRACT

After reviewing some of the important differences between compositeness and elementarity, in these lectures I detail what are the dynamical requirements for constructing composite models of quarks and leptons. Particular attention is paid to the notion of chiral protection and various dynamical criteria for ensuring that some chiral symmetries survive in the binding are discussed. Many of these ideas are exemplified via two useful toy models. The latter part of the lectures is devoted to understanding dynamically how to generate light, rather than massless, bound state fermions. Here the idea of vector-like gauging of some preons in a chiral gauge theory is explained and it is again illustrated by a simple example. Features of a more realistic model are also discussed and the relative ease of generating an up-down mass splitting is contrasted with the difficulty of getting any fermion mixing at all. At the end, difficulties and some toy model hopes for producing families are discussed.

---

* Lectures given at the XXXI International School of Subnuclear Physics, Erice, Italy, July 1993.

# INTRODUCTION

One of the profound mysteries of particle physics is the origin of the family replication for quarks and leptons, along with the reasons for the peculiar mass spectrum observed. As Fig. 1 illustrates, there are enormous mass ratios between the various charged excitations. For instance, $m_t/m_e > 10^5$ while, even within the 3rd generation, one has already $m_t/m_b > 40$. Furthermore, all neutrinos are very much lighter than their charged constituents, with the bound $m_{\nu_i} \leq 10^2$ eV following from cosmological considerations of the neutrino's contribution to the Universe's energy density.[1]

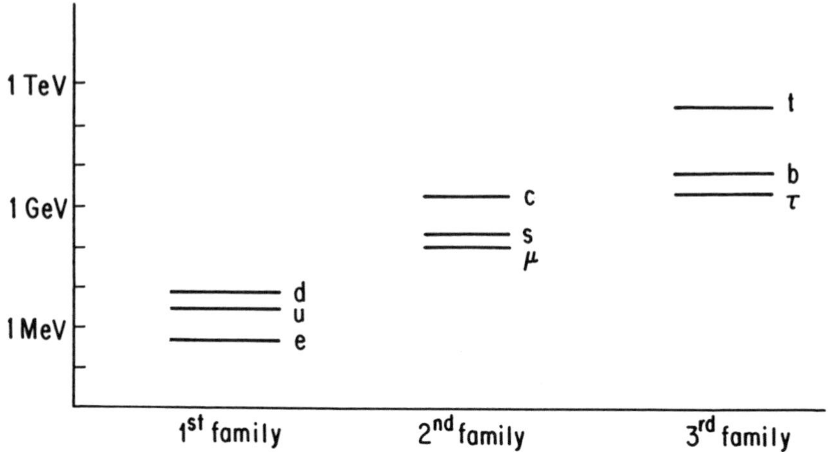


**Fig. 1:** Schematic of the mass spectrum of quarks and leptons.

The natural tendency when one is confronted with such a spectrum is to imagine that it is a reflection of the **composite nature** of quarks and leptons. This idea has clear historical precedents, since as one has probed deeper into the structure of matter one has found that the elementary constituents at one level then proved to be composite at the next. Molecules are composed of atoms, while atoms are bound states of electrons and nuclei. Nuclei themselves are bound states of nucleons which, in turn, are made up of quarks. The idea that quarks and leptons may themselves be composite, however, runs into an immediate serious problem: all experimental data we have on quarks and leptons is consistent with their being elementary objects, since they do not seem to possess any of the characteristic features one associates with bound state systems! Therefore to contemplate composite models of quarks and leptons, one has to find sufficient reasons why, despite first impressions, there are grounds for supposing that quarks and leptons are composite. Indeed, to speak about composite models of quarks and leptons, one must first motivate the reasons why compositeness may be desirable for these excitations and then describe and explain the dynamics which manages to endow quarks and leptons with the features of (nearly) elementary particles.

The most compelling motivation for supposing that quarks and leptons may be composite is related to the dynamical breakdown of SU(2)× U(1). The argument is as follows. The dynamical breakdown of the electroweak theory necessitates introducing into the theory some new fermions, $T$, which are subject to some other underlying interaction which causes the formation of SU(2)× U(1) breaking condensates, $\langle \bar{T}T \rangle$. Although the presence of these condensates suffices to generate non-zero masses for the W and Z bosons, no masses can be produced for the quarks and leptons (which I shall generically denote by $f$), unless there exists some bridge between the $f$ and $T$ fermions. Traditionally, this bridge has been built by assuming that both $T$ and $f$ fit in a representation of some other non-Abelian gauge group, which itself is spontaneously broken (ETC theory).[2] Mass for the fermions $f$ then ensues as a result of the combined exchange of these superheavy gauge bosons between $f$ and $T$ and the formation of $\langle \bar{T}T \rangle$ condensates. This is illustrated schematically in Fig. 2(a). However, this is not the only possible scenario. In particular, if $f$ and $T$ were made of the same stuff one would expect some effective 4-fermion interaction between $f$ and $T$ to exist. These effective interactions, when $\langle \bar{T}T \rangle$ condensates form, can also give rise to masses for the $f$ fermions, through the diagram shown in Fig. 2(b).

## COMPOSITENESS VERSUS ELEMENTARITY

Quarks and leptons fail to show any of the typical features of composite systems. Therefore, if these states are in fact bound states, their dynamics must be quite peculiar. To see this, it is useful to contrast typical bound states, like hadrons, with what we know about quarks and leptons. To begin with, consider the issue of the **mass spectrum**. We know of the existence of literally hundreds of hadronic resonances, excited states of the lowest quark-antiquark, or tri-quark, bound states of QCD. For these excitations, typically the difference between the excited state masses, $M^*$, and the ground state mass, $M$, is itself of order $M$:

$$\Delta M = M^* - M \sim M \ . \tag{1}$$

**Fig. 2:** Schematics of mass generation for the quarks and leptons, $f$, in the case of: (a) an ETC theory or (b) if $f$ and $T$ are both made of the same subconstituents.

For quarks and leptons, on the other hand, we know from LEP that no excitated quarks and leptons exist with masses $m_q^*$ or $m_\ell^*$ less than about 45 GeV.[3] Thus the mass gap–at least for the light quarks and leptons–is much greater than the mass of the presumed ground states:

$$\Delta m = m_f^* - m_f \gg m_f \ . \tag{2}$$

Similar differences exist when one compares the **scattering properties** of hadrons and those of quarks and leptons. Hadrons scatter strongly, leading to scattering cross sections which reflect the geometric size of the hadrons themselves, which are in turn simply related to their masses:

$$\sigma_{\text{had}} = \pi \langle r_s \rangle^2 \sim \frac{1}{M^2} \ . \tag{3}$$

On the other hand, we know that quarks and leptons interact weakly at short distances, leading to cross sections which fall in a scale invariant fashion as the CM energy E increases

$$\sigma_f \sim \frac{\alpha_f}{E^2} \ . \tag{4}$$

High energy data at both LEP and the Fermilab Collider is consistent with the above scale invariant behavior and suggests that if quarks and leptons have any scale at all, the corresponding mass scale $\Lambda_f \sim 1/\langle r_f \rangle$ is greater than a TeV.[4]

Similar qualitative information follows from considering how hadrons differ from quarks and leptons under **electromagnetic probing**. The scattering of electrons on protons revealed long ago[5] that these particles possessed a form factor, reflecting their size

$$\langle r_{\rm em}\rangle \sim \frac{1}{\Lambda_{\rm em}} \,. \tag{5}$$

No similar form factors, or an electromagnetic charge radius, have been found as yet for quarks and leptons, indicating that for these excitations $\Lambda_{\rm em} > \text{TeV}$.

It is perhaps worthwhile commenting here how the parameter $\Lambda_{\rm em}$ is determined for leptons, for it illustrates an important lesson. The anomalous magnetic moment for the electron and muon is predicted to incredible accuracy by QED and these predictions agree with experiment to high precision. One has

$$\delta(a) = \frac{1}{2}\left[(g-2)_{\rm exp} - (g-2)_{\rm QED}\right] = \begin{cases} 3.2\times 10^{-10} & e \\ 1.5\times 10^{-8} & \mu \end{cases} \tag{6}$$

If leptons had some substructure, one would expect that they would exhibit an additional anomalous interaction with the EM field, typified by the effective Lagrangian

$$\mathcal{L}_{\rm comp} = \frac{e}{\Lambda_{\rm em}}\left[\frac{m_\ell}{\Lambda_{\rm em}}\right]\bar{\ell}_L\sigma_{\mu\nu}\ell_R F^{\mu\nu} \,. \tag{7}$$

The extra factor of $m_\ell/\Lambda_{\rm em}$ in the above is included because the magnetic moment operator is **chirality breaking**, since it connects $\ell_L$ with $\ell_R$. One expects that no such operators are induced until masses for the leptons are generated and hence that $\mathcal{L}_{\rm comp} \to 0$ as $m_\ell \to 0$. The presence of this factor gives[6]

$$\delta a \sim \left[\frac{m_\ell}{\Lambda_{\rm em}}\right]^2 \tag{8}$$

and leads to the bounds

$$\Lambda_{\rm em} \geq \begin{cases} 50 \text{ GeV} & e \\ 900 \text{ GeV} & \mu \end{cases} \tag{9}$$

Without the chiral protection factor, then $\delta a \sim m_\ell/\Lambda_{\rm em}$ and the bounds on $\Lambda_{\rm em}$ would be very much tighter:

$$\Lambda_{\rm em} \geq 10^6 - 10^7 \text{ GeV} \,. \tag{10}$$

To summarize, there are many different ways by which one can determine that hadrons are composite: by looking at their spectrum; from the geometrical form of their scattering cross section; and by probing electromagnetically for their size. For hadrons, all these three methods reveal scales which, although not identical, are all of the same order of magnitude

$$\langle r_s \rangle \sim \langle r_{\rm em} \rangle \sim \frac{1}{\Delta M} \; . \tag{11}$$

That this must be the case is easily understood from QCD. In QCD the strong interaction coupling constant $\alpha_s$, through its running, determines only one scale in the theory $\Lambda_{\rm QCD}$–the scale where, roughly speaking, $\alpha_s(\Lambda_{\rm QCD}) \simeq 1$. This is shown schematically in Fig. 3. Because $\Lambda_{\rm QCD}$ is the only scale in the theory, clearly all parameters which have the dimension of length must be inversely proportional to $\Lambda_{\rm QCD}$. Thus

$$\langle r_s \rangle \sim \langle r_{\rm em} \rangle \sim \frac{1}{\Delta M} \sim \frac{1}{\Lambda_{\rm QCD}} \; . \tag{12}$$

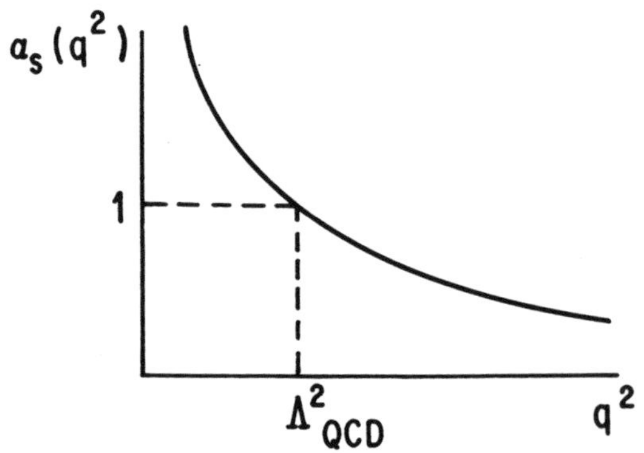


**Fig. 3:** The running of the strong coupling constant $\alpha_s$, provides a definition for $\Lambda_{\rm QCD}$.

If quarks and leptons are composite and their dynamics is generated by some non-Abelian gauge theory like QCD, then our discussions tells us that the scales we discussed: $\langle r_f \rangle \sim 1/\Lambda_f$, $\langle r_{\rm em} \rangle \sim 1/\Lambda_{\rm em}$ and $m_f^* - m_f$ should be all of the order of the dynamical scale $\Lambda_c$ of this theory:

$$\Lambda_c \sim \Lambda_f \sim \Lambda_{\rm em} \sim m_f^* - m_f \; . \tag{13}$$

However, there is a strong constraint in the dynamics, since we have seen already that quarks and leptons appear essentially elementary. This near elementarity requires that the dynamical scale $\Lambda_c$ obey

$$\Lambda_c \gg m_f \ . \tag{14}$$

It is this constraint which constitutes the real physical challenge for constructing composite models of quarks and leptons.

## DYNAMICAL REQUIREMENTS OF MODELS OF COMPOSITE QUARKS AND LEPTONS

We saw that the near elementarity of the quarks and leptons requires a dynamics which produces bound states whose size $\langle r_f \rangle \sim 1/\Lambda_c$ is much less than their Compton wavelength $1/m_f$:

$$\langle r_f \rangle \ll 1/m_f \ . \tag{15}$$

Our experience does not encompass bound state systems which are so tightly bound. For instance, in the case of positronium, the size of the bound atom is very much bigger than the electron's Compton wavelength:

$$\langle r_{\text{positronium}} \rangle \sim \frac{1}{\alpha m_e} \gg \frac{1}{m_e} \ . \tag{16}$$

For hadrons, on the other hand, as we discussed

$$\langle r_{\text{hadrons}} \rangle \sim \frac{1}{M_{\text{hadrons}}} \ . \tag{17}$$

This reflects the fact that the strong coupling constant $\alpha_s$ for these bound states is indeed of $O(1)$, leading to hadronic sizes which are comparable to the hadron's Compton wavelength.

Up to now there has been really only one idea put forward for the dynamics of composite models–preon dynamics–which could give rise to such tight bound states, so that $m_f \ll \Lambda_c$. Namely, that the preon theory is some confining Yang Mills theory which has some (approximate) **protective global symmetries**. In such a theory, $\Lambda_c$ is the scale where the preon coupling constant squared $\alpha_{\text{preon}}$ becomes of $O(1)$. Necessarily, therefore, one expects all bound states to have masses $M_{\text{BS}} \sim \Lambda_c$, **except** possibly for a few states which some (approximate) protective symmetries in the theory force to be (nearly) massless. These latter states are then identified with the quarks and leptons, for which indeed $m_f \ll \Lambda_c$.

QCD has a bosonic example of this phenonemon–the pseudoscalar pions $\{\pi_i\}$. In QCD the dynamical masses for the $u$ and $d$ quarks are much less than $\Lambda_{\text{QCD}}$:

$$m_u,\ m_d \ll \Lambda_{\rm QCD}\ . \tag{18}$$

If these masses can be neglected, then QCD has an exact global symmetry

$$G_{\rm global} = SU(2)_{\rm L} \times SU(2)_{\rm R} \times U(1)\ . \tag{19}$$

This global symmetry is not preserved by the QCD dynamics, since non-vanishing quark condensates $\langle \bar{u}u\rangle$, $\langle \bar{d}d\rangle$ form. As a result, $G_{\rm global}$ is spontaneously broken to

$$H_{\rm global} = SU(2)_{\rm L+R} \times U(1)\ . \tag{20}$$

Because of the spontaneous breakdown of $G_{\rm global} \to H_{\rm global}$, a set of 3 Goldstone bosons–the pions $\{\pi_i\}$–appear in the theory. These particles are massless if $G_{\rm global}$ is exact ($m_\pi^2 = 0$). However, if one includes the explicit breaking of $G_{\rm global}$ caused by the presence of the small $u$ and $d$ quark masses, then

$$m_\pi^2 \sim m_q \Lambda_{\rm QCD} \qquad \{q = u, d\} \tag{21}$$

One sees that, in this example, the approximate global $SU(2)_{\rm L} \times SU(2)_{\rm R} \times U(1)$ symmetry of QCD protects some states–the $\{\pi_i\}$–from getting masses of $O(\Lambda_{\rm QCD})$. Indeed, for small quark masses one has a hierarchy

$$m_\pi \sim (m_q \Lambda_{\rm QCD})^{1/2} \ll \Lambda_{\rm QCD}\ , \tag{22}$$

which singles out the pions as being special states in QCD.

Two different protective mechanisms have been suggested in preon models to obtain dynamically (nearly) massless fermions. The first of these uses global chiral symmetries, which are preserved in the binding, to guarantee the presence of massless bound state fermions [**chiral protection**].[7] The other mechanism uses global symmetries which are spontaneously broken in supersymmetric preon models and obtains the desired massless bound state fermions as superpartners of the Nambu-Goldstone bosons resulting from the symmetry breakdown [**Quasi Goldstone Fermion mechanism**].[8] Both mechanisms require also that a number of other dynamical conditions hold. Thus, it behooves us to discuss the general aspects of each of these suggestions in some more detail in what follows.

### Chiral Protection

Massless fermionic bound states may ensue in a preon theory if this theory possesses some **global** chiral symmetries. However, this is not absolutely guaranteed, since it is possible that these symmetries are broken in the binding. Indeed, QCD, as we just discussed, provides a nice counterexample. Two-flavor ($m_q = 0$)

QCD is invariant under chiral transformations in which the right- and left-handed fundamental fermions are rotated in an independent fashion:

$$\begin{pmatrix} u \\ d \end{pmatrix}_{\rm L,R} \to e^{i\vec{\alpha}_{\rm L,R}\cdot\vec{\tau}} \begin{pmatrix} u \\ d \end{pmatrix}_{\rm L,R} . \tag{23}$$

If this $SU(2)_{\rm L} \times SU(2)_{\rm R}$ symmetry were preserved in the binding then nucleons, transforming as quarks do under this symmetry, would also be massless. However, the global $SU(2)_{\rm L} \times SU(2)_{\rm R}$ symmetry at the quark level, is not preserved in the binding due to the formation of $\langle \bar{u}u \rangle = \langle \bar{d}d \rangle \neq 0$ condensates. As a result, the only symmetry preserved in QCD is a diagonal isospin symmetry $[SU(2)_{\rm L+R}]$ times baryon number and this symmetry allows nucleon masses to exist. So, in QCD, one ends up with massive nucleons and, in the limit where $m_q = 0$, massless pions:

$$M_{\rm N} \sim \Lambda_{\rm QCD} \; ; \quad m_\pi = 0 \, . \tag{24}$$

Thus, of one wants to adduce chiral global symmetries in preon theories as the root cause for the existence of massless quarks and leptons, one must **dynamically guarantee** that these chiral symmetries are not broken in the binding. Necessary conditions for this to obtain were first enunciated by G. 't Hooft[7] and will be discussed below.

**Quasi Goldstone Fermions**

One can produce massless bound state fermions in a supersymmetric preon theory provided this theory has some global symmetry $G$ which is spontaneously broken down to another global symmetry $H$. The breakdown $G \to H$ necessarily produces a number of massless Nambu-Goldstone bosons in the theory–one each for each generator in the coset $G/H$. Supersymmetry (SUSY) then requires that these massless bosonic excitations have an appropriate number of massless fermionic partners. These, so-called, Quasi Goldstone Fermions (QGF),[8] however, will only be massless in the **exact** SUSY limit. Once supersymmetry is broken, the QGF are no longer protected and they can acquire a mass. If the global symmetry $G$ is not broken explicitly, one could then end up with the phenomenologically undesirable solution of having still some massless Nambu-Goldstone bosons in the theory, but having massive Quasi Goldstone Fermions! To avoid these problems, in more realistic models[9] one tries to incorporate a **double protection**,[10] where the remaining global group $H$ is itself a chiral group. In this case, even after supersymmetry breaking, the chiral symmetry of $H$ can serve to keep the Quasi Goldstone Fermions massless: $m_{\rm QGF} = 0$.

## ANOMALY MATCHING AND OTHER DYNAMICAL CONSTRAINTS

Because chirality, in the end, plays a fundamental role in trying to construct models of composite quarks and leptons, it is clearly important to try to understand better under what conditions chirality is preserved in the binding. The conditions which are necessary for this to occur were spelled out about 15 years ago by 't Hooft.[7] In what follows, I shall try to review and explain what these conditions are.

Suppose one has some preon theory which has a group $G$ of chiral symmetries. That is imagine, for example, that one has a set of chiral fermionic preons which are invariant under global $G$-transformations (we shall soon see some explicit examples). Then for each generator $g_a$ of the group $G$ there is an associated chiral current $J_a^\mu$. Even though $G$ is a symmetry, the existence of Adler-Bell-Jackiw[11] anomalies spoils the conservation of the chiral currents $J_a^\mu$. In particular, the three-point function involving these currents

$$\begin{aligned}(2\pi)^4\delta(\Sigma q_i)\ \Gamma_a^{\mu\nu\lambda}(q_1,q_2,q_3)\\ = \int\prod_i d^4x_i\ e^{iq_ix_i}\langle 0|T(J_a^\nu(x_1)\ J_a^\mu(x_2)\ J_a^\lambda(x_3)|0\rangle\end{aligned} \tag{25}$$

has an anomalous divergence

$$q_{3\lambda}\Gamma_a^{\mu\nu\lambda}(q_1,q_2,q_3) = A_{\rm preon}^{J_a}\epsilon^{\mu\nu\alpha\beta}q_{1\alpha}q_{2\beta}\ . \tag{26}$$

This anomaly at the preon level originates from the triangle graph shown in Fig. 4 and

$$A_{\rm preon}^{J_a}\sim {\rm Tr}\ \lambda_a^3\ , \tag{27}$$

where the $\lambda_a$ are the corresponding generator matrices for the preons.

't Hooft's[7] observation was that the chiral symmetry $G$ will **only** be preserved in the binding if there are $m=0$ fermionic bound states such that their contributions to the divergence of $\Gamma_a^{\mu\nu\lambda}$ leads also to an anomaly with

$$A_{\rm bound\ states}^{J_a} = A_{\rm preon}^{J_a}\ . \tag{28}$$

That is, chirality preservations necessitates that the massless bound states contribute precisely the same anomaly as the preons do. This is essentially a consistency condition on the theory. It says that, irrespective of which level one calculates the 3-point functions $\Gamma_a^{\mu\nu\lambda}$, its divergence must reproduce the **same** anomaly.

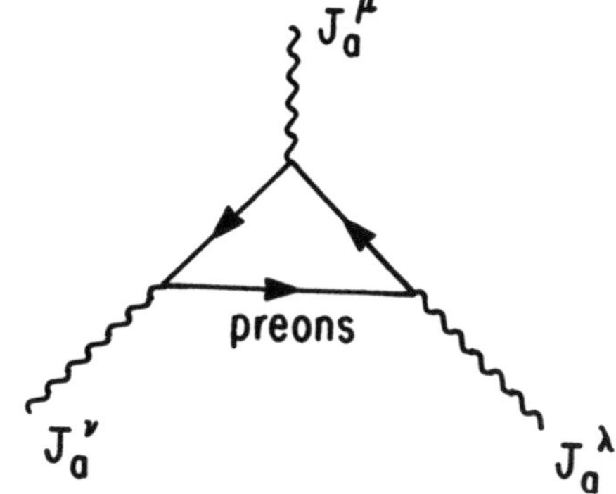


**Fig. 4:** Graph leading to the chiral anomaly at the preon level.

The logic of the 't Hooft condition (28) is quite non-trivial and was elucidated in twin papers by Banks, Frishman, Schwimmer and Yankielowicz,[12] and by Coleman and Grossmann.[13] Although the anomaly equations, Eq. (26), is a short-distance effect, it turns out[12,13] that it also fixes the form of the 3-point function $\Gamma_a^{\mu\nu\lambda}$ at long distances. In particular, it fixes the singularity structure of $\Gamma_a^{\mu\nu\lambda}$ at the symmetric point $q_i^2 = q^2 \to 0$. To be consistent with Eq. (26), it is necessary that at the symmetric point

$$\Gamma_a^{\mu\nu\lambda}(q_1, q_2, q_3)\bigg|_{q_i^2=q^2} = \frac{A_{\rm preon}^{J_a}}{q^2} C^{\mu\nu\lambda} + \text{non} - \text{singular terms} \tag{29}$$

where

$$C^{\mu\nu\lambda} = \epsilon^{\mu\nu\alpha\beta} q_{1\alpha} q_{2\beta} q_3^\lambda + \text{cyclic terms} . \tag{30}$$

Because the 3-point function can be computed either at the preon level or by using the bound states of the theory, consistency requires that also in this latter computation one finds the singularity structure displayed in Eq. (29).

What Banks *et al.*[12] and Coleman and Grossmann[13] showed, was that the required $1/q^2$ singularity in $\Gamma_a^{\mu\nu\lambda}$ at the bound state level arises only as a result of two circumstances. Either

i) there are massless bound state Nambu-Goldstone bosons in the theory

or

ii) there are massless spin-1/2 fermions in the theory.

In the first case, obviously the group $G$ is spontaneously broken and chirality is not preserved in the binding. For chirality to be preserved in the binding, the second circumstance must obtain. In this case one needs really to have massless spin-1/2 bound states, which lead to the same anomaly coefficient. Spin-3/2 massless states, for instance, would not do. Neither would massless spin-1/2 bound states which do not reproduce the anomaly. So consistency in this latter case requires that Eq. (28) hold.

It is interesting to comment briefly on the case when $G$ is not preserved in the binding, so that there are now Nambu-Goldstone bosons in the theory. The contributions of these Nambu-Goldstone bosons is shown schematically in Fig. 5. If one takes their contribution to $J^{\mu}_{a}$ to be given by

$$J^{\mu}_{a} = f_{\pi}\partial^{\mu}\pi_{a} + \dots \ , \tag{31}$$

one sees that in this case

$$\Gamma^{\mu\nu\lambda}_{a}(q_1, q_2, q_3)\bigg|_{q_i^2=q^2} = \frac{f_{\pi}g_{\pi JJ}C^{\mu\nu\lambda}}{q^2} \ . \tag{32}$$

Here $f_{\pi}$ is the coupling of the Nambu-Goldstone boson $\pi_a$ to the broken current $J^{\mu}_{a}$ and $g_{\pi JJ}$ is the coupling of this state to the two currents $J^{\mu}_{a}$. One sees, from Eq. (32) that when chirality is broken, the presence of the chiral anomaly $A^{J_a}_{\text{preon}}$ at the fundamental level fixes the coupling $g_{\pi JJ}$ in terms of $f_{\pi}$ and this anomaly coefficient

$$g_{\pi JJ} = \frac{A^{J_a}_{\text{preon}}}{f_{\pi}} \ . \tag{33}$$

It is precisely through a similar set of considerations that the coupling of the neutral pion to two photons, $g_{\pi\gamma\gamma}$, is determined in QCD.[14]

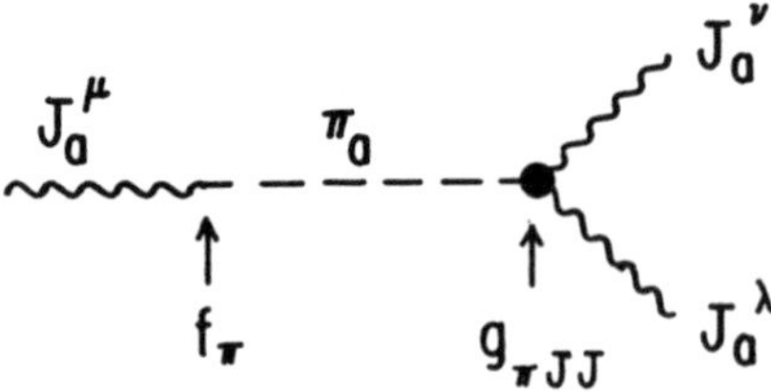


**Fig. 5:** Nambu-Goldstone contribution to $\Gamma^{\mu\nu\lambda}_{a}$ when $G$ is spontaneously broken.

It is important to remark that just because the preon and bound state anomalies match, it does not follow that chirality is preserved in the binding. The 't Hooft ondition (28) is a **necessary** condition for chirality preservation in the binding. It is, however, not a sufficient condition. Thus, in any given preon model, one must look for further dynamical arguments to bolster the assumption that certain chiral symmetries remain unbroken in the binding, leading then to massless bound states whose specific transformation properties under the unbroken group allow anomaly matching. Perhaps the most useful of these auxiliary dynamical indications is provided by the notion of **complementarity**.[15]

Roughly speaking, complementarity allows one to study the $m = 0$ bound state spectrum of a confining theory by looking at which fermions remain massless in a broken phase of the underlying theory The equivalence between the symmetries (and low lying states in the spectrum) in the confining phase and in the broken phase of a theory is not a general property. Rather, it holds only in confining theories where there are scalars (either true or effective) transforming according to the fundamental representation of the confining group. In this case, one can show[15] that there is no phase boundary between the confining and broken phase. Thus one can typify the massless fermions in the confining theory by looking at the massless fermions in the broken theory.

To illustrate these ideas, and for future use in these lectures, I will consider next two very nice toy models. These models will exemplify some of the abstract concepts just discussed, and hopefully will render them more transparent.

## TWO USEFUL TOY MODELS

The first model I want to discuss was studied originally by Bars and Yankielowicz.[16] It is an example of a **chiral gauge theory**. This is a confining gauge theory with Weyl fermions * transforming according to representations $R$ of the gauge group which:

i) Do not admit mass terms. Thus $R \otimes R \not\supset 1$.

ii) Are free of gauge anomalies, and thus give rise to perfectly consistent quantum field theories.[16]

### The Bars-Yankielowicz Model

In the Bars-Yankielowicz model, the preon gauge group is $SU(N)$ and the

* Weyl fermions are 2-component fermions and are equivalent to 4-component Dirac fermions of only one- handedness.

theory has $(N+4)$ Weyl fermion preons in the fundamental representation and one Weyl fermion preon transforming according to the conjugate two-rank symmetric tensor representation of $SU(N)$. That is, one has a chiral $SU(N)$ gauge theory with preons transforming according to

$$R = \{(N+4)F_\alpha \oplus S^{\alpha\beta}\} \; . \tag{34}$$

Because of the above representation content, it is clear that the theory has, at the classical level, a global chiral symmetry

$$G_{\rm class} = SU(N+4) \times U(1)_F \times U(1)_S \; , \tag{35}$$

where the last two $U(1)$ symmetries correspond to the fermion numbers of the $F$ and $S$ preons, respectively. At the quantum level, however, these $U(1)$ symmetries are not preserved since they have $SU(N)$ gauge anomalies.[11] The anomaly coefficients corresponding to the triangle graphs of Fig. 6 are easily seen to be

$$A_F = (N+4) \; ; \quad A_S = (N+2) \; . \tag{36}$$

Even though both $U(1)_F$ and $U(1)_S$ are anomalous symmetries, it is clear that the linear combination of these fermion numbers, $U(1)_Q$, with charge

$$Q = \frac{1}{N}\left[(N+4)Q_S - (N+2)Q_F\right] \; , \tag{37}$$

is anomaly free. Thus, quantum mechanically, the Bars-Yankielowicz[16] model has the global symmetry

$$G_{\rm quantum} = SU(N+4) \times U(1)_Q \; . \tag{38}$$

Using the set of preons in $R$, one can readily construct appropriate gauge-singlet fermion bound states by contracting two $F$ preons with an $S$ preon. This contraction can involve different Lorentz combinations of the Weyl fermions and, in general, will involve different members of the $(N+4)$ preons of the $F$ type. If we denote the different $F$ preons by $F_{\alpha a}$, with $a = 1, \cdots, (N+4)$, then the Pauli principle allows two different bound states to form, $B_{[a,b]}$ and $C_{\{a,b\}}$. These states are, respectively, antisymmetric and symmetric in the $SU(N+4)$ indices:

$$\begin{aligned} B_{[a,b]} &\sim (F^T_{\alpha a}\ \sigma_2 \sigma_\mu F_{\beta b})\sigma^\mu S^{\alpha\beta} \\ C_{\{\alpha,\beta\}} &\sim (F^T_{\alpha a}\ \sigma_2 F_{\beta b}) S^{\alpha\beta} \end{aligned} \tag{39}$$

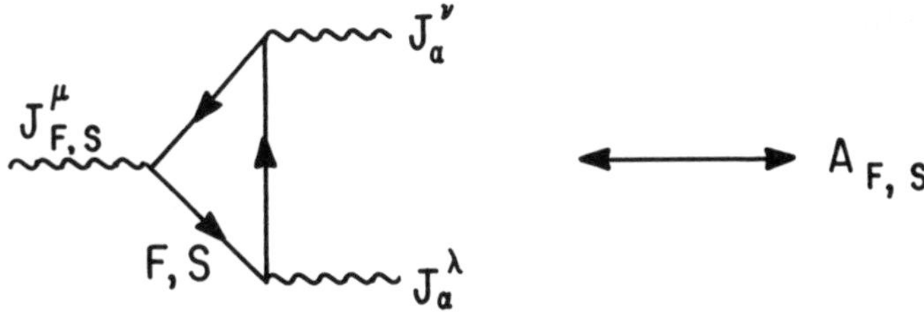


**Fig. 6:** Triangle graphs leading to the anomaly coefficients $A_F$ and $A_S$.

Both of these bound states have $Q = -1$, but obviously transform differently under the global $SU(N+4)$ group.

It is easy to check that only the $B$-states, but not the $C$-states, are suitable bound states if we want the symmetry $G_{\text{quantum}}$ to be preserved in the binding. That is, only the $B$-states have the same anomalies for the $G_{\text{quantum}}$ currents as those found at the preon level. I will not demonstrate this in detail here, but only consider the $[U(1)_Q]^3$ anomaly to illustrate how this goes. At the preon level this anomaly is computed from the graphs in Fig. 7. The anomaly coefficient from any one of the $F$ preons, using Eq. (37), is $Q_F^3 = -[(N+2)/N]^3$ while that for any of the $S$ preons is $Q_S^3 = +[(N+4)/N]^3$. However, there are $(N+4)\cdot N$ preons of the $F$ type, where the second factor of $N$ counts the gauge degrees of freedom, and $N(N+1)/2$ preons of the $S$ type. Hence the $[U(1)_Q]^3$ anomaly at the preon level is given by

$$\begin{aligned} A[U(1)_Q^3] &= (N+4)\cdot N\cdot\left[-\frac{N+2}{N}\right]^3 + \frac{N(N+1)}{2}\cdot\left[\frac{N+4}{N}\right]^3 \\ &= -\frac{(N+4)(N+3)}{2} \end{aligned} \tag{40}$$

This is clearly also the result of the anomaly computed at the bound state level with the $B$-fermions, since there are precisely $(N+4)(N+3)/2$ of them and each has a charge $Q = -1$. The $C$-fermions do not match this anomaly since there are $(N+4)(N+5)/2$ of them.

Just because the $B$ bound states match the preon anomalies of the global $G_{\text{quantum}}$ group, it does not necessarily follow that this symmetry is preserved in the binding. It is therefore necessary to find other dynamical indications that this is the case. I indicate here how **complementarity** can help. To apply complementarity we must deal with a confining theory in which there are scalars which transform

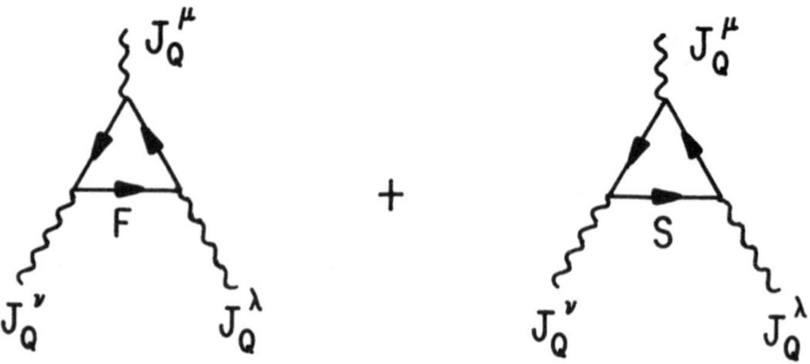


**Fig. 7:** Graphs contributing to the $[U(1)_Q]^3$ anomaly at the preon level.

according to the fundamental representation of the gauge group. Even though in the Bars-Yankielowicz model one has only fermionic preons, one can construct an **effective** Higgs scalar out of the $F$ and $S$ preons which indeed transforms according to the (conjugate) of the fundamental representation of $SU(N)$

$$\phi_a^\alpha = F_{\beta a} S^{\alpha\beta} \ . \tag{41}$$

Using this effective scalar field we can look at the model, equivalently, in a broken phase where the effective scalar field of Eq. (41) acquires a non-zero vacuum expectation value *

$$\langle \phi_a^\alpha \rangle = \langle F_{\beta a} S^{\alpha\beta} \rangle = V \ \delta_a^\alpha \ . \tag{42}$$

The above VEV breaks the $SU(N)$ gauge symmetry and the global quantum symmetry $G_{\text{quantum}}$ down to a diagonal subgroup of both:[19]

$$SU(N) \times \left[ SU(N+4) \times U(1)_Q \right] \longrightarrow SU(N)_{\text{diag}} \times SU(4) \times U(1)_{Q'} \ . \tag{43}$$

Here the charge $Q'$ is a linear combination of the original charge $Q$ and the diagonal generators $I_{N,4}$ of $SU(N+4)$. One has

* One can check that this condensate is actually the most attractive condensate in the model.[18]

$$Q' = Q - 2I_{N,4} \; , \tag{44}$$

where $I_{N,4}$ is given by the block matrix

$$I_{N,4} = \frac{1}{N} \begin{bmatrix} 1 & 0 \\ 0 & -N/4 \end{bmatrix} \; , \tag{45}$$

with the diagonal blocks being $N \times N$ and $4 \times 4$, respectively. Under $Q'$, the preons $F$ and $S$ have the following assignments

$$\begin{aligned} F_{a\alpha} &= \begin{cases} -(N+4)/N & a \subset SU(N) \\ -(N+4)/2N & a \subset SU(4) \end{cases} \\ S^{\alpha\beta} &= (N+4)/N \; , \end{aligned} \tag{46}$$

so that indeed the VEV (42) preserves the $U(1)_{Q'}$ symmetry.

Effective interactions of the Higgs scalar $\phi_a^\alpha$ with the preons can generate mass for these states when $\phi_a^\alpha$ acquires a VEV. However, even in this case there are a number of fermion states which remain massless. It is easy to check that $\langle \phi_a^\alpha \rangle$ **cannot** give mass to two preson combinations:

$$\begin{aligned} F_{a\alpha} - F_{\alpha a} \quad &\text{with} \quad a \subset SU(N) \\ F_{a\alpha} \quad &\text{with} \quad a \subset SU(4) \end{aligned} \tag{47}$$

The first of these states transforms as a second rank antisymmetric tensor under $SU(N)_{\text{diag}}$ and a singlet under $SU(4)$, while the second transforms as the product of the two fundamental representations of these groups.

The unbroken chiral symmetry in the Higgs phase of the Bars-Yankielowicz model

$$H = SU(N)_{\text{diag}} \times SU(4) \times U(1)_{Q'} \tag{48}$$

is a subgroup of the chiral global symmetry $G_{\text{quantum}}$ in the confining phase. Furthermore, the fermions which remain massless in the Higgs phase, given by Eq. (47), have the quantum numbers corresponding to those of a subset of the bound state fermions $B_{[a,b]}$ which we found earlier matched the chiral anomalies. Indeed, if one decomposes the second rank antisymmetric representation of $SU(N+4)$ into representations of $SU(N) \times SU(4)$ one finds

$$(N+4)(N+3)/2 = \big(N(N-1)/2, 1\big) \oplus (1,6) \oplus (N,4) \; . \tag{49}$$

The first and last of the above representations are those we have found in the Higgs phase.

The analysis of the Bars-Yankielowicz model in the Higgs phase provides important dynamical information. Since there is no phase boundary between this phase and the confining phase, one expects that the surviving symmetry and the massless excitations found in the Higgs phase should correspond to those of the confining phase. This analysis therefore suggests that in the confining phase the full global symmetry $G_{\text{quantum}}$ does not survive in the binding, but that a slightly smaller symmetry $H$ actually remains a good chiral symmetry. The chiral anomalies of this symmetry at the preon level are matched not by the full set of bound states $B_{[a,b]}$ we discussed earlier, but by the smaller subset which emerged in our study of the complementary phase of the model. Indeed, it is easy to check that the bound states

$$\begin{aligned} B_{[a,b]} &\sim (F^T_{\alpha a}\sigma_2\sigma_\mu F_{\beta b})\sigma^\mu S^{\alpha\beta} \quad a,b \subset SU(N) \\ B'_{ai} &\sim (F^T_{\alpha a}\sigma_2\sigma_\mu F_{\beta i})\sigma^\mu S^{\alpha\beta} \quad a \subset SU(N);\ i \subset SU(4) \end{aligned} \tag{50}$$

precisely match all the anomalies of the $H$-group at the preon level.

For future use, I note also that the following 6-fermion effective interaction is $H$ invariant:

$$\mathcal{L}_{\text{eff}} = \frac{1}{\Lambda_c^5} BB\bar{B}'\bar{B}'\bar{B}'\bar{B}' \ . \tag{51}$$

In the above, all unnecessary indices are suppressed and $\Lambda_c$ is the dynamical scale associated with the $SU(N)$ preon theory. This interaction can be traced to the presence of underlying effective interactions at the preon level which preserve $U(1)_Q$ but violate individually $U(1)_F$ and $U(1)_S$.[20]

### The Novino Model

As a second interesting illustrative model, I want to discuss the, so called, novino model.[9] This model is based on an $SU(2)$ supersymmetric gauge theory in which, as a result of the breakdown of a global symmetry, massless Quasi Goldstone Fermions (QGF) are produced as bound states. To be more specific, the model has six preons superfields $\Phi^p_\alpha$ $(p = 1, \cdots, 6)$, with each superfield

$$\Phi^p_\alpha = (\varphi^p_\alpha,\ \psi^p_\alpha) \tag{52}$$

containing a complex scalar $\varphi^p_\alpha$ and a Weyl fermion $\psi^p_\alpha$. The classical global symmetry of the model also here contains two $U(1)$ factors:

$$G_{\text{classical}} = SU(6) \times U(1)_\Phi \times U(1)_R \ , \tag{53}$$

where the first factor is just the particle number associated with the superfield $\Phi$, while the other is an $R$ symmetry associated with the supersymmetry.* Again only a linear combination of $U(1)_\Phi$ and $U(1)_R$ will be free of gauge anomalies and one finds that the quantum symmetry for the model is:[9]

$$G_{\text{quantum}} = SU(6) \times U(1)_\chi \tag{54}$$

with

$$Q_\chi = Q_\Phi + 3Q_R \; . \tag{55}$$

In the novino model the chiral global symmetry (54) can be broken down by condensate formation:

$$\langle \epsilon^{\alpha\beta} \Phi^5_\alpha \Phi^6_\beta \rangle \neq 0 \tag{56}$$

resulting in a smaller surviving symmetry for the theory

$$G_{\text{quantum}} \longrightarrow H = SU(4) \times SU(2) \times U(1)_{\chi'} \; . \tag{57}$$

The $U(1)_{\chi'}$ surviving symmetry here can be seen to involve the particle number for the first four superfields $\Phi_4$ and the $R$ symmetry:

$$Q_{\chi'} = Q_{\Phi_4} + 2Q_R \; . \tag{58}$$

This breakdown produces 17 Nambu-Goldstone bosons, since this is the dimension of the coset space $G_{\text{quantum}}/H$. It is easy to see that these states decompose with respect to $SU(4) \times SU(2)$ as

$$NGB = (4,2) \oplus (\bar{4},\bar{2}) \oplus (1,1) \; . \tag{59}$$

If the underlying dynamics of the model preserves supersymmetry in the binding, then this set of NGB must be matched by some massless fermionic partners (QGF) and possibly some other scalar companions, which we shall denote as QGB–quasi Nambu-Goldstone bosons. Supersymmetry requires that the number of degrees of freedom of the scalars match those of the fermions. Thus one must have that

$$2n_{\text{QGF}} = n_{\text{NGB}} + n_{\text{QGB}} \; . \tag{60}$$

---

* The $R$ charge of $\psi$ is -1, while that of the $SU(2)$ gauginos, $\tilde{g}$, is +1.

The minimal solution of the above equation, given that $n_{\rm NGB} = 17$, is to have 9 QGF and only 1 additional QGB. This is precisely the dynamical solution which will emerge from the novino model.[9] This pattern provides a semirealistic model for one family of left-handed fermions. These states transform as

$$\mathrm{QGF} \sim (4,2) \oplus (1,1) \ , \tag{61}$$

and the bound state fermions

$$\Psi^{ai}_{BS} \sim \epsilon^{\alpha\beta}\left[\psi^a_\alpha\varphi^i_\beta - \psi^i_\alpha\varphi^a_\beta\right] \quad (a = 1, \cdots, 4;\ i = 5, 6) \tag{62}$$

comprise one lepton doublet and a (color triplet) quark doublet. In addition, there is an $SU(2)$ singlet state, the novino,

$$\Psi^{56}_{BS} \sim \epsilon^{\alpha\beta}\left[\psi^5_\alpha\varphi^6_\beta - \psi^6_\alpha\varphi^5_\beta\right] \ . \tag{63}$$

One can check dynamically by two different means that this minimal solution is consistent. First, the group $H$ is chiral, and thus one can check that the anomalies of $H$ at the preon level are matched by those of the bound states we have identified. Full matching is readily found. For instance, at the preon level the $[U(1)^3_{\chi'}]^3$, anomaly contributes

$$A[U(1)^3_{\chi'}] = 3(2)^3 + 8(-1)^3 + 4(-2)^3 = -16 \ , \tag{64}$$

where the first contribution is that of the gauginos, the second is that of the $SU(4)$ preons and the last that of the $SU(2)$ preons. This same anomaly computed at the bound state level, on the other hand, gives

$$A[(U(1)^3_{\chi'}] = 8(-1)^3 + 1(-2)^3 = -16 \ , \tag{65}$$

where the first contribution is that from the quarks and leptons and the second is that from the novino. Because this anomaly, and all the others match, it is at least consistent to imagine that the set of QGF of Eq. (61) indeed forms.

Because the scalar preons of the model $\varphi^p_\alpha$ transform according to the fundamental representation of $SU(2)$, to reinforce this result one can use, as a second dynamical tool, complementarity. Thus one can look at the broken phase of the theory and try to deduce, through complementarity, both what is the surviving symmetry and which are the $m = 0$ fermions in the spectrum. If one assumes that $\varphi^p_\alpha$ gets a VEV:

$$\langle\varphi^p_\alpha\rangle = v\delta^p_\alpha \ , \tag{66}$$

it is easy to check that the gauge times global symmetry of the model breaks down precisely to $H$! That is,

$$SU(2) \times \left[SU(6) \times U(1)_\chi\right] \longrightarrow SU(2)_{\text{diag}} \times SU(4) \times U(1)_{\chi'} \sim H \ . \tag{67}$$

This VEV, however, can only give masses to 3 fermions out of the 12 fermions in $\psi^p_\alpha$, namely those included in

$$\chi^i_j = \psi^i_j - \frac{1}{2}\ \delta^i_j \psi^k_k \quad i,j,k = 5,6\ . \tag{68}$$

One is left over, therefore, with 9 massless fermions and these transform under $H$ precisely as the QGF we identified in Eq. (61).

Having ascertained that the dynamics of this model is consistent and leads to an interesting set of massless fermions, it is useful to discuss here one other aspect of this toy model which will be of some use later. Because the model is supersymmetric, the low energy interactions of the QGF are (almost) fixed by the coset space $G_{\text{quantum}}/H$. For Nambu Goldstone bosons, of course, it is known that their low energy interactions are entirely fixed by the coset space metric.[21] One has

$$\mathcal{L}_{\text{eff}} = -\partial_\mu \pi_i\ g_{ij}(\pi/v)\partial^\mu \pi_i\ , \tag{69}$$

where $g_{ij}$ is the metric of the coset space and $v$ is a scale associated with the breakdown. There is a supersymmetric generalization of this that involves the Kähler potential for the associated Nambu- Goldstone superfields.[22] However, if in addition to having purely Nambu-Goldstone bosons in the theory there are also some quasi Nambu-Goldstone bosons, then the low energy interactions are no longer uniquely fixed by the coset space.[23] This is precisely the case here due to the presence of 1 QGB, associated with the novino.

Let $\phi$ denote the novino superfield and $\phi_{ai}$ denote the superfield connected with the $(4,2)$ bound state representation. Then one can show[9] that the low energy effective interactions of the model are given by

$$\begin{aligned}\mathcal{L}_{\text{eff}} = \Big[ & v_1^2 \bar{\phi}_{ai}\phi_{ai} + v_2^2 \bar{\phi}\phi - \frac{v_1^2}{2}\bar{\phi}_{ai}\bar{\phi}_{bj}\phi_{aj}\phi_{bi} \\ & + \frac{v_2^2}{4}\bar{\phi}_{ai}\bar{\phi}_{bj}\phi_{ai}\phi_{bj} + \cdots \Big]\Big|_{\theta\theta\bar{\theta}\bar{\theta}}\end{aligned}\ , \tag{70}$$

where $v_1, v_2$ are scales associated with the breakdown. A simple calculation then gives the following 4-fermion effective interaction for the doublet fermions $\psi_{aL}$ ($a = 1, \cdots, 4$) contained in the $(4,2)$ superfield:

$$\begin{aligned}\mathcal{L}_{\text{Fermi}} = & \frac{1}{8v_1^2}(\bar{\psi}_{aL}\gamma^\mu\vec{\tau}\psi_{aL})\cdot(\bar{\psi}_{bL}\gamma_\mu\vec{\tau}\psi_{bL}) \\ & - \frac{(v_1^2 - v_2^2)}{8v_1^4}(\bar{\psi}_{aL}\gamma^\mu\psi_{aL})(\bar{\psi}_{bL}\gamma_\mu\psi_{bL})\end{aligned} \quad . \tag{71}$$

## A FIRST ATTEMPT AT MASS GENERATION

Although protective symmetries are a necessary ingredient of any composite model of quarks and leptons, for a realistic theory one must find a way to generate small masses for the bound states which the protective symmetries forced to zero mass. One can accomplish this by gauging a subset of the preons in a **vector-like** way, exploiting the fact that global chiral symmetries in these theories are spontaneously broken. This is a rather old idea,[24] which has been examined recently by Khlebnikov and me.[25] Before discussing some of the details behind this idea, it is useful to illustrate its qualitative features. In models of the type we are considering there are two operating scales. One is the very high scale $\Lambda_c$, which gives rise to very tight preon bound states of size $\langle r\rangle \sim 1/\Lambda_c$. The other is the scale associated with the range of the vector-like interactions, typified by the dynamical scale of these theories, $\Lambda_v$, and one assumes that $\Lambda_v \ll \Lambda_c$. Masses for the states which were massless when the vector-like interactions are neglected then turn out to be small on the scale of $\Lambda_c$, because they are suppressed by the scale mismatch between $\Lambda_c$ and $\Lambda_v$. Typically one obtains, as a result,

$$m_f \sim \Lambda_c\left[\frac{\Lambda_v}{\Lambda_c}\right]^n \ll \Lambda_c \ , \tag{72}$$

with $n$ some small integer.

To get a first orientation, I want to begin by illustrating how the idea of vector-like gauging works in the case of the toy Bars- Yankielowicz model[16] discussed earlier. Recall that in this model the global quantum symmetry $G_{\text{quantum}}$ was broken to $H = SU(N)\times SU(4)\times U(1)_{Q'}$ and that two kinds of $m = 0$ bound states ensued in the model, which we denoted as $B$ and $B'$ respectively (cf. Eq. (50)). The $B$ states are $SU(4)$ singlets, while the $B'$ states transform as the fundamental of $SU(4)$.

Imagine now that in addition to the chiral $SU(N)$ gauging of the preons, 4 of the $F$ preons also felt an $SU(2)$ gauge symmetry, with the 4 states acting as two doublets of $SU(2)$. That is, under $SU(2)$ these preons transform as *

* Even though $\bar{2}$ is equivalent to 2 in $SU(2)$, we find it convenient to carry this

$$4 = 2 \oplus \bar{2} \ . \tag{73}$$

Assuming that the $SU(4)$ in $H$ coincides with the one defined through the 4 preons which get this vectorial gauging, it follows that the $B'$ states also split into two sets of $SU(2)$ doublets

$$B' = B'_2 \oplus B'_{\bar{2}} \ . \tag{74}$$

The strong $SU(2)$ gauge group can form condensates of $B'_2$ and $B'_{\bar{2}}$ much in the same way as quark-antiquark condensates form in QCD. Since the $B'$ states transform non-trivially under the global $SU(N) \times U(1)_{Q'}$ symmetry, the presence of these condensates, whose scale is fixed by the dynamical scale $\Lambda_2$ of the $SU(2)$ theory,

$$\langle B'_2 \ B'_{\bar{2}} \rangle \sim \Lambda_2^3 \ , \tag{75}$$

will break this symmetry down. One can check[25] that Eq. (75) causes the breakdown of $SU(N) \times U(1)_{Q'}$ to $O(N)$. Most importantly, the presence of these condensates can generate a mass for the otherwise massless bound states $B$, through the action of the $H$- invariant effective interaction given in Eq. (51). Replacing pairs of $B'$ states by their corresponding condensates, one sees that

$$\mathcal{L}_{\rm eff} = \frac{1}{\Lambda_c^5} BB\bar{B}'\bar{B}'\bar{B}'\bar{B}' \longrightarrow \mathcal{L}_{\rm mass} = \frac{\Lambda_2^6}{\Lambda_c^5} BB \ . \tag{76}$$

Thus the $B$ states acquire a (Majorana) mass of order

$$m_B = \frac{\Lambda_2^6}{\Lambda_c^5} \ll \Lambda_c \ , \tag{77}$$

assuming that indeed $\Lambda_2$ is much smaller than $\Lambda_c$.

Let me recapitulate what we have accomplished. Recall that the Bars-Yankielowicz model produces naturally two kinds of bound states. Heavy states with mass $M \sim \Lambda_c$ and, before vector-like gauging, massless states $B$ and $B'$ protected by the global chiral symmetry $H$. After vectorially gauging an $SU(2)$ at the preon level, the erstwhile massless bound states $B'$ are confined into some $SU(2)$ singlet bound states of intermediate mass $M_I \sim \Lambda_2$. However, in addition, through the formation of the condensate (75), the remaining $m = 0$ bound states $B$ now get the small mass, given in Eq. (77). So the result of this dynamics is a theory where there are three distinct scales associated with the bound states:

---

notational distinction in what follows.

$$\begin{aligned} \text{Heavy states} &\leftrightarrow M \sim \Lambda_c \\ \text{Intermediate states} &\leftrightarrow M_I \sim \Lambda_2 \\ \text{Light states} &\leftrightarrow m_B \sim \Lambda_2^6/\Lambda_c^5 \end{aligned} \tag{78}$$

Unfortunately, the model itself is not very realistic since we have only obtained masses for **vector-like fermions**, with the final group being an $O(N)$ not an $SU(N)$ group. What one is interested in is to be able to give masses to the quarks and leptons, which are **chiral fermions**.

## ELEMENTS OF A SEMIREALISTIC MODEL

To construct a semirealistic composite model of quarks and leptons one needs to allow for the possibility that, after vector-like gauging of some of the preons, the light fermions which ensue in the model are in **chiral representations**. Only if this is the case, it is possible to introduce the weak interactions as gauge interactions, where the left-handed fermions are organized in doublets and the right-handed fermions are singlets under the gauge group. To accomplish this, the simplest scheme is to imagine that mass generation occurs in two stages. In the first stage, as a result of condensate formation, left- and right-handed fermions are tied together. This can occur either through some appropriate vector-like gauging or through interactions already present at the preon level. In the language of a dynamical symmetry breaking theory, this first stage is very much like an ETC theory which ties left- and right-handed fermions together. Here, as in the ETC case, some of the $m = 0$ fermions will play the role of techniquarks. In the second stage mass generation really comes about. This follows here in ways quite akin to Technicolor[26] in which $\langle \bar{T}T \rangle$ condensates team up with the effective interactions of fermions and techniquarks to give mass to the ordinary quarks and leptons.

I want to illustrate these ideas by means of a nice, albeit uneconomical model for one generation of quarks (leptons need a separate theory). This model, which I developed in collaboration with Khlebnikov,[25] has many features in common with models developed ten years ago by Yamawaki and Yokota[24] and shares in the spirit of the earlier (supergroup) preon models put forward by Bars.[27]

The main idea that Khlebnikov and I pursued,[25] was to build in *ab initio* the asymmetry that the weak interactions impose on quarks and leptons, in which $f_L \sim 2$ while $f_R \sim 1$ under the electroweak group. To achieve this we started with 3 distinct preon models. The first of these models serves to build up the left-handed quark doublets and is already organized so that also some additional fermions ensue. The latter two preon theories exist essentially to provide the theory with the full complement of right-handed quarks. Taken together these models can produce a set of light, but nearly elementary, quarks.

The gauge group acting on the preons in the theory is $SU(6)_{\text{preon L}} \times SU(6)_{\text{preon u}_R} \times SU(6)_{\text{preon d}_R}$, where each of the separate preon structures serves to build the corresponding quark states. The $SU(6)_{\text{preon u}_R}$ and $SU(6)_{\text{preon d}_R}$ theories have the Bars-Yankielowicz[16] structure we discussed earlier, while for the $SU(6)_{\text{preon L}}$ theory this structure is doubled. In more detail, for each of these preon gauge groups one has the following set of fermions:

$$
\begin{aligned}
SU(6)_{d_R,u_R}: R_{d,u} &= \left\{\bar{3}(F^c_{u,d})_R \oplus \bar{3}(F^T_{u,d})_R \oplus 4(F^M_{u,d})_R + (S_{u,d})_R\right\} \\
SU(6)_L: R_L &= \left\{\bar{3}\begin{pmatrix} F^c_u \\ F^c_d \end{pmatrix}_L \oplus \bar{3}\begin{pmatrix} F^T_u \\ F^T_d \end{pmatrix}_L \oplus 4F^{M_1}_L + 4F^{M_2}_L + \begin{pmatrix} S_u \\ S_d \end{pmatrix}_L\right\}
\end{aligned} \tag{79}
$$

Here the $F$ preons are either right-handed or left-handed Dirac states (2-component Weyl preons) which transform according to the fundamental representation of the appropriate $SU(6)$ group, while the $S$ preons, either right-handed or left-handed, transform according to the second rank conjugate symmetric tensor representation of $SU(6)$. The various labels $c, T$ and $M$ (or $M_1$ and $M_2$) identify future vectorial interactions which will act on these specific preons. For the case of $SU(6)_L$, the preons are also organized in terms of future doublets, or singlets, under the weak $SU(2)$ group. Each of these preons also have specific $U(1)_Y$ assignments. These, however are left unspecified in Eq. (79) to ease the notation. These $U(1)_Y$ assignments forbid one to generate both quarks and leptons in the model.[25]

Given the representation assignments of Eq. (79), one can easily deduce the bound states which will ensue for each preon theory. The results here follow precisely our earlier discussion in the toy models. In particular, for each preon theory, one expects two $m = 0$ preon bound states to form, analogous to the $B$ and $B'$ states found in the previous section. More specifically, one finds the following set of bound states[25]

$$
\begin{aligned}
B &= \left\{B_{u_R} \sim (15,1); \quad B_{d_R} \sim (15,1); \quad B_L \sim \begin{pmatrix} 15,1 \\ 15,1 \end{pmatrix}\right\} \\
B' &= \left\{B'_{u_R} \sim (6,4); \quad B'_{d_R} \sim (6,4); \quad B'_{1L} \sim (6,4); \quad B'_{2L} \sim (6,4)\right\}
\end{aligned} \tag{80}
$$

In the above, I indicated the transformation of the bound states under the natural $SU(6) \times SU(4)$ global chiral group of each preon theory. Each of the $SU(6)$ states decomposes into appropriate representations of the two $SU(3)$ subgroups which eventually are gauged at the preon level, with the first $SU(3)$ group corresponding to ordinary color (c) and the other to technicolor (T). One has

$$\begin{aligned} 6 &= (\bar{3},1) \oplus (1,\bar{3}) \\ 15 &= (3,1) \oplus (1,3) \oplus (\bar{3},\bar{3}) \end{aligned} \tag{81}$$

The $(3,1)$ states in the 15 representation above are the desired quark states with the $(1,3)$ and $(\bar{3},\bar{3})$ states playing the role of techniquarks.

The $M$ preons, like the $c$ and $T$ preons, are also subject to a vector-like interaction (metacolor) whose purpose in the model is to link the states produced by the "right" preon theories ($SU(6)_{\text{preon u}_R}$ and $SU(6)_{\text{preon d}_R}$) with those of the "left" preon theory ($SU(6)_{\text{preon L}}$). In particular, after gauging the $SU(4)$ metacolor group two different condensates will form, each of size $\Lambda_4^3$:

$$\langle \bar{B}'_{1L} B'_{uR} \rangle \sim \Lambda_4^3 \; ; \quad \langle B'_{2L} B'_{dR} \rangle \sim \Lambda_4^3 \; . \tag{82}$$

As a result of this condensate formation, and of the residual interactions between $B$ and $B'$ bound states produced by each preon theory, an effective ETC interaction will ensue. Schematically, the three preon theories–whose dynamical scales will be denoted by $\Lambda_L$, $\Lambda_{u_R}$ and $\Lambda_{d_R}$, respectively–will give rise to residual contact terms between the $B$ and $B'$ bound states of the form*

$$\begin{aligned} \mathcal{L}^{\text{preon}}_{\text{residual}} = & \frac{1}{\Lambda_L^2}(\bar{B}_L B_L)(\bar{B}'_{1L} B'_{1L}) + \frac{1}{\Lambda_L^2}(\bar{B}_L B_L)(\bar{B}'_{2L} B'_{2L}) \\ & + \frac{1}{\Lambda_{u_R}^2}(\bar{B}_{u_R} B_{u_R})(\bar{B}'_{u_R} B'_{u_R}) + \frac{1}{\Lambda_{d_R}^2}(\bar{B}_{d_R} B_{d_R})(\bar{B}'_{d_R} B'_{d_R}) \end{aligned} \tag{83}$$

Using these interactions and the condensates of Eq. (82), produced as a result of gauging the $SU(4)$ metacolor group, will give rise to an effective interaction among the $B_L$ and $B_{u_R}$, or $B_{d_R}$, states. This mechanism is sketched schematically in Fig. 8. It leads to the following effective Lagrangian describing these interactions[25]

$$\begin{aligned} \mathcal{L}_{\text{ETC eff}} \simeq & \frac{\Lambda_4^2}{\Lambda_L^2 \Lambda_{u_R}^2}(\bar{B}_L B_L)(\bar{B}_{u_R} B_{u_R}) \\ & + \frac{\Lambda_4^2}{\Lambda_L^2 \Lambda_{d_R}^2}(\bar{B}_L B_L)(\bar{B}_{d_R} B_{d_R}) \end{aligned} \tag{84}$$

Since the $B$ states contain both quarks and techniquarks, the above effective interactions can produce mass for the quarks once the technicolor group is gauged, precisely as it occurs in ordinary ETC theories.[2] One finds in this way

* We suppress below all $\gamma$ matrices and group indices in the interactions.

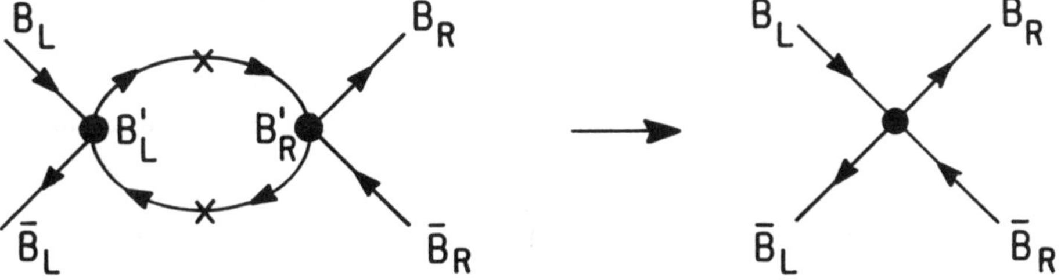


**Fig. 8:** Origin of $\mathcal{L}_{\text{ETC eff}}$ in the model.

$$\mathcal{L}_{\text{mass}} = \frac{\langle \bar{T}_L T_R \rangle}{\Lambda_{\text{eff}}^2} \bar{f}_L f_R \ . \tag{85}$$

Here, as in Technicolor theories,[26] the scale of the technicolor condensates is related to the mass of the $W$:

$$\langle \bar{T}_L T_R \rangle \sim \Lambda_T^3 \ ; \quad M_W \sim g\Lambda_T \ . \tag{86}$$

In our model, furthermore, we have two different $\Lambda_{\text{eff}}$ depending on whether the effective ETC interactions involve up-like or down-like quarks. Since

$$(\Lambda_u^2)_{\text{eff}} = \frac{\Lambda_L^2}{\Lambda_4^2} \Lambda_{u_R}^2 \ ; \quad (\Lambda_d^2)_{\text{eff}} = \frac{\Lambda_L^2}{\Lambda_4^2} \Lambda_{d_R}^2 \ , \tag{87}$$

for the model the masses of the up and down quarks will be different:

$$m_u \sim \frac{\Lambda_T^3 \Lambda_4^2}{\Lambda_L^2 \Lambda_{u_R}^2} \sim K \langle r_u \rangle^2 \ ; \quad m_d \sim \frac{\Lambda_T^3 \Lambda_4^2}{\Lambda_L^2 \Lambda_{d_R}^2} \sim K \langle r_d \rangle^2 \ . \tag{88}$$

This result informs us that the $u - d$ mass difference (or probably, more realistically the $t - b$ mass difference) is set by the different extent of the preonic bound states. That is, the dynamical scales of the different $SU(6)_{\text{preon } u_R}$ and $SU(6)_{\text{preon } d_R}$ theories determine the different masses of the corresponding composite quarks. However, to the extent that the various scales in the theory obey the hierarchy:

$$\Lambda_T < \Lambda_4 \ll \Lambda_L \lesssim \Lambda_{d_R} < \Lambda_{u_R} \ , \tag{89}$$

then the resulting quarks will be massive but so tightly bound that, for all practical purposes, they will look essentially elementary.

It is possible to arrive at quite similar results by a simple extension[28] of the novino model[9] discussed earlier. Instead of considering an $SU(2)$ preon theory with 6 superfields $\Phi^p_\alpha$ $(p = 1, \cdots, 6)$, one considers instead[28] an $SU(2)_{\text{preon L}} \times SU(2)_{\text{preon R}}$ theory with three different kinds of preon superfields. These are: a set of 8 superfields $\Phi^p_\alpha$ $(p = 1, \cdots, 8)$ of the $SU(2)_{\text{preon L}}$ theory which will build up the left-handed quarks and leptons, plus two doublets of left-handed techniquarks; a set of another 8 superfields $\Phi'^{\alpha'}_{p'}$ $(p' = 1, \cdots, 8)$ of the $SU(2)_{\text{preon R}}$ theory which will build up the analogous right-handed states; and, finally, a bridging preon superfield $\chi^\alpha_{\alpha'}$ which is a doublet under both confining preonic groups and whose purpose is to tie the left-handed states with the right-handed states.

As in the novino model, one can argue that the natural quantum chiral global symmetry of the model

$$G_{\text{quantum}} = SU(8) \times SU(8)' \times U(1)_\chi \times U(1)_{\chi'} \ , \tag{90}$$

is broken down by the condensates

$$\langle \epsilon^{\alpha\beta} \Phi^7_\alpha \Phi^8_\beta \rangle \neq 0 \ ; \quad \langle \epsilon_{\alpha'\beta'} \Phi'^{\alpha'}_7 \Phi'^{\beta'}_8 \rangle \neq 0 \ . \tag{91}$$

These condensates break down the respective $SU(8)$ groups to $SU(6) \times SU(2) \times U(1)$. In addition, the model becomes interesting, if one assumes that two further **left-right connecting** condensates form which are also alligned in the 7-8 direction. Namely

$$\langle \Phi'^{\alpha'}_7 \chi^\alpha_{\alpha'} \Phi^7_\alpha \rangle = \langle \Phi'^{\alpha'}_8 \chi^\alpha_{\alpha'} \Phi^8_\alpha \rangle \neq 0 \ . \tag{92}$$

These latter condensates break the two $SU(2)$ global symmetries remaining to their diagonal sum $\left[SU(2) \times SU(2)' \to SU(2)_{\text{diag}}\right]$ and serve to eliminate one of the remaining $U(1)$ symmetries.

The condensates (91) and (92), therefore, cause the breakdown of $G_{\text{quantum}}$ to a smaller group $H$, with

$$H = SU(6) \times SU(6)' \times SU(2)_{\text{diag}} \times U(1)_{\tilde{\chi}} \ . \tag{93}$$

The coset space $G_{\text{quantum}}/H$ is 54-dimensional and the associated Nambu-Goldstone bosons of the breakdown transform in the following way under $H$:*

* We do not detail here the $U(1)_{\tilde{\chi}}$ quantum numbers.

$$\begin{aligned} \text{NGB} = &(6,1,2) \oplus (\bar{6},1,2) \oplus (1,6,2) \oplus (1,\bar{6},2) \\ &\oplus (1,1,3) \oplus 3(1,1,1) \end{aligned} \tag{94}$$

By arguments quite analogous to the ones we discussed in connection with the novino model, one can show that[28] there are 30 QGF and 6 QGB in the theory.

The QGF of the model contain a full set of left- and right-handed quarks and leptons for one generation, along with 2 doublets of technifermions. Specifically, these states are given by

$$\begin{aligned} \Psi_{\rm L}^{ai} &= \epsilon^{\alpha\beta}[\psi_\alpha^a \varphi_\beta^i - \psi_\alpha^i \varphi_\beta^a] \sim (6,1,2) \quad (a = 1,\cdots,6; i = 7,8) \\ \Psi'_{\rm R\ ai} &= \epsilon_{\alpha\beta}[\psi_a'^\alpha \varphi_i'^\beta - \psi_i'^\alpha \varphi_a'^\beta] \sim (1,6,2) \quad (a = 1,\cdots,6; i = 7,8) \end{aligned} \tag{95}$$

In addition, this model has 3 novinos and an $SU(2)$ triplet of fermions belonging to the superfield

$$\Pi_j^i = \Xi_j^i - \frac{1}{2}\delta_j^i\ \Xi_k^k \tag{96}$$

with

$$\Xi_j^i = \Phi_j'^{\alpha'} \chi_{\alpha'}^{\alpha} \Phi_\alpha^i \ . \tag{97}$$

The effective low energy interactions of the $\Psi$ and $\Psi'$ bound states can be computed analogously to what was done in the novino toy model. Of interest for mass generation are interactions which connect Left $\leftrightarrow$ Right. For these interactions specifically, one finds,[28]

$$\mathcal{L}_{\rm residual} = \left[\frac{v_\pi^2}{8v_1^2 v_{1'}^2}\right](\bar{\Psi}_{\rm aL}\gamma^\mu \vec{\tau}\Psi_{\rm aL}) \cdot (\bar{\Psi}_{\rm bR}\gamma_\mu \vec{\tau} \cdot \Psi_{\rm bR}) \ . \tag{98}$$

Here $v_1, v_1'$ are the scales of the $SU(8) \to SU(6) \times SU(2) \times U(1)$ and $SU(8)' \to SU(6)' \times SU(2)' \times U(1)'$ breakdowns, respectively, while $v_\pi$ is the scale of the breakdown of $SU(2) \times SU(2)' \to SU(2)_{\rm diag}$.

Since the fermion fields $\Psi_{\rm aL}$ and $\Psi_{\rm bR}$ contain, in addition to the quarks and leptons, also two doublets of technifermion fields, if one were to gauge the $SU(2)$ associated with the two preons $p = 5,6$; $p' = 5,6$ then the quarks and leptons would get a mass of size

$$m_f = \frac{v_\pi^2 \Lambda_T^3}{v_1^2 v_{1'}^2} \ , \tag{99}$$

as a result of the formation of the technicolor condensates (of size $\Lambda_T^3$). However, at this stage, because of the L-R symmetry in the model (and the overall $SU(4)$ symmetry of quarks and leptons), the mass $m_f$ is a **common mass** for all excitations. That is,

$$m_f \equiv m_u = m_d = m_e = m_{\nu_e} \ . \tag{100}$$

Fortunately, this large degeneracy can be split by gauging the color $SU(3)$ and providing a Majorana mass for $\nu_R$, using the residual 't Hooft interactions in the theory. Nevertheless, there remains an embarassing up-down mass degeneracy. In this respect the model discussed earlier is better suited, since this degeneracy never occurs.

I should note that in this extension of the novino model the weak ineractions are incorporated by gauging an $SU(2) \times U(1)_Y$ symmetry at the preon level, using $\Phi_\alpha^p$ with $p = 7, 8$ as a weak doublet. This preon superfield is assumed not to carry any hypercharge but the other sueprfield $\Phi'^{\alpha'}_{p'}$ for $p' = 7, 8$ obeys:

$$Y\Phi'^{\alpha'}_{p'} = \begin{pmatrix} -1/2 \\ 1/2 \end{pmatrix} \Phi'^{\alpha'}_{p'} \ . \tag{101}$$

Because of these assignments under the electroweak group, the trilinear condensates of Eq. (92)–connected to the scale $v_\pi$ and the breakdown of $SU(2) \times SU(2)' \to SU(2)_{\rm diag}$–also break the gauge symmetry $SU(2) \times U(1)_Y$. Thus in the model one has really two distinct sources of $SU(2) \times U(1)_Y \to U(1)_{\rm em}$ breaking; the trilinear preonic condensates associated with the scale $v_\pi$ and the technicolor condensates, associated with the scale $\Lambda_T$. *A priori*, it is not clear whether $\Lambda_T > v_\pi$ or *vice versa*. However, one naively expects that the scale of the 3-body condensates, $v_\pi$, should be less than that associated with the 2-body condensates, $v_1$ and $v_1'$. Thus, the resulting quark and lepton masses (99) should be very light.

## FAMILY PROBLEMS AND IDEAS

Up to now I have discussed only one-family models. This is not an accident, since the issue of families and of the origin of interfamily mass splittings is not well understood in substructure models. Even so, it is worthwhile to discuss some of the failures and lessons one encounters in trying to build in families in preon models.

In the models discussed earlier, it is rather easy to mechanically introduce families by making copies of the one generation preon models. For instance, to get a 3-generation model, one can consider instead of the preon group $SU(6)_{\rm preon\ L} \times SU(6)_{\rm preon\ u_R} \times SU(6)_{\rm preon\ d_R}$ a preon theory based on the group

$$\left[SU(6)_{\mathrm{preon\ L}} \times SU(6)_{\mathrm{preon\ u_R}} \times SU(6)_{\mathrm{preon\ d_R}}\right]^3 . \tag{102}$$

In such a theory one would get again mass formulas of the type

$$m_f \simeq K\langle r_f^2\rangle , \tag{103}$$

which–if true–would tell us that top is heavy because it has the **largest** size compared to all the other bound states. However, these mechanical extensions have considerable difficulties.[25] Since some of these troubles are generic, it is worth discussing them briefly here. For definiteness, I will focus on the model based on the large preon group of Eq. (102).

This model has two types of problems: theoretical problems and phenomenological problems. In the former category fall issues connected to the fact that this theory really has a very large number of preons. In fact, with the gauge group of Eq. (102), there are so many preons that, at the compositeness level, color, technicolor and metacolor are all **not** asymptotically free! So there is a real question of how one then controls the dynamics.* Even if one were to, somehow, resolve this issue, there are more immediate phenomenological problems. Because of the mechanical way in which we have introduced family replication, the model in general has a variety of protective family symmetries. These symmetries, prevent the generation of a quark-mixing matrix: $V_{\mathrm{CKM}} = 1$. Of course, this is a real phenomenological disaster!

It is worthwhile to elaborate a little more on the last point, to understand better how, even though one is able to generate a spectrum of quarks, one cannot get any mixing among them. Using the gauge group of Eq. (102), it is easy to check[25] that each generation preserves a generalized vectorial $U(1)$ symmetry, $U(1)_V$ in which the colored preons have $Q_V = +1$, while the technicolor preons have $Q_V = -1$. Normally, such vectorial symmetries are never broken by condensates and one is immediately in trouble, because the presence of these $U(1)_V$ symmetries automatically forces $V_{\mathrm{CKM}}$ to be diagonal. However, because the states that carry metacolor, $B'$, have strong chiral interactions, the metacolor condensates may not respect the Vafa-Witten theorem.[29] Indeed, to have some hope that any family mixing ensues, one must assume that it is possible to form off-diagonal metacolor condensates of the type analogous to Eq. (82):

$$\langle \bar{B}'_{1\mathrm{L\ i}} B'_{\mathrm{u_R j}}\rangle = \Lambda_4^3 \Sigma_{ij}^{(1)} \ ; \quad \langle \bar{B}'_{2\mathrm{L\ i}} B'_{\mathrm{d_R j}}\rangle = \Lambda_4^3 \Sigma_{ij}^{(2)} , \tag{104}$$

---

* In this sense the SUSY preon model of the novino type fare much better. But, in this latter case, one has to worry about how SUSY is ultimately broken.

where $i$ and $j$ are family indices and

$$\Sigma_{ij}^{(1)} \neq \delta_{ij} \; ; \quad \Sigma_{ij}^{(2)} \neq \delta_{ij} \; . \tag{105}$$

Even if Eq. (105) holds, this is not sufficient to generate some family mixing, because the effective ETC interactions (analogous to those given in Eq. (84)) still preserve B family number, since they always involve $\bar{B}_i B_i$ terms:

$$\mathcal{L}_{\mathrm{ETC\ eff}} = \frac{\Lambda_4^2}{\Lambda_{\mathrm{L}i}^2 \Lambda_{\mathrm{u_R,d_R}j}^2} (\bar{B}_{\mathrm{L}i} B_{\mathrm{L}i})(\bar{B}_{u_R,d_R j} B_{\mathrm{u_R,d_R}j}) \; . \tag{106}$$

However, the above is not the only effective interaction present at the bound state level. Each $SU(6)$ preon gauge theory will have its own 't Hooft interaction, analogous to Eq. (51). For the model in question, one has[25]

$$\begin{aligned} \mathcal{L}_{\mathrm{eff}} = & \frac{1}{\Lambda_{\mathrm{u_R,d_R}i}^5} (B_{\mathrm{u_R,d_R}}^i B_{\mathrm{u_R,d_R}}^i)(\bar{B}_{\mathrm{u_R,d_R}}^{\prime i} \bar{B}_{\mathrm{u_R,d_R}}^{\prime i} \bar{B}_{\mathrm{u_R,d_R}}^{\prime i} \bar{B}_{\mathrm{u_R,d_R}}^{\prime i}) \\ & + \frac{1}{\Lambda_{\mathrm{L}j}^{14}} (B_L^{\prime j} B_L^{\prime j} B_L^{\prime j} B_L^{\prime j} B_L^{\prime j} B_L^{\prime j} B_L^{\prime j} B_L^{\prime j}) \cdot (\bar{B}_L^j \bar{B}_L^j \bar{B}_L^j \bar{B}_L^j) \end{aligned} \tag{107}$$

These interactions, obviously do not conserve the individual $B_i$-fermion number and, after metacolor is gauged and the metacolor condensates (104) are formed, one will have $B_i$-violating interactions of the type:

$$\mathcal{L}_{B_i \ \mathrm{violating}} = \frac{\Lambda_4^{16}}{\Lambda_{\mathrm{L}k}^{14} \Lambda_{\mathrm{u_R}j}^5 \Lambda_{\mathrm{d_R}i}^5} (B_{\mathrm{u_R}}^j B_{\mathrm{u_R}}^j B_{\mathrm{d_R}}^i B_{\mathrm{d_R}}^i)(\bar{B}_{\mathrm{L}}^k \bar{B}_{\mathrm{L}}^k \bar{B}_{\mathrm{L}}^k \bar{B}_{\mathrm{L}}^k) \; . \tag{108}$$

If $\Lambda_4 < \Lambda_{\mathrm{L}k}$, $\Lambda_{\mathrm{u_R}j}$, $\Lambda_{\mathrm{d_R}i}$, this is a small perturbation which breaks the natural residual family symmetry $[U(1)_B]^3$ down. However, this perturbation cannot really generate family mixing since it preserves a discrete $(Z_2)^2 \; Z_4$ symmetry![25]

There is no theorem which prevents the technicolor interactions, in principle, to break down this final discrete family symmetry. However, it is difficult to see how the remaining chiral interactions are strong enough at the technicolor scale to induce the formation of non- vector-like condensates. Thus these remnant discrete family symmetries will survive and the dynamics will prevent the appearance of any family mixing, leading to $V_{\mathrm{CKM}} = 1$. This circumstance is quite generic for "mechanical" family models, where in the end always some protective family symmetry–either discrete or continuous–survives. There are no problems in these models to get arbitrary fermion masses and mass splittings, but these theories are not phenomenologically viable since there is no quark mixing!

It may be possible to avoid this disaster in more sophisticated models for family generation. However, even in these cases there are difficulties in trying to reproduce the interfamily mass hierarchy seen in nature. It is perhaps useful here to indicate the structure of a more sophisticated family model, even though the model is not quite realistic. The theory I want to briefly discuss is based on an $SU(8n+k)$ preon theory, first discussed some time ago by Preskill.[30] The model uses a combination of two $SU(N)$ chiral gauge theories which are free of gauge anomalies containing, respectively, the following representation content:

$$R_S = \{(N+4)\ F_\alpha \oplus S^{\alpha\beta}\} \tag{109}$$

and

$$R_A = \{(N-4)\ F_\alpha \oplus A^{\alpha\beta}\}\ . \tag{110}$$

Preskill's model involves the difference of the above representations and hence has the preon content

$$R = \{8F_\alpha \oplus S^{\alpha\beta} \oplus A_{\alpha\beta}\}\ . \tag{111}$$

The global symmetry of this $SU(N)$ preon theory,

$$G_{\text{quantum}} = SU(8) \times U(1)_Q \times U(1)_{Q'}\ , \tag{112}$$

in general is not preserved in the binding. The actual breakdown pattern and the $m=0$ bound states in the theory depend on the value of $N$ in the preon gauge symmetry.[19] If one writes

$$N = 8n + k\ , \tag{113}$$

then the pattern of the breakdown depends on $k$, while the number of $m=0$ fermion bound states depends on the value of $n$. By studying the model using complementarity, one finds[19]

$$G_{\text{quantum}} \longrightarrow H = SU(k) \times SU(8-k) \times U(1)_{\tilde{Q}_1} \times U(1)_{\tilde{Q}_2}\ . \tag{114}$$

The massless fermions which match the $H$ anomalies come in repetitive patterns related to the size $n$ of the gauge group, with three distinct bound states appearing: $B, \tilde{B}$ and $B'$. The quantum numbers of these states under $H$, and their multiplicities, are as follows:

$$\begin{aligned} B &\sim (k(k-1)/2,\ 1,\ 2(8-k),\ N-2-4s) \text{ with } s = 0,\cdots,2n+1 \\ \tilde{B} &\sim (1,\ (8-k)(7-k)/2,\ -2k,\ N-2-4s) \text{ with } s = 0,\cdots,2n-1 \\ B' &\sim (k,\ 8-k,\ 8-2k,\ N-2-4s) \text{ with } s = 0,\cdots,2n \end{aligned} \tag{115}$$

We see from the above that the second $U(1)$ symmetry, $U(1)_{\tilde{Q}_2}$, acts as a family number, with the multiplicity of states closely correlated with the size of the preon gauge group, $N = 8n + k$. In this model the family symmetry carried by $U(1)_{\tilde{Q}_2}$ is related to the number of $(AS)$ pairs found in the wave-functions of the $B, \tilde{B}$ and $B'$ states. One can check[19] that the "valence" combination of preons (FFS) present in all three of these states has $\tilde{Q}_2 = N - 2$, while an $(AS)$ pair has $\tilde{Q}_2 = -4$. Thus the number of $(AS)$ pairs distinguishes the various family replications in the $B, \tilde{B}$ and $B'$ states.

To make this model semirealistic, one can play the same tricks we did before, constructing separately left- and right-handed states via a combined set of preon gauge groups

$$SU(8n+k)_{\text{preon L}} \times SU(8n+k)_{\text{preon } u_R} \times SU(8n+4)_{\text{preon } d_R} \,. \tag{116}$$

The separate left- and right-handed bound states, as before, communicate with each other via the gauging of a metacolor group in $SU(8-k)$. The condensates formed through this metacolor gauging, in turn, can break the $U(1)_{\tilde{Q}_2}$ family symmetry. So, in principle, this model should have a non-vanishing CKM matrix. However, here it appears to be difficult to generate any inter-family mass-hierarchy since there are only 3 confining scales in the theory, associated with each of the three preon groups in Eq. (116). While $\Lambda_{u_R} \neq \Lambda_{d_R}$ can give a $t-b$ mass difference, what provides a $t-c$ or $t-u$ mass difference?*

## CONCLUDING REMARKS

The preon models discussed in these lectures have all various kinds of difficulties. These troubles tell us that we are still far from achieving a true dynamical model of composite quarks and leptons. Nevertheless, in my opinion, these models are very useful theoretical exercises and more people should be encouraged to think hard about this issue. We will know eventually whether the breakdown of $SU(2) \times U(1)_Y$ to $U(1)_{\text{em}}$ is triggered by a scalar VEV or comes about from the formation of some dynamical condensates. If the latter possibility is the way of nature, understanding how fermions get their mass will be the central theoretical issue of the day. I hope to have shown in these lectures that preon models provide

---

* As can be seen from Eq. (115), the $B$ and $\tilde{B}$ states have always an even number of families, providing yet another obstacle to surmount if one wants to make contact with reality!

an interesting framework for considering this question. Although present attempts at understanding the origin of quark and lepton masses are toy exercises, the dynamics of chiral plus vectorial gauge interactions is very rich.[31] It is my hunch that it may well hold the key to answering this deep question.

## ACKNOWLEDGMENTS

I am grateful to Nino Zichichi for the always perfect hospitality at Erice and to Sergei Khlebnikov for discussions on the topics of these lectures. This work was supported in pat by the Department of Energy under Grant No. FG03-91ER40662, Task C.

## REFERENCES

1. See, for example, E. W. Kolb and M. Turner, **The Early Universe** (Addison Wesley, Redwood City, CA, 1990).
2. S. Dimopoulos and L. Susskind, Nucl. Phys. B**155** (1979) 237;
   E. Eichten and K. D. Lane, Phys. Lett. **90B** (1980) 237.
3. For a discussion of these bounds see, for example, F. Thompson, Proceedings of the XXVI International Conference of High Energy Physics, Dallas, Texas, 1992, J. Sanford ed. (AIP, New York, 1993).
4. For a discussion see, for example, M. Gold, Proceedings of the XXVI International Conference on High Energy Physics, Dallas, Texas, 1992, J. Sanford. (AIP, New York, 1993).
5. For an interesting collection of reprints see, for example, R. Hofstadter **Electron Scattering and Nuclear and Nucleon Structure** (W. A. Benjamin, New York, 1963).
6. S. Brodsky and S. Drell, Phys. Rev. D**22** (1980) 2236;
   G. L. Shaw, D. Silverman and R. Slansky, Phys. Lett. **94B** (1980) 343.
7. G. 't Hooft in **Recent Developments in Gauge Theories**, ed. G. 't Hooft *et al.* (Plenum Press, New York, 1980).
8. W. Buchmüller, S. T. Love, R. D. Peccei, and T. Yanagida, Phys. Lett. **115B** (1982) 233;
   W. Buchmüller, R. D. Peccei, and T. Yanagida, Phys. Lett. **124B** (1983) 67.
9. W. Buchmüller, R. D. Peccei, and T. Yanagida, Nucl. Phys. B**231** (1984) 53.
10. R. Barbieri, A Masiero, and G. Veneziano, Phys. Lett. **128B** (1983) 493;

W. Buchmüller, R. D. Peccei, and T. Yamagida, Nucl. Phys. **B227** (1983) 503.

11. S. Adler, Phys. Rev. **177** (1969) 2426;
    J. Bell and R. Jackiw, Nuovo Cimento **60A** (1969) 47.
12. T. Banks, Y. Fishman, A. Schwimmer, and S. Yankielowicz, Nucl. Phys. **B177** (1981) 157.
13. S. Coleman and B. Grossmann, Nucl. Phys. **B203** (1982) 205.
14. See, for example, S. Adler in **Lectures on Elementary Particles and Quantum Field Theory**, ed. S. Deser *et al.* (MIT Press, Cambridge, MA, 1970).
15. K. Osterwalder and E. Seiler, Ann. Phys. **110** (1978) 440;
    E. Fradkin and S. Shenker, Phys. Rev. **D19** (1979) 3682;
    T. Banks and E. Rabinivici, Nucl. Phys. **B160** (1979) 349.
16. I. Bars and S. Yankielowicz, Phys. Lett. **101B** (1981) 159.
17. D. Gross and R. Jackiw, Phys. Rev. **D6** (1972) 477;
    C. Bouchiat, J. Iliopoulos, and P. Meyer, Phys. Lett. **38B** (1972) 519.
18. S. Dimopoulos, S. Raby, and L. Susskind, Nucl. Phys. **B169** (1980) 373; Nucl. Phys. **B173** (1980) 208.
19. J. Goity, R. D. Peccei, and D. Zeppenfeld, Nucl. Phys. **B262** (1985) 95.
20. G. 't Hooft, Phys. Rev. **D14** (1976) 3432.
21. S. Coleman, J. Wess, and B. Zumino, Phys. Rev. **177** (1969) 2239;
    S. Coleman, C. Callan, J. Wess, and B. Zumino, Phys. Rev. **177** (1969) 2247.
22. B. Zumino, Phys. Lett. **87B** (1979) 203.
23. W. Buchmüller, R. D. Peccei, and T. Yanagida, Nucl. Phys. **B227** (1983) 503.
24. K. Yamawaki and T. Yokota, Phys. Lett. **113B** (1982) 293; Nucl. Phys. **B223** (1983) 143.
25. S. Khlebnikov and R. D. Peccei, Phys. Rev. **D48** (1993) 361.
26. L. Susskind, Phys. Rev. **D20** (1979) 2619;
    S. Weinberg, Phys. Rev. **D13** (1976) 974.
27. I. Bars, Nucl. Phys. **B208** (1982) 27.
28. W. Buchmüller, R. D. Peccei, and T. Yanagida, Nucl. Phys. **B244** (1984) 186.
29. C. Vafa and E. Witten, Nucl. Phys. **B234** (1984) 173.
30. J. Preskill in **Particles and Fields 1981: Testing the Standard Model**, C. A. Heusch and W. T. Kirk, eds. (AIP, New York, 1982).
31. R. S. Chivukula and H. Georgi, Phys. Lett. **188B** (1987) 99. For a more extensive discussion, see for example, H. Georgi in **SCGT 90**, ed. T. Muta and K. Yamawaki (World Scientific, Singapore, 1991).

## CHAIRMAN: R.Peccei

*Scientific Secretaries: R.Budzynski, H. Petrarche*

## DISCUSSION

– *McPherson:*

Is it true that the compositeness scale, the inverse of the size of, for example, the electron, leads to energy splittings of about the same order? Why can't the muon be an excited electron?

– *Peccei:*

The second question is easier to answer. The muon cannot be an excited electron because in that case you would expect the muon to transmute into an electron and you would have processes like $\mu \to e\gamma$ at a very rapid rate. So it has to be a different species. We know that flavor-changing processes like that are very rare.

– *McPherson:*

Is that model dependent?

– *Peccei:*

It isn't model dependent. If you do not have a conservation law for muon number separate from electron number you get, generally speaking, into trouble. What I tried to say is that in a theory where you have only one scale, then in fact all measures of compositeness give you the same thing roughly, whether you are talking about mass, splittings, scattering or electromagnetic probing. It may well be that the theories you start building these things from, are in fact theories where more than one scale appears, so it could be that there is no reason why the compositeness scale should give you the splittings exactly, even if it does so as an order of magnitude.

– *Giannakis :*

Usually composite models have some problem with flavor-changing neutral currents. How do you deal with this problem?

– *Peccei:*

What I did today was basically to give just an introduction to the subject. What you will see tomorrow is that our main problem is not actually the one of eliminating flavor–changing neutral currents. Rather, in the kind of models I will discuss one obtains a Kobayashi–Maskawa matrix that is identically equal to one. That is, it is very difficult to get flavor–mixing. It is true that generically these are problems that exist for any kind of model in which you do not build in the GIM mechanism to begin with. My impression is that if you eventually succeed in understanding the nature of families in these models, you will also understand how to get an effective GIM mechanism.

– *Giannakis:*

A few years ago there were some attempts to combine compositeness with supersymmetry in the context of "walking technicolor". Is there any resemblance between this approach and yours?

– *Peccei:*

There are some similarities. But you have to be careful. By combining too many things you usually don't make a good meal, unless you're an exceedingly good cook. There are some advantages in using supersymmetry. One of them is that it is very simple to make composite fermions in supersymmetric theories, because you have many scalars around. On the other hand, you then have all the problems of breaking supersymmetry. So they do have some nice structures, but technically they're more complex.

– *Kaplunovsky:*

I'd like to add a comment. In supersymmetric theories it is indeed easy to achieve compositeness, and to the best of my knowledge the only example where it's been proven that the theory does create massless composite fermions and does not break chiral symmetry is a supersymmetric model with more flavors than colors.

– *Peccei:*

In fact what Professor Kaplunovsky says is quite correct. In some supersymmetric models you can actually show that the kind of chiral protection of massless fermions which I discussed really is obtained. I think that all that I have shown today, and what I will discuss tomorrow, are qualitative and semi-qualitative arguments rather than exact arguments. That's where supersymmetry is very helpful.

– *Vassilevskaya:*

Is it possible to explain such a difference between fermion masses using your preon model?

– *Peccei:*

There are two answers to that. One answer is yes: I will show you a model which is exceedingly ugly, where I can get any mass that I want in a technically natural way. However, if the question is does one reproduce the observed spectrum in an intuitive way, the answer is no.

– *Altarelli:*

One can have compositeness near the weak scale or at the Planck scale. For compositenesss near the weak scale, all talks start by mentioning motivations related to family replication, or to continuation of the tradition that goes from atoms down to quarks. But these motivations are immediately contrasted with the fact that different families must have different constituents, (because of $\mu \to e\gamma$ etc.) and the problem of scale of energy and size of the bound states shows that quarks and leptons are a different case than atoms etc. So the question is what is left as a motivation for compositeness at the weak scale?

– *Peccei:*

Let me tell you what my motivation is. There's only one that I consider reasonable, but may in fact be totally unreasonable physically. If you try to think of a theory that breaks $SU(2) \times U(1)$ not by the Higgs mechanism, then you must have some underlying theory that gives you condensates, and you must be able, in that theory to get fermion masses. If you just replace the Higgs mechanism by these nice condensates, and these condensates do nothing else, then you might as well stay with an elementary Higgs and be happy. I think it is important you also think about fermion masses if you want to break $SU(2) \times U(1)$ dynamically. It may be that the best proof that this is not the right way to go, is that nobody has succeeded in solving the problem of how to give masses to the fermions. What happens if you take the other route, like most people do? Then you basically postpone this problem. You say that there are some Yukawa couplings that are set by some magic manifold, or something like that. You basically bypass the fermion mass problem by saying that this problem is solved at the Planck scale – and it may very well be that this is a Planck scale problem, and you're just barking in the wrong direction. On the other hand, I don't find it sensible to worry about dynamical symmetry breaking and not to worry at the same time about the fermions. I think that is contradictory. And if you worry about the fermions, masses in these schemes, the kind of theories that you build by using extended technicolor and more and more groups, are in fact no less complicated and no more successful than theories where things are made of some kind of substructure. To that extent, I think it's worthwhile to explore these models. That is the clear motivation for compositeness, not saying " Look, these things repeat, and they are all alike so we should be able to make them of the same stuff." If you follow this route, you really have to worry about the dynamics, and do the best you can to figure out whether you have some sensible dynamics. The fact that basically, you are unsuccessful and I will show you an unsuccessful attempt tomorrow may tell you that you should do something else. But if you want to think this way, I think it is still worthwhile, understanding what the real problems are.

– *Kawall:*

In this picture, will you necessarily generate a non-zero magnetic moment for the neutrino?

– *Peccei:*

This is a very advanced question. The neutrinos are very difficult states. In this scheme they have to be very much more massless than anything else, and it's very difficult to give mass to some things and not to give mass to the neutrinos. If you eventually give them mass, you also give them some magnetic moments. There's no reason why they shouldn't have a magnetic moment. But, as you will see, we have enough troubles that worrying about the magnetic moment of the neutrino is not the biggest problem.

– *Beneke:*

In the discussion of your SU(N) toy model, you mentioned an effective six-fermion interaction as a remnant of the 'tHooft interaction of the instantons of the $SU(N)$. In QCD the role of instantons chiral symmetry breaking is still somewhat obscure. In your model you do not have an anomaly. Can you elaborate on the role of instantons in your scheme?

– *Peccei:*

For these purposes, the instantons are just a way of keeping track of what is going on. In this theory, you end up with the two kinds of bound states which are masses, and these are the $B$'s and the $B'$'s. They have certain quantum numbers with respect to the remaining chiral global group, and you can ask yourself: what is the simplest operator involving these states that allows you to combine them? – and you find that this is the simplest operator that you can write down. Now if you ask: why is this particular operator allowed? Nnaïvely you can trace it back to certain instantonic interactions that preserve, at the preon level some original U(1). If you look at these interactions, they will give rise, at the bound state level, to an operator like the one I wrote down. This doesn't mean, however, that it comes from that, it's just that given these states – and composite bound states – this is the simplest operator one can write down with these states, and so it will scale like $1/\Lambda_c^5$ where $\Lambda_c$ is the dynamical scale of the theory. If you have a more complex theory, it's sometimes more difficult to figure out what are the effective operators that you can write down, and if you trace some of these to t'Hooft operators, that's a simple way to get them. That such an effective operator will exist is certainly true, because it's allowed by all the remaining symmetries of the theory, but how does it actually get produced? You use the instantons just to get the quantum numbers right.

– *Trocsanyi:*

Could you recall the gap between the ground and first excited states you showed us in the morning? I am asking for this because I would like to know what sort of experiment would show the first signs of compositeness in quarks and leptons and how possible is such an experiment in the next 10 or 20 years.

– *Peccei:*

I can tell you what people do. They look for deviations from elementary behaviour. The kind of deviations to look for are deviations from the scattering cross-sections that go like $\alpha/s$. For example, H.Montgomery showed some bounds today from Fermilab, on some effective extra terms. Basically, residual interactions arising from the preon theory will modify the $\alpha/s$ behaviour and you look for some interference. So you look for deviations from what you would expect from just quark scattering or lepton scattering, as a result of hitting some sort of a hard core. There have been many analyses of LEP data and Fermilab data, looking for these kind of interactions – effective 4-fermion interactions and higher-dimensional interactions. Essentially, at low energy, compositeness would just give you some extra non–renormalisable interactions. In certain cases, you are also looking for direct excitations, for example, you look for an excited electron which would then disintegrate into an electron and a photon – a process which would occur quite rapidly. My guess is that it is highly unlikely that we will see any such excitations in present experiments — I think that the energy is too low. But you might see some remnant interactions.

–*Budzynski:*

What is your opinion at present on attempts to marry preons with superstrings? For instance, might the extra gauge sector which is generically present in 4D string vacua be used to provide the preon–confining forces instead of being dumped into a hidden sector?

– *Peccei:*

This is a little bit like Guido's question. I think you have to distinguish two things: whether you think that quarks and leptons are composite at a scale which is, for example, in the TeV or multi–TeV scale, or whether you think there are composite entities that come at a Planckian scale. Indeed, Dmitri, in one of his 1001 transparencies, was constructing some states by summing up and multiplying things, using precisely the same kind of algebra that I do. However, something which is a strong excitation at the Planck scale, whether you want to call it composite or call it preonic (so that the things that come out of the string are not the quarks and leptons, but the quarks and leptons are composite things) is quite a different game. Then you've moved everything up to an unreachable scale and that could have some advantage. However, it seems to me that things are sufficiently complex at the Planck scale that you should just get the excitations you get and not try to make bound states or funny combinations of these bound states. I think there's a difference between low-energy compositeness, where the object is to try to replace the Higgs sector by something more dynamical, and trying to do things at the Planck scale.

– *Budzynski:*

Would it then be fair to summarize your answer by saying that it would be premature to think of where the preons come from (e.g. the superstring), at least before you have a viable preon model?

– *Peccei:*

Yes, I think you have to do one problem at a time. You cannot solve all problems. In fact, you probably cannot solve any problem.

– *Budzynski:*

Considering the level of constraints provided by present data on departures from the Standard Model in the electroweak gauge sector, would you say that the idea of composite $W$ and $Z$ is definitely dead and buried?

– *Peccei:*

Yes, I think that idea was always very difficult to implement in a reasonable dynamical way. Although you can make bosons massless by a Goldstone-like phenomenon, or fermions massless by using chirality, there was never any good idea of how to keep the $W$'s light except by gauge symmetry. Once you give up the gauge symmetry it becomes very difficult. But now I think that, in particular with the kind of data you get out of LEP for the $Z^0$, to imagine that these states are not elementary up to enormous scales (I don't know what happens with them at the Planck scale) that you have some nearby compositeness, it seems to me that this is not tenable at all. The models were never very reasonable to begin with, and now I don't think there's practically anybody that thinks along these lines.

– *Lu:*

I am trying to understand your philosophy here. On the one hand, you want to explain why the up quark is so light, and $m_t/m_b$ is so large, within some composite model. But then on the other hand you have some superheavy preons forming a bound state which is very light. Isn't that very unnatural?

– *Peccei:*

First of all, whatever I will or will not do will not be very successful. Nevertheless, it still will be perfectly natural. That is, one of the problems of the standard model, when people talk about naturalness, is that you cannot control the scale in the scalar sector unless you have something like a supersymmetry. If all the scales are dynamical, scales like $\Lambda_{QCD}$ and mass ratios are ratios of these dynamical scales of different groups, the results that you get are perfectly natural – in the sense that they are a dynamical result. What I will try to show you tomorrow is that it is possible to get small masses out of very large scales. What I did today in the very simple examples discussed, was to show you that if one had a preonic theory with no other scale except for its dynamical scale, one could get some states which have zero mass, and they are independent of this scale. The next trick is to try to figure out how to give these states a small mass, not a TeV mass, but something that is MeV or GeV. I will show you how to do this tomorrow.

– *Lu:*

Another technical point: your approach relies on the assumption of complementarity. I thought that this was just a hypothesis. Can you comment on that?

– *Peccei:*

Complementarity is proven in some simplified theories. For a theory with many groups, as I will have, it certainly has not been proven.

– *Lu:*

For say $SU(2)$, is it proven?

– *Peccei:*

If you have many fields, you cannot prove it exactly. But we have so little guidance on what the dynamics of these theories is, it is important to use whatever we can.

**CHAIRMAN: R.Peccei**

*Scientific Secretaries: M. Giovannini, A.Papa*

## DISCUSSION

– *Langacker:*

You have made a heroic attempt to construct realistic compositeness models. However, the cynic might argue that the enormous complications of such models is a hint that it is the wrong approach. Would you care to comment?

– *Peccei:* My attitude on the subject is the following: I think that it is interesting to pursue models with dynamical symmetry breaking. I prefer breaking symmetry by dynamical condensates: we have at least an example of such a theory which is the BCS theory of superconductivity where there are no Landau-Ginzburg fields, but correlated pairs of electrons which break the symmetry. Moreover, I think that it is dangerous when everybody follows the same direction of research. I think the direction I follow contains many interesting theoretical problems.

– *Arnowitt:*

Important in the models you have been considering is the formation of condensates. Has there been any progress in showing the existence of these condensates?

– *Peccei:*

There is a toy-model (supersymmetric QCD, with one more flavour than the number of colours) for which it has been proven more or less rigorously by Kaplunovsky and Louis that such condensates exist. Unfortunately, the models I study are chiral theories which cannot be studied on the lattice so one has to resort to less theoretically secure arguments. However, I hope that precisely because these models are not like QCD there may be some surrounding pieces of physics, such as, for example, asymptotic freedom was for vector gauge theory.

– *Beneke:*

In some recent talks Nambu has proposed a mechanism for intrafamily mass splittings that is based on minimising the vacuum energy with respect to the parameters of the Lagrangian. Can you comment on this scenario?

– *Peccei:*

No, I can't. I have not seen the paper.

– *Kawall:*

Apart from looking for new hard cores in very, very deep inelastic scattering, what experimental signatures should we be looking for?

– *Peccei:*

One thing one could look for is excited electrons, but this may not be very practical. A better candidate to study would be the top. Since it is the most massive, it is also the most extended and I expect it may have different properties than an elementary top.

– *Lu:*

If Higgs is found to be responsible for electroweak symmetry breaking, what do you think about the prospects of understanding fermion masses and mixings?

– *Peccei:*

There are two possible scenarios. In the first one, you can find a light Higgs boson but no SUSY matter. So the Standard Model would be a theory perfectly finite, but without any explanation of the hierarchy problem. On the other hand, my guess is that if you find a light Higgs you will also find SUSY matter; even in this case the problem of fermion masses is still open, this problem is here pushed to the Planck scale.

# PRESENT LEP DATA AND ELECTROWEAK THEORY °

**V.A. Novikov,**
**L.B. Okun**
and
**M.I. Vysotsky**

ITEP, Moscow 117259, Russia

## ABSTRACT

The Born approximation, based on $\bar{\alpha} \equiv \alpha(m_Z)$ instead of $\alpha$, reproduces all electroweak precision measurements within their $(1\sigma)$ accuracy. The astonishing smallness of the loop corrections results from the cancellation of a large positive contribution from the heavy top quark and large negative contributions from all other virtual particles. It is precisely the non-observation of electroweak radiative corrections that places stringent upper and lower limits on the top mass.

° **Lecture presented by M.I. Vysotsky**

The precision measurements of $Z$ decays at LEP are usually considered as providing evidence for non-vanishing electroweak radiative corrections (see e.g. [1], where a representative list of references is given). The aim of this letter is to stress that present LEP data [2]-[5] are in perfect agreement [6]-[8] with the Born approximation and that no genuine electroweak corrections (involving loops with heavy virtual bosons, neutrinos and top quarks) have as yet been observed. The disagreement with statements to the contrary stems from the different definitions of the Born approximation being used. Usually it is defined in terms of electric charge at zero momentum transfer, i.e.

$$\alpha \equiv \alpha(0) = e^2/4\pi = 1/137.0359895(61) \ , \tag{1}$$

while we argue that the true Born approximation should be defined in terms of

$$\bar{\alpha} = \alpha(m_Z) = 1/128.87(12) \ , \qquad [9],[10] \ . \tag{2}$$

By using $\bar{\alpha}$ instead of $\alpha$ one automatically takes into account the only purely electromagnetic correction (polarization of vacuum by the photon), which has not already been allowed for by the experimentalists.

While $\alpha(q^2)$ is running, the other two gauge couplings

$$\alpha_W = g^2/4\pi \ , \ \alpha_Z = f^2/4\pi \tag{3}$$

are "frozen" for $|q^2| \leq m^2_{Z,W}$ and start to run only for $|q^2| \gg m^2_{Z,W}$. Therefore it is natural to consider the electroweak Born approximation at the Fermi scale, i.e. at $q^2 \simeq m_Z^2$. In a sense, $\alpha$, with all its accuracy, is irrelevant to electroweak physics; what is relevant is $\bar{\alpha}$. Hence, if Glashow, Weinberg and Salam [11]-[13] had thought about actually calculating electroweak radiative corrections they would have used $\bar{\alpha}$ from the beginning. Then they would have defined the weak angle $\theta$ through the equations

$$\alpha_W = \alpha_Z c^2 \ , \ \bar{\alpha} = \alpha_W s^2 \ , \tag{4}$$

where $c \equiv \cos\theta$, $s \equiv \sin\theta$. (We do <u>not</u> use $\theta_W$, $s_W$, $c_W$ here, because in the literature they are associated with $\alpha$, not $\bar{\alpha}$.) Hence

$$c^2 s^2 = \bar{\alpha}/\alpha_Z \ . \tag{5}$$

According to the Minimal Standard Model [11]-[13]:

$$m_W = g\eta/2 \ , \ m_Z = f\eta/2 \tag{6}$$

where $\eta$ is the vacuum expectation value of the Higgs field, so that in the Born approximation:

$$m_W/m_Z = c \ . \tag{7}$$

To obtain $\eta$ we consider, as usual, the four-fermion coupling of $\mu$-decay (see, for instance, [14])

$$\frac{G_\mu}{\sqrt{2}} = \frac{g^2}{8m_W^2} \ . \tag{8}$$

Then it follows from (6):

$$\eta^2 = 1/\sqrt{2}G_\mu \ . \tag{9}$$

The value

$$G_\mu = 1.16639(2) \times 10^{-5} \ GeV^{-2} \tag{10}$$

gives:

$$\eta = 246.2185(21) \ GeV \ . \tag{11}$$

In the pre-LEP era, both $m_Z$ and $m_W$ were poorly known. This justifies the definition [15] $s_W \equiv m_W/m_Z$ from the historical point of view. At present, however, $m_Z$ is known with much higher accuracy than $m_W$ [16]:

$$m_Z = 91.187(7) \tag{12}$$

$$m_W = 80.22(26) \tag{13}$$

It is reasonable therefore to express $s$ and $c$ in terms of $m_Z$:

$$f^2 = 4m_Z^2/\eta^2 = 4\sqrt{2}G_\mu m_Z^2 = 0.548636(84) \tag{14}$$

$$\alpha_Z = \frac{\sqrt{2}}{\pi} \cdot G_\mu m_Z^2 = 1/22.9047(35) = 0.0436592(66) \tag{15}$$

$$\frac{1}{4}\sin^2 2\theta = c^2 s^2 = \bar{\alpha}/\alpha_Z = \frac{\pi\bar{\alpha}}{\sqrt{2}G_\mu m_Z^2} = 0.177735(16) \tag{16}$$

$$s = 0.48081(33) \ , \ c = 0.87682(19) \tag{17}$$

$$s^2 = 0.23118(33) \ , \ c^2 = 0.76881(33) \ . \tag{18}$$

Now we are ready to derive the electroweak Born predictions for various observables. We first compare Eq. (7) with the experimental ratio

$$m_W/m_Z = 0.8797(29) \ . \tag{19}$$

The agreement is within $1\sigma$. Next we consider decays of the $Z$-boson. The amplitude of the decay into pairs of charged leptons, $e^+e^-$, $\mu^+\mu^-$, $\tau^+\tau^-$, has the form:

$$M_l = \frac{1}{2} f\bar{l}[g_A\gamma_\alpha\gamma_5 + g_V\gamma_\alpha]lZ_\alpha \ , \tag{20}$$

where $l$ and $Z_\alpha$ are the wave functions of lepton and $Z$-boson.

The corresponding width $\Gamma_l$ is given by the expression:

$$\Gamma_l = \left(1 + \frac{3\bar{\alpha}}{4\pi}\right) \times 4(g_A^2 + g_V^2)\Gamma_0 \ , \tag{21}$$

where

$$\Gamma_0 = \frac{f^2 m_Z}{196\pi} = \frac{\sqrt{2}G_\mu m_Z^3}{48\pi} = 82.941(19)\ MeV \ . \tag{22}$$

The forward-backward asymmetry in the channel $f\bar{f}$ is

$$A_{FB} = \frac{3}{4} A_e A_f \ , \tag{23}$$

where

$$A_i = 2g_A^i g_V^i / (g_A^{i^2} + g_V^{i^2}) \ , \quad (i = e, \mu \ldots) \tag{24}$$

and the longitudinal polarization of the $\tau$-leptons

$$P_\tau = -A_\tau \ . \tag{25}$$

As is well known, the Born aproximation gives, for charged leptons

$$g_A = T_3 = -1/2 = -0.5000 \ , \tag{26}$$

$$g_V/g_A = 1 - 4s^2 = 0.0753(12) \ , \tag{27}$$

which should be compared with corresponding experimental values [2]-[5]:

$$g_A^{exp} = -0.4999(9) \tag{28}$$

and

$$(g_V/g_A)^{exp} = 0.0728(28) \ , \tag{29}$$

(the latter was obtained from the measurements of $A_{FB}$ for leptons and hadrons, and from $\tau$-polarization). Again we see agreement to within $1\sigma$.

The decays into hadrons may be considered as decays into quark + anti-quark pairs. In this case, as before,

$$g_A = T_3 \ , \tag{30}$$

but the fractional charges and the colour degrees of freedom of the quarks must be taken into account:

$$g_V/g_A = 1 - 4|Q|s^2 \ , \tag{31}$$

$$\Gamma_q = 3\Big(1 + \frac{3}{4\pi}Q^2\bar{\alpha}\Big)G\Gamma_0 \ , \tag{32}$$

where the factor $G$ describes the final state exchange and emission of gluons [17]-[19]

$$G = 1 + \bar{\alpha}_s/\pi + 1.4(\bar{\alpha}_s/\pi)^2 - 13(\bar{\alpha}_s/\pi)^3 + ... \tag{33}$$

Here $\bar{\alpha}_s \equiv \alpha_s(m_Z)$ is the gluonic coupling constant at the Fermi scale. For further estimates we will assume that

$$\bar{\alpha}_s = 0.12 \pm 0.01 \ , \tag{34}$$

which agrees with the global analysis of all pertinent data: $\bar{\alpha}_s = 0.118 \pm 0.007$ [20],[21]. Then

$$G(\bar{\alpha}_s = 0.12 \pm 0.01) = 1.0395(33) \tag{35}$$

LEP data are compared with the Minimal Standard Model predictions in the Table.

Note that in the Table the "Born" values of hadronic observables are obtained by using the gluonic factor $G$, given by Eq. (33) for all quark flavours. The specific gluonic corrections to $\Gamma_b$ [21]-[27] caused by the non-vanishing $\bar{m}_b = m_b(m_Z) = 2.3$ GeV and the large $m_t$ are included in the MSM corrections. Allowing for them in the $G$-factor of $b\bar{b}$ decay would give $G_B \simeq G - 0.01$; the new central Born values ($\Gamma_h$1739 MeV, $\Gamma_Z = 2487$ MeV, $\sigma = 41.46$ nb, $R_\ell = 20.82$ and $R_b = 0.218$) would still preserve the $1\sigma$ agreement with experimental data.

The agreement between "Born" and experiment is stunning, even if one allows for the fact that not all the observables in the Table are independent: $\Gamma_\ell$ can be expressed in terms of $g_A$ and $g_V/g_A$, $R_l = \Gamma_h/\Gamma_\ell$ and

$$\sigma_h = 12\pi\Gamma_e\Gamma_h/m_Z^2\Gamma_Z^2 \ . \tag{36}$$

The coincidence of the central experimental and Born values is amazing: their differences are in some cases smaller than the experimental uncertainties. This fact must be considered a rare statistical fluctuation.

What is much more interesting from the physics point of view is the smallness of the electroweak radiative corrections as compared with naïve estimates ($\sim \alpha_W/\pi \sim \bar{\alpha}$). This originates from the compensation of two large contributions: a positive one from the heavy top quark ($m_t \sim 150$ GeV), and a negative one from all other virtual particles (light quarks, higgs, $W, Z$-bosons).

Were the top quark much lighter, the agreement with the Born approximation would be destroyed. This is shown in detail in [7],[8]. It may seem paradoxical, but it is precisely the non-observation of electroweak radiative corrections that places stringent upper and lower limits on the top's mass. (Note that in the usual approach based on $\alpha$, not $\bar{\alpha}$, the same limits appear as a result of precision measurements of non-vanishing radiative corrections, some of which are very large.)

The results of the "low-energy" electroweak experiments complement the above picture of "Born"–experiment agreement. From $\nu_\mu e$ scattering experiment [28]:

$$g_A^{e\nu_\mu} = -0.5030(81) \ , \quad g_V^{e\nu_\mu}/g_A^{e\nu_\mu} = 0.0500(380) \ , \tag{37}$$

which should be compared with $-0.5000$ and 0.0753(12), respectively. The three experiments on deep inelastic neutrino scattering (CHARM, CDHS, CCFR) give for $m_W/m_Z = 0.8785(30)$ (see [16]), where the uncertainty seems to be less reliable than in the case of direct $m_W$ measurement (UA2, CDF). Still, it would be interesting to check whether the electroweak Born approximation (with due allowance for strong interactions) would describe deep inelastic scattering also to within $1\sigma$. The most promising seems to be the experiment on atomic parity violation in $^{133}$Cs. The experimental value of the weak charge $Q_W(^{133}\text{Cs})$=-71.04(1.81) is $1.5\sigma$ away from its "Born" value: $-73.9$.

The reduction of the one-loop corrections by a factor 3 to 5 (through partial cancellation) is important in the light of the experimental uncertainties. Even if these were reduced by, say, a factor of 3, this would still not make the electroweak two-loop contribution essential. Hence one can safely limit oneself to the one-loop electroweak approximation.

The fact that one can confine oneself from the beginning to the one-loop approximation not only makes the calculations quite simple but also extremely transparent. It is convenient to organize them in five steps:

**Step 1.** Start with a Lagrangian, which contains only bare couplings ($e_0, f_0, g_0$) and bare masses ($m_{W0}, m_{Z0}\ldots$). Substitute $m_t, m_b$ and $m_H$ for $m_{t0}, m_{b0}$, and $m_{H0}$, since two loops are neglected.

**Step 2.** Calculate one-loop Feynman diagrams for the three most accurately known observables – $G_\mu, m_Z, \bar{\alpha}$ – in terms of the bare quantities $e_0, f_0$ ... and $1/\varepsilon$, where $\varepsilon$ is the parameter used in dimensional regularization: $2\varepsilon = D - 4$ and $D$ is the dimension in which Feynman integrals are calculated.

**Step 3.** Invert the equations resulting from Step 2 by expressing all bare quantities in terms of $G_\mu, m_Z, \bar{\alpha}\ , m_t, m_b, m_H$ and $1/\varepsilon$.

**Step 4.** Calculate Feynman integrals for $m_W/m_Z, g_A, g_V/g_A$ or any other electroweak observable in terms of bare quantities $e_0, g_0$ ... and $1/\varepsilon$.

**Step 5.** Express $m_W, g_A, g_W/g_A$, etc, in terms of $G_\mu, m_Z, \bar{\alpha}, m_t, m_b, m_H$. At this step all terms proportional to $1/\varepsilon$ cancel each other and the resulting relations contain no infinities in the limit $\varepsilon \to 0$.

Each of the "gluon-free" observables, $m_W/m_Z, g_A, g_V/g_A$, is conveniently presented as the sum of the Born term and the one-loop term [7]:

$$\begin{aligned} m_W/m_Z &= c + \bar{\alpha}\frac{3c}{32\pi s^2(c^2 - s^2)}V_m(t,h) = \\ &= 0.8768 + 0.00163 V_m \ , \end{aligned} \tag{38}$$

$$\begin{aligned} g_A &= -\frac{1}{2} - \bar{\alpha}\frac{3\alpha}{64\pi s^2 c^2}V_A(t,h) = \\ &= -0.5000 - 0.00065 V_A \ , \end{aligned} \tag{39}$$

$$g_V/g_A = 1 - 4s^2 + \bar{\alpha}\frac{3}{4\pi(c^2 - s^2)}V_R(t,h) = \\ = 0.0753(12) + 0.00345V_R \ , \quad (40)$$

where

$$t = (m_t/m_Z)^2 \ , \quad h = (m_H/m_Z)^2 \ . \quad (41)$$

All three functions $V_i$ are normalized in such a way that they behave similarly for $t \gg 1$, i.e.:

$$V_i \simeq t \ \text{ for } t \gg 1 \ .$$

By comparing Eqs. (38), (39), (40) with the corresponding experimental values (see column 1 of the Table ) one obtains the experimental values of the $V_i$'s:

$$V_i^{exp} = \bar{V}_i \pm \delta V_i \ . \quad (42)$$

They are:

$$V_m^{\text{exp}} = 1.78 \pm 1.78 \quad (43)$$

$$V_A^{\text{exp}} = -0.15 \pm 1.38 \quad (44)$$

$$V_R^{\text{exp}} = -0.73 \pm 0.81 \ . \quad (45)$$

The fact that experiments are, to within $1\sigma$, described by the Born approximation means that

$$|\bar{V}_i| \leq |\delta V_i| \ . \quad (46)$$

The one-loop approximation leads to an important property of the functions $V_i(t,h)$, namely, they may be presented in the form [7]:

$$V_i(t,h) = t + T_i(t) + H_i(h) \ . \quad (47)$$

Thus $V_i(m_t)$ for different values of $m_H$ differ only by a shift.

Simple analytical expressions and numerical tables for functions $T_i(t)$ and $H_i(h)$ are given in Ref. [7]. It is important to emphasize that Eqs. (38)-(40) are *exact* in the one-loop approximation, unlike the so-called "improved Born approximation" (see, e.g., [29]), which starts with $\alpha$ and includes terms proportional to $\Delta\alpha = \bar{\alpha} - \alpha$ and $t$.

The functions $V_i(t,h)$ form a surface over the plane $m_t, m_H$. The intersection with a plane orthogonal to the axis $m_H$ gives a curve describing the

$m_t$-dependence of $V_i$ at given $m_H$. Similarly, the intersection with a plane orthogonal to the axis $m_t$ gives a curve describing the $m_H$-dependence of $V_i$ at given $m_t$. Horizontal planes at $V_i = \bar{V}_i \pm \delta V_i$ give isolines corresponding to a central value $\bar{V}$ and $1\sigma$ uncertainties. The crossing point of central-value isolines determines in principle the values of $m_t$ and $m_H$. Unfortunately the $\delta V_i$'s are so large that reliable limits may be obtained only for $m_t$.

As for $m_H$, the minimum $\chi^2$ lies at $m_H = 10$ GeV, which is much below the lower experimental LEP bound (62 GeV). This false minimum corresponds to the crossing point of central-value isolines. It is evident that this contradiction is statistically not significant.

A similar analysis may be performed for the hadronic decays of $Z$-bosons. As basic observables one can choose $\Gamma_Z$, $R_l = \Gamma_h/\Gamma_l$ and $\sigma_H$, which depend on $\bar{\alpha}_s$, and also $R_b$, whose dependence is less pronounced. From these observables, stringent limits can be obtained not only on $m_t$, but also on $\bar{\alpha}_s$ [8]. Our results for $m_t$ and $\bar{\alpha}_s$ are in qualitative agreement with the results of $\chi^2$ fits published by other authors (see, for instance: [1],[4], [16],[30], [31]).

In the last three columns of the Table, we give the values of MSM radiative corrections, which illustrate the sensitivity of various observables to the values of $\bar{\alpha}, \bar{\alpha}_s$ and $m_t$. (These corrections also include the effects of virtual gluons in electroweak quark loops [32]).

When the data from LEP and the SLC have a better accuracy, and when the top quark is discovered and its mass is measured, the MSM corrections may become non-adequate. This would signal the existence of New Physics. A convenient parametrization of it has already been worked out [33].

M.Vysotsky is grateful to Professor Antonino Zichichi for kind hospitality at Ettore Majorana Centre for Scientific Culture.

| 1 | 2 | 3 | 4 | 5 | 6 |
|---|---|---|---|---|---|
| Observable | Exp. value | "Born" | MSM corrections | | |
| $m_W/m_Z$ | 0.8798(29) | 0.8768(2) | 33(7) | 25(7) | 17(7) |
| $g_A$ | -0.4999(9) | -0.5000 | -9(2) | -7(2) | -5(2) |
| $g_V/g_A$ | 0.0728(28) | 0.0753(12) | -31(12) | -56(12) | -75(12) |
| $\Gamma_\ell$ (MeV) | 83.51(28) | 83.57(2) | 25(9) | 15(9) | 7(9) |
| $\Gamma_h$ (MeV) | 1740(6) | 1741.5(5) | -1(1.5) | -3.0(1.5) | -6(1.5) |
| $\Gamma_Z$ (MeV) | 2487(7) | 2489.8(5.2) | 3.6(2.1) | 0.3(2.1) | -3.0(2.1) |
| $\sigma_h$ (nb) | 41.45(17) | 41.44(5) | -0.4(0.7) | -1(0.7) | -0.4(0.7) |
| $R_\ell \equiv \Gamma_h/\Gamma_\ell$ | 20.83(6) | 20.84(6.4) | -6.1(0.3) | -7.5(0.3) | -8.2(0.3) |
| $R_b \equiv \Gamma_b/\Gamma_h$ | 0.2201(32) | 0.2197(1) | -31(3) | -31(3) | -31(3) |

**Table -** Comparison of experimental values of various LEP observables with the electroweak "Born" approximation (column 3). The quotation marks indicate that for the hadronic decays the virtual gluons are taken into account by the universal factor $G$ [see Eq. (33) and the discussion following Eq. (35)]. The last three columns (4,5,6) give the values of the MSM one-loop electroweak corrections to the "Born" approximation (and also of specific gluonic corrections depending on $m_b$ and $m_t$) for $m_t = 150 \pm 10$ GeV and for three values of $m_H$: 100, 300 and 750 GeV, respectively. The $1\sigma$ uncertainties in columns 2 and 3 are quoted in brackets, with the customary convention regarding digits. The figures in columns 4, 5 and 6 are again given in the corresponding units with their respective "uncertainties", which are actually shifts induced by $\Delta m_t = \pm 10$ GeV. The Born uncertainties for $m_W/m_Z$ and $g_V/g_A$ derive mainly from $\Delta\bar{\alpha}/\bar{\alpha} = \pm 9.3 \times 10^{-5}$. The uncertainty of $g_A$ is "hidden" in $\Gamma_0$ [Eq. (22)]. The uncertainties of hadronic observables in column 3 correspond to $\Delta\bar{\alpha}_s = \pm 0.01$. The figure in brackets is underlined when the coefficient in front of $\Delta\bar{\alpha}_s$ or $\Delta m_t$ is negative.

# References

[1] Review of Particle Properties, *Phys.Rev.* **D15** N11, Part II (1992), p. III.59.

[2] C. DeClercq, Talk given at XXVIIIth Rencontres de Moriond "Electroweak Interactions and Unified Theories", Les Arcs, 1993.

[3] V. Innocente, Talk given at XXVIIIth Rencontres de Moriond "Electroweak Interactions and Unified Theories", Les Arcs, 1993.

[4] M. Pepe Altarelli, Talk given at Les Rencontres de Physique de la Vallée d'Aoste "Results and Perspectives in Particle PHysics, La Thuile, 1993.

[5] R. Tenchini, Talk given at XXVIIIth Rencontres de Moriond "Electroweak Interactions and Unified Theories", Les Arcs, 1993.

[6] V.A. Novikov, L.B. Okun and M.I. Vysotsky, CERN Preprint TH. 6053/91 (1991); TPIMINN-91/14-T (1991).

[7] V.A. Novikov, L.B. Okun and M.I. Vysotsky, *Nucl.Phys.* **B397** (1993) 35.

[8] V.A. Novikov, L.B. Okun and M.I. Vysotsky, CERN Preprint TH. 6855/93 (1993)

[9] F. Jegerlehner, Villigen Preprint PSI-PR-991-08 (1991).

[10] H. Burkhardt, F. Jegerlehner, G. Renso and C. Verzegnassi, *Z.Phys.* **C43** (1989) 497.

[11] S.L. Glashow, *Nucl.Phys.* **22** (1961) –.

[12] S. Weinberg, *Phys.Rev.Lett.* **19** (1967) 1264.

[13] A. Salam, in Elementary Particle Theory, ed. N. Svartholm (Almquist and Wiksells, Stockholm, 1969), p. 367.

[14] L.B. Okun, Leptons and Quarks (North Holland, Amsterdam, 1982).

[15] A. Sirlin, *Phys.Rev.* **D22** (1980) 971.

[16] L. Rolandi, Plenary Talk at XXVIth International Conference on High Energy Physics, Dallas, CERN Preprint PPE/92-175 (1992).

[17] S. Gorishny, A. Kataev and S. Larin, *Phys.Lett.* **B259** (1991) 144.

[18] L. Surguladze and M. Samuel, *Phys.Rev.Lett.* **66** (1991) 560.

[19] A. Kataev, *Phys.Lett.* **B287** (1992) 209.

[20] G. Altarelli, Rapporteur Talk at the Conference "QCD-20 Years Later", Aachen, 1992, CERN Preprint TH. 6623/92 (1992).

[21] S. Bethke, Plenary Talk at XXVIth International Conference on High Energy Physics, Dallas, 1992. Preprint HD-PY 92/13, OPAL-CP093 (1993).

[22] B. Kniehl and J. Kühn, *Nucl.Phys.* **B329** (1990) 547.

[23] B. Kniehl and J. Kühn, *Nucl.Phys.* **B224** (1989) 229.

[24] J. Kühn, in Proceedings XXVth International Conference on High Energy Physics, Singapore, 1990, eds. K.K. Phua and Y. Yamaguchi (World Scientific, Singapore, 1991), Vol. II, p. 1451.

[25] K. Chetyrkin and J. Kühn, *Phys.Lett.* **B248** (1990) 359.

[26] A. Kataev and V. Kim, Preprint ENSLAPP-A-407/92 (1992).

[27] A. Kataev, Talk given at Quark '92 Conference, Zvenigorod, to be published in "Quark '92", Singapore, 1993.

[28] CHARM II Collaboration, P. Vilain et al., *Phys.Lett.* **B281** (1992) 159.

[29] G. Altarelli, R. Kleiss and C. Verzegnassi, ets., Physics at LEP 1, Vol. 1, CERN Report 89-08 (CERN, Geneva, 1989).

[30] G. Altarelli, CERN Preprint TH. 6867/93 (1993).

[31] J. Ellis, G. Fogli and E. Lisi, *Phys.Lett.* **B292** (1992) 427.

[32] N. Nekrasov, V.A. Novikov and M. Vysotsky, CERN Preprint TH. 6696/92 (1992), to be published in Yadernaya Fizika.

[33] G. Altarelli, R. Barbieri and F. Caravaglios, CERN Preprint TH. 6770/93/Rev. (1993).

[34] V.A. Novikov, L.B. Okun and M.I. Vysotsky, CERN Preprint TH. 6849/93 (1993).

**CHAIRMAN: M.Vysotsky**

*Scientific Secretary: R.McPherson*

# DISCUSSION

– *Khoze:*

I have a question stemming from the title of your talk. What is the proper way to understand the EW radiative correction?

– *Vysotsky:*

The title was not entirely serious. You have to calculate the tree diagram contribution, avoid the introduction of all the additional parameters like $\sin^2\theta_{\bar{MS}}$, $\sin^2\theta_{eff}$, etc., and then add loop terms.

– *Ellis:*

a) Are the most important $M_t$-dependent 2-loop contributions to the vacuum polarization taken into account?

b) Do you have the $\chi^2$ contour lines for the full set of data including gluon-dependent variables, and are the $\Delta\chi^2 = 1$ etc. contours significantly contracted?

– *Vysotsky:*

a) Yes.

b) I think that the curves are not significantly contracted. The most important constraint is $g_V/g_A$, which is included.

– *Beneke:*

You analyzed two independent parameter sets, both of which favoured top mass around 110 GeV. Combining them, you obtain $M_t = 157$ GeV. How does this happen?

– *Vysotsky:*

Gluon free parameter set for a Higgs mass of 300 GeV gives $M_t = 170$ GeV.

– *Lopez:*

What happens to the top mass if the Higgs mass is less than 150 GeV, as is the case in some sense in supersymmetry? Do you agree with Langacker?

– *Vysotsky:*

The central value of $M_t$ decreases and agrees with Langacker. You must include all supersymmetric contributions in polarization operators, of course.

– *Montgomery:*

In your response to an earlier question, you suggest that the preferred fit values suggest $M_H$ below the experimental bound. Does this mean that there is an inconsistency even without knowledge of $M_t$?

– *Vysotsky:*

No, not really. The dependence on $M_H$ is very weak. Perhaps there is some information at the $1\sigma$ level, but not at 90% or 95% confidence level.

– *Langacker:*

I would like to make two comments.

a) The upper limit on the Higgs mass is rather slippery. You get very different results depending on whether or not you include a priori knowledge that the Higgs mass is above 60 GeV.

b) There is a common mistake in the literature which you made. The confidence level on a region in two parameter space is determined by the $\chi^2$ distribution for 2 degrees of freedom, so that $\Delta\chi^2 = 4.6$ for 90% confidence limit. The quantity

#of data points - # of parameters

is only relevant for goodness-of-fit determination.

– *Ellis:*

I have a comment about the value of $M_H$. Your contour lines indicate incompatibility with the direct limit on $M_H$ at the $\Delta\chi^2 = 1$ level, indicating compatibility with the direct limit at about the 10% level. It is not possible to bound $M_H$ at the 99, 95, or 90% level, but it is possible at the 68% confidence level, corresponding to the $1\sigma$ level that you quote for $M_t$.

– *Trocsanyi:*

Is it correct to say that your approach is identical to the improved Born Approximation, which is just the usual tree level calculation with running coupling $\alpha$?

– *Vysotsky:*

No, we take loop corrections into account as well.

– *Vassilevskaya:*

a) Is it possible to determine the minimum number of physical scalars which are necessary to cause spontaneous breakdown in gauge theory?

b) No Higgs has been found experimentally. What do you think of attempts to construct EW theory without the Higgs sector?

– *Vysotsky:*

a) At least one.

b) I personally prefer a theory with an elementary Higgs.

– *Anselm:*

You have said that if there were not radiative corrections, it would be OK experimentally. Can you imagine a theory with smaller radiative corrections than the Standard Model?

– *Vysotsky:*

No, I cannot imagine such a theory.

– *Khoze:*

I am a little confused with the absence of EW radiative corrections. One example is $B^0 - \bar{B}_0$ mixing, where W induced loop contributions are experimentally observed.

– *Vysotsky:*

I agree that in $B^0 - \bar{B}_0$ mixing we see the top loop contribution, but the accuracy of testing the Standard Model is not high in this particular place.

– *Forshaw:*

Why has the central experimental value of the top mass been rising over the last few years?

– *Vysotsky:*

Given the error bars surrounding the central value, there is no inconsistency.

# Precision Tests of the Standard Model

G. Altarelli
CERN- Geneva

## Content



The electroweak theory originally developed by Glashow, Salam and Weinberg[1], together with QCD, is the Standard Model of fundamental particle interactions that provides us with the basic framework for the description of all observed particle properties. I think that presenting the updated evidence that so far completely supports the intuition of the inventors of the Standard Model is a very appropriate way to honour the great achievements of Abdus Salam in particle physics. In the following I shall discuss the analysis of the available data in the Standard Model and beyond. In writing this article I decided to include all the present information, that is, also those data that became available after the Meeting in honour of Abdus Salam.

## 1. Precision Electroweak Data and the Standard Model

For the analysis of electroweak data in the Standard Model[1-3] one starts from the input parameters: some of them, $\alpha$, $G_F$ and $m_Z$, are very well measured, some other ones, $m_{flight}$ and $\alpha_s(m_Z)$ are only approximately determined while $m_t$ and $m_H$ are largely unknown. Then one computes the radiative corrections[4,5] to a sufficient precision to match the experimental capabilities, compares the theoretical predictions for the numerous observables which have been measured, checks the consistency of the theory and derives constraints on $m_t$ and hopefully also on $m_H$. The light quark masses $m_{flight}$ enter in the large logs that determine the running of $\alpha$ from the very small momenta where it is defined up to $m_Z$. By using a dispersion relation, the relevant contribution to the vacuum polarisation function of the photon can be directly obtained[6] from the experimental data on the hadronic cross-section in $e^+e^-$ annihilation at energies below $m_Z$. The resulting value[7] of $\alpha(m_Z)$, in a specified definition for this quantity, is $\alpha(m_Z) = (128.87 \pm 0.12)^{-1}$. The error quantifies the theoretical ambiguity associated to $m_{flight}$ for precision tests of the Standard Model (the largest source of error apart from our relative ignorance on $m_t$, $m_H$ and $\alpha_s(m_Z)$).

Recently there has been some additional progress[8] in the control of radiative corrections by new computations of some potentially dominant higher-loop effects: terms of order $(G_F m_t^2)^2$ in the $Z \to b\bar{b}$ vertex and, for all values of $m_H$, in $\Delta\rho$ [9]; terms of order $\alpha_s G_F m_t^2$ in the $Z \to b\bar{b}$ vertex[10] and, for some refinements, in $\Delta r$ and $\Delta\rho$ [11]; terms of order $(\alpha_s m_b/m_Z)^2$ in $\Gamma(Z \to \text{hadrons})$[12]. The impact of these new additions is a shift upward of 2-3 GeV in the fitted central value of $m_t$.

The experimental values, presented at the Marseille Conference[13-15], of the most relevant observables are collected in table 1, together with, for comparison, their values at the '93 winter meetings[19]. Not all LEP measurements listed in table 1 are independent. From resonance scanning $m_Z$ and $\Gamma_T$ are obtained. At the peak one measures the ratios $\sigma_h$, $R_h$, $R_{bh}$ ... and the asymmetries. Thus, for example, $\Gamma_l$ and $\Gamma_h$ are derived quantities.

In figs.1-6 we present a direct comparison of the data and the Standard Model predictions (as functions of $m_t$, for $m_H$=50-1000 GeV and $\alpha_s$= 0.111-0.125) for the six most relevant observables, assuming lepton universality: $m_W/m_Z$, $\Gamma_T$, $\sigma_h$, $R_h$, $R_{bh}$ and $g_V/g_A$, the last quantity, which refers to charged leptons, being derived from all the asymmetries listed in table 1.

Complete Standard Model fits are discussed in ref.14. Here I present the results of a simplified fit [20] based on the six most important quantities discussed above. Fitting at fixed $m_H$ in terms of $m_t$ and $\alpha_s(m_Z)$ one obtains:

$$
\begin{aligned}
m_t &= 141 \pm 20 \text{ GeV}, \ \alpha_s(m_Z) = 0.118 \pm 0.006 \text{ for } m_H = 60 \text{ GeV} \\
m_t &= 162 \pm 18 \text{ GeV}, \ \alpha_s(m_Z) = 0.120 \pm 0.006 \text{ for } m_H = 300 \text{ GeV} \\
m_t &= 180 \pm 18 \text{ GeV}, \ \alpha_s(m_Z) = 0.123 \pm 0.006 \text{ for } m_H = 1 \text{ TeV}
\end{aligned}
\tag{1}
$$

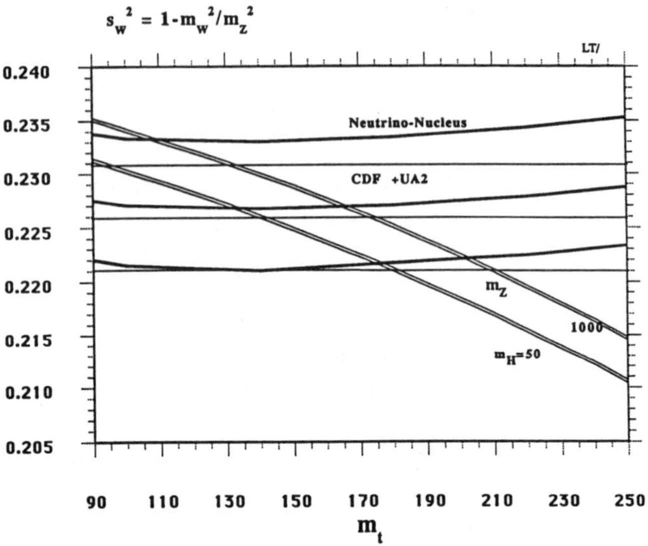


Fig. 1 . Labelled by $m_Z$ is the prediction for $s_W^2 = 1 - m_W^2/m_Z^2$ obtained in the Standard Model as a function of $m_t$ for $m_H$=50-1000 GeV. The 1$\sigma$ error bands implied by the CDF and UA2 measurements of $m_W/m_Z$ [16] and by the data on $R_\nu$ [17] are also shown.

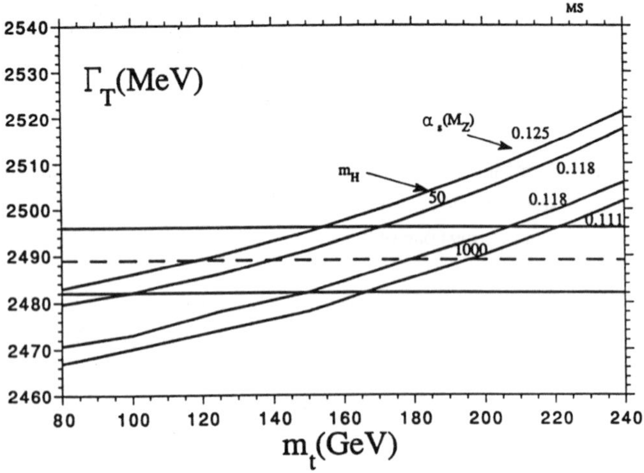


Fig. 2. The Standard Model prediction for $\Gamma_T$ as function of $m_t$ for $m_H$=50-1000 GeV and $\alpha_s(m_Z)$ = 0.111-0.125 is compared with the LEP result.

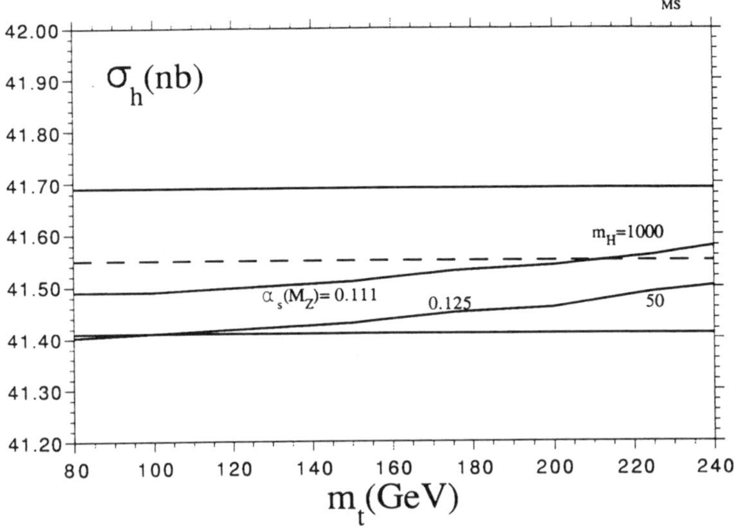


Fig. 3. The Standard Model prediction for $\sigma_h$ as function of $m_t$ for $m_H$=50-1000 GeV and $\alpha_s(m_Z)$ = 0.111-0.125 is compared with the LEP result.

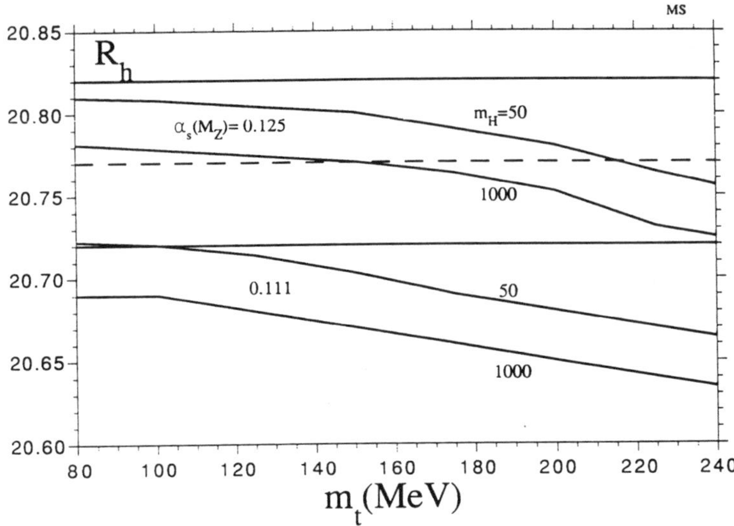


Fig. 4. The Standard Model prediction for $R_h$ as function of $m_t$ for $m_H$=50-1000 GeV and $\alpha_s(m_Z)$ = 0.111-0.125 is compared with the LEP result.

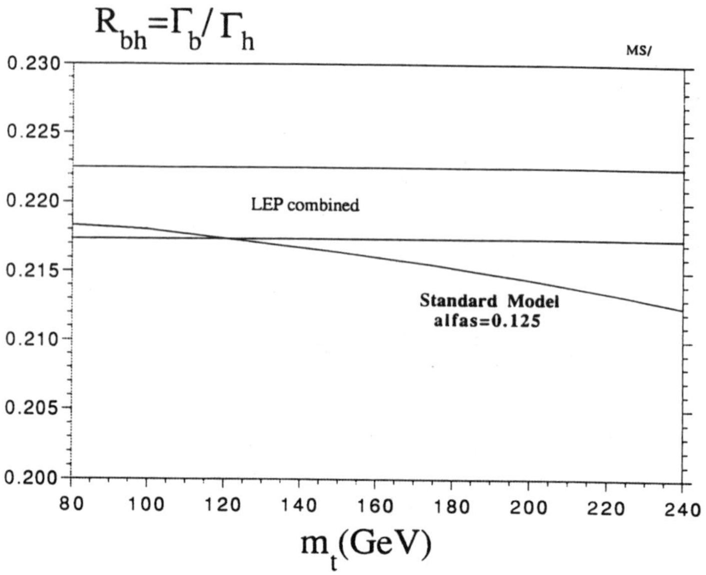


Fig. 5. The Standard Model prediction for $R_{bh}$ as function of $m_t$ for $m_H$=50-1000 GeV and $\alpha_s(m_Z)$ = 0.125 (that leads to the maximum prediction in the range 0.111-0.125) is compared with the LEP result.

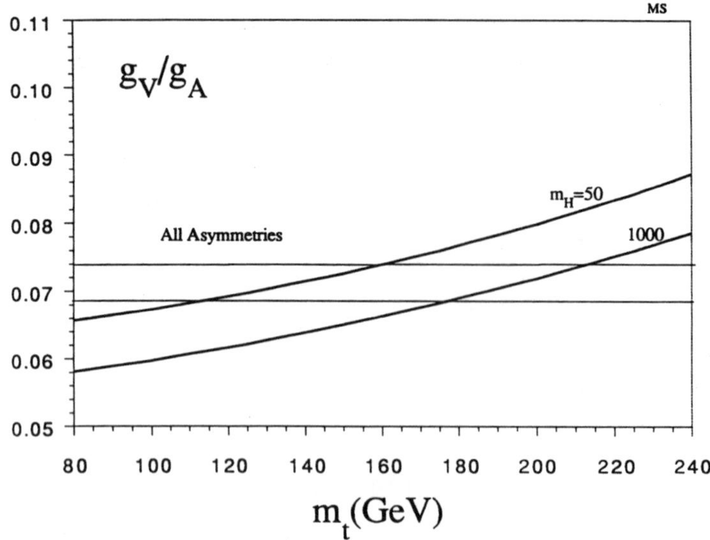


Fig. 6. The Standard Model prediction for $g_V/g_A$ as function of $m_t$ for $m_H$=50-1000 GeV is compared with the LEP result obtained by combining all the measured asymmetries (see table 1).

| | LATHUILE-MORIOND [19] | MARSEILLE[14] |
|---|---|---|
| $m_Z$(GeV) | $91.187 \pm 0.007$ | |
| $\Gamma_T$(MeV) | $2488 \pm 7$ | $2489 \pm 7$ |
| $R_h=\Gamma_h/\Gamma_l$ | $20.83 \pm 0.06$ | $20.77 \pm 0.05$ |
| $\sigma_h=12\pi\Gamma_e\Gamma_h/m_Z^2\Gamma_T^2$ (nb) | $41.44 \pm 0.17$ | $41.55 \pm 0.14$ |
| $\Gamma_l$(MeV) | $83.52 \pm 0.28$ | $83.79 \pm 0.28$ |
| $\Gamma_h$(MeV) | $1739.7 \pm 6.3$ | $1740 \pm 6$ |
| $\Gamma_b$(MeV) | $383 \pm 6$ | $383 \pm 5$ |
| $R_{bh}=\Gamma_b/\Gamma_h$ | $0.220 \pm 0.003$ | $0.220 \pm 0.0027$ |
| $A_{FB}^l$ | $0.0164 \pm 0.0021$ | $0.0162 \pm 0.0020$ |
| $A_{pol}^\tau$ | $0.142 \pm 0.017$ | $0.138 \pm 0.014$ |
| $A^e$ | $0.130 \pm 0.025$ | |
| $A_{FB}^b$ | $0.098 \pm 0.009$ | $0.098 \pm 0.006$ |
| $A_{FB}^c$ | $0.090 \pm 0.019$ | $0.075 \pm 0.015$ |
| $A_{LR}$ (SLD) | $0.100 \pm 0.044$ | |
| $g_V/g_A$ (all asymmetries) | $0.0725 \pm 0.0033$ | $0.0712 \pm 0.0028$ |
| $m_W/m_Z$ [16] | $0.8798 \pm 0.0028$ | |
| $R_{neutrino}=\sigma_{NC}/\sigma_{CC}$ [17] | $0.312 \pm 0.003$ | |
| APV( Cs) $Q_W$ [18] | $-71.04 \pm 1.81$ | |
| $\alpha_s(m_Z)$ [21,22] | $0.118 \pm 0.007$ | |

Table 1
The results presented at the Marseille Conference compared with the situation last winter.

No important differences are found if $\alpha_s(m_Z)$ is fixed in the range shown in table 1. Comparing these results with the complete fits of ref.14 one indeed realises that the above six observables completely dominate the fit. The previous results imply significant upper and lower limits on $m_t$: at $1.64\sigma$ one has:

$$106 \text{ GeV} < m_t < 208 \text{ GeV} \qquad (2)$$

The fitted value of $\alpha_s(m_Z)$, also including the uncertainty from $m_H$ is given by $\alpha_s(m_Z) = 0.120 \pm 0.007$, in perfect agreement with the value obtained from jet studies: $\alpha_s(m_Z) = 0.123 \pm 0.006$ [21] and with the world average of table 1 [21,22].The $\chi^2$ prefers

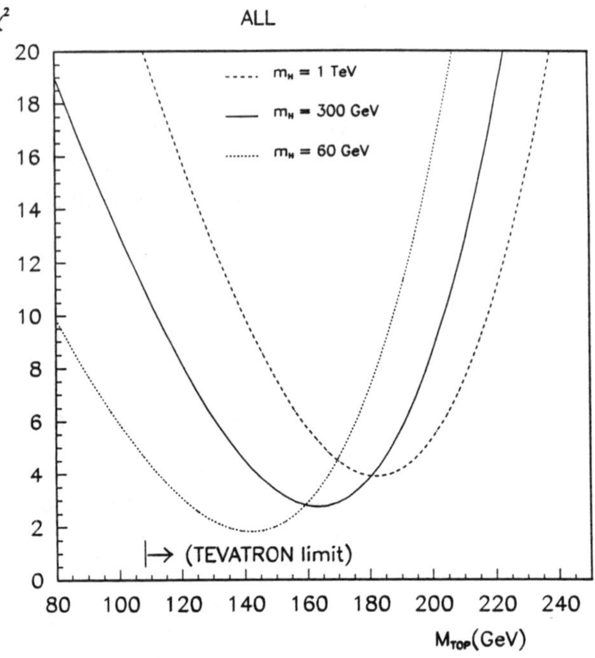


Fig. 7 The $\chi^2$ vs $m_t$ of the Standard Model fit for different fixed values of $m_H$ [20].

smaller values of $m_H$ (fig.7), but within $2\sigma$ no significant upper bound on $m_H$ is obtained [20] (i.e. one below 1 TeV which is already too large according to the bound from avoiding the Landau pole [23]). Actually a 2-variable fit (fig.8), i.e. using $\chi^2(m_t,m_H)$, indicates a value for $m_H$ below [20,24] the direct experimental lower limit $m_H$>63.5 GeV [25]. From the fit one can also obtain the best estimates for any other quantity. For example, one gets $s_W^2 = 1- m_W^2/m_Z^2 =$ $0.2259\pm 0.0019$, which, given $m_Z$ from LEP, corresponds to $m_W$=80230± 100 MeV. The effective $\sin^2\theta_W$ defined from the $Z\rightarrow l\bar{l}$ vertex via $g_V/g_A$ [15] is given by $\sin^2\theta_W=\bar{s}_W^2 =$ $0.2325\pm 0.00055$ (while from $g_V/g_A$ given in table 1 we have $\bar{s}_W^2 = 0.2322\pm 0.0007$).

The Standard Model fits the data better now than at the winter conferences: the $\chi^2$/degrees of freedom was 4.2/5 and is now 2.8/5 [20]. The major change was that $\Gamma_l$ went up by about $1\sigma$ (fig.9), because of comparable shifts in $\sigma_h$ and $R_h$ (see table 1). As a consequence, the value of $\alpha_s(m_Z)$, which used to be large, went down. Similarly, we shall see that the value of $\varepsilon_3$, which used to be low, went up.

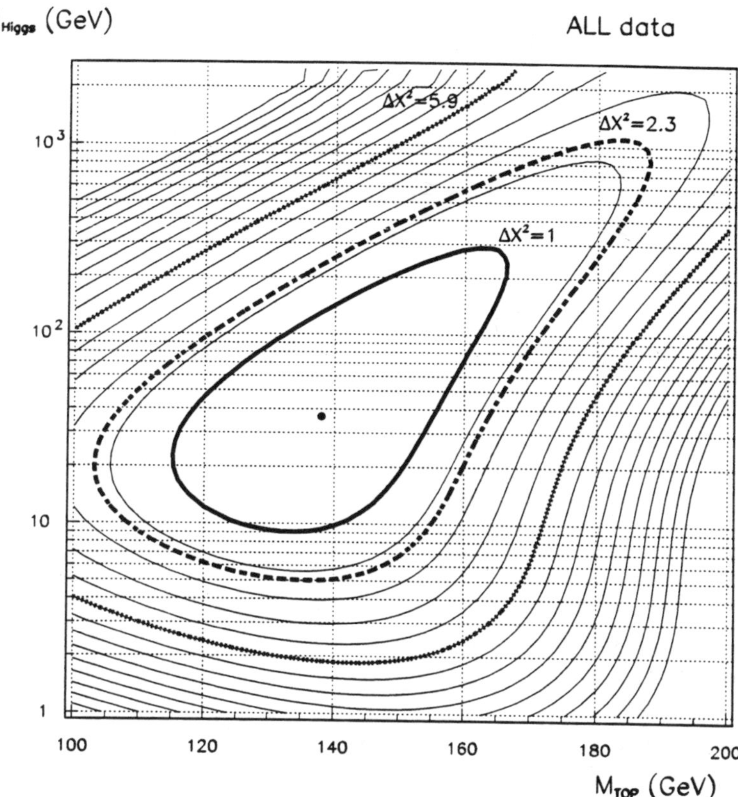


Fig.8. Contour plot of the function $\chi^2(m_t,m_H)$ for the Standard Model fit [20].

In a sense my presentation could well end here. The LEP experiments have already reached a remarkable accuracy of about or below the 0.5% level, yet the Standard Model passes all precision tests. Concerning the search for new physics, that we discuss in the following, the consequence is that, by now, only those extensions of the Standard Model can survive that very delicately perturb its basic framework.

## 2. Strategy for a (More) Model Independent Analysis

Recently we have proposed [26,27] a general strategy for the analysis of precision electroweak tests in view of the search for new physics beyond the Standard Model. Our method is more complete and less model dependent than the similar approach based on the

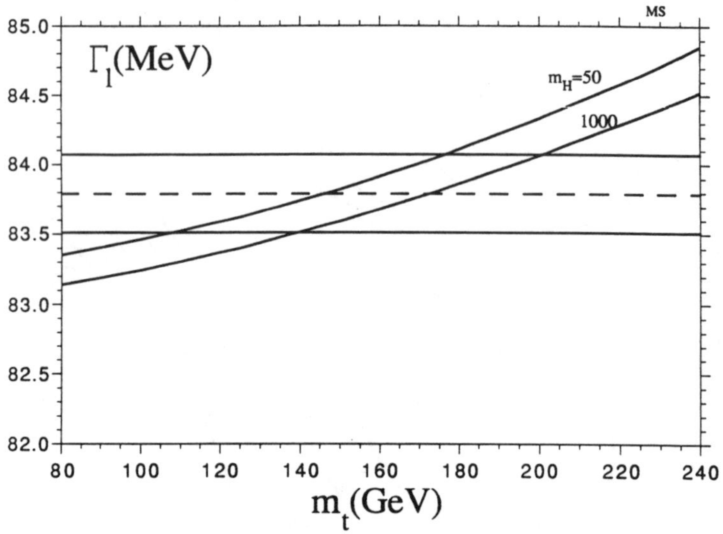


Fig 9. The Standard Model prediction for $\Gamma_l$ as function of $m_t$ for $m_H$=50-1000 GeV is compared with the LEP result.

variables S, T and U, [28-31] which, from the start, necessarily assumes dominance of vacuum polarisation diagrams from new physics and truncation of the $q^2$ expansion of the corresponding amplitudes. In a completely model independent way we define [27] four variables, called[f1] $\varepsilon_1$, $\varepsilon_2$, $\varepsilon_3$ and $\varepsilon_b$, that are precisely measured and can be compared with the predictions of different theories. The quantities $\varepsilon_1$, $\varepsilon_2$, $\varepsilon_3$ and $\varepsilon_b$ are defined in ref.27 in one to one correspondence with the set of observables $m_W/m_Z$, $\Gamma_l$, $A^l_{FB}$ and $\Gamma_b$. The four epsilons are defined without need of specifying $m_t$ (and $m_H$). In the Standard Model, for all observables at the Z pole, the whole dependence on $m_t$ arising from one-loop diagrams only enters through the epsilons. The same is true for any extension of the Standard Model such that all possible deviations only occur through vacuum polarisation diagrams and/or the $Z \to b\bar{b}$ vertex. As discussed in detail in ref.27, for such a model one can compare the theoretical predictions with the experimental determination of the epsilons as obtained from

---

[f1] Here we resume the notation $\varepsilon_i$ for exactly the same quantities as defined in ref.27, where they were denoted $\varepsilon_{Ni}$ (the index N, for "new", had been inserted to signal some small differences with respect to the original definitions in refs.26,29.

the whole set of LEP data. If a model does not satisfy this requirement then the comparison is to be made with the epsilons determined from the defining variables only, or with some more limited enlargement of the same set, depending on the particular case. For example, if lepton universality is maintained, then the data on $A_{FB}^{l}$ can be replaced by the combined result on $g_V/g_A$ from all lepton asymmetries (see, e.g., the following section on extended gauge models).

The epsilons represent an efficient parameterisation of the small deviations from what is solidly established in a way that is unaffected by our ignorance of $m_t$. The variables S, T, U depend on $m_t$ because they are defined as deviations from the complete Standard Model prediction for specified $m_t$ (and $m_H$). Instead the epsilons are defined with respect to a reference approximation which does not depend on $m_t$. In fact the epsilons are defined in such a way that they are exactly zero in the Standard Model in the limit of neglecting all pure weak loop-corrections (i.e. when only the predictions from the tree level Standard Model plus pure QED and pure QCD corrections are taken into account). This very simple version of improved Born approximation is amazingly well supported by the data [27,32,33]. In table 2 we compare the Born predictions, computed for $\alpha(m_Z)=1/128.87$ and $\alpha_s(m_Z)=0.118$ (the values used in the definitions of the epsilons) with the experimental data. We see that there is agreement to better than 1-1.5 $\sigma$. This does not mean that LEP experiments were badly designed and their accuracy is not sufficient to detect the pure weak corrections. Rather it reflects a very important a-posteriori dynamical result that the experimental values of the radiative corrections are smaller than their a-priori estimate, signalling the presence of important cancellations in the key quantities like $\Delta\rho$, $\Delta r$, $\Delta k'$ [3,4] that determine the whole pattern of radiative corrections. This information is very constraining both within and beyond the Standard Model. For example, in the Standard Model the lower limit on $m_t$ in eq.2 arises from the necessity of cancelling in $\Delta\rho$ the negative residual corrections with the positive contribution of $m_t$.

| | Born [27] | Experiment |
|---|---|---|
| $m_W/m_Z$ | 0.8768 | $0.8798 \pm 0.0028$ |
| $\Gamma_T$(MeV) | 2488 | $2489 \pm 7$ |
| $R_h=\Gamma_h/\Gamma_l$ | 20.81 | $20.77 \pm 0.05$ |
| $\sigma_h$ (nb) | 41.425 | $41.55 \pm 0.14$ |
| $R_{bh}=\Gamma_b/\Gamma_h$ | 0.218 | $0.220 \pm 0.0027$ |
| $g_V/g_A$ (all asym.s) | 0.0753 | $0.0712 \pm 0.0028$ |

Table 2

The QED and QCD improved Born approximation, obtained with $\alpha(m_Z)=1/128.87$ and $\alpha_s(m_Z) = 0.118$, is compared to the data.

By combining the value of $m_W/m_Z$ [16] with the LEP results on the charged lepton partial width and the forward-backward asymmetry, all given in table 1, and following the definitions of ref.27, one obtains:

$$g_A^2 = 0.2508 \pm 0.00085$$

$$x = g_V/g_A = 0.0738 \pm 0.0046 \quad \text{or} \quad \bar{s}_W^2 = 0.2316 \pm 0.0012 \tag{3}$$

and:

$$\begin{aligned} \varepsilon_1 &= \Delta\rho = (3.0 \pm 3.4)\ 10^{-3} \\ \varepsilon_2 &= (-5.3 \pm 7.6)\ 10^{-3} \\ \varepsilon_3 &= (3.2 \pm 4.1)\ 10^{-3} \end{aligned} \tag{4}$$

Finally, by adding the value of $\Gamma_b$ listed in table 1 and using the definition of $\varepsilon_b$ given in ref.27 one finds:

$$\varepsilon_b = (3.0 \pm 5.8)\ 10^{-3} \tag{5}$$

In fig.10 the experimental $1\sigma$ ellipse in the $\varepsilon_1$-$\varepsilon_3$ plane is compared with the Standard Model predictions for different $m_t$ and $m_H$ values. We recall that $\varepsilon_1$ and $\varepsilon_3$ are completely determined by $\Gamma_l$ and $A_{FB}^l$. We see that the value of $\varepsilon_3$ went up in the latest data and is by now well compatible within $1\sigma$ with the Standard Model. This is because $\varepsilon_3$ fine-tunes the value of $\Gamma_l$ once $\bar{s}_W^2$ is given from $A_{FB}^l$ [27]. In the data presented at Marseille $\Gamma_l$ went up with $A_{FB}^l$ essentially unchanged (see table 1). In fig.11 the experimental value of $\varepsilon_2$ is compared with the Standard Model prediction as a function of $m_t$ and there is consistency at all practical values of $m_t$. Note that $\varepsilon_2$ also depends on $m_W/m_Z$ and better measurements of this quantity are needed in order to make this test more stringent. Finally, in fig.12 we compare the experimental value of $\varepsilon_b$ with the Standard Model prediction. Here we see that $\varepsilon_b$ prefers small values of $m_t$. This result is a simple and direct consequence of the fact that the measured value of $R_{bh}$ is a bit high ($\Gamma_b$ went up over the last year and the error considerably decreased).

To proceed further, and include other measured observables in the analysis we need to make some dynamical assumptions. The minimum amount of model dependence is introduced by including other purely leptonic quantities at the Z pole such as $A_{pol}^{\tau}$, $A_e$ (measured [15] from the angular dependence of the $\tau$ polarisation) and $A_{LR}$ (measured by SLD [34]). At this stage, one is simply relying on lepton universality. With essentially the same assumptions one can also include the data on the b-quark forward backward asymmetry $A_{FB}^b$. In fact it turns out that $A_{FB}^b$ is almost unaffected by the $Z \to b\bar{b}$ vertex correction.

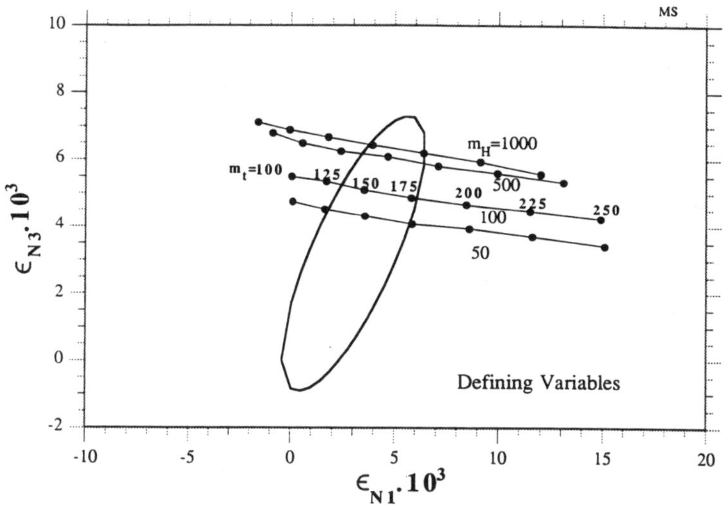


Fig. 10. The 1σ ellipse in the plane $\varepsilon_1$-$\varepsilon_3$ obtained from the data on the defining variables $\Gamma_l$ and $A^{l}_{FB}$ compared with the Standard Model predictions for the indicated values of $m_t$ and $m_H$.

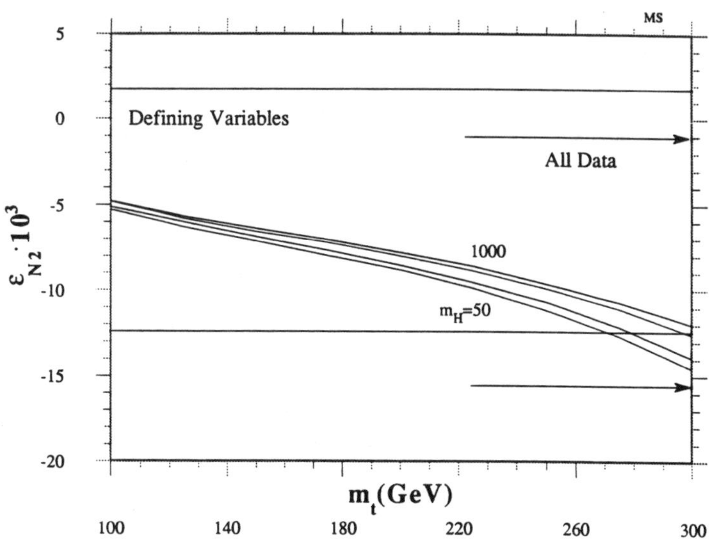


Fig. 11. The 1σ data on $\varepsilon_2$ obtained from the data on the defining variables $\Gamma_l$, $A^{l}_{FB}$ and $m_W/m_Z$ compared with the Standard Model predictions as functions of $m_t$ for the indicated values of $m_H$. The arrows indicate the experimental 1σ band from the fit in eq.8 to all electroweak data.

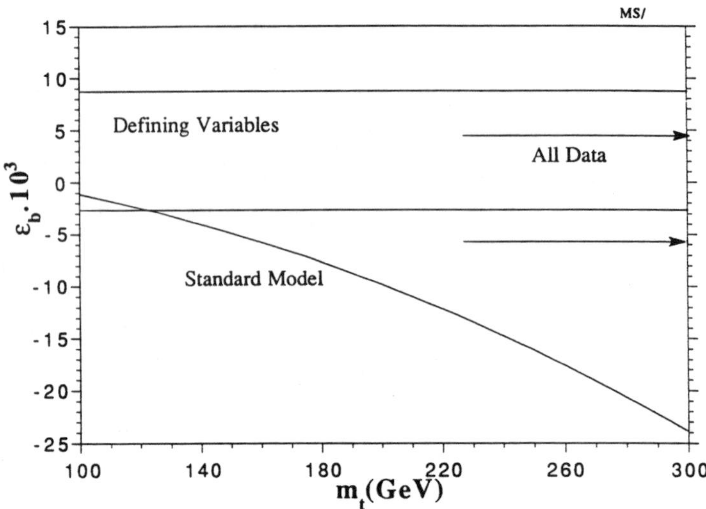


Fig. 12. The $1\sigma$ data on $\varepsilon_b$ obtained from the data on the defining variables $\Gamma_l$, $A^l_{FB}$, $m_W/m_Z$ and $\Gamma_b$ compared with the Standard Model predictions as functions of $m_t$ for $\alpha_s(m_Z)$=0.118. The arrows indicate the experimental $1\sigma$ band from the fit in eq.8 to all electroweak data.

As a result of the previous discussion, we can combine the values of $x = g_V/g_A$ obtained from the whole set of asymmetries measured at LEP and we get the value reported in table 1. At this stage the best values of $\varepsilon_1$, $\varepsilon_2$ and $\varepsilon_3$ are modified according to

$$\begin{aligned} \varepsilon_1 &= \Delta\rho = (3.4 \pm 3.4)\ 10^{-3} \\ \varepsilon_2 &= (-6.3 \pm 7.0)\ 10^{-3} \\ \varepsilon_3 &= (5.0 \pm 3.2)\ 10^{-3} \\ \varepsilon_b &= (3.2 \pm 5.8)\ 10^{-3} \end{aligned} \tag{6}$$

From fig.13 we see that the agreement with the Standard Model is even improved by this step: the center of the $1\sigma$ ellipse of the data in the $\varepsilon_1$-$\varepsilon_3$ plane is very close to the point $m_t$=150 GeV, $m_H$=100 GeV.

All observables measured on the Z peak at LEP can be included in the analysis provided that we assume that all deviations from the Standard Model are only contained in vacuum polarisation diagrams (without demanding a truncation of the $q^2$ dependence of the corresponding funtions) and/or the $Z\to b\bar{b}$ vertex. Note that this is true for whatever partition of the new effect between $g_{bV}$ and $g_{bA}$, because only one combination of them is

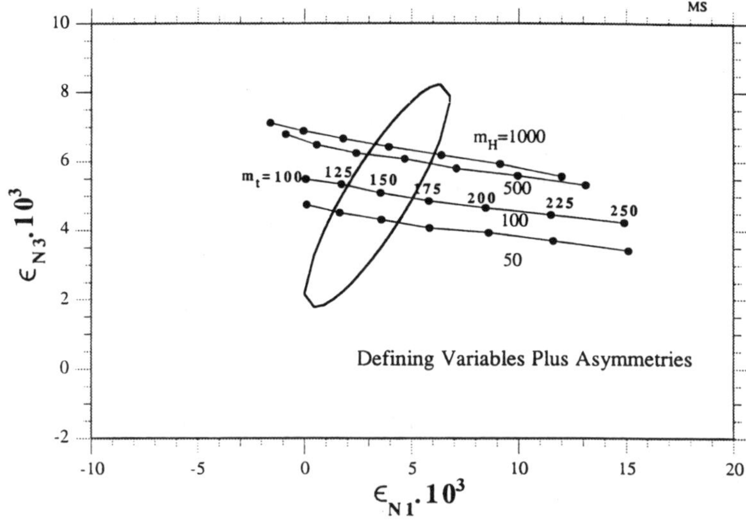


Fig. 13. The 1σ ellipse in the plane $\varepsilon_1$-$\varepsilon_3$ obtained from the data on $\Gamma_l$ and $g_V/g_A$ derived from all the asymmetries (see table 1) compared with the Standard Model predictions for the indicated values of $m_t$ and

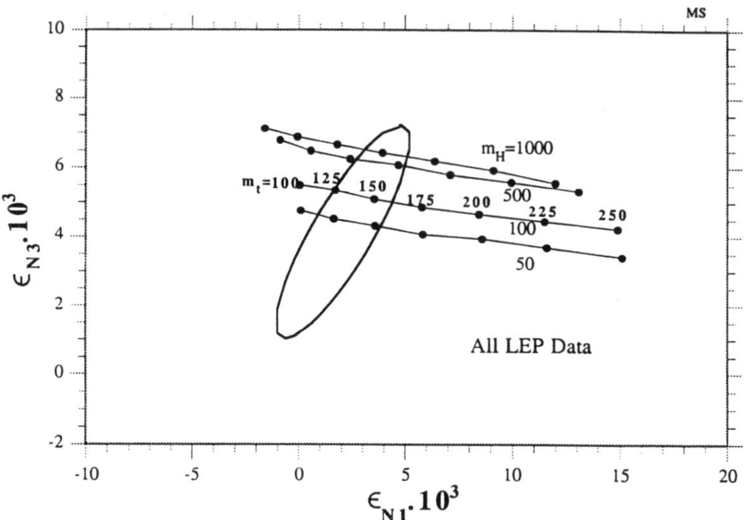


Fig. 14. The 1σ ellipse in the plane $\varepsilon_1$-$\varepsilon_3$ obtained from the data on $m_W/m_Z$, $\Gamma_T$, $\sigma_h$, $R_h$, $R_{bh}$ and $g_V/g_A$ derived from all the asymmetries (see table 1) compared with the Standard Model predictions for the indicated values of $m_t$ and $m_H$.

measured in $\Gamma_b$, while, as already mentioned, $A^b_{FB}$ is nearly independent of the $Z \to b\bar{b}$ vertex.

From a global fit of the data on $m_W/m_Z$, $\Gamma_T$, $R_h$, $\sigma_h$, $R_{bh}$ and $x=g_V/g_A$ given in table 1 (for LEP data, we have taken the correlation matrix for $\Gamma_T$, $R_h$ and $\sigma_h$ given in ref.15, while we have considered the additional information on $R_{bh}$ and x as independent) we obtain :

$$\begin{aligned} \varepsilon_1 &= \Delta\rho = (1.8 \pm 3.1)\ 10^{-3} \\ \varepsilon_2 &= (-7.7 \pm 6.9)\ 10^{-3} \\ \varepsilon_3 &= (4.0 \pm 3.1)\ 10^{-3} \\ \varepsilon_b &= (-0.5 \pm 5.1) 10^{-3} \end{aligned} \qquad (7)$$

The comparison of theory and experiment in the planes $\varepsilon_1$-$\varepsilon_3$, $\varepsilon_b$-$\varepsilon_3$ and $\varepsilon_b$-$\varepsilon_1$ is shown in figs. 14, 15 and 16, respectively. We see that, except for $\varepsilon_b$ which noticeably moves in the direction of the Standard Model prediction, the inclusion of all LEP quantities does not change the epsilons very much. This is because $\Gamma_T$, $\sigma_h$ and $R_h$ (or equivalently the ratios of $\Gamma_T$, $\Gamma_h$ and $\Gamma_l$) are normal.

To include in our analysis lower energy observables as well, a stronger hypothesis needs to be made: only vacuum polarization diagrams are allowed to vary from the Standard Model ones and only in their constant and first derivative terms in a $q^2$-expansion [28-30], a likely picture, e.g., in technicolour theories [35,36]. In such a case, one can, for example, add to the analysis the ratio $R_\nu$ of neutral to charged current processes in deep inelastic neutrino scattering on nuclei [17],the "weak charge" $Q_W$ measured in atomic parity violation experiments on Cs [18] and the measurement of $g_V/g_A$ from $\nu_\mu e$ scattering [38] (the final result of CHARM-II corresponds to $\bar{s}^2_W$ =0.2324 ± 0.0086). In this way one obtains the global fit:

$$\begin{aligned} \varepsilon_1 &= \Delta\rho = (1.5 \pm 2.6)\ 10^{-3} \\ \varepsilon_2 &= (-7.8 \pm 6.8)\ 10^{-3} \\ \varepsilon_3 &= (3.5 \pm 2.8)\ 10^{-3} \\ \varepsilon_b &= (-0.5 \pm 5.0)\ 10^{-3} \end{aligned} \qquad (8)$$

With the progress of LEP the low energy data, while important as a check that no deviations from the expected $q^2$ dependence arise, play a lesser role in the global fit. The $\varepsilon_1$-$\varepsilon_3$ plot for all data is shown in fig.17. Note that the present ambiguity on the value of $\alpha(m_Z) = (128.87 \pm 0.12)^{-1}$ [7] corresponds to an uncertainty on $\varepsilon_3$ (the other epsilons are not much affected) given by $\Delta\varepsilon_3 \cdot 10^3 = \pm 0.6$ [27]. Thus the theoretical error is still confortably less than the experimental error but the two will become close at the end of the LEP1 phase. The values of $\varepsilon_2$ and $\varepsilon_b$ in eq.8 are compared with the Standard Model predictions in figs. 11 and 12. The fitted value of $\varepsilon_b$ analysed in the Standard Model

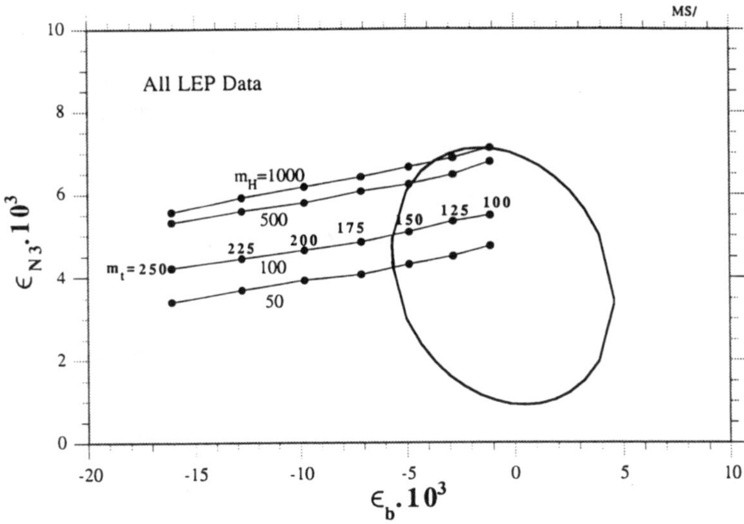


Fig. 15. The $1\sigma$ ellipse in the plane $\varepsilon_b$-$\varepsilon_3$ obtained from the data on $m_W/m_Z$, $\Gamma_T$, $\sigma_h$, $R_h$, $R_{bh}$ and $g_V/g_A$ derived from all the asymmetries (see table 1) compared with the Standard Model predictions for the indicated values of $m_t$ and $m_H$.

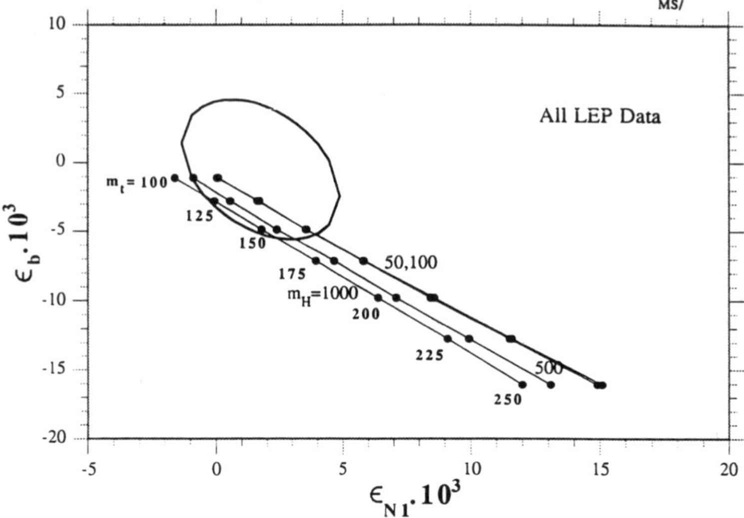


Fig. 16. The $1\sigma$ ellipse in the plane $\varepsilon_1$-$\varepsilon_b$ obtained from the data on $m_W/m_Z$, $\Gamma_T$, $\sigma_h$, $R_h$, $R_{bh}$ and $g_V/g_A$ derived from all the asymmetries (see table 1) compared with the Standard Model predictions for the indicated values of $m_t$ and $m_H$.

implies $m_t$<190 GeV at 1.64σ if the CDF limit $m_t$>113 GeV [39] is not taken into account. Note that the limit from $\varepsilon_b$ is independent of $\varepsilon_1 = \Delta\rho$, i.e. of a large class of possible large vacuum polarisation effects. It however depends on the assumed value of $\alpha_s(m_Z)$ that we have taken as in table 1. If we repeat the fit with $\alpha_s(m_Z)$ free, $\varepsilon_b$ moves up to $\varepsilon_b \cdot 10^3 = 3.1 \pm 7.1$ (to fix $R_{bh}$ which is nearly independent of $\alpha_s(m_Z)$) while $\alpha_s(m_Z)$ goes down to $\alpha_s(m_Z) = 0.105 \pm 0.013$. Finally, $\varepsilon_1$, $\varepsilon_2$ and $\varepsilon_3$ are quite insensitive to $\alpha_s(m_Z)$ and closely keep their values in eq.8.

## 3. Specific Examples

We now concentrate on a number of well known extensions of the Standard Model which not only are particularly important per se but also are interesting in that they clearly demonstrate the constraining power of the present level of precision tests.

3.1 Technicolour. It is well known that technicolour models [35-37] tend to produce large and positive corrections to $\varepsilon_3$. As the central value of $\varepsilon_3$ went up one might imagine that the experimental problems of technicolour with respect to electroweak tests are now solved. This is not the case. First, by glancing at fig.18 [37], where the data on $\varepsilon_3$ and $\varepsilon_1$ are compared with the predictions of a class of simple versions of technicolour models, one realises, that the experimental errors on $\varepsilon_3$ are by now small enough that these models remain clearly disfavoured with respect to the Standard Model.

Second, recently it has been shown [40] that the data on $\varepsilon_b$ also produce evidence against technicolour models. The same mechanism that in extended technicolour generates the top quark mass also leads to large corrections to the $Z \to b\bar{b}$ vertex that have the wrong sign. For example, in a simple model with two technidoublets, ($N_{TC}$=2), the Standard Model prediction is decreased by the amount [40]:

$$\Delta\varepsilon_b = -\ 16 \cdot 10^{-3} \left| \frac{\xi}{\xi'} \left(\frac{m_t}{100\ \mathrm{GeV}}\right) \right| \tag{9}$$

where ξ and ξ' are Clebsch-like coefficients, expected to be of order 1. The effect is even larger for larger $N_{TC}$. In a more sophisticated version of the theory, the so called "walking" technicolour [41], where the relevant coupling constants walk (i.e. they evolve slowly) instead of running, the result is somewhat smaller [42] but still gigantic (fig.19).

In conclusion, it is difficult to really exclude technicolour because it is not a completely defined theory and no realistic model could be built sofar out of this idea. Yet, it is interesting that the most direct realisations tend to produce $\Delta\varepsilon_3 \gg 0$ and $\Delta\varepsilon_b \ll 0$ which are both disfavoured by experiment.

3.2 Minimal Supersymmetric Standard Model (MSSM). Contrary to technicolour the MSSM [43-45] is a completely specified, consistent and computable theory. There are too many parameters to attempt a direct fit of the data to the most general framework. So in

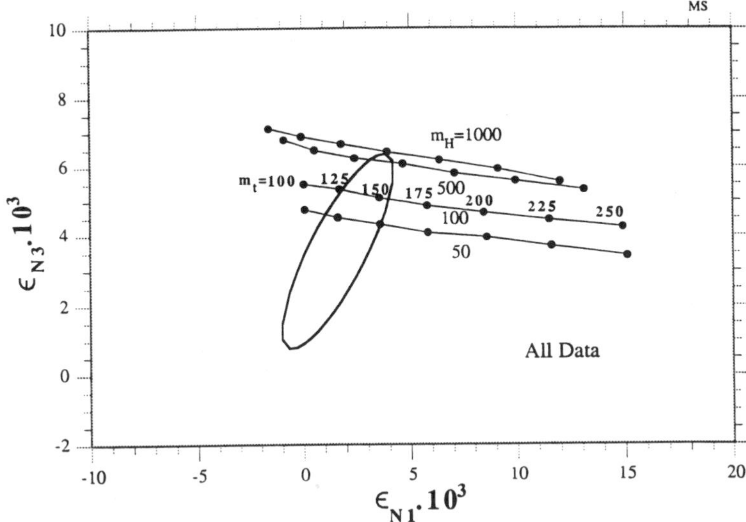


Fig. 17. The $1\sigma$ ellipse in the plane $\varepsilon_1$-$\varepsilon_3$ obtained from all the data also including the low energy data compared with the Standard Model predictions for the indicated values of $m_t$ and $m_H$

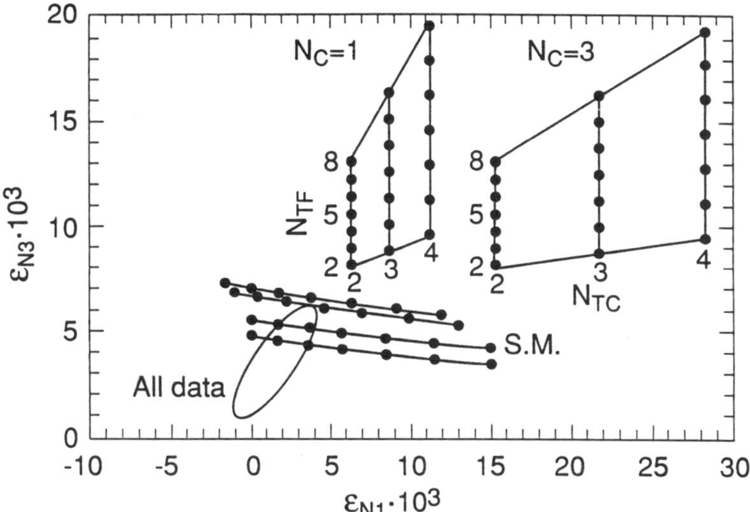


Fig. 18. The same as fig.17 but with the predictions of simple technicolour models also shown [37] , where $N_C$, $N_{TC}$ and $N_{TF}$ are the numbers of colours, technicolours and techniflavours, respectively.

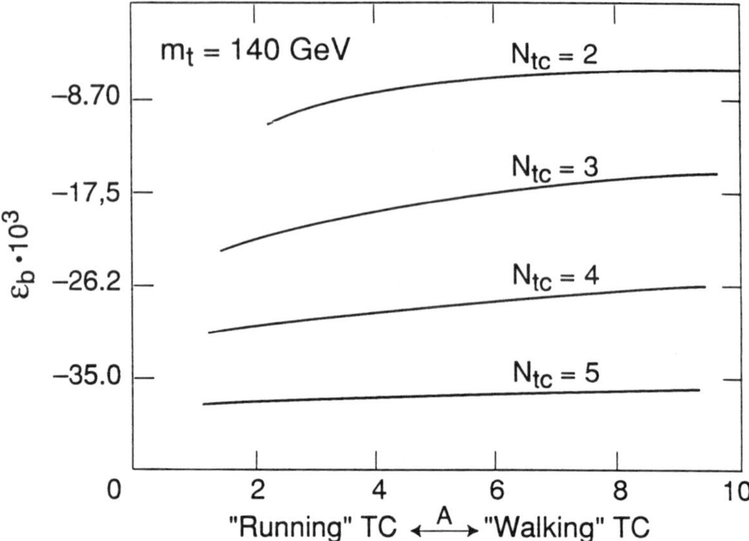


Fig. 19. Evolution of the contributions to $\Delta\varepsilon_b$ in moving from running (A small) towards walking (A large) tecnicolour, for $m_t$=140 GeV and different values of $N_{TC}$ (from ref.42).

ref.46 we restricted ourselves to two significant limiting cases: the "heavy" and the "light"MSSM.

The "heavy" limit correspond to all s-particles being sufficiently massive, still within the limits of a natural explanation of the weak scale of mass. In this limit a very important result holds [47]: for what concerns the precision electroweak tests, the MSSM predictions tend to reproduce the results of the Standard Model with a light Higgs, say $m_H \lesssim 100$ GeV.

In the "light" MSSM option some of the superpartners have a relatively small mass, close to their experimental lower bounds. In this case the pattern of radiative corrections may sizeably deviate from that of the Standard Model. The most interesting effects occur in vacuum polarisation amplitudes and/or the $Z \to b\bar{b}$ vertex and therefore are particularly suitable for a description in terms of the epsilons (because in such a case, as explained in ref.27, the predictions can be compared to the experimental determination of the epsilons from the whole set of LEP data). They are:

i) a threshold effect in the Z wave function renormalisation [47] mostly due to the vector coupling of charginos and (off-diagonal) neutralinos to the Z itself. Defining the vacuum polarisation functions by $\Pi_{\mu\nu}(q^2)=-ig_{\mu\nu}[A(0)+q^2F(q^2)]+q_\mu q_\nu$ terms, this is a positive contribution to $e_5=m_Z^2 F'_{ZZ}(m_Z^2)$ [47],the prime denoting a derivative with respect to $q^2$ (i.e. a contribution to a higher derivative term not included in the usual S,T,U formalism [28-31]). The $e_5$ correction shifts $\varepsilon_1$, $\varepsilon_2$, $\varepsilon_3$ by $-e_5$, $-c^2e_5$ and $-c^2e_5$ respectively, where $c^2=\cos^2\theta_W$, so that all of them are reduced by a comparable amount. Correspondingly all the Z widths are reduced without affecting the asymmetries. This effect can be significant but requires the lightest chargino to have a mass close to the experimental lower limit of 45 GeV.

ii) a positive contribution to $e_1$ from the virtual exchange of the scalar top and bottom superpartners [48], analogous to the contribution of the top-bottom quark doublet. The needed isospin splitting requires one of the two scalars (in the MSSM the s-top) to be light.

iii) a negative contribution to $\varepsilon_b$ due to the virtual exchange of a charged Higgs [49]. If one defines, as customary, $tg\beta=v_2/v_1$ ($v_1$ and $v_2$ being the vacuum expectation values of the Higgs doublets giving masses to the down and up quarks, respectively), then, for negligible bottom Yukawa coupling or $tg\beta << m_t/m_b$, this contribution is proportional to $m_t^2/tg^2\beta$.

iv) a positive contribution to $\varepsilon_b$ due to virtual chargino--s-top exchange [50] which in this case is proportional to $m_t^2/\sin^2\beta$. This effect again requires the chargino and the s-top to be light in order to be sizeable.

As an example, in fig. 20-22, we give the pair correlations $\varepsilon_3$- $\varepsilon_1$, $\varepsilon_3$- $\varepsilon_b$ and $\varepsilon_b$- $\varepsilon_1$ in the MSSM in the form of scatter plots [46]. The ellipses are the $1\sigma$ contours obtained from the present combined LEP experimental data (with the addition of the measurements of $m_W/m_Z$) [16]. The theoretical points in each plot are for fixed $m_t$=130 GeV. The Standard Model prediction as a function of $m_H$ is shown by two black stars corresponding to $m_H$=50 GeV and $m_H$=1000 GeV connected by a line. The MSSM scatter plot are obtained for $tg\beta>1$ (which is relevant to the effects described in iii) and iv) above), the charged Higgs mass $m_{H^+}>100$ GeV (relevant for iii)) the lightest s-top and the approximately degenerate s-bottom masses (relevant to ii) and iv)), $m_{stop}>50$ GeV and $m_{sbottom}>150$ GeV. This last constraint requires an additional qualification. In the mass matrix of the two top superpartners we limit the off diagonal term by taking $m_A< 3m_{sbottom}$. This constrains the relative s-top--s-bottom splitting when $m_{sbottom}$ gets large and consequently the size of the effect on $\varepsilon_1$. A particularly important role is played by the lightest chargino mass $m_{\chi^+}$, relevant to i) and iv). The white bullets refer to very light charginos: $60 > m_{\chi^+} > 48$ GeV, while the white stars, that by superposition form a dark area, refer to $m_{\chi^+}>60$ GeV. In figs.20-22 the dark area leans towards the heavy MSSM while the white bulletts are from the (very) light MSSM. As apparent from figs.20-22, the differences between the theoretical predictions of the Standard Model and of the MSSM are not large in comparison with the present accuracy of the experiments, but they will become more important with the increase of statistics at LEP.

In conclusion, the present electroweak data are well consistent with the MSSM. As the present data are very constraining this statement is highly non trivial. In its heavy version, the pattern of radiative corrections predicted by the MSSM is practically

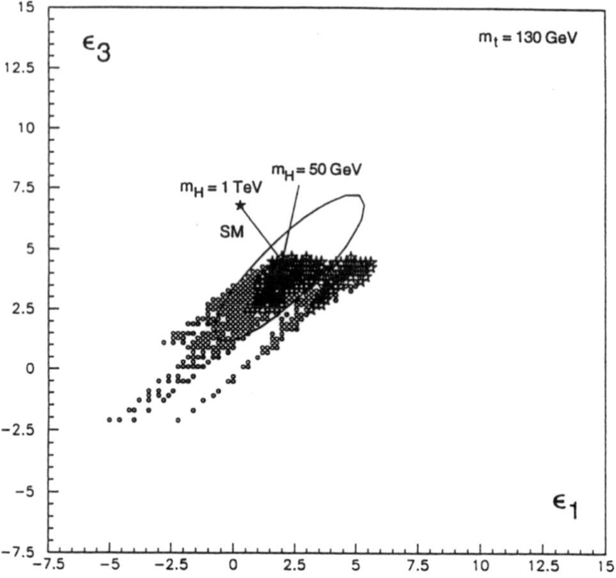


Fig. 20. ε3 vs ε1 at fixed $m_t$: $m_t$=130 GeV (from ref.46). The ellipse is the 1σ contour from all LEP data.The Standard Model prediction as a function of $m_H$ is shown by two black stars corresponding to $m_H$=50 GeV and $m_H$=1000 GeV connected by a line. The MSSM scatter plot are obtained for tgβ>1, the charged Higgs mass $m_{H^+}$>100 GeV, the lightest s-top and the approximately degenerate s-bottom masses $m_{stop}$>50 GeV and $m_{sbottom}$>150 GeV.The white bullets refer to very light charginos: 60 >$m_{\chi^+}$>48 GeV, while the white stars, that by superposition form a dark area, refer to $m_{\chi^+}$>60 GeV. The figures are actually realised by taking tgβ=1 and 4, $m_{H^+}$=100 GeV or ∞ (i.e. its contributions set to zero), $m_{stop}$=50 or 200 GeV, $m_{sbottom}$>150 GeV and varying the chargino masses in the mentioned ranges.

indistinguishable from that obtained from the Standard Model with a light Higgs. Sizeable departures from the Standard Model arise in the MSSM if a light gaugino and a light s-top exist, with masses close to their experimental bounds and such as to make them visible at LEP200. The resulting effects are still consistent with the data on $\varepsilon_3$ and $\varepsilon_b$.

3.3 Models with an Extended Gauge Group. In the simplest models with an extra U(1), for a specified definition of that U(1), i.e. for given couplings to fermions of the associated neutral vector boson $Z_N$, irrespective of assumptions on the new Higgs sector (namely on the heavy $Z_H$ mass) two new parameters are introduced, tgξ and $\delta\rho_M$ [51]. The angle ξ defines the mixtures of the standard ($Z_S$) and the new ($Z_N$) vector boson that make up the light ($Z_L$), observed at LEP, and the heavy ($Z_H$) mass eigenstates, e.g. $Z_L = \cos\xi Z_S + \sin\xi Z_N$. $\delta\rho_M$ describes the shift induced at tree level in the $\rho$ parameter by the mixing: $\cos^2\theta_W = m_W^2/(m_{Z_L}^2 \rho)$ with $\rho = 1 + \delta\rho_M$ (we assume that the breaking of the ordinary SU(2) is induced by Higgs doublets). One has $\delta\rho_M > 0$ because the mixing pushes the lowest state down. If the $Z_H$ mass was known $\delta\rho_M$ and $\sin^2\xi$ would be related: $\delta\rho_M = (m_{Z_H}^2/m_{Z_L}^2 - 1)\sin^2\xi$. More details can be found in ref.[52]. Here for lack of space we go directly to the results for the class of models based on E6 [53] (with arbitrary orientation of the extra U(1) in the group space, determined by an angle $\theta_2$) and for the left-right symmetric model (assuming heavy $W_R$) [54].The constraints on $\xi$ and $\delta\rho_M$ are very strong [52]. In fig.23 the 90% c.l. bounds on $\xi$ are plotted for $m_t$=100-200 GeV, $m_H$=100 GeV and $\alpha_s(m_Z)$=0.118. The dependence on $m_t$ of the limits on $\xi_0$ is very moderate, except in the region near $\theta_2$=0. One finds that in most cases $|\xi| < 0.5\%$ is implied for all values of $m_t$, $m_H$ and $\alpha_s(m_Z)$, with $|\xi| < 1\%$ in the least favourable instances. Similarly the 90% c.l. bounds on the quantity $\Delta\rho = (\varepsilon_1)_{SM} + \Delta\rho_M$ are displayed in fig.24, for the same values of $m_t$, $m_H$ and $\alpha_s(m_Z)$. Note that considering $\Delta\rho$ is useful because the corresponding bounds are nearly independent of $m_t$. On the contrary the allowed range for $\Delta\rho_M$ rapidly decreases with increasing $m_t$ because of the quadratic $m_t$ dependence of $(\varepsilon_1)_{SM}$ and the fact that $\Delta\rho_M$ >0. A typical value for the upper bound on $\Delta\rho$ (at $\alpha_s$ = 0.118) of about 0.006 is already saturated by the contribution from $\Delta\rho_t$ with $m_t \sim 175$ GeV. We have also studied the dependence on $\alpha_s$ in the range $0.111<\alpha_s<0.125$. It turns out that, for $m_H$ = 100 GeV, $m_t$=150 GeV, by increasing $\alpha_s$ of 0.01 the allowed region for $\Delta\rho_M$ is shifted down by about 0.0015.

One can obtain a pictorial impression of the comparison with the Standard Model by using the epsilon variables . As discussed in ref.[26], in extended gauge models $\varepsilon_1$, $\varepsilon_2$ and $\varepsilon_3$ receive the additional contributions (with respect to the Standard Model). In figs.25 we plot for $m_t$=$m_H$=100 GeV and $\alpha_s$ =0.118 a set of dots in the plane $\varepsilon_3$ vs $\varepsilon_1$ that correspond to the best fit for each $\theta_2$ value obtained [52] by considering the whole set of LEP data, plus $m_W/m_Z$, given in table 1. The solid lines, for each dot, show the 1$\sigma$ variation due to $\Delta\rho$, while the dashed lines span 1$\sigma$ in $\xi$. The result for the LR model also appears in the figures, marked by a box symbol. The corresponding Standard Model point is also shown, together with the ellipse of the experimental data (with 1$\sigma$ projections on both axis). The data correspond to a fit to the defining variables for $\varepsilon_1$, $\varepsilon_2$ and $\varepsilon_3$ , i.e. $m_W/m_Z$, $\Gamma_l$ and the ratio $g_V/g_A$ obtained by combining all the asymmetries. In general, it is at small values of $m_t$ that there is a substancial improvement in going from the Standard Model to the extended gauge models, mostly because $\Delta\rho_M$ increases $\Delta\rho$. Finally, the variable $\varepsilon_b$ is not particularly relevant in the present context because, in all extended gauge

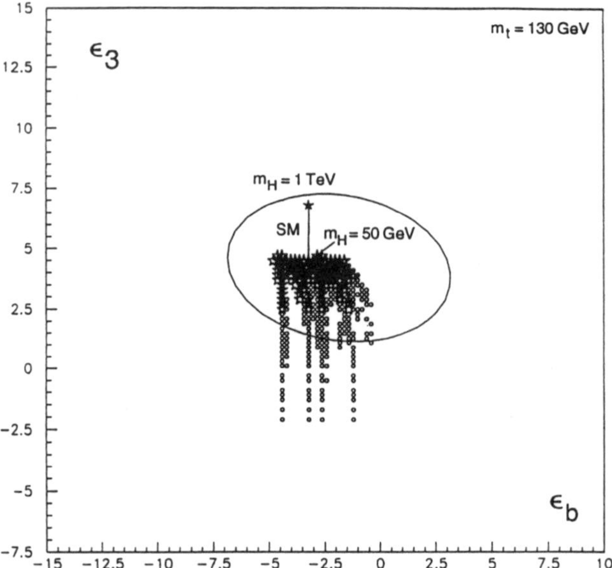


Fig. 21. The same as fig.20, but for $\varepsilon_3$ vs $\varepsilon_b$.

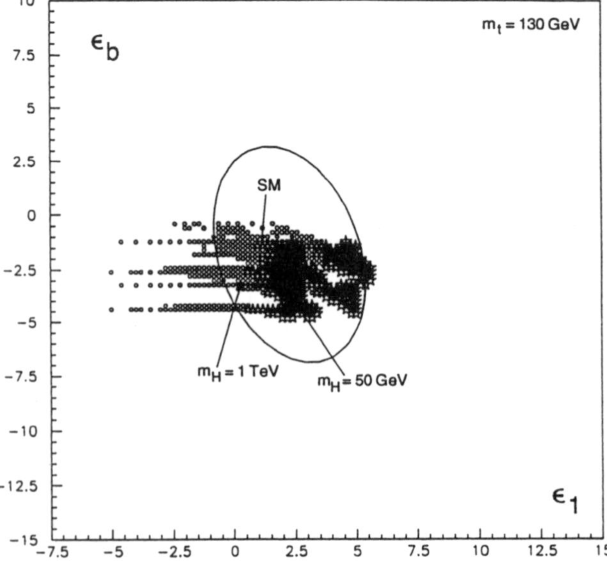


Fig. 22. The same as fig.20, but for $\varepsilon_b$ vs $\varepsilon_1$.

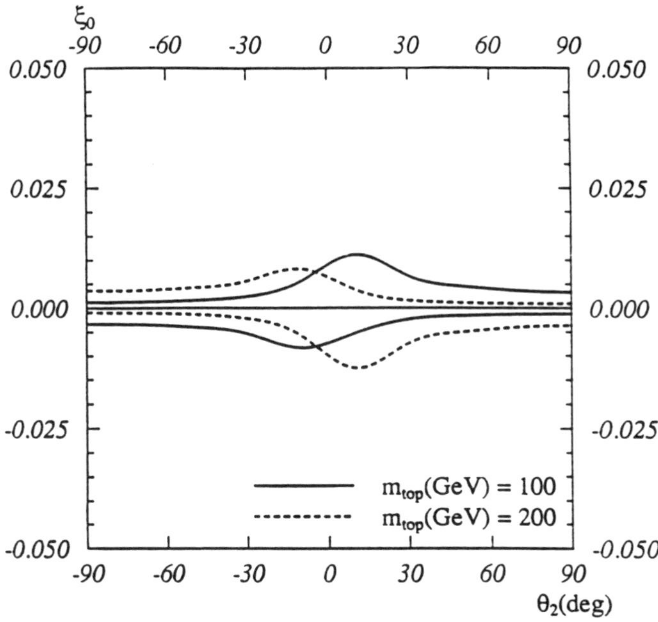


Fig. 23. Allowed range (90% c.l.) for $\xi$ versus $\theta_2$, for $m_H$ =100 GeV, $\alpha_s$=0.118 (from ref.52).

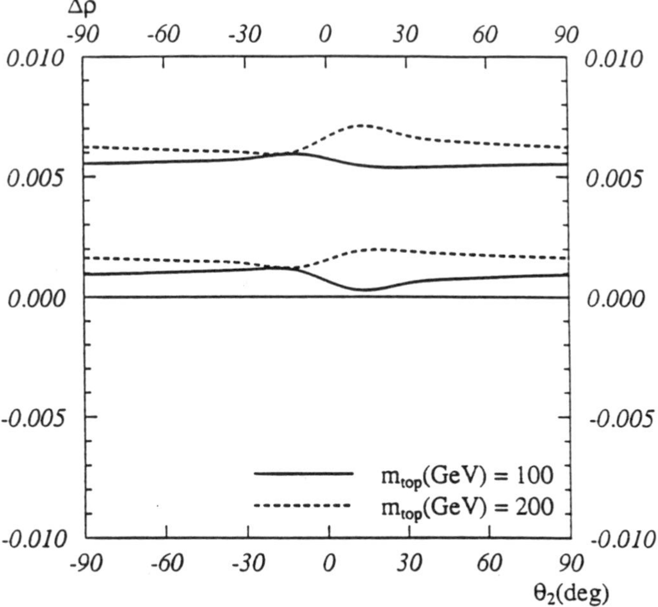


Fig. 24. Allowed range (90% c.l.) for $\Delta\rho$ versus $\theta_2$, for $m_H$ =100 GeV, $\alpha_s$=0.118 (from ref.52).

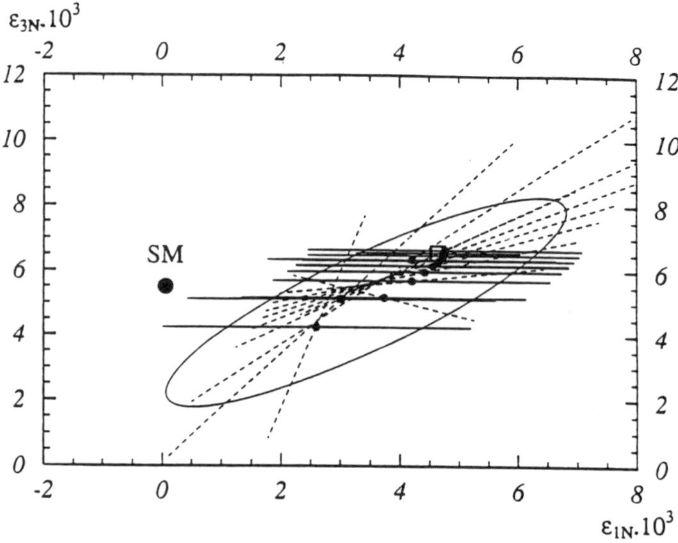


Fig. 25. In the $\varepsilon_{N1}$, $\varepsilon_{N3}$ plane the $1\sigma$ ellipse corresponding to the experimental data on $m_W/m_Z$, $\Gamma_l$ and the ratio $g_V/g_A$ obtained by combining all the asymmetries is shown together with the Standard Model point . The set of dots correspond to the best fit for different $\theta_2$ values. The box symbol corresponds to the LR model.The solid lines, for each dot, show the $1\sigma$ variation due to $\Delta\rho$, while the dashed lines span $1\sigma$ in $\xi$. Here we have assumed $m_H$=100 GeV, $\alpha_s$ =0.118, with $m_t$=100 GeV (from ref.52).

models we consider, all the couplings are family independent. Thus, whatever correction to the $Z \to b\bar{b}$ vertex would also affect all down quark vertices by the same amount . Since the accuracy of the data on $\Gamma_b$ is relatively modest, we cannot adjust the value of $\Gamma_b$ without at the same time spoil the agreement with the other hadronic observables.

In summary, the data collected at LEP on the Z peak impose very severe constraints on the mixings of a non standard Z'. At present, the Standard Model is in such good agreement with the data that no indication for additional corrections is found, at least for unspecified $m_t$. In fact, only a very small amount of mixing is allowed, with $\xi$ always less than 1%.

## 4.Conclusion

With the progress of the LEP programme the electroweak precision tests have become extremely constraining. There is at the moment perfect agreement with the Standard Model. A quite restricted range of $m_t$ is indicated by the data: for example, if we take the central value obtained in sect.1 for m=300 GeV, and we enlarge the error so as to include the uncertainty on $m_H$ (a procedure of merely indicative value), we get

$$m_t = 162 \pm 27 \text{ GeV} \tag{10}$$

Similarly, tight bounds on all conceivable forms of new physics are also obtained. We have considered a few particularly important examples. Technicolour theories are not in better shape now in spite of the fact that the $\varepsilon_3$ deficit has disappeared from the data, because of the smaller errors and the problem with $\varepsilon_b$. The MSSM is the most hard to beat model of new physics because it only makes the Standard Model more stable and robust, without disrupting its structure. It predicts corrections that are in most cases smaller than present errors, unless some of the spartners are very light. In models with extended gauge structure the amount of allowed mixing of the observed Z with the non standard component must be below 1%.

In the '93 and '94 runs of LEP one aims at collecting ~50 $pb^{-1}$ of additional integrated luminosity per experiment with resonance scanning. The errors on the epsilons will then go down by about a factor of two. Another big step in probing the Standard Model is ahead of us !

Acknowledgements. It is pleasure for me to thank R. Casalbuoni, F. Caravaglios, S. De Curtis, M. Martinez, M. Pepe-Altarelli for their qualified help in preparing the material for this talk, and R. Barbieri, D. Bardin, L. Okun , G. Passarino for important interactions.

**References**

1. S.L. Glashow, *Nucl. Phys.* **22** (1961) 579;
S. Weinberg, *Phys. Rev. Lett.* **19** (1967) 1264;
A. Salam, Proc. 8th Nobel Symposium, Aspenägården, 1968, ed. N. Svartholm (Almqvist and Wiksell, Stockholm, 1968) p. 367.

2. See, for example, the following textbooks:
I. Aitchison and A. Hey, *Gauge theories in particle physics: a practical introduction* (A.~Hilger, Bristol, 1982);
D. Bailin, *Introduction to gauge field theory* (A.~Hilger, Bristol, 1986);
E.D. Commins and P.H. Bucksbaum, *Weak interactions of leptons and quarks*, (Cambridge Univ. Press, Cambridge, 1983);
L.B. Okun, *Leptons and quarks* (North Holland Publ. Co., Amsterdam, 1982).

3. G. Altarelli, Proceedings of the 32nd Int. Winter School, Schladming,Austria, CERN preprint TH.6867/93 (1993).

4. G. Altarelli, R. Kleiss and C. Verzegnassi (eds.), Z Physics at LEP 1 (CERN 89-08, Geneva, 1989), Vols. 13.

5. V. Novikov, L. Okun and M. Vysotsky, *Nucl. Phys.* **B397** (1993) 35; CERN preprint CERN-TH.6696/92, CERN-TH.6855/93 (1993); *Phys. Lett.* **299** (1993) 329; ITEP 104-92 (1992).

6. H. Burkhardt, F. Jegerlehner, G. Penso and C. Verzegnassi, *Z.Phys.* **C43** (1989) 497.

7. F. Jegerlehner, in Proceedings of the 1990 Theoretical Advanced Study Institute in Elementary Particle Physics, ed. by P. Langacker and M. Cvetic (World Scientific, Singapore, 1991), p. 476.

8. B. Kniehl, Proceedings of the EPS Conference on High Energy Physics, Marseille, France, 1993.

9. R. Barbieri, M. Beccaria, P. Ciafaloni, G. Curci and A. Viceré, *Phys. Lett.* **B288** (1992) 95 and CERN preprint CERN-TH.6713/92 (1992);
A. Denner, W. Hollik and B. Lampe, CERN preprint CERN-TH.6874/93 (1993).

10. J. Fleischer, F. Jegerlehner, P. Raczka and O.V. Tarasov, *Phys.Lett.* **B293** (1992) 437;
G. Buchalla and A.J. Buras, Munich preprint MPI-PTh/111-92 (1992), TUM-T31-36/92 (1992);
G. Degrassi, Padua preprint DFPD 93/TH/03 (1993).

11. S. Fanchiotti, B. Kniehl and A. Sirlin, CERN preprint CERN-TH.6749/92 (1992).

12. K.G. Chetyrkin and J.H. Kühn, *Phys. Lett.* **B248** (1990) 359;
K.G. Chetyrkin, J.H. Kühn and A. Kwiatkowski, *Phys. Lett.* **B282** (1992) 221.

13. G. Quast, Proceedings of the EPS Conference on High Energy Physics, Marseille, France, 1993.

14. J. Lefrançois, Proceedings of the EPS Conference on High Energy Physics, Marseille, France, 1993.

15. The LEP Electroweak Working Group, CERN preprint CERN/PPE/93-157 (1993).

16. CDF Collaboration, F. Abe et al., *Phys. Rev. Lett.* **65** (1990) 2243; *Phys. Rev.* **D43** (1991) 2070;
UA2 Collaboration, *Phys. Lett.* **B276** (1992) 354; the value shown in Table 1 is from the Particle Data Group 1992: *Phys. Rev.* **D45** (1992).

17. CHARM Collaboration, J.V. Allaby et al., *Phys. Lett.* **B177** (1986) 446; *Z. Phys.* **C36** (1987) 611;
CDHS Collaboration, H. Abramowicz et al., *Phys. Rev. Lett.* **57** (1986) 298;
A. Blondel et al., *Z. Phys.* **C45** (1990) 361;
CCFR Collaboration, A. Bodek, Proceedings of the EPS Conference on High Energy Physics, Marseille, France, 1993.

18. M.C. Noecker et al., *Phys. Rev. Lett.* **61** (1988) 310;
M. Bouchiat, M.A. Proceedings of the 12th International Atomic Physics Conference (1990).

19. M. Pepe-Altarelli, Proceedings of the Rencontres de la Vallée d'Aoste, La Thuile, Italy, 1993.

20. I thank M. Martinez for providing me with the results of this simplified fit from exactly the same code used for the complete fit presented in Ref. [14].

21. S. Catani, Proceedings of the EPS Conference on High Energy Physics, Marseille, France, 1993.

22. G. Altarelli, in QCD-20 years later, ed. by P.M. Zerwas and H.A. Kastrup, World Scientific 1993;
S. Bethke, Proceedings of the XXVI International Conference on High Energy Physics, Dallas, 1992.

23. M.A.B. Bég et al., *Phys. Rev.* **52** (1984) 883;
D.J. Callaway, *Nucl. Phys.* **B233** (1984) 189;
R. Dashen and H. Neuberger, *Phys. Rev. Lett.* **50** (1983) 189;
K.J. Babu and E. Ma, *Phys. Rev.* **D31** (1984) 2861;
E. Ma, *Phys. Rev.* D31 (1985) 322;
M. Lindner, *Z. Phys.* **C31** (1986) 295;
P. Hasenfratz, *Nucl. Phys.* (Proc. Suppl.) **B9** (1989) 3;
J. Kuti, ibid., p. 55;
H. Neuberger, Proc. Symposium on Lattice field theory, Capri, 1989 (to appear as Nucl. Phys. B. Proc. Suppl. 17, 1990);
M. Lüscher and P. Weisz, *Nucl. Phys.* **B290** (1987) 5; **B295** (1988) 65 and **B318** (1989) 705.

24. M. Martinez, private communication;
V. Novikov et al., *Phys. Lett.* **B308** (1993) 123.

25. D. Treille, Proceedings of the EPS Conference on High Energy Physics, Marseille, France, 1993.

26. G. Altarelli, R. Barbieri and S. Jadach, *Nucl. Phys.* **B369** (1992) 3.

27. G. Altarelli, R. Barbieri and F. Caravaglios, *Nucl. Phys.* **B405** (1993) 3; CERN preprint CERN-TH.6859/93 (1993).

28. M.E. Peskin and T. Takeuchi, *Phys. Rev. Lett.* **65** (1990) 964 and *Phys. Rev.* **D46** (1991) 381.

29. G. Altarelli and R. Barbieri, *Phys. Lett.* **B253** (1990) 161;
B.W. Lynn, M.E. Peskin and R.G. Stuart, SLAC-PUB-3725 (1985); in Physics at LEP, Yellow Book CERN 86-02, Vol. I, p. 90.

30. B. Holdom and J. Terning, *Phys. Lett.* B247 (1990) 88;
D.C. Kennedy and P. Langacker, *Phys. Rev. Lett.* **65** (1990) 2967 and preprint UPR-0467T;
B. Holdom, Fermilab 90/263-T (1990);
W.J. Marciano, as in Ref. [73];
A. Ali and G. Degrassi, DESY preprint DESY 91-035 (1991);
E. Gates and J. Terning, *Phys. Rev. Lett.* **67** (1991) 1840;
E. Ma and P. Roy, *Phys. Rev. Lett.* **68** (1992) 2879;
G. Bhattacharyya, S. Banerjee and P. Roy, *Phys. Rev.* **D45** (1992) 729.

31. M. Golden and L. Randall, *Nucl. Phys.* **B36**1 (1991) 3;
M. Dugan and L. Randall, *Phys. Lett.* **B264** (1991) 154;
A. Dobado et al., *Phys. Lett.* **B255** (1991) 405;
J. Layssac, F.M. Renard and C. Verzegnassi, Preprint UCLA/93/TEP/16 (1993).

32. V. Novikov, L. Okun and M. Vysotksy, *Mod. Phys. Lett.* **A8** (1953) 2529.

33. M. Bilenky et al., preprint BI-TP93/46 (1993);
D. Schildknecht, Proceedings of the EPS Conference on High Energy Physics, Marseille, France, 1993.

34. The SLD Collaboration, SLAC-PUB-6030 (1993);
W. Ash, Proceedings of the EPS Conference on High Energy Physics, Marseille, France, 1993.

35. S. Weinberg, *Phys. Rev.* **D13** (1976) 974 and *Phys. Rev.* **D19** (1979) 1277;
L. Susskind, *Phys. Rev.* **D20** (1979) 2619;

E. Farhi and L. Susskind, *Phys. Rep.* **74** (1981) 277.

36. R. Casalbuoni et al., *Phys. Lett.* **B258** (1991) 161;
R.N. Cahn and M. Suzuki, LBL-30351 (1991);
C. Roisnel and Tran N. Truong, *Phys. Lett.* **B253** (1991) 439;
T. Appelquist and G. Riantaphylou, Yale Univ. preprint YCTP-p49-91 (1991);
T. Appelquist, Proceedings of the Rencontres de la Vallée d'Aoste, La Thuile,Italy, 1993.

37. J. Ellis, G.L. Fogli and E. Lisi, CERN preprint CERN-TH.6383/92 (1992).

38. The CHARM II Collaboration, R. Berger, Proceedings of the EPS Conference on High Energy Physics, Marseille, France, 1993.

39. The CDF Collaboration, A. Barbaro-Galtieri, Proceedings of the EPS Conference on High Energy Physics, Marseille, France, 1993.

40. R.S. Chivukula, S.B. Selipsky and E.H. Simmons, *Phys. Rev. Lett.* **69** (1992) 575.

41. B. Holdom, *Phys. Lett.* **105** (1985) 301;
K. Yamawaki, M. Bando and K. Matumoto, *Phys. Rev. Lett.* **56** (1986) 1335;
V.A. Miransky, *Nuov. Cim.* **90A** (1985);
T. Appelquist, D. Karabali and L.C.R. Wijewardhana, *Phys. Rev.* **D35** (1987) 389; 149;
T. Appelquist and L.C.R. Wijewardhana, *Phys. Rev.* **D35** (1987) 774;
*Phys. Rev* .**D36** (1987) 568.

42. R.S. Chivukula et al., Preprint BUHEP-93-11 (1993).

43. H.P. Nilles, *Phys. Rep.* **C110** (1984) 1;
H.E. Haber and G.L. Kane, *Phys. Rep.* **C117** (1985) 75;
R. Barbieri, *Riv. N. Cim.* **11** (1988) 1.
E. Eliasson, *Phys. Lett.* **147** (1984) 67;
S. Lim et al., *Phys. Rev.* **D29** (1984) 1488;
J. Grifols and J. Sola, *Nucl. Phys* **B253** (1985) 47;
B. Lynn et al., in Physics at LEP, ed.by J. Ellis and R. Peccei, CERN86-02, Vol. 1 (1986);
R. Barbieri et al., *Nucl. Phys.* **B341** (1990) 309;
P. Gosdzinsky and J. Sola, preprint UAB-FT-247/90 (1990);
M. Drees, K. Hagiwara and A. Yamada, DESY preprint DTP/91/34 (1991);
J. Ellis, G. Fogli and E. Lisi, CERN preprint CERN-TH.6642/92 (1992);
J. Lopez et al., Texas Univ. preprint CTP.TAMU/19/93 (1993).

44. R. Barbieri, S. Ferrara and C. Savoy, *Phys. Lett.* **119B** (1982) 343;
P. Nath, R. Arnowitt and A. Chamseddine, *Phys. Rev. Lett.* **49** (1982) 970.

45. G. Altarelli, R. Barbieri and F. Caravaglios, *Phys. Lett.* **B314** (1993) 357.

46. R. Barbieri, F. Caravaglios and M. Frigeni, *Phys. Lett.* **B279** (1992) 169.

47. R. Barbieri and L. Maiani, Nucl. Phys. B224 (1983) 32;
L. Alvarez-Gaumé, J. Polchinski and M. Wise, Nucl. Phys. B221 (1983) 495.

48. W. Hollik, *Mod. Phys. Lett.* **A5** (1990) 1909.

49. A. Djouadi et al., *Nucl. Phys.* **B349** (1991) 48;

M. Boulware and D. Finell, *Phys. Rev.* **D44** (1991) 2054. The sign discrepancy between these two papers appears now to be solved in favour of the second one.

50. G. Altarelli, R. Casalbuoni, S. De Curtis, F. Feruglio and R. Gatto, *Mod. Phys. Lett.* **A5** (1990) 495; *Nucl. Phys.* **B342** (1990) 15;
G. Altarelli, R. Casalbuoni, S. De Curtis, F. Feruglio and R. Gatto, *Phys.Lett.* **B235** (1990) 669;
G. Altarelli, R. Casalbuoni. S. De Curtis, N. Di Bartolomeo, F. Feruglio and R. Gatto, *Phys. Lett.* **B263** (1991) 459;
A. Chiappinelli, *Phys. Lett.* **B263** (1991) 287;
M.C. Gonzalez-Garcia and J.W.F. Valle, *Phys. Lett.* **B259** (1991) 365;
F. Del Aguila, W. Hollik, J.M. Moreno and M. Quiros, *Nucl. Phys.* **B372** (1992) 3;
P. Langacker and M. Luo, *Phys. Rev.* **D45** (1992) 278;
J. Layssac, F.M. Renard and C. Verzagnassi, *Z. Phys.* **C53** (1992) 114;
A. Leike, S. Riemann and T. Riemann, *Phys. Lett.* **B291** (1992) 187;
E. Nardi, E. Roulet and D. Tommasini, *Phys. Rev.* **D46** (1992) 3040.

51. G. Altarelli, R. Casalbuoni, S. De Curtis, N. Di Bartolomeo, R. Gatto and F. Feruglio, CERN preprint CERN-TH.5947/93 (1993).

52. R.W. Robinet and J.L. Rosner, *Phys. Rev.* **D26** (1982) 2396;
E. Witten, Nucl. Phys. B258 (1985) 75;
M. Dine, V. Kaplunovsky, M. Mangano, C. Nappi and N. Seiberg, Nucl. Phys. B259 (1985) 519;
S. Cecotti, J.-P. Derendinger, S. Ferrara, L. Girardello and M. Roncadelli, *Phys. Lett.* **156B** (1985) 318;
J.D. Breit, B.A. Ovrut and G. Segré, *Phys. Lett.* **158B** (1985) 33;
E. Cohen, J. Ellis, K. Enqvist and D.V. Nanopoulos, *Phys. Lett.* **165B** (1985) 76;
J. Ellis, K. Enqvist, D.V. Nanopoulos and F. Zwirner, *Nucl. Phys.* **B276** (1986) 14; *Mod. Phys. Lett.* **A1** (1986) 57;
F. Del. Aguila, G. Blair, M. Daniel and G.G. Ross, *Nucl. Phys.* **B272** (1986) 413;
D. London and J.L. Rosner, *Phys. Rep.* **34** (1986) 1530;
G. Belanger and F. Godfrey, *Phys. Rev.* **D35** (1987) 378;
L. Ibànez and J. Mas, *Nucl. Phys.* **B286** (1987) 107.

53. J.C. Pati and A. Salam, *Phys. Rev.* **D10** (1974) 275;
R.N. Mohapatra and J.C. Pati, *Phys. Rev.* **D11** (1975) 566 and ibid. 2559;
G. Senjanovic and R.N. Mohapatra, *Phys. Rev.* **D12** (1975) 152;
G. Senjanovic, *Nucl. Phys.* **B153** (1979) 334;
R.W. Robinet and J.L. Rosner, *Phys. Rev.* **D25** (1982) 3035;
C.N. Leung and J.L. Rosner, *Phys. Rev.* **D29** (1982) 2132.

**CHAIRMAN: G. Altarelli**

*Scientific Secretaries: M. Giovannini, A. Ianni*

## DISCUSSION

– *Morpurgo:*

1) In the transition of $\alpha^{-1}$ from 137 to $\sim 128$, what is the contribution of the electrons?

2) Then in calculating the contribution of the quarks to the transition, must one also include the effects of the strong interactions, including for instance the creation of $q\bar{q}$ aggregates live pions?

3) What are (approximately) the masses of the light quarks to be used in the calculations?

– *Altarelli:*

1) I have shown that:

$$\frac{1}{\alpha(0)} - \frac{1}{\alpha(t)} = \frac{1}{3\pi} \sum_f Q_f^2 t_f$$

where $t_f = \ln \frac{M^2}{m_f^2}$ $(t = \ln \frac{M^2}{m^2})$. In the case of electrons $t_f$ is $\simeq 24$ over a total of $\simeq 84$.

2) I think that in practice the question is completely bypassed by the method that I described. This method relates the imaginary part of the vacuum polarization to the data of the actual cross section $\sigma$ $(e^+e^- \rightarrow \gamma \rightarrow \text{hadrons})$. So of course you evaluate the imaginary part of the vacuum polarization through the physical hadrons that are produced.

3) But, interestingly enough, the results that you obtain correspond to the results that you would obtain from leptons, quarks ( plus QCD connection ) with values of light quark masses of the order of $100 - 200\, MeV$.

– *Arnowitt:*

The LEP data average for $\alpha_S(M_Z)$ was given as $0.128 \pm 0.008$. How much variation in $\alpha_S(M_Z)$ is there between the different LEP groups?

– *Altarelli:*

Unfortunately I do not have the separate data of each group with me.

– *Gibilisco:*

1) What test can be performed in the future at LEP in order to investigate the way of going beyond the Standard Model?

2) Can you explain to me in a more detailed way the resonant spin depolarization method used to calibrate the energy?

– *Altarelli:*

1) To your first question I can answer that all the LEP program is devoted to test the Standard Model. In the first phase (LEPI) a number of precise electroweak tests are performed on Z decays. In the phase of LEPII the scenario could be completely different because at LEPII we could hopefully be in the situation of producing new particles such as, for example, charginos, neutralinos, charged Higgs and sleptons. One also hopes to find a Higgs particle (of the Standard Model or else). We want also to test, at LEPII, the three gauge bosons vertex.

2) This method is based on the fact that these are electrons with transverse polarization that circulate. One can put a small magnetic field at a certain point of the collider. Every time the electrons pass they get a kick from the magnetic field which tends to decrease the polarisation. One varies the frequency of the magnetic field until the polarisation disappears. By measuring the frequency of depolarization you can measure the period of the electrons on the orbit and their energy.

– *Beneke:*

Since the allowed range for $m_{top}$ is getting narrow, what is the definition of $m_{top}$ used in LEP analysis. And, even if it is too early to worry about this, can you quantify the effect of defining $m_{top}$, say at $m_Z$ or $m_{top}$?

– *Altarelli:*

The definition of $m_{top}$ used in radiative corrections is as the position in the pole in the propagator, i.e. $m_{top}(m_{top})$. $m_{top}$ is so close to $m_Z$ that practically I do not see at the moment any difference using $m_{top}$ at $m_Z$ or $m_{top}$. A different question may be the following: as in the near future we hope to measure the top mass and because we hope to make precision measurements of the top mass at the Tevatron, LHC and SSC, which are the theoretical uncertainties of $m_{top}$ calculated in different processes?

The answer to this question is that various definitions of $m_{top}$ differ by terms of $O(\alpha_S)$. So the accuracy with which you can define top mass depends on the feasible theoretical computations of such correction terms and is, I would say, around a few hundred $MeV$.

- *McPherson:*

1) Can you remind me of the best method for finding $m_W$ at LEPII? Do you scan the machine energy across threshold?

2) Will the experimental error on $m_W$ be competitive with the $\pm 120\, MeV$ you showed from EW fit?

- *Altarelli:*

1) The most promising method in order to measure the mass of the $W$ at LEPII is from the invariant mass of the 2-jets from $W$ decay.

2) The errors needs to be $\leq 100\, MeV$ to be a good test. The claim is that the experimental error will be between $\pm 50\, MeV$ and $\pm 100\, MeV$ with $500\, pb^{-1}$ of integrated luminosity.

- *Khoze:*

I would like to make a comment concerning the accuracy of $W$ mass determination at LEP via the reconstruction of 4 hadrons jets. This problem is not so clear to me at the moment since this accuracy is connected with the structure of the existing Monte Carlo models. I hope the problem will be solved in the near future. We are working on it together with Torbjon Sjostrand from CERN.

- *Anselm:*

Sometime ago several exciting events were reported by the L3 group. I mean the decay of Z-boson into a lepton pair and two photons with a fixed invariant mass of $60\, GeV$. What has happened with these events? Are they still alive?

- *Altarelli:*

No further signals of this kind were found either by L3 scanning in the 1992 data sample or in other experiments. If you take the collected statistics by now and if you compute the QED background you find no problems.

- *Lenti:*

Is it possible to measure $\alpha(M_Z)$ (the QED coupling) directly at LEP?

- *Altarelli:*

It is not possible with the precision that you would need, i.e. $\frac{\delta\ alpha}{\alpha} \simeq 10^{-4}$.

- *Lu:*

1) You showed LEP data on $Z \to b\bar{b}$ decay mode from the point of view of extracting top and Higgs masses. How about using those $b$ hadrons to test predictions of heavy quark symmetry?

2) Have there been any signals for charmed B mesons, $B_c = b\bar{c}$?

– *Altarelli:*

1) You are talking about the use of LEPI as a $b$ factory for the study of $b$ decays and the spectrum of $b$ hadrons. This is currently being done. About $5-6\times 10^6$ Z have been produced by the four experiments all together. The branching ratio of $Z\to b\bar{b}$ is 15%. So you have a number of $b$ quarks which is close to $2\times 10^6$. This number of $b$ is currently being studied. At the moment every experiment has installed vertex detectors and really interesting results are starting to come: not only the measurements of the $b$ lifetime, but also the first results are coming of new hadrons (for example the $\Lambda_b$). Also there is an attempt to separately measure the lifetime of the different hadrons.

2) No, $B_c$ has not been discovered.

– *Hoang:*

1) You mentioned that the model independent lower bound for $m_t$ is $45\,GeV$. It seems to me that this limit comes just from the fact that top is not seen in $Z$ decay. Now, the particle data book proposes $55\,GeV$ as the model independent lower bound. Can you say something about this discrepancy?

2) Not long ago there was a consistency problem, the so called $\tau$- puzzle, where about $2\,\sigma$ inconsistency of the Standard Model was measured. Does this puzzle still exist?

– *Peccei:*

The limit $m_{top} > 55Gev$ is from the W width at CDF. It is not completely model independent but is independent of the top decay modes.

– *Altarelli:*

From time to time there are some $\tau$ problems that are invoked. The most serious one was the problem of universality: is the Fermi coupling for the $\tau$ lepton the same as the Fermi coupling for the muon? The new measurement of $\tau$ mass at Beijing showed that the previous measurement by Delco was off by $3\,\sigma$. That has reduced the discrepancy. Moreover the LEP data do not show the discrepancy.

– *Vassilevskaya:*

At the moment Z-boson decay is a subject of constant experimental interest. On the other side, if the masses of supersymmetric particles are below the mass of Z-boson they are produced in the Z-decay and consequently could be detected at LEP. No signal has been obtained. What resolution on parameters of supersymmetry model can be obtained from the experimental investigation of Z-decay?

– *Altarelli:*

In the Higgs sector you can derive bounds on the masses of supersymmetric Higgs' and also on the parameter $\tan\beta$ which is the ratio of VEV's for Higgs' that give mass to the up and down fermions. By now we know that the SUSY Higgs has to be heavier than $43\,GeV$, and in the plane $\tan\beta$-$m_{Higgs}$ you have a region which is by now forbidden. You can look for sleptons, sneutrinos, neutralinos and then you also get limits in the plane of the parameters M, $\mu$ that are also significant limits. That was done at LEP. At the moment the improving of this limit is very slow and we have to wait for LEPII.

– *Forshaw:*

For the QED radiative corrections, is an all order sum of the leading logarithms performed?

– *Altarelli:*

Yes, leading and next-to-leading logarithms are resummed.

– *Langacker:*

As you know there is a small tendency for values of $\alpha_S(M_Z)$ extrapolated from low energy experiments to be smaller that the LEP value. Recently the lattice Fermilab group has done an intense study of the charmonium spectrum, which also gives a low value with a small error claimed. Would you care to comment?

– *Altarelli:*

I think that in this respect the situation is now better than it used to be a year ago. The determination from $\tau$ decay went up and the central value is now 0.122. On the other hand the value from the lineshape went down. All numbers coming from high energies that I've shown this morning are really on top of the determination with $\tau$. What is true is that D.I.S. is low and also $\Upsilon$ decay is low. But first of all is not so low: the difference between 0.127 and 0.112 is 0.015 which is of the order of the sum of the 2 errors. From the point of view of the experiments I do not see a big problem.

I want to make a comment: in D.I.S. if you put yourself at fixed $x$ and you vary $Q^2$ at some point you start producing heavy quarks ( $c$ and $b$ quarks). Then there is a distortion of the apparent slope in $Q^2$ at fixed $x$ because of the fact that the structure functions change their regime. This is an effect which is not considered. It is very well possible that D.I.S. has some bias. The lattice calculations claim a small theoretical error. I am sceptical at present. But even if, at the moment, we are not in a condition to be so confident in lattice calculation of $\alpha_S$, in the future the lattice can certainly be very useful for precise determination of $\alpha_S$.

– *Montgomery:*

With respect to the determination of $\alpha_s$ from DIS and effect of the thresholds. I recall that charm gives about 30% of the scale breaking at $x$ of about 0.05. To the extent that the data at low $x$ are not the dominant contributions; then the effect of heavy quark thresholds is diminished. I'm considering for example the EMC experiment.

– *Altarelli:*

I am considering the determination by both BCDMS and CCFR which are in a range of energy where this small $x$ is pushed a little bit above. I would like to have a more precise quantification of these questions.

**CHAIRMAN: G. Altarelli**

*Scientific Secretaries: C.Acerbi, F.Piccinini*

## DISCUSSION

– *Morpurgo:*

If the top does not exist could it still be possible to construct a divergence-free theory? This was in some sense stated by an earlier speaker and I would like to know your point of view on this because it certainly contradicts the notion that cancellation of anomalies can take place only if complete families are present.

– *Altarelli:*

First of all there is also another problem. By now we know without doubt that the $b$ quark has weak isospin which is not zero but rather $-1/2$. This we know, for example, from the measurement of $\Gamma_b$ (the vector and axial vector couplings of the $b$ depend on $T_3$). So if $T_3$ were zero we would have got something many more standard deviations away in the measurement of $\Gamma_b$. Similarly for the forward-backward asymmetry of the b quark. So by now, the anomaly appears at a secondary level. The first thing is that the $b$ quark is part of a weak isospin multiplet so in the absence of a partner the gauge symmetry would be spoiled and the renormalizability would be destroyed even without considering the anomaly.

– *Wegrzyn*

You are looking for some four parameters which depend, as you said, more on the leptonic than the hadronic part of the theory. Suppose that these parameters are very small but non-zero when measured with great accuracy. Is it possible that the electroweak sector of the theory is correct and the hadronic one should be changed?

– *Altarelli:*

If I understand correctly you are asking if it is not possible that the leptonic part of the theory is correct but the hadronic sector is wrong. You are afraid that the variables that I define would not reveal a problem in the hadronic sector because they are too much oriented to the leptonic sector. I don't think so. In fact this is why we introduced $\epsilon_b$ and this is why we have considered a global fit including not only the leptonic variables but also the hadronic variables There are no signals in the data for a violation of lepton–quark universality i.e. there is no pronounced new physics effect that affects in an asymmetric way, leptons and quarks to within the present accuracy.

– *Peccei:*

You showed that in the $\epsilon_3 - \epsilon_b$ space there is a deviation of one sigma with respect to the data. If you remember there was a change that came from Dallas in $\epsilon_b$. In that case was the deviation worse or better?

– *Altarelli:*

In Dallas it was better. $\epsilon_b$ was consistent. The error was larger but the central value was smaller. It was perfect.

– *Peccei:*

Thus in the new data there was a tendency away from the SM.

– *Altarelli:*

Yes, it's also true, however, as I said, that the measuring technique has considerably evolved. At Dallas time, only the leptonic signal was considered.

– *Lenti:*

Is it important to measure $A_{FB}$ out of the Z-peak, where its absolute value is greater?

– *Altarelli:*

Yes, it is useful to measure $A_{FB}$ as a function of the energy and this is in fact done. Actually, the common procedure is to make a fit of the cross-section and the angular distribution using a program like Z–FITTER. The cross-section is written down with all the radiative corrections in terms of the effective parameters $g_V$ and $g_A$. By measuring the cross-section as a function of energy and angle they get a fit of $g_V$ and $g_A$ which are the parameters that give the width and asymmetry at the peak. This year they are going to do a scan again so they will collect statistics on the cross-section and angular distribution outside the peak.

– *McPherson:*

On Thursday you pointed out the importance of $Z \to f\overline{f}\ b\overline{b}$ with low energy $b\overline{b}$, for the $\Gamma_h$ measurements. It seems that that the three different methods of B tagging should be very anticorrelated, so if this is a significant effect the three methods should give different answers but they are the same. Doesn't that mean this is not significant?

– *Altarelli:*

Yes, if you remember, I said yesterday that I expect an effect that would be at the most of the order of one sigma if the $b\overline{b}$ pair pass all the cuts although much less energetic than the $b\overline{b}$ pair from direct production. So, in other words, the prediction of the standard model will move up by a fraction of one sigma. This is the maximum impact that I can imagine for such an effect and I think it should be diluted by a factor of 3, so I agree that it is not significant.

*–McPherson:*

As Professor Zichichi pointed out $\chi^2 << 1$ for one degree of freedom is very unlikely. Doesn't this mean the errors are badly overestimated?

*– Altarelli:*

If you have a theory that passes through all central points, then the resulting minimum in the $\chi^2$ would be zero. It is unlikely for that to happen even for a true theory. But if a theory does that, you certainly cannot conclude that the theory is wrong because it passes through all central points.

*– McPherson:*

But I conclude that your sigmas are wrong.

*– Altarelli:*

But that is not our (the theorists) business !!

There is a serious aspect in what you say. It has been repeatedly observed that if you take the various LEP results they tend to underfluctuate from one experiment to another. That means that systematics are dominant. That would be the normal conclusion. In making the average they separate the statistical and systematic error, extract the common systematic error and then combine the residual statistical error, which eventually is smaller than the systematic error. That means that the experimentalists are very conservative. In other words indeed the error appears at face value to be too large in some cases.

*– Trocsanyi:*

During the discussion after the first lecture you said that the results for $\alpha_s$ obtained from DIS and $\Upsilon$–decay on one hand and from line shape measurements on the other hand are compatible. Today, however, you hinted that the difference could be due to an anomaly in $Z \rightarrow b\bar{b}$ vertex. How seriously do you take this latter statement?

*– Altarelli:*

I don't think there is an incompatibility because it's an effect of order one sigma at most. However, it is true that all experiments in DIS tend to give a central value which is less than all experiments on the line shape at LEP. So I was simply pointing out that all measurements of $\alpha_s$ from the electroweak quantities of course, assume the validity of the electroweak theory. In particular, if you find that the central value of $\frac{\Gamma b}{\Gamma_h}$ is larger than in the SM then this effect cannot be explained by $\alpha_s$, because $\alpha_s$ drops in the ratio almost precisely. This simply would indicate a small electroweak problem in the $Z \to b\bar{b}$ vertex. Then I was simply noticing that if this is the case, then the determination of $\alpha_s$ from the line shape should be reduced by a corresponding amount and this is sufficient to bring it down from 0.128 to 0.115. My opinion is that while the $\epsilon_3$ measurement is very much consolidated (I don't expect it to change unless the experimentalists find a bug in their program or something like that) the $\Gamma(Z \to b\bar{b})$ measurement is still at the beginning and so it will evolve. Only I use it as it is given in order to make an application of the proposed method to analyse the data.

– *Ng:*

The $\epsilon$ parameters are correlated; is this correct?

– *Altarelli:*

Yes.

– *Ng:*

Then does this mean that the errors are also correlated?

– *Altarelli:*

Yes, sure.

– *Ng:*

When you put the error on each $\epsilon$ parameter what does the error mean?

– *Altarelli:*

You see that I've drawn ellipses that are not parallel to the co-ordinate axis; they are inclined. This is the effect of the correlation.

– *Ng:*

Are there no problems when you use $\epsilon_3$ for example to exclude some non–standard models? Do the errors on the $\epsilon$ parameters mean 68% of confidence level?

– *Altarelli:*

I'm not giving a confidence level against some model. I used another language. I said that the fact that the central value of $\epsilon_3$ is below the SM , means that models that add corrections to $\epsilon_3$ in the positive direction are disfavoured with respect to the ones that add corrections in the wrong direction.

– *Lopez:*

In the SM, what is the relative importance of "non–oblique" radiative corrections in the predictions for $\epsilon_{1,2,3}$?

– *Altarelli:*

In the SM the oblique corrections become larger if $m_t$ is very heavy, and they dominate. Now, with the present limits on $m_t$ there is not such a pre-eminence of oblique corrections with respect to vertex corrections. Since the $\epsilon$s are very small quantities of order $10^{-3}$ they are sensitive to small details of radiative corrections. So there is no simple recipe to evaluate $\epsilon_{1,2,3}$ in the SM in the mass top range supported by the data.

– *Piccinini:*

You use in your analysis the data corrected for the $\gamma - Z$ interference and (in the case of the electrons) t–channel contributions. These contributions are calculated in the Standard Model. The question is whether all these contributions are well under the accuracy required to determine the $\epsilon$s without any bias and when the errors on the widths are of the order 3 MeV.

– *Altarelli:*

The question is essentially what theoretical precision we have in this game because it's clear that if experiments were to become more accurate than the theoretical precision it would be useless to continue. So it is important to know what the theoretical precision is. Now, I have indicated that the ignorance of the light quark masses or ignorance of the precise value of $\alpha(M_Z)$ is one source of uncertainty which is really serious because it cannot be lowered by more calculations. With time more and more calculations are done; for example, in the last year, people have computed a number of mixed weak–strong corrections. However, the problem of light quark masses is something that cannot be cured by calculations. One would need new measurements. We observed that the error on $\epsilon_3$ is $\pm 0.6 \times 10^{-3}$ which is induced by this ignorance and that the current experimental error is $\pm 3 \times 10^{-3}$ and when it becomes $\pm 1.5 \times 10^{-3}$ it will be a factor of 3 away from the theoretical accuracy. The general opinion is that the present status of knowledge of radiative corrections is adequate for the LEP 100 program. So for the run of this year and for the run of next year there is still space for the experimentalists to improve their measurements without reaching the limit of the present theoretical accuracy.

**CHAIRMAN: G. Altarelli**

*Scientific Secretaries: C.Acerbi, F.Piccinini*

## DISCUSSION

– *Lopez:*

Isn't it premature to say that MSSM with light sparticles fits the LEP data better than the SM?

– *Altarelli:*

I didn't talk of best fits. You don't have to fit anything because the Standard Model works very well. I only tried to point out from a semi-quantitative point of view some features of different models. Some of them go in the wrong direction e.g. Technicolor, some models are indifferent e.g. $Z'$, while some others go at least partly in the right direction.

– *Lopez:*

But there was a stronger statement than that in the paper you presented: "..the fit especially to $\epsilon_3$ and $\epsilon_b$ becomes better than for the Standard Model if the chargino and the scalar top are light, that is close to their experimental limits and within reach of LEP200 ".

– *Altarelli:*

Yes, this is true.

– *Lopez:*

I agree that the central points become better, but if you consider a 95% confidence level the central point could be anywhere in the parameter space.

– *Altarelli:*

Sure. Actually the most likely situation is that Supersymmetry gives results for radiative corrections which are practically indistinguishable from Standard Model with light Higgs.

– *Arnowitt:*

Does the LEP data produce any constraints on the $t - \bar{t}$ condensate scenario as a replacement of the Higgs?

– *Altarelli:*

I don't think so because in that scenario you need a sufficiently heavy top, let us say 200 GeV, and LEP doesn't exclude this value. The apparent Higgs would be presumably a little bit heavier, let us say 250 – 300 GeV and this is perfectly compatible with the Standard Model. Actually the main objection to this scenario is that at low energies it is indistinguishable from the Standard Model.

– *Hoang:*

You explained in detail at the beginning of your talk that expanding $\pi$ (vacuum polarisation) needs also the second term of order $q^2$. I'm confused because you have extracted only six form factors out of the four vacuum polarisations $\pi_{WW}, \pi_{ZZ}, \pi_{\gamma\gamma}$ and$\pi_{\gamma Z}$. I think there should be eight. Can you comment on this point?

– *Altarelli:*

If you choose a physical gauge, the $\gamma Z$ and the $\gamma\gamma$ polarisation diagrams start from $q^2$ and their values at $q^2 = 0$ are zero. Then you have two for $\pi_{WW}$, two for $\pi_{ZZ}$ , one for $\pi_{\gamma\gamma}$ and one for $\pi_{\gamma Z}$. Six.

– *Hoang:*

You told us that by pushing all SUSY particles to high scales you can obtain the Standard Model. I'm confused about that because we learned that Higgs self-couplings are proportional to their masses, so that decoupling does not occur if the Higgs masses are made equal to infinity.

– *Altarelli:*

This is a misunderstanding. The Higgs is not a spartner. In the MSSM you have a Higgs structure made by two doublets and there is a constraint that the lightest Higgs is below or nearby the $Z$. When I'm sending the spartners to $\infty$ I'm not sending the Higgs to $\infty$.

– *Gibilisco:*

1) What could be the decay channels of the heavy $Z'$ boson and what decays could be observable at LEP200?

2) How could a very heavy $Z'$ affect other decays, for instance those appearing in penguin diagrams?

– *Altarelli:*

1) I don't think there is any chance of producing a $Z'$ at LEP200. First of all because of the CDF limit and also because most of the models get constraints from LEP data which put the mass of the $Z'$ boson above at least several hundred GeV.

2) Maybe we will have indirect evidence of $Z'$ in fermion-antifermion or $WW'$ channels. But you have to consider that the LEP200 accuracy is much lower than LEP100. They will only have 5000 $W$'s. If you want evidence for $Z'$ at LEP200

you will have to be particularly lucky. Of course it is possible to find it at hadron colliders (LHC, SSC) where the energies are much higher.

– *Werner:*

Suppose a Higgs is found at LEP200 or LHC. How do we know whether this is the Standard Model or the MSSM one?

– *Altarelli:*

I think it could be difficult to decide if it is Standard or non-Standard unless you find charged Higgs or candidates for sparticles.

– *Abu Leil:*

1) You have mentioned that at LEP200 there will be about 5000 pairs in 4 years running. Can you please tell me whether LEP200 could investigate the anomalous couplings at the $WW\gamma$ –$WWZ$ vertex?

2) Is there any possibility of doing that at the quartic vertex $WW\gamma\gamma$ and $WWZZ$?

– *Altarelli:*

1) Yes, it will be possible and this is one main part of the program. There has been in the last few months a discussion about the possibility of probing this vertex at LEP200. There is a paper (de Rujula et al.) which correctly points out that all modifications must preserve the $SU(2)\bigotimes U(1)$ symmetry. From this statement they initially deduced that this vertex is so constrained that you cannot expect to get any new constraint from LEP200. But this statement was not exact, as stated by the authors and others, and actually LEP200 will give us useful information.

2) I don't think there is any possibility because they are high order effects in the weak couplings.

– *Lu:*

We understand little about Physics at the Planck scale because it is so much higher than the electroweak scale that there are virtually no effects. However, in cases where decoupling theories do not hold there may be some hope in understanding ultra–high energy physics. Since you mentioned violation of the decoupling theorem in your lecture could you be more specific by giving an example?

– *Altarelli:*

The hierarchy problem mentioned by Professor Arnowitt in his lecture, is related to the Planck mass, which is very high. This scale implies constraints on low energy physics.The real problem is not of a mathematical nature but of naturalness. If you consider the procedure of regularisation as a temporary trick you apply to get rid of divergencies then one hopes to find a mechanism that makes the cut-off a physical entity. If you assume the Standard Model valid up to the Planck mass you have to explain why the $W$ mass is so small compared to the Planck scale.

– *McPherson:*

In the graph you showed of $\epsilon_i$ versus $m_t$ showing the LEP data and a best fit (A) and extreme fit (B) from the MSSM, was the same global fit used for all $\epsilon_i$?

– *Altarelli:*

I don't want to talk about best fits. Among the points shown in the scatter plot, we picked up by eye those very close to the ellipse of the data and those which were far from that. The lines A and B refer to the same set of parameters in all figures.

# Some Recent Experimental Results from Fermilab

**H.E.Montgomery**

Fermilab
P.O.Box 500
Batavia, IL 60510

February 15, 1994

**Abstract**

A selection of recent experimental results from experiments at Fermilab is presented and discussed. In some cases the information is updated from that available at the time of the presentation of the talk.

*Lecture given at the International School of Sub-nuclear Physics, Erice, Trapani, Sicily, Italy, July, 1993.*

**Introduction, Accelerator, Experiments**

This lecture in contrast to many of the others at this school is a presentation of data. The Tevatron accelerator at Fermilab has been running for physics in both fixed target and collider modes, most recently the latter. The breadth of the results from the different experiments is very impressive and in order to give this talk I have been forced to be selective. Nevertheless, I have chosen to give many samples rather than just a few results with in depth discussion. In this way it is hoped that an impression of the potential and the contributions to the advancement of the science are given.

In 1992-3 the Tevatron operated in collider mode with 900 GeV anti-protons and 900 GeV protons. The accelerator surpassed many records for anti-proton production and for luminosity, exceeding 9 $10^{30}$ $cm^{-2}.\ sec^{-1}$ in peak luminosity. The integrated luminosity delivered to the experiments was more than 30 $pb^{-1}$, CDF was able to log data corresponding to 21 $pb^{-1}$ and DØ, the new Tevatron collider experiment, 16 $pb^{-1}$. The difference between the two is largely accounted for by the fact that the conventional Main Ring accelerator, which is used to produce anti-protons, actually passes through the outer part of the DØ calorimeter. Thus far this has been handled primarily by gating off the experiment when beam in the Main Ring is passing.

While the talk will cover many experiments I would like to illustrate the actual apparati with just four examples. First, in Fig 1, the E687, photoproduction experimental apparatus is shown. This is typical of the "800 GeV" program fixed target experiments. It is many tens of metres long and as in all cases where the study of heavy quarks is concerned there is an elaborate silicon micro-strip detector, which has sufficient tracking precision to measure the secondary vertices associated with the decays of charm and bottom particles.

A contrast is offered by the E760 apparatus, Fig 2., which is installed in the anti-proton accumulator where it uses low energy anti-protons and a hydrogen gas jet to study charmonium states in formation. The technique evades the restriction in quantum numbers associated with electron positron formation experiments and bore fruit with the first observation[1] of the $^1P_1$, $c\bar{c}$ state a year or more ago. Given the low energy, the apparatus is compact, but the asymmetric kinematics mean it is somewhat different from a low energy collider experiment.

The largest experiments at Fermilab are the two Tevatron collider experiments, DØ, shown in Fig. 3. and CDF in Fig. 4. DØ is new, it operated for the first time in the 1992-3 collider run. It has a hermetic calorimeter, good muon coverage and a compact non-magnetic central tracking system. In contrast, CDF has magnetic tracking and the new feature of the detector for the recent run was the addition of a 45000 channel silicon microstrip barrel. This introduced precision determination of heavy quark decay vertices to high energy hadron colliders.

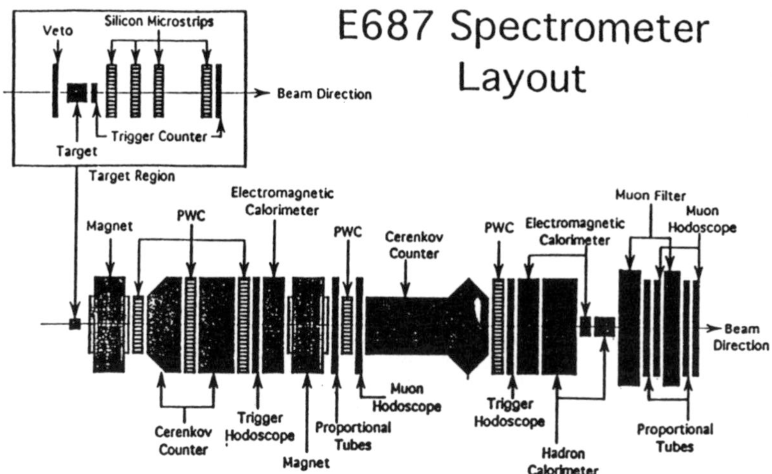


Figure 1: The E687 Apparatus.

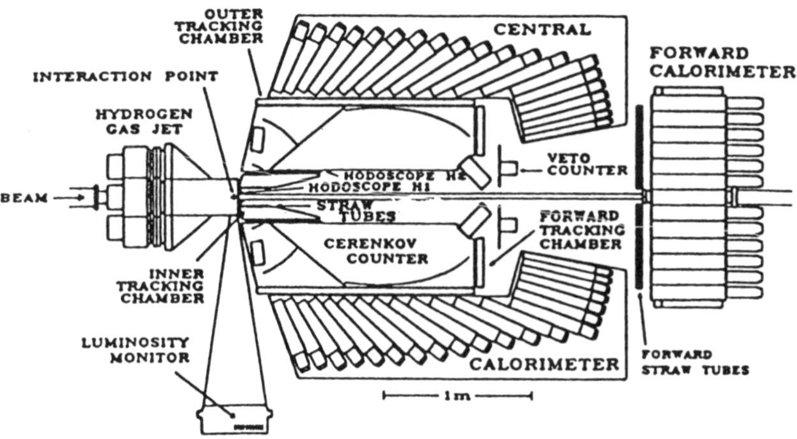


Figure 2: The E760 Apparatus.

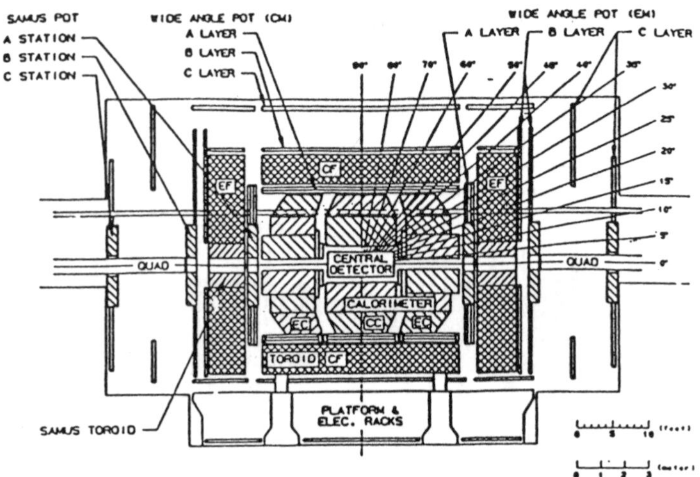


Figure 3: The DØ Experiment Apparatus.

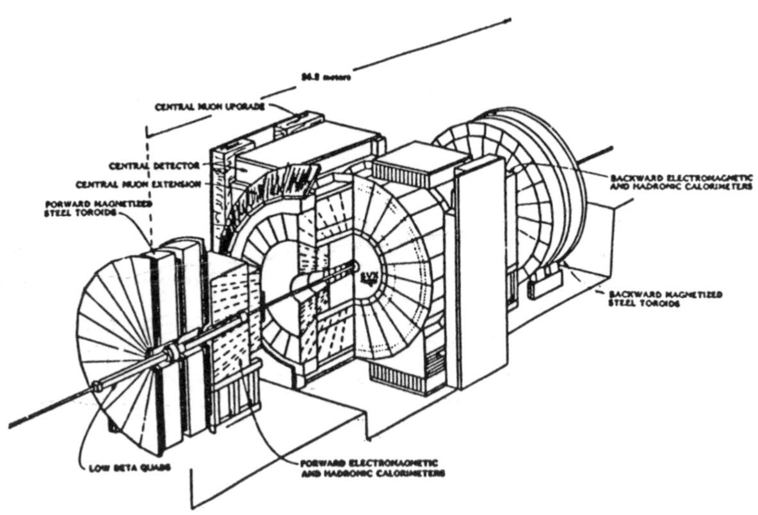


Figure 4: The CDF Experiment Apparatus.

**Cross-Sections, QCD Measurements**

Because of the possibility that in certain circumstances they can interact directly with partons, photons figure prominently as probes of the strong interaction.

Real photon interactions contain both soft interactions, where the photon behaves as a hadronic vector meson, and hard interactions involving the photon and individual partons. Jets are produced through a QCD Compton process and through the photon gluon fusion graph. In addition, there may be higher order resolved photon process. Typically these components lead to one, two and three jets respectively, in addition to the target recoil jet. The simplest parton diagrams produce no beam spectator jet, whereas the resolved photon process does. The expected topology with two recoiling high $p_T$ jets is observed by E683[2], working with a 250 GeV broad band photon beam, in the azimuthal angle between jets shown in Fig. 5a. The forward calorimeter energy distribution shown in Fig. 5b is well described by the expected proportions of these different processes.

Using the transverse energy distribution of resolved forward jets in deep inelastic scattering of muons, E665 has determined[3] the strong coupling constant, $\alpha_S$, see Fig 6a. As with real photons there are kinematic regions where muon scattering exhibits a hadron like component of the virtual photon. Studies[4] of the hadronic final state in this *shadowing* region have so far shed little light on the question of which component of the final state is being shadowed. E665 has extended its jet studies to nuclear targets and they show preliminary evidence[5] that the muli-jet final states are more shadowed than the single parton jet final states. Figure 6 b) shows the ratio of yields of events of different topologies compared to that for the elementary deuteron target as a function of $x_{Bj}$. $x_{Bj} < 0.01$ is where shadowing is observed in the total virtual photon cross-section in the same data[6].

Single photons in the final state can be produced through QCD compton scattering, quark-antiquark annihilation and quark bremsstrahlung. The experimental challenge is to distinguish single photons from the decay photons from $\pi^0$ s. E706 working with a 530 GeV beam has measured[7] single photon cross-sections as a function of $p_T$ for incident pions and protons. The results are shown in Fig 7. The data can be described by next to leading order QCD calculations and are of sufficient precision to distinguish different parton distribution sets. However, at next to leading order the predictions still show a dependence on the definition of scale. The collider experiments are attempting to extend their measurements to large $\eta$. The sensitivity to the calculations then extends into low effective $x_{Bj}$, where the calculations are not well tested. Figure 8 shows the ratio between the measured[8] yields in two different ranges of $\eta$ from CDF. It appears as though the theoretical calculations have the wrong trend at low $p_T$.

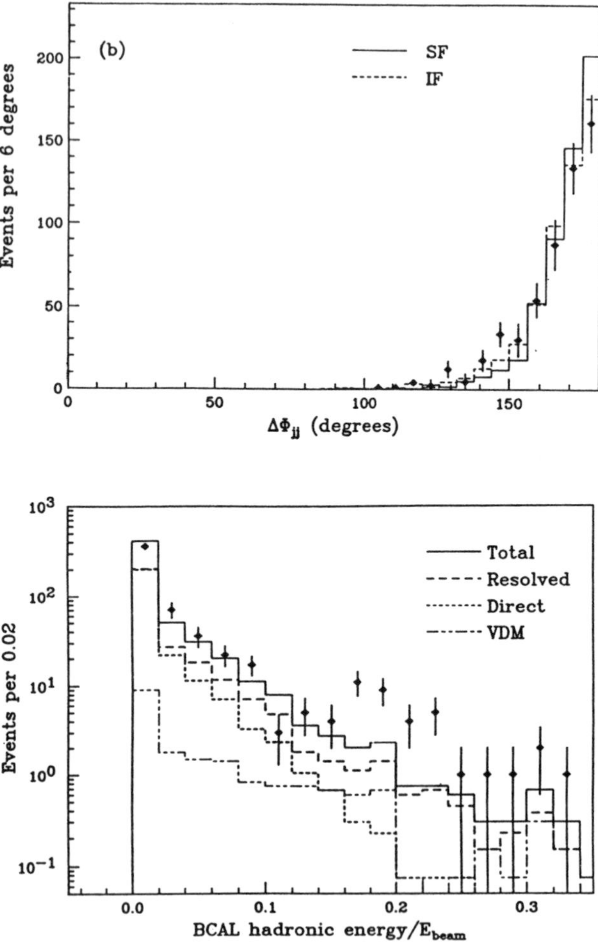


Figure 5: E683 measurements of a) the azimuthal separation between jets in photon interactions, b) the observed energy deposition in a small angle calorimeter compared with the various components of the photon interaction.

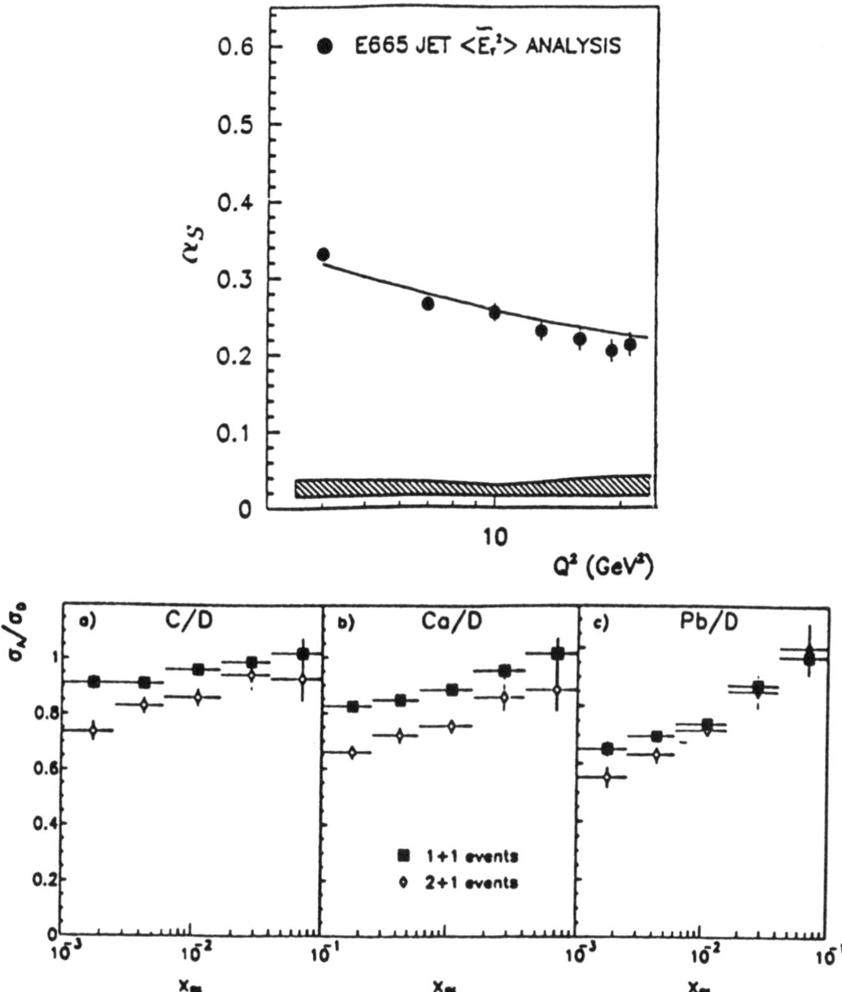


Figure 6: a) $\alpha_S$ as a function of $Q^2$ in muon scattering, b) the relative yields of 2+1 and 1+1 jet topologies in deep inelastic muon scattering, as a function of $x_{Bj}$ from different nuclear targets.

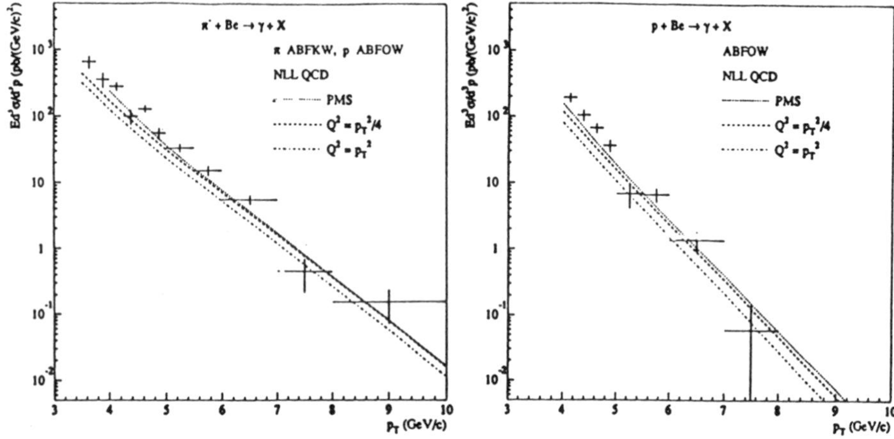


Figure 7: E706 single photon cross-sections compared with a variety of QCD calculations, a) for $\pi^-$, b) for protons.

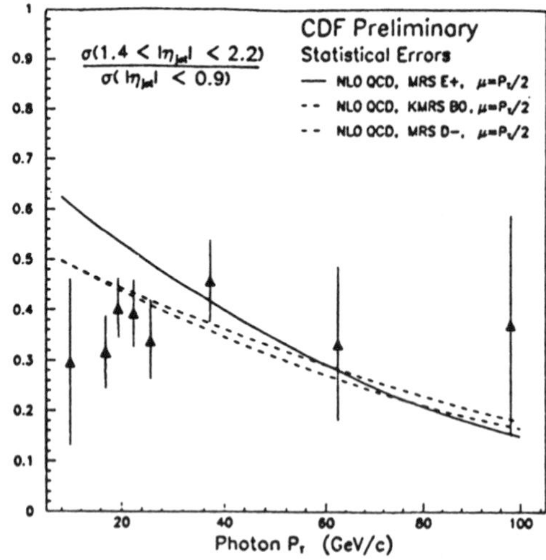


Figure 8: CDF, ratio of single photon cross-section in different $\eta$ ranges.

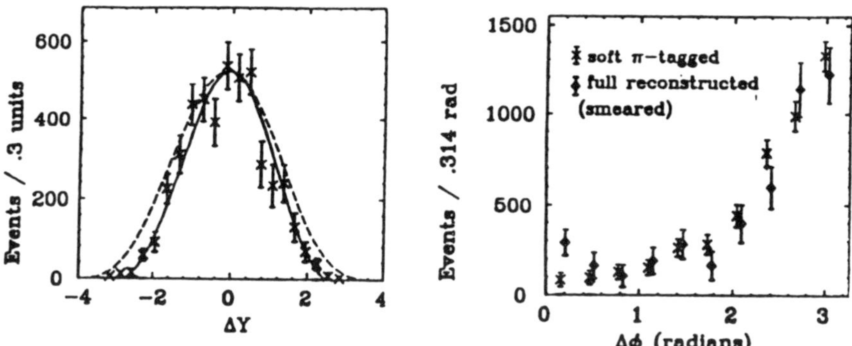


Figure 9: E687 measurements of charm pair production, a) the rapidity difference between pairs, b) the azimuthal angle difference between pairs.

The production cross-sections for mesons containing heavy quarks, charm and bottom, are expected to be amenable to perturbative QCD description. One of the generally accepted tests of this hypothesis is that the cross-section for production of charm mesons should be independent of atomic number, exhibiting no shadowing. Until recently there has been some inconsistency between different data sets. E769 [9] has studied the issue extensively. Their data shows no evidence for shadowing and now dominates the world data. The cross-section per nucleus is linear with atomic number at the few % level. The $x_F$ dependence shows[10] little difference when the produced meson contains quarks in common with the beam particle compared to when it does not. There is no evidence for a strong leading particle effect in contrast to earlier data. Recently there are measurements[11] which show modest differences when ratios of cross-sections are constructed as a function of $x_F$.

E687 has sufficient charm meson pairs to examine the production mechanism for photoproduction of charm and finds that the data[12] exhibit, see Fig. 9, the correlations between the charm pairs expected of the photon-gluon fusion mechanism. They use both fully reconstructed mesons and mesons in which a $D^*$ is tagged by a soft charged pion.

Within the last year, two fixed target experiments have reported cross-sections for bottom production. One, E672[14], bases the measurement on the rate of detached $J/\psi$ vertices. E653[13] used a hybrid emulsion spectrometer and attempted to fully reconstruct the bottom hadron decays. Based on nine pairs of bottom particles reconstructed they are able to look at

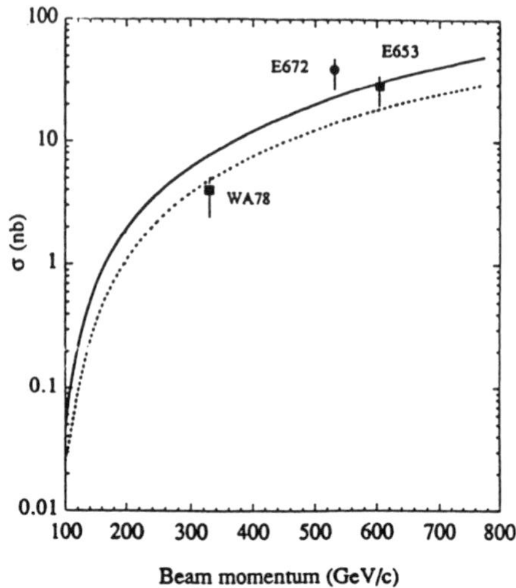


Figure 10: Bottom hadron production cross-section measurements at low energies.

the $x_F$ distribution as well as the cross-section. The results from both experiments are shown in Fig. 10 compared with the theoretical expectations.

In collider experiments the signal to background for heavy quark production is much more favorable. The semi-leptonic decays of charm mesons are characterized by a relatively low $p_T$ muon. As the $p_T$ rises, but the muon is still within a jet, bottom production becomes the dominant process. Fig 11 shows recent measurements[15] from DØ in which the different components of the single muon cross-section as predicted by theory are also shown. The agreement is good.

CDF, with the silicon vertex detector and a magnetic field, are able to reconstruct, partially or fully, a limited number of bottom particles. They also deduce the bottom cross-section based on the yield of detached $J/\psi$ particles. The results[16] are shown in Fig. 12. The measurements appear to be high compared to the theoretical expectations. This is a possible area of disagreement between the two collider experiments which has not been resolved.

As mentioned earlier, there is a premium on measurements with a wide range of $\eta$ when the goal is to test QCD. DØ has a rather good calorimetric coverage and in Fig. 13, the jet cross-section[17] as a function of $p_T$ is shown for two different ranges out to $\eta = 3.0$ and extending in $p_T$ to several hundred GeV. This $\eta$ coverage is parlayed into a measurement of

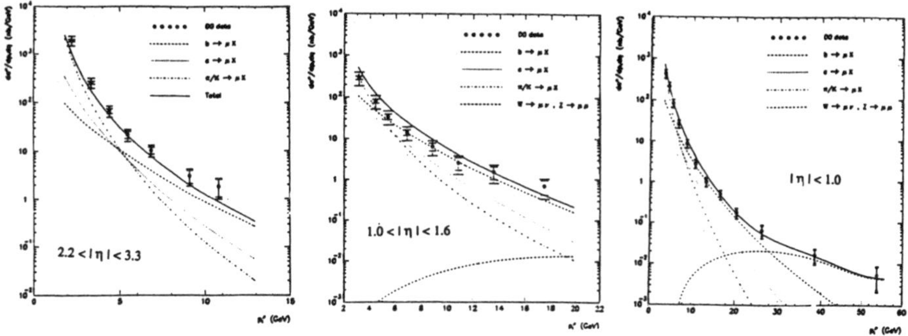


Figure 11: Measurement of the single muon cross-section at 1800 GeV from DØ.

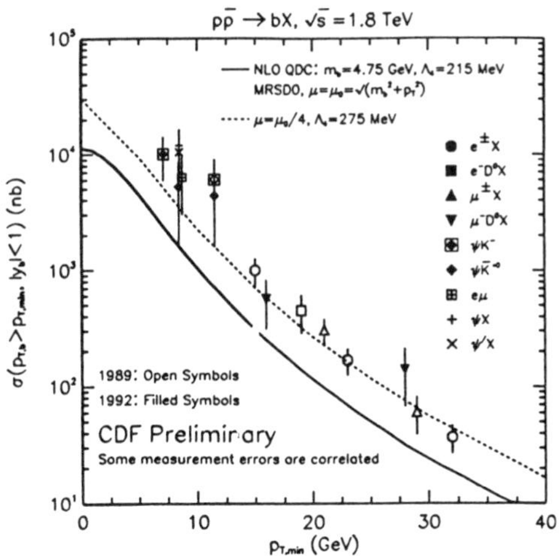


Figure 12: CDF B cross-section measurement.

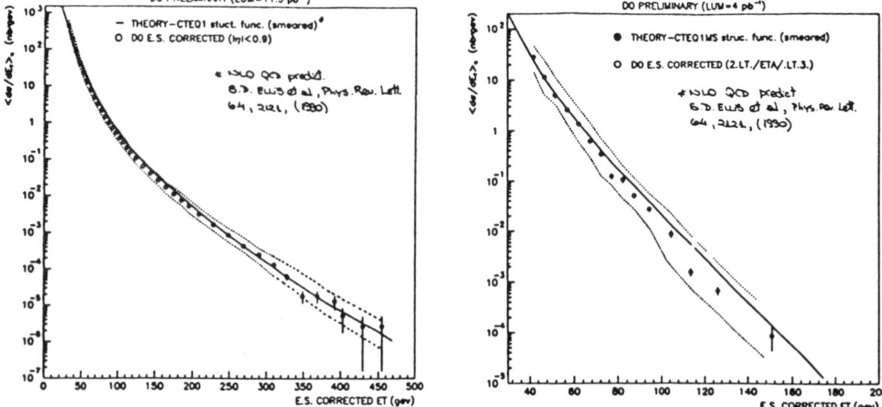


Figure 13: Jet production cross-sections in two different $\eta$ ranges from DØ.

the dijet angular distribution as shown in Fig. 14. If the underlying process were Rutherford scattering, the $\chi = (1 + cos\theta^*)/(1 - cos\theta^*)$, where $\theta^*$ is the parton center of mass scattering angle, distribution shown would be flat. Scale breaking in the parton distributions lead to the slope from low to high $\chi$. These data are a dramatic extension of the measured range.

Recently an old topic has been revitalised and new experimental results are appearing. In the 1970s rapidity gaps were a signal of the exchange of a pomeron. With the current theoretical understanding the more general case of the exchange of any colorless object has been considered[19] and a search has been executed[20] by DØ. They find that if they plot the probability of finding *nothing* between two jets separated by some distance $\Delta\eta$ in pseudo-rapidity, then this probability decreases over a range $\Delta\eta \simeq 2$ and then reaches a plateau. As shown in Fig. 15, for 30 GeV jets and where the threshold for *something* is defined by towers of 200 MeV in the electromagnetic calorimeter, the plateau probability is a few per thousand. The characteristic of the plot is very similar to that expected for a pomeron (a colorless combination of gluons) exchange.

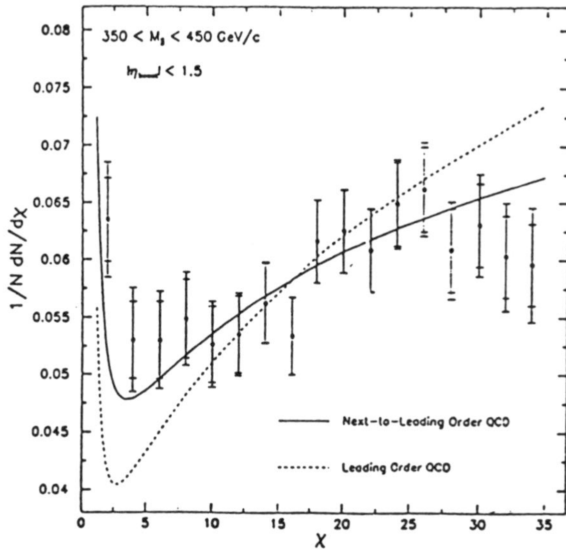


Figure 14: Dijet angular distributions.

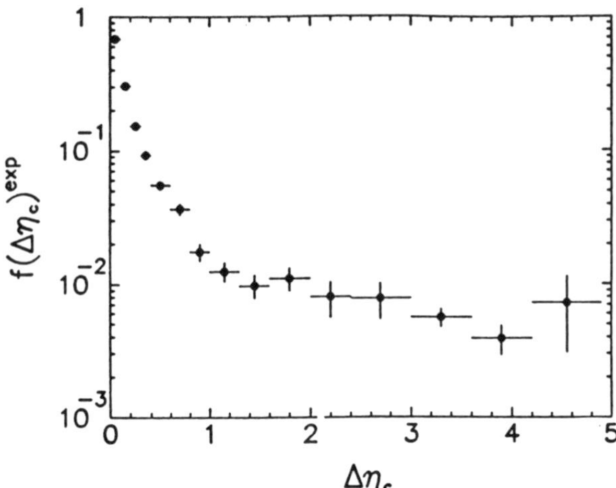


Figure 15: Fraction of events that have a tagged particle between the two leading jets as a fraction of their separation in pseudo-rapidity.

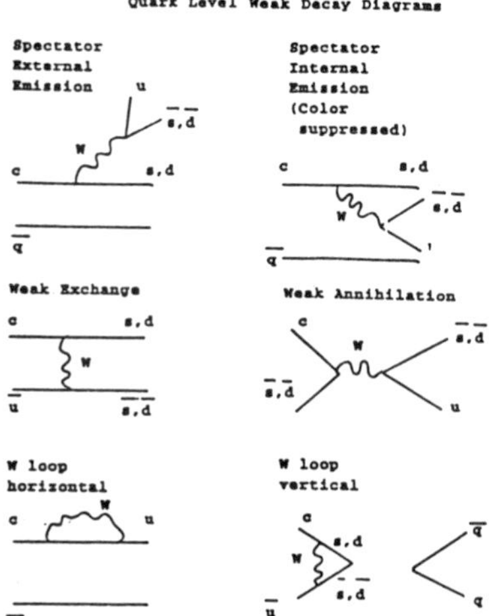


Figure 16: Quark level weak decay diagrams

## Decay Physics

In the previous section we have referred to heavy quark production as a means of obtaining a handle on some of the more rigorous calculations of perturbative QCD. In the following we turn to the physics associated with the characteristics of the particular heavy flavor and its couplings to other flavors, the domain of the Cabibbo-Kobayashi-Maskawa(CKM) matrix. At least that will be the goal. As we shall see effects of the strong interaction are difficult to escape. Much of the content of this section comes from recent reviews[21][22].

First we consider charm decays, some relevant diagrams are shown in Fig. 16. The simplest are the spectator diagrams and the expectation is that as the mass of the heavy quark increases from strange to charm and to bottom, these diagrams should increasingly dominate. The understanding is that the semi-leptonic decays of the charmed mesons $D^0$ and $D^+$ are equal but that complications in the hadronic decays lead to an enhancement of those of the $D^0$ by a factor 2 and hence make the $D^0$ lifetime correspondingly shorter. The $D_s^+$ has a similar lifetime to that of the $D^0$. The data of Fig. 17, from E687, show the change in the fractions of $D^+$ and $D_s^+$ as the cut on the significance of the detachment of the decay vertex from the primary vertex is increased. The lifetime results for these cases suggest that the W exchange graph is not the dominant source of the charged-neutral difference. Rather, strong interaction interference effects are playing a big role.

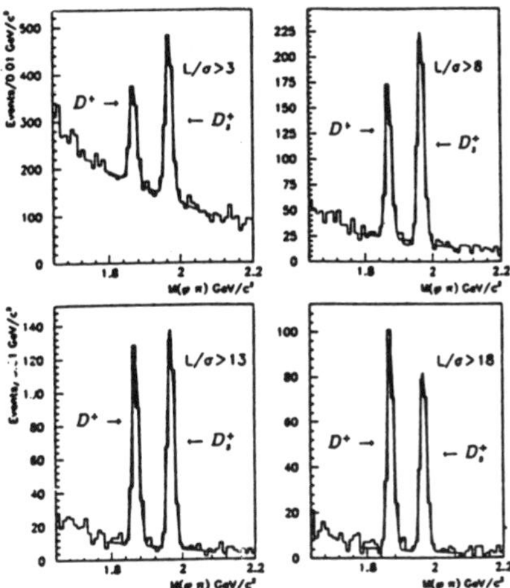


Figure 17: $D^+$ and $D_s^+ \rightarrow \phi\pi$ for increasing $L/\sigma$.

The charm baryons have lagged behind the charm mesons both in terms of number detected and in understanding. E687 are building an impressive array of results on this subject and one can compare the results for the lifetimes of the $\Lambda_c$, the $\Omega_c$ and the $\Xi_c$. The competing models[23][24] differ slightly in that the former would expect a hierarchy $\tau(\Omega_c) < \tau(\Xi^0{}_c) < \tau(\Lambda^+{}_c) \simeq \tau(\Xi^+{}_c)$ while the latter predicts $\tau(\Omega_c) \simeq \tau(\Xi^0{}_c) < \tau(\Lambda^+{}_c) < \tau(\Xi^+{}_c)$. The data have $\tau(\Xi^0{}_c)/\tau(\Lambda^+{}_c) = 0.47 \pm 0.12$ from E687, which favours the Guberina et al.[24] prediction at the level of $2\sigma$ but which is not yet conclusive. Figure 18 shows the relevant mass peaks in the data from which the determinations are made. In contrast to the meson case, the W exchange is not expected to be helicity suppressed, and within the models, for some baryons, it plays a more significant role.

The B hadron lifetime has been measured by CDF[25] using their new silicon vertex detector to measure inclusive $J/\psi$ decay lengths. Figure 19 shows the proper length associated with the observed $J/\psi$ mass peak. It contains a component of prompt production with a symmetric distribution about zero. A background is determined from the sidebands, which is attributed to dimuons from the decay of two B mesons, or at least due to B meson decays, which results in a finite positive lifetime. There is also the real signal from $B \rightarrow J/\psi$. A combined fit, taking into account the scaling to the full B meson momentum, results in the measurement $\tau(B) = 1.46 \pm 0.06(stat) \pm 0.06(syst)ps$.

CDF have also detected a number of B meson exclusive decay modes involving the $J/\psi$ and by comparing $J/\psi K^+$ with $J/\psi K^{*0}$ they obtain a ratio between the lifetimes for charged

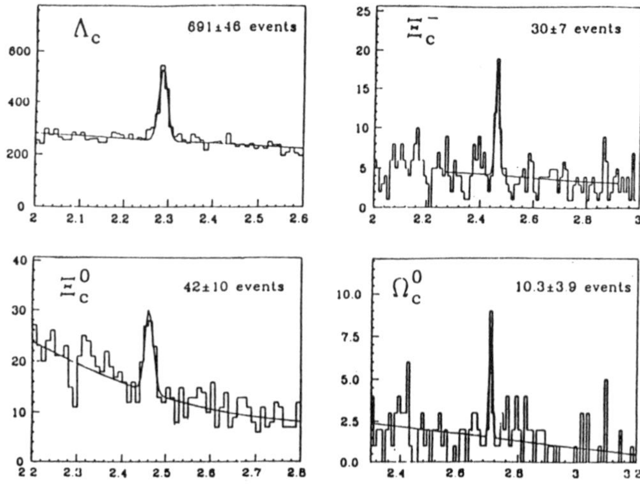


Figure 18: Charmed baryon mass peaks from E687.

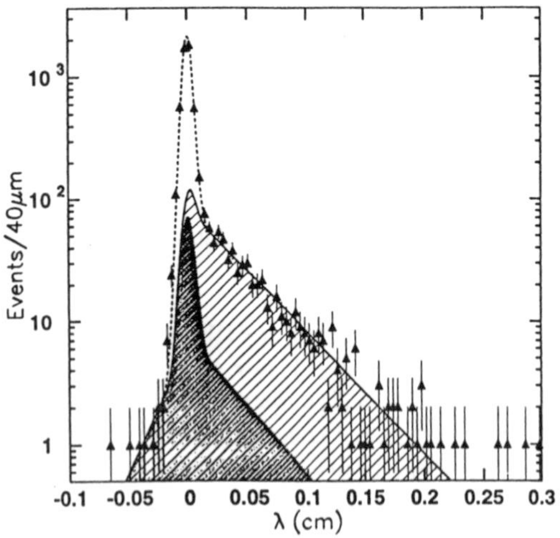


Figure 19: Proper decay length, $B \to J/\psi$.

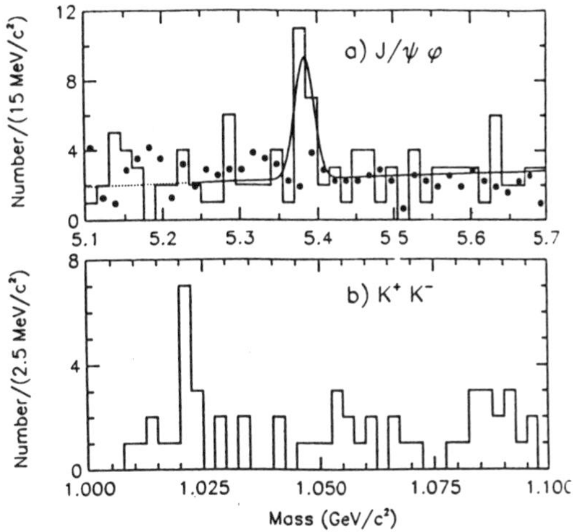


Figure 20: Mass peak for a) $B_s$ and b) $\phi$ in the decay.

and neutral B mesons of $\tau(B^+)/\tau(B^0) = 1.06 \pm 0.15(stat) \pm 0.05(syst)$; this result is still preliminary. Finally, CDF has observed[26] the charmed strange meson through the decay $B_s \to J/\psi\phi$; the mass peak for the $B_s$ and for the $\phi$ are shown in Fig. 20. The mass value they quote[26] is $m_{B_s} = 5383.3 \pm 4.5(stat) \pm 5.0(syst) MeV$ and the lifetime they obtain is $\tau(B_s) = 1.54\ ^{+0.42}_{-0.34}(stat)\ ^{+0.11}_{-0.12}(syst) ps$.

$B^0 - \overline{B^0}$ mixing has been measured in several $e^+e^-$ and $p\overline{p}$ experiments. The most recent measurement comes from DØ[27]. They have examined the ratio of yields of like-sign and unlike-sign muons. The mass spectrum in the two cases is shown in Fig. 21. Cuts are applied to avoid the low mass region where the $J/\psi$ is evident in the unlike-sign spectrum. No isolation cuts are applied since the muons from the decays of B mesons are typically contained within the $b$ quark jets. The problem with the measurement is understanding the alternative sources of dimuons, for example sequential decays of the B mesons to charm which decay semi-leptonically. The result obtained after Monte Carlo correction for these and other effects is $\chi = 0.13 \pm\ 0.02(stat) \pm\ 0.05(syst)$, where $\chi$, the mixing parameter does not separate the $B_d$ from $B_s$ mixing. This result confirms similar measurements from CDF and the LEP experiments with comparable precision.

Although there is hope for the B system, currently the only open window on CP violation is the $K^0 - \overline{K^0}$ system. The produced states, $K^0$ and $\overline{K^0}$ are superpositions of the CP eigenstates $K_1^0$ and $K_2^0$. However, the states $K_L^0$ and $K_S^0$ observed through their decays appear to be themselves also superpositions of the CP eigenstates. The extent of the mixing is parameterised

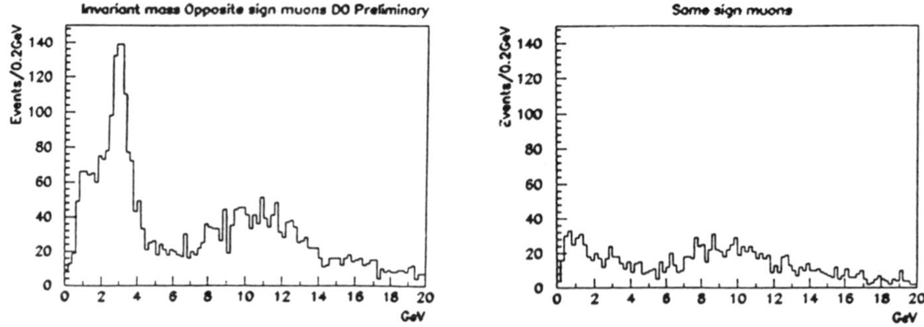


Figure 21: Mass spectra of Unlike-sign and Like-sign muon pairs.

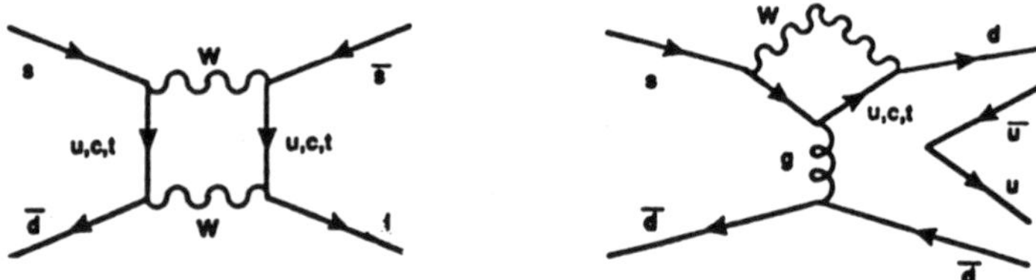


Figure 22: CP violating $K^0$ decay diagrams.

by the quantity $\epsilon$ which is determined by the inequalities of the decay rates of the $K^0_L$ and $K^0_S$ to two pions. Within the Standard Model this CP violation could have two origins depicted by the diagrams in Fig. 22a) and Fig 22b). The first, a box diagram containing a virtual top quark, would be a $\Delta S = 2$ transition and is labeled *indirect CP violation*; the second, a penguin diagram, would be $\Delta S = 1$ and is termed *direct CP violation*. The contribution of the *direct CP violation* is measured by another parameter $\epsilon'$. Experimentally its existence would lead to a difference between the ratio $K^0_L/K^0_S$ decaying to two charged pions as compared to two neutral pions. The Tevatron experiment E731 has recently published a result [28] of $\epsilon'/\epsilon = (7.4 \pm 5.2(stat) \pm 2.9(syst))\ 10^{-4}$. Other experiments, E799 and E773, use modifications of the E731 apparatus to look for CP violation in other channels and in the latter case to look for CPT violation. CPT studies are particularly interesting in the context of the theoretical lecture at this school by Nick Mavramatos[30][31]. The E773 measurement is of the phase difference between the charged and neutral two pion decays, currently the precision, from E731[32] is slightly more than 1 deg., the E773 measurement[29] will reduce that error by more than a factor two.

**W/Z Physics**

The study of vector bosons is a primary domain for the $p\bar{p}$ collider and there are some new measurements from the Tevatron. New cross-section measurements have been presented by DØ[18] and CDF[33] based on their 1992-93 data. The ratio of W and Z cross-section times branching ratio is an interesting quantity. It can be related through LEP measurements and some well established theoretical cross-section calculations, to the ratio of the leptonic and total widths of the W boson. In turn this ratio is sensitive to the existence of a low mass top quark. The measured values for the experimental ratios are CDF(electrons) $10.65 \pm 0.36(stat) \pm 0.27(syst)$, CDF(muons) $12.38 \pm 0.63(stat) \pm 0.45(syst)$, DØ(electrons) $10.57 \pm 0.60(stat) \pm 0.50(syst)$ and DØ(muons) $10.0 \pm 1.1(stat) \pm 2.4(syst)$. For the CDF measurement this translates into a decay mode independent top quark mass limit of 62 GeV at 95% confidence level and for DØ, 56 GeV.

The couplings between the vector bosons, the photon, W and Z, are sensitive to the structure of the standard model and with the high energy of the Tevatron, there has been progress with the experimental measurements, particularly the measurement of the W-$\gamma$ coupling. The signal for the events is a single photon in conjunction with the decay lepton from the W and the missing energy which is the signature of the partner neutrino. DØ[34] has shown results in both the electron and muon channels and finds limits on the anomalous magnetic moment of the W of $-2.5 \leq \Delta\kappa \leq 2.7$ at 95% confidence level. For the electric quadrupole moment the limits are $-1.2 \leq \lambda \leq 1.1$. These are significant improvements on the previous limits.

Finally, the primary goal of current W and Z physics is the precise measurement of the mass of the W. At the time of the Erice School there were no results from the 1992-93 data. Since that time both CDF and DØ have presented measurements and I give the status as of the Tsukuba $p\bar{p}$ Workshop. The measurement is very sensitive to the systematic errors associated with understanding the detector. Considerable effort is expended by both collaborations to understand their momentum scales whether they be those of magnetic tracking or those of the calorimetry. The result from DØ[35] is $M_W = 79.86 \pm 0.16(stat) \pm 0.31(syst)$ GeV $\pm 0.30(scale)$ GeV and from CDF[36] $M_W = 80.47 \pm 0.15(stat) \pm 0.25(syst)$ GeV.

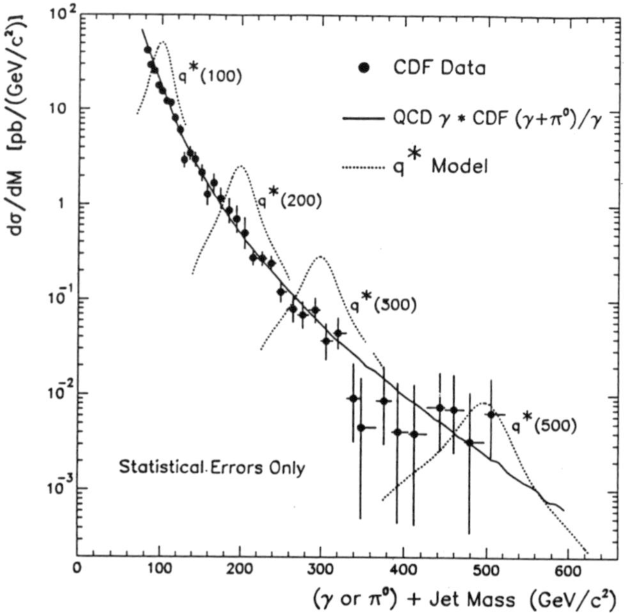


Figure 23: Excited quark production. a) shows the observed spectrum of quark-photon effective masses.

**Searches for New Physics**

In the category of searches for new physics we briefly consider searches for excited quarks and searches for leptoquarks, composites of leptons and quarks. Such searches are potentially the most important physics at the highest energy machine of the day. However, as yet all results are negative.

A first example[37] comes from the CDF group which has searched for excited quarks decaying to a quark, signalled by a jet, and a photon. In practice the latter objects may include some contamination from unresolved neutral pions. The results are shown in Fig. 23, where the observed yield is compared with that expected from background and putative resonance signals superimposed. When the background is subtracted, the result is compared to the theoretical expectations[38] in which the branching ratio for the decay is taken to be unity. In that case the result reduces to a mass limit of 470 GeV at 95% confidence level for this and the analogous quark - boson possibility.

DØ has searched[39] for pair produced leptoquarks decaying into an electron and a quark

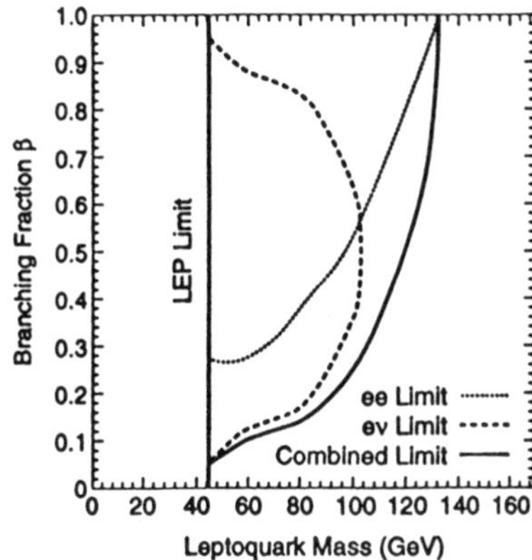


Figure 24: DØ Lepto-quark mass limits as a function of branching ratio, $\beta$.

or a neutrino and a quark. The electron is detected in the calorimeter and the neutrino would result in significant missing transverse energy. The primary signals are then two electrons or one electron plus missing transverse energy. Since the particles, if they existed, would dominantly be pair produced at the Tevatron, such a method is relatively insensitive to the quark-lepton-leptoquark coupling. The measurement leads to the limits shown in Fig. 24 for a range of decay branching ratios, $\beta$. Note that the use of both decay modes results in a limit which is about 3 times higher than the LEP limit except for values $\beta < 0.2$.

**The Search for the Top Quark**

The search for the top quark is a major activity within both of the Tevatron collider experiments, CDF and DØ. The analysis of the data is rapidly evolving. While the description given here follows that of the lecture, the quantitative limits are updated to correspond to the values shown by the two groups at the $p\bar{p}$ Workshop in Tsukuba in October, 1993,[40],[41],[42], [43].

Using data taken in 1989, CDF has shown[44] that it is unlikely that the mass of the top is less than that of the W boson. The discussion here therefore concentrates on the search for the top quarks where they are assumed to be produced in pairs($t\bar{t}$) which then decay, each to a W boson and a $b$ quark. Schematically the possibilities are shown in Fig. 25. For the range of relatively high masses expected, the top quarks and their decay products populate the central range of rapidity, $|\eta| < 2.5$. A potentially clean signal would be two high transverse momentum, isolated leptons $(ee, e\mu, \mu\mu)$ with associated missing transverse momentum and two hadronic jets. Although the branching ratio is low, this is a primary search channel. A signal of one high $p_T$ isolated lepton, missing transverse energy and four hadronic jets covers the case where one of the W bosons decays hadronically. This has a higher branching ratio but also has backgrounds associated with QCD W+ multi-jet production. In both experiments multiple triggers with different thresholds and combinations of requirements ensure relatively high efficiencies for the relevant modes.

The DØ analysis of the dilepton channels concentrated on the $ee$ and $e\mu$ modes. For the $ee$ channel each electron had to have $p_T > 20\ GeV$, and the missing $E_T$ had to be greater than 25 GeV, unless the dielectron mass was in the region of the Z peak when 40 GeV of missing energy was required. Finally at least one other jet with $E_T > 15 GeV$ was required. One event was found. In the $e\mu$ analysis the analagous cuts were 15 GeV in lepton $p_T$ and 20 GeV in calorimeter missing $E_T$ and also the missing $E_T$ calculated taking into account the muon $p_T$. Separation of the electron and muon in pseudo-rapidity and azimuthal angle was also required and the muon required to be isolated, that is to say, not buried in a jet. Again one event was found.

Backgrounds were estimated using a combination of monte carlo studies and estimates for processes using ancillary data samples. The estimates of the backgrounds are about a half and one event in the $ee$ and $e\mu$ channels respectively. The expected $t\bar{t}$ yields are at the single event level for masses in the range 130 to 140 GeV. With these numbers no claim of discovery was made. The scatter plot of the muon $p_T$ versus the electron $p_T$ is shown in Fig. 26a) for data and in Fig 26b) for a large 160 GeV top quark monte carlo sample. One sees that the event that

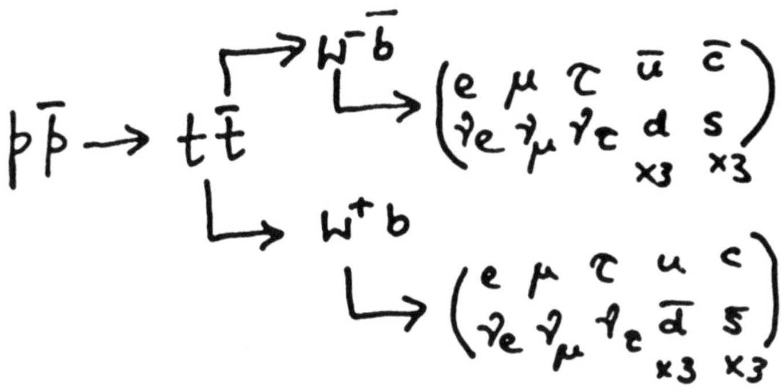

$$p\bar{p} \to t\bar{t}$$

$$t \to W^-\bar{b},\quad W^- \to \begin{pmatrix} e & \mu & \tau & \bar{u} & \bar{c} \\ \nu_e & \nu_\mu & \nu_\tau & d & s \\ & & & \times 3 & \times 3 \end{pmatrix}$$

$$\bar{t} \to W^+ b,\quad W^+ \to \begin{pmatrix} e & \mu & \tau & u & c \\ \nu_e & \nu_\mu & \nu_\tau & \bar{d} & \bar{s} \\ & & & \times 3 & \times 3 \end{pmatrix}$$

Figure 25: Top quark production and decay

the analysis cuts is, in fact, well above the cuts and it has generated considerable interest. Using the measurements of the leptons, the jets and the missing transverse momentum, the event is underconstrained. When the leptons and jets are organised in a particular combination and the W mass constraints and the requirement $m_t = \overline{m_t}$, are imposed, a kinematic fit is still not constrained. However, using knowledge of the incident parton distributions and the probabilities for different decay configurations, an analysis of the probability for different top masses can be performed[45]. Such an analysis, extended to account for detector resolutions and other effects, suggests a mass in the range around 145 GeV [46] if this event is assumed to be from $t\bar{t}$.

As mentioned above the single lepton channels are expected to have significant background. The problem is that conventional W boson production has higher order QCD contributions which lead to multiple jets which mock up the top quark signal. For this reason both CDF and DØ are examining more sophisticated approaches using either topological cuts or more promisingly, attempting to tag the b jets which are expected to be more likely in the top events than in the W plus jets background. A $b$ jet has a good probability to contain a soft lepton from the semi-leptonic decay of either the $b$ or $c$ quarks in the chain. Also the decay length for the $b$ quark offers the possibility of tagging its presence by observing evidence for a displaced vertex. This is where the silicon vertex detector in CDF plays a role. They have several vertex tagging methods and for the moment the details are unimportant. Fig. 27 shows the observed jet multiplicity distribution in the W plus jets (top candidate) events for a subsample of the data compared with the expectations from top of different masses. Even with the requirement of greater than 3 jets the data lie well above the signal expectations. If a vertex tagging requirement is applied, the situation shown in Fig 28 results. The data before the vertex tagging are replotted along with the data after applying the vertex tag. There are

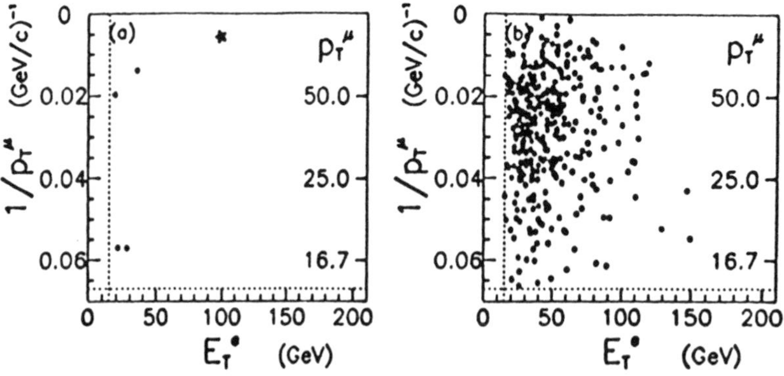


Figure 26: Scatter plot of lepton $p_T$ a) for the $e\mu$ data and b) for a top quark monte carlo with $m_t$= 160 GeV.

events with a jet multiplicity greater than three, two have a multiplicity of three. Further, the estimated background, as shown, is significantly less than one event. The analogous situation for a soft (>4 GeV) muon tagging analysis is depicted in Fig 29. For a range of top masses the data, the background expectations and the signal expectations are all shown. With a greater than 3 jet requirement there appear to be no events in the data, a background expectation of about 0.2 events, and a signal plus background which is slightly greater than one event at $m_t = 140\ GeV$. This analysis is fascinating but has not yet converged to the point where a statement is made about finding top nor to where the data are included in a limit estimation. It is worth repeating that these data represent about half of the total collected by CDF during the 1992-3 running. DØ using a $b$ tag analysis did include the lepton plus jets channel in a limit presented at Tsukuba[43]; however, here also the analysis has not yet matured to publication quality.

With the evolution in the analyses the situation with respect to the current quotable, 95% confidence limit on a lower mass limit is in flux and by the time this report appears, it may well have changed. Be that as it may, the situation was summarised at Tsukuba by Grannis[47]. He obtained 129 GeV based on the channels presented by CDF and DØ at Cornell[48] and Tsukuba respectively. He used a NNLO cross-section calculation for the purpose.

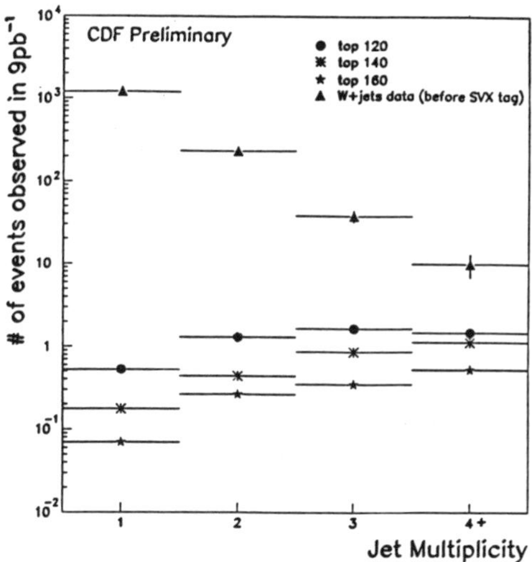


Figure 27: CDF lepton plus jets with no b tagging requirements.

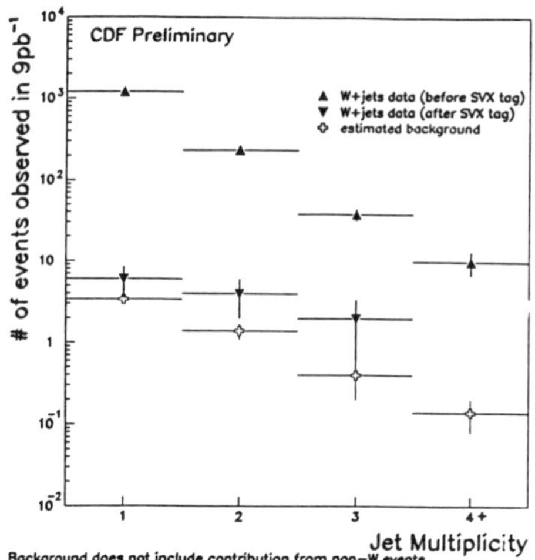


Figure 28: CDF lepton plus jets with detached vertex b tagging requirements.

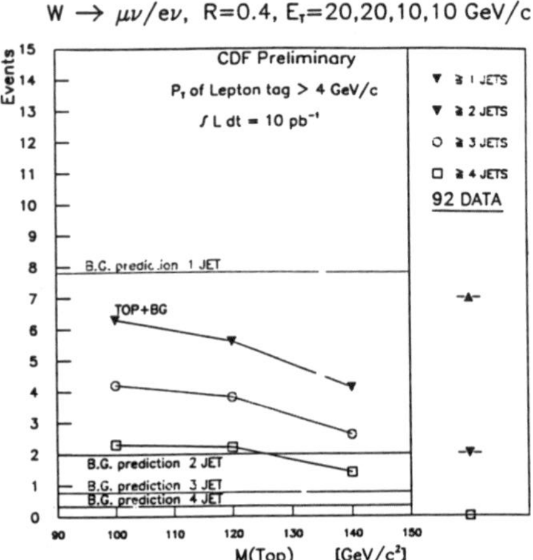


Figure 29: CDF lepton plus jets, soft muon b tagging requirements.

## Conclusions

The aim of this talk was to give an impression of the tremendous range and depth of the data being produced by experiments at Fermilab, both fixed target and collider. Despite the generous allotment of time it was not possible to do more than scratch the surface of some subjects. The collider experiments, using the measurements of the W mass and with the top search and mass limits, are approaching the situation where a statement about the Higgs mass, or a sensitive test of the consistency of the standard model become a possibility.

I would like to thank the Director of the Erice School and all its organizing staff for a very pleasant week and the students for the stimulating discourse. In addition, I would like to acknowledge the assistance of members of the Fermilab experiments who so generously gave their time and results. Finally, I would like to thank Sonya Wright for her assistance with the manuscript.

# References


[1] E760,T.Armstrong et al., Phys. Rev. Lett.**68**, 1468(1992).

[2] E683,D.Adams et al., Rice U. Preprint DE/F605/92ER40717-6, July 1993, FERMILAB-Pub-94/025-E, submitted to Phys. Rev. Lett.

[3] E665,M.R.Adams et al., Phys. Rev. Lett. **72**,466(1994).

[4] E665, M.R.Adams et al., FERMILAB-Pub-93/245-E, July 1993, submitted to Phys. Rev. D.

[5] H.Melanson, FERMILAB-Conf-93/165, presented at the 28th Rencontres de Moriond, Les Arces, Savoie, France, March 20-27, 1993.

[6] E665, T.Carrol, FERMILAB-Conf-93/165, presented at the 28th Rencontres de Moriond, Les Arces, Savoie, France, March 20-27, 1993.

[7] G.Alverson et al., FERMILAB-Pub-93/007-E, January 1993, submitted to Phys. Rev. D.

[8] CDF,F.Abe et al., Phys. Rev. Lett. **71**,679(1993); F.Abe et al., FERMILAB-Conf-93/206-E, Submitted to the 16th International Symposium on Lepton and Photon Interactions, Ithaca, Aug. 10-15, 1993.; F.Abe et al., FERMILAB-Conf-93/202-E, Submitted to the 16th International Symposium on Lepton and Photon Interactions, Ithaca, Aug. 10-15, 1993.

[9] E769, G.A.Alves et al., Phys. Rev. Lett. **70**,722(1993).

[10] E769, G.A.Alves et al., FERMILAB-Pub-93/081-E, April 1993, submitted to Phys. Rev. Lett.

[11] J.Appel, XVI Int. Symp. on Lepton and Photon Interactions, Cornell U., Ithaca, New York, Aug. 1993.

[12] E687, P.L.Frabetti et al., Phys. Lett. **B308**, 193(1993).

[13] E653, K.Kodama et al., Phys. Lett. **B305**, 359(1993); K.Kodama et al., Prog. Theor. Phys., **89**, 679(1993).

[14] E672, R.Jesik: Ph.D. Thesis, Univ. of Illinois at Chicago, August, 1993; R.Jesik et al., Proc. XXVI Int. Conf. on High Energy Physics, Dalas, Texas, 1992, AIP Conf. Proc. No 272, ed. J.R.Sandford, page 284.

[15] DØ, A. Maciel, Presented at the 9th Topical Workshop on $p\bar{p}$ Collider Physics, Tsukuba, Japan, October 17-22 1993.

[16] CDF, F.Abe et al., Phys. Rev. Lett. **71**, 500(1993) F.Abe et al., FERMILAB-Pub-93/145-E, June 1993, submitted to Phys. Rev. Lett.

[17] H.Weerts, Presented at the 9th Topical Workshop on $p\bar{p}$ Collider Physics, Tsukuba, Japan, October 17-22 1993.

[18] DØ, N.Graf, Presented at the 9th Topical Workshop on $p\bar{p}$ Collider Physics, Tsukuba, Japan, October 17-22 1993.

[19] J.D.Bjorken, Phys. Rev. **D47**, 101(1992).

[20] DØ, F.Borcherding, Presented at the 9th Topical Workshop on $p\bar{p}$ Collider Physics, Tsukuba, Japan, October 17-22 1993. S.Abachi et al., FNAL-Pub-94/005-E, 1994, to be published in Phys. Rev. Lett.

[21] P.Garbincius, Invited Talk presented at Hadron '93 Conference, Centro di Cultura scientifica "A.Volta", Villa Olmo, Como, Italy, June 21-25,1993 .

[22] Joel Butler and Peter H.Garbincius, Fermilab Report, Summer, 1993.

[23] M.B.Voloshin and M.A.Shifman, Sov. Phys, JETP 64,698(1987).

[24] B.Guberina,R.Ruckl and J.Trampetic, Z. Phys. **C33**, 297(1986).

[25] CDF, F.Abe et al., FERMILAB-Pub-93/158-E, June 1993, submitted to Phys. Rev. Lett.

[26] CDF,F.Abe et al.,Phys. Rev. Lett. **71**,1685(1993).

[27] DØ, S.Igarashi, Presented at the 9th Topical Workshop on $p\bar{p}$ Collider Physics, Tsukuba, Japan, October 17-22 1993.

[28] E731, L.K. Gibbons, Phys. Rev. Lett. **70**, 1203 (1993).

[29] E773, G. Gollin, Private Communication.

[30] Talk by Nick Mavramatos at this school.

[31] The relevance of the measurement to CPT violation was discussed by M.Gourdin in the discussion session associated with this lecture at this school.

[32] E731, L.K.Gibbons, Phys. Rev. Lett. **70**, 1199 (1993).

[33] CDF, W. Badgett, Presented at the 9th Topical Workshop on $p\bar{p}$ Collider Physics, Tsukuba, Japan, October 17-22 1993.

[34] DØ, A.Spadafora, Presented at the 9th Topical Workshop on $p\bar{p}$ Collider Physics, Tsukuba, Japan, October 17-22 1993.

[35] DØ, Q. Zhu, Presented at the 9th Topical Workshop on $p\bar{p}$ Collider Physics, Tsukuba, Japan, October 17-22 1993.

[36] CDF, D.Saltzberg, Presented at the 9th Topical Workshop on $p\bar{p}$ Collider Physics, Tsukuba, Japan, October 17-22 1993.

[37] CDF, FERMILAB-Pub-93/341-E, Nov. 1993, Submitted to Physical Review Letters.

[38] U.Baur, I.Hinchliffe and D.Zeppenfeld, Int. Journal of Mod. Phys. A2(1987)1285.

[39] DØ, S. Abachi et al, FERMILAB-Pub-93/340-E, 1993, to be published in Phys. Rev Lett.

[40] CDF, T.Chikamatsu, Presented at the 9th Topical Workshop on $p\bar{p}$ Collider Physics, Tsukuba, Japan, October 17-22 1993.

[41] CDF, M.Contreras, Presented at the 9th Topical Workshop on $p\bar{p}$ Collider Physics, Tsukuba, Japan, October 17-22 1993.

[42] DØ, M.Fatyga, Presented at the 9th Topical Workshop on $p\bar{p}$ Collider Physics, Tsukuba, Japan, October 17-22 1993.

[43] DØ, H.Greenlee, Presented at the 9th Topical Workshop on $p\bar{p}$ Collider Physics, Tsukuba, Japan, October 17-22 1993.

[44] CDF, F.Abe, et al., Phys. Rev. D **45**, 3921 (1992).

[45] R.H.Dalitz and G.R.Goldstein, Phys. Lett. **B287**, 225 (1992); K.Kondo, T.Chikamatsu and S.Kim, Journal of the Phys. Soc. of Japan, **62**,1177(1993).

[46] DØ, M.Strovink, Proceedings of the International Europhysics Conference on High Energy Physics, Marseille, France, 1993, eds. J.Carr and M.Perrottet.

[47] P.Grannis, Presented at the 9th Topical Workshop on $p\bar{p}$ Collider Physics, Tsukuba, Japan, October 17-22 1993.

[48] P.Tipton, XVI Int. Symp. on Lepton and Photon Interactions, Cornell U., Ithaca, New York, Aug. 1993.

**CHAIRMAN: H.Montgomery**

*Scientific Secretaries: A.Ianni, R.McPherson*

## DISCUSSION

– *Gourdin:*

a) CPT invariance does not predict the equality $\phi_{00} - \phi_{+-}$ of the phases of the CP violating parameters $\eta_{+-}$ and $\eta_{00}$ of the $K \to 2\pi$ decay. It predicts that the phase of $\epsilon'$ is $\pi/2 + \delta_2 - \delta_0$ where $\delta_0$ and $\delta_2$ are strong interactions phases for the scattering I=0,2 isospin states. However, the difference $\delta\phi = \phi_{00} - \phi_{+-}$ can be computed and using the experimental facts $|\eta_{+-}| \approx |\eta_{00}|$ and $\phi_{+-} \approx \phi_{00}$ we can obtain

$$\delta\phi = \left[1 - |\frac{\eta_{00}}{\eta_{+-}}|\right] + g\left[\phi_{+-} + \delta_0 - \delta_2 - \frac{\pi}{2} + \delta_\omega\right]$$

where $\delta\omega$ is the phase of $(1+\omega)(1-2\omega)$ due to a violation of the $\Delta I = 1/2$ rule. With $|\omega|^{-1}$=31.68 and $\delta_0 - \delta_2 \approx \pi/4$ we get $\delta_\omega \approx 1.4^o$

b) Do you have any plan for measuring the pure leptonic decay $D^+ \to l^+ + \nu_e$ and $B^+ \to l^+ + \nu_e$. This will clarify the somewhat confused situation for the decay constants $f_D$ and $f_B$, which are theoretically estimated with large uncertainties.

– *Montgomery:*

a) I'm not qualified to answer this question.

b) No, I don't know of any experimental program that really gets at that. Part of the aim of the charm program is to get at some of the measurements in the charm system and calibrate the theory.

– *Peccei:*

Actually, the reason that $\phi_{+-} - \phi_{00}$ tends to measure the possible CPT violation is precisely the fact that the phase of $\epsilon'$ and the phase of $\epsilon$ are very close to each other, such that the imaginary part of $\epsilon'/\epsilon$ which measures the phase difference is in fact, to a very good approximation, a purely CPT test.

– *McPherson:*

You showed a couple of dozen $B^0 \to J/\psi K_S$ events. With detector upgrades and more data, how many will you get? Will you make a meaningful CP violation measurement with CDF and D0, or will a dedicated B collider detector be needed?

– *Montgomery:*

With D0 or CDF and 1 $fb^{-1}$ of data, or one year of running with the main injector which is scheduled to be completed in 1999, we expect 1000 to 2000 $J/\psi K_S$ events. The errors that we quote on $\sin 2\beta$ are around 0.15.

– *Hoang:*

You mentioned the two direct CP violation experiments NA31 and E731 that seem to give different answers to the question "does direct CP violation exist". Are there any hints that one of these experiments made a mistake?

– *Montgomery:*

Not that I know of. The two used somewhat different techniques. By the way as an experimentalist, I don't see anything that I call a major discrepancy between those two. There is a difference, but it's not $3\sigma$, and if you do a lot of experiments a $3\sigma$ effect is perfectly natural.

– *Gibilisco:*

a) Have you determined the form factors for B and D meson decays, in particular in the vector case? If yes, do these results confirm the on lattice calculations, in particular the D form factor $A_2$, which on lattice is very near to 0 and in other models is different from 0?

b) In your opinion, what is the effective importance of the annihilation and W-exchange contributions for D decays? Are they helicity suppressed or not?

– *Montgomery:*

a) There is some information on these form factors. There are three experiments: E687 with $0.78 \pm 0.18 \pm 0.10$, E691 with $0.0 \pm 0.5 \pm 0.2$, and E653 with $0.82 \pm 0.23 \pm 0.11$. That's not resolved yet.

b) There are certainly suppressions in these, but these might be offset by some strong interaction loops, and the problem becomes a strong interaction calculation.

– *Langacker:*

If the main injector and reasonable improvement in the CDF and D0 detectors are completed what will be the reach for new physics, in particular SUSY?

– *Montgomery:*

I don't have a number on SUSY, but I did show expected errors on $M_W$ of $\pm 50 MeV$ and on $M_t$ of $\pm 5 GeV$.

– *Skenderis:*

You have shown in the morning the various channels for the top quark and the various cuts you use to select events. How do you choose these cuts, or in other words how arbitrary is the choice?

– *Montgomery:*

The present cuts were more-or-less decided before the run began. There is an industry about how one sets cuts in an impartial way. It's important to answer only one question at a time; for example, in doing the background calculation you should be conservative in one direction if you want to put a lower a limit on the mass of the top, while you should be conservative in the other direction when deciding whether or not you have a signal.

– *Arnowitt:*

a) A suggestion has been made to use the 6-jet signal to detect the top quark; i.e. 2 jets would be b-quarks tagged with vertex detectors and the 4 remaining jets should have 2 pair of two jets consistent with W kinematics. Is this a realistic suggestion?

b) What are the current CDF bounds on SUSY particles? (the CDF paper last year analyzed $\mu$ and one $\tan\beta$ and thus didn't give a clear picture of bounds.)

– *Montgomery:*

a) There are people working on it. You did see the W+jet signal I showed this morning, and it had a rather significant background in it. One can imagine that if one could come up with sufficiently high efficiency on the tagging then one might have a good chance. I consider that a long shot.

b) I can give you a preprint afterwards. It shows a hyperbola in mass of the gluino vs the mass of the squark, which is around 120 GeV in each.

– *Vassilevskaya:*

There are several sufficiently definite theoretical predictions about the top quark mass in comparison with the Higgs mass. What are the main experimental difficulties to discovering the top?

– *Montgomery:*

As I pointed out this morning, the whole problem is to reduce the background down to essentially zero, and still have a signal coming through. This requires a certain luminosity, which goes up with the mass of the top because the production cross-section goes down. We have not actually had the discussion about how many candidates are sufficient to claim discovery in the zero background case.

– *Mavromatos:*

Can the top quark be missed at FNAL experiments?

– *Montgomery:*

There is certainly the possibility of detector effects creating background. In my opinion, if you only had one experiment then making a statement that something is excluded is not very sensible. You need to confirm experimental results.

– *Kawall:*

I have a question regarding E769, the charm production experiment. You showed a plot of cross section vs. Atomic number. Was the slope exactly one? Do you know the $Q^2$? I'm interested to know if this is a reasonable place to look for colour transparency.

– *Montgomery:*

The slope is $1.00 \pm 0.02$, where I have made an eyeball estimate of the error bar shown. It was $\pi^- N$, so I'm not sure how we should define $Q^2$.

– *Ng*

a) Have rapidity gap events like Wiik discussed from Hera been seen at FNAL? b) W, Z masses were not shown. Why? c) Are there new $W \to l\nu$ lepton assymetry results? d) How well is data distribution handled? e) What's the FNAL run plan.

– *Montgomery:*

a) I'm not sure the subject is entirely the same; however, there has been a fair amount of noise in the last year or so from, among others, Bjorken (Dokshitzer, Khoze, Troyan) which has to do with rapidity gaps from the exchange of colourless objects. They were hypothesizing that, for the SSC, this was a nice prospect for looking for the Higgs. At the ISR, we looked for rapidity gaps indicating double pomeron exchange. In $\bar{p}p$ collisions, if you look for another jet as a function of separation then you would see something flattening out which would indicate the exchange of a colourless object. D0 now has some data on this subject, which may indicate a flattening out at $10^{-2}$ to $10^{-3}$.

b) D0 has not fully understood to their satisfaction our calorimeter calibration in absolute terms. We use the LEP Z mass as our final adjustment, which right now is too large. I think that CDF has not completed their analysis.

c) I don't know of any new results.

d) D0 wrote data at around 2Hz, CDF at 6Hz. Basic event reconstruction is performed centrally at FNAL. Summary tapes are distributed to the institutions.

e) We're in a shutdown now. There is an increase in energy of the LINAC. There will then be a run accumulating 75 $pb^{-1}$. It is possible there will be a shutdown before the fixed target program in 1995, and then a further run of the Tevatron in 1996. We're going to get around 100 $pb^{-1}$ in the next year.

– *Skenderis*

You didn't say many things about the two CDF events. Is there any special reason for this?

– *Montgomery:*

I didn't have information about the event they had in their sample last year, and so the compromise I reached was to show the lepton analysis from D0, and the lepton + jets from CDF since they have the Silicon Vertex detector.

– *Petrache:*

In your plot of charm production, it seems that you have nice agreement between the theoretical prediction and the experimental data. But, there are two problems. First of all, I don't think you can put the theoretical predictions in a single line because your predictions depend on the model you choose. Then, for the experimental data, you have to reconstruct charm from the particles you detect, and for this you assume a certain model of charm fragmentation. So the plot is not a comparison between "pure" theoretical prediction and "pure" experimental data.

– *Montgomery:*

I think I agree with most of what you say. You are looking at a differential cross-section, so at least you don't have to integrate over an unknown production mechanism to produce the total cross-section.

– *Zichichi:*

Why is there no apparent leading particle contribution to the D production cross-section?

– *Montgomery:*

This experiment addressed that by looking for $\pi^- \rightarrow D^+$ and $\pi^- \rightarrow D^-$, fitted to $x_F$ distributions, and they find that there is very little difference. So it seems that this new high precision data is not seeing any leading particle effect in $D^+$, $D^-$ production.

– *Zichichi:*

I would like to point out that in addition to our experiment, we have done a world analysis of all processes, including $\nu$ production. The existence of a leading particle effect is well established, if you look for it. For example, $\nu$ production of $\Lambda$ has a very clear leading effect, shown by us in our world data analysis.

– *Montgomery:*

I have a reprint which I will give to you.

# HERA: ACCELERATOR PERFORMANCE AND PHYSICS RESULTS

B. H. Wiik
II. Institut für Experimentalphysik und
Deutsches Elektronen-Synchrotron DESY
Hamburg, Germany

## 1. INTRODUCTION

The electron-proton collider HERA and its large multipurpose detectors H1 and ZEUS has now made the transition from a virtual to a real source of data on electron-proton interactions in a new, greatly expanded kinematic region.

The HERA project is a truly international effort. It was built within the framework of a collaboration where institutions in 10 countries contributed either components built at home or delegated skilled manpower to work on the project at DESY. Also the large detectors H1 and ZEUS have been built and are exploited by international collaborations. Only some 25% of the 750 physicists presently involved in the programme are from German institutions while the remainder comes from some 90 institutions in 19 countries.

In 1992 a total integrated luminosity of 58 $nb^{-1}$ was delivered to each of the experiments and data corresponding to about 25$nb^{-1}$ were recorded.

The winter shutdown 92/93 was used to upgrade the HERA control system and to prepare HERA for multibunch operation. Due to repair work on the H1 jet chamber luminosity running only started towards the end of June. By the end of October a total integrated luminosity of more than 1000$nb^{-1}$ had been delivered to the experiments and data corresponding to 600$nb^{-1}$ had been recorded on tape.

In this talk I'll first focus on the present performance of the accelerator and the plans for 1994 and then discuss the physics results obtained by H1 and ZEUS.

HERA offers the intriguing possibility of performing high luminosity, high duty cycle fixed target experiments by installing internal targets in the electron and the proton beam. In the last part of my talk I'll discuss physics motivation and status of this programme.

## 2. HERA

### 2.1 Overview

HERA [1] is made of two independent accelerators designed to accelerate and to store, respectively 820 GeV protons and 30 GeV electrons and to collide the two counterrotating beams head on in four interaction regions spaced equidistant around its 6.4 km long circumference.

The layout of the accelerator complex is shown in Fig. 1.

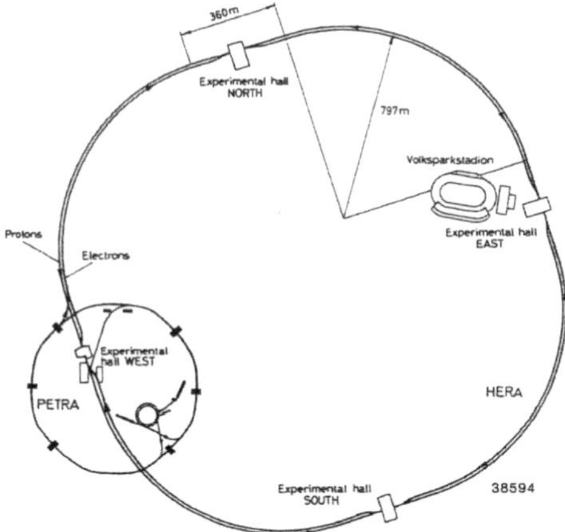


Fig. 1: The layout of the HERA accelerator complex.

The collider general purpose detectors H1 and ZEUS are installed in straight section North, respectively South. In straight section East the HERMES experiment, designed to scatter longitudinally polarized electrons on polarized H, D and $He^3$ targets will be installed. An experiment, designed to measure the CP violation in the $b\bar{b}$ systems, using an internal wire target in the halo of the proton beam, is being considered. If approved, this experiment will be installed in straight section West.

## 2.2 The HERA Rings

The operation of the HERA electron and proton rings has been greatly eased by the stability and the reproducibility of the two rings.

The conventional RF system of the electron ring has been augmented by 16 four cell superconducting 500 MHz cavities, assembled pairwise into 8 cyostats. These cavities have been industrially produced from high purity Niobium (RRR=300). At the design current of 58 mA the gradient is limited to 2.05 MV/m due to the 100 kW power rating of the input coupler. Without beam all cavities reached 5 MV/m. Initially the design Q-value of $2 \cdot 10^9$ could not be reached in all cases, due to the presence of $Nb_xH_y$ precipitation on the Nb surface.

The electron beam has been accelerated to 30.3 GeV with both the normal and the superconducting RF systems working in parallel. However, during physics runs the accelerator is operated at 26.7 GeV which permits the stored beam to remain in the ring even if an RF station trips off.

An attractive feature of HERA is the option of colliding longitudinally polarized electrons or positrons with protons.

The spins of the electrons in a stored beam will become aligned [2] antiparallel to the direction of the magnetic guide field through the emission of synchrotron radiation. The transverse polarization P(t) builds up exponentially:

$$P(t) = P_0(1 - e^{-t/\tau_0}).$$

The equilibrium polarization $P_0 = 92{,}4\%$ in a perfect, flat machine and the time constant $\tau_0$ decreases proportional to $\gamma^{-5}$, where $\gamma$ is the electron energy measured in units of its rest mass.

In the presence of depolarizing effects with a build up time $\tau_D$ the equilibrium polarization $P_m$ is reduced to

$$P_m = P_0 \frac{\tau_D}{\tau_0 + \tau_D}$$

and the effective build up time is

$$\tau_m = \tau_0 (P_m / P_0).$$

Here $\tau_D$ is the time constant of the depolarizing effect.

The transverse electron beam polarization is determined [3] by measuring the up-down asymmetry of backscattered polarized laser light using a tungsten scintillator sandwich counter. The observed electron beam polarization is shown in Fig. 2 as a function of time. Both the build up time

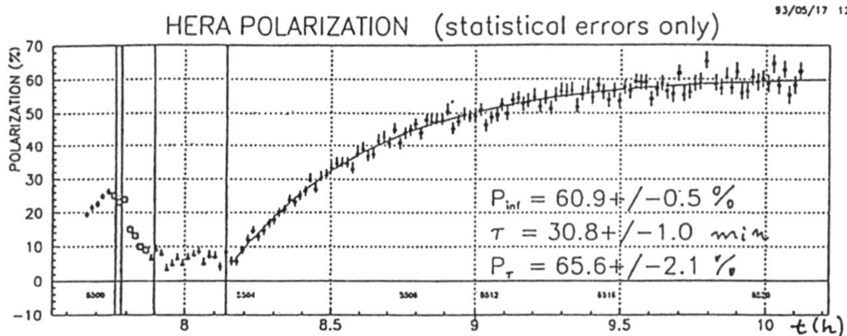


Fig. 2: a) The transverse electron beam polarization as a function of time.

of 31 min and the measured polarization are consistent with a transverse electron polarization of more than 60%. Furthermore this polarization is both robust and reproducible and it coexists with luminosity operation, at least for proton currents up to 30% of the design value. Tuning using beam based alignment may increase the degree of polarization even further.

In preparation for the HERMES experiment a pair of 60 m long spin rotator magnets designed to turn the transversely polarized electron beam in the arcs into states of well defind helicity at the collision point and to restore the transverse polarization before the beam enters the downstream arc will be installed in straight section East in the 93/94 winter shut down.

Two sets of spin rotators for the H1 and ZEUS experiments are on order, and are scheduled to be installed in the 95/96 shutdown.

The cryogenic system and the 2156 superconducting magnets and correction coils are the most challenging components of the HERA proton ring.

The central refrigerator is located on the DESY site. It is subdivided into 3 identical plants with each plant capable of providing 6.6 kW isothermally at 4.3K, 20.4g liquid helium per second and 20 kW at 40K to 80K. The heat loads of the whole ring, including transferlimes and feed boxes are 5.1 kW at 4.4K and 28.5 kW at the shield level. At 820 GeV an additional head load of 170W is observed. It thus requires some 2.7 MW from the mains to operate HERA at 820 GeV. For comparison 65 MW from the mains are needed to run the SPS at 310 GeV.

Recently the temperature of the proton ring has been lowered from 4.4K down to 3.7K. The availability of the cryogenic plant is of order 99%.

The superconducting magnet system has been very reliable and none of the magnets had to be removed during 3 years of operation.

Adjusted to an operating temperature of 4.4K the average quench current is (6900±130)A for the dipoles and (7840±160)A for the quadrupoles. Only magnets with a quench current above 6500A, compared to the nominal current of 5020A are installed in HERA. The design energy of 820 GeV is thus easily reached. To increase the energy beyond the nominal energy may require changes to the quench protection system.

The energy stored in the magnets at 820 GeV is of order 270 MJA and a sophisticated quench detection and protection system is installed to dump this energy in a safe way once a quench occurs.

Only few beam induced quenches are observed during operation. This is presumably due to the large safety margin which exist both in field (30%) and in temperature (1.5K).

The field quality of superconducting magnets is seriously affected by persistent magnetization currents. Any change of the field in a magnet induces magnetization currents within the superconducting filaments with a strength which is proportional to filament diameter and to the critical current density at ambient temperature and field strength. The persistent current changes the value of all allowed multipoles. At HERA, with a ratio of peak magnetic field to injection field of 20 and a filament diameter of 14μm, persistent current effects are rather strong.

The persistent current multipoles are compensated using local correction elements. In order to determine the required strength of the correction elements at injection and during acceleration, the dipole and sextupole fields are measured continuously in two superconducting reference magnets, powered in series with the ring magnets.

The proton beam lifetime at injection is on the order of 10hrs after a careful cycling of the magnets and after correction of persistent current multipoles. Thus the beam lifetime is long compared to the time it takes to fill the proton ring.

During the 1992 run and the initial part of the 1993 run protons were accelerated and stored using only the 52 MHz RF system yielding a bunch length ($\sigma$) of 18 to 20 cm. With the 208 MHz system now in use, the 1$\sigma$ bunch length at 820 GeV is 11.6 cm in agreement with the design value. The ensuing event vertex distribution is shown in Fig. 3.

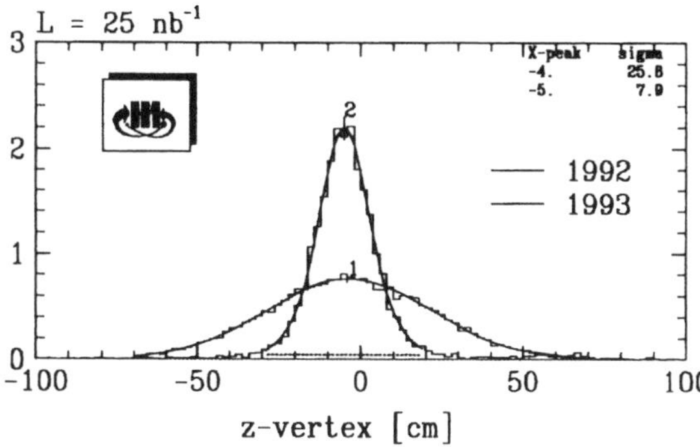


Fig. 3.: The event vertex distribution as observed by the H1 Collaboration during the 1992 data taking periode (wide distribution) and in 1993 (narrow distribution).

## 2.3 LUMINOSITY

The following strategy is used to bring the two counterrotating beams into collisions. The proton R.F. frequency is locked to the electron R.F. frequency to insure the same revolution time for electron and proton bunches and then the arrival time of the bunches measured at a monitor located at interaction point East is used to adjust the phase such that the bunches arrive simultaneously in interaction points North and South. The final step is then to make the beams overlap transversely at the interaction point.

The luminosity is measured [4,5] using the reaction $e + p \rightarrow e + \gamma + p$ with the electron and the proton detected in coincidence. The arrangement of the luminosity monitors is shown in Fig. 4.

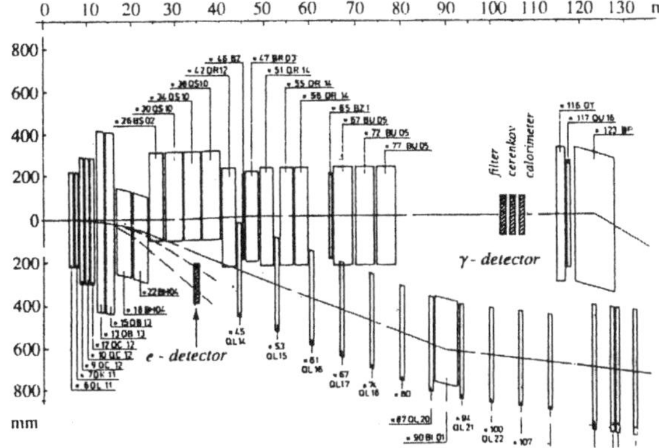


Fig. 4: The layout of magnets and luminosity detectors in the straight section.

The electron-proton luminosity is given by:

$$L = \frac{N_p \cdot N_e \cdot n_b \cdot f_R}{\left(\sigma_{x,e}^2 + \sigma_{x,p}^2\right)^{1/2} \cdot \left(\sigma_{y,e}^2 + \sigma_{y,p}^2\right)^{1/2}}$$

$N_p$ and $N_e$ denotes the number of protons, respectively electrons per bunch, $n_b$ is the number of colliding bunch pairs and $f_R$ is the revolution frequency. The transverse electron and proton beam size at the interaction point are denoted by $\sigma_{x,e}$, $\sigma_{y,e}$, $\sigma_{x,p}$ and $\sigma_{y,p}$ and are given by $\sigma_j = \sqrt{\beta_j \varepsilon_j}$ with $j = x, y$.

Here $\beta_j$ is the value of the $\beta_j$-function at the interaction point and $\varepsilon_j$ is the beam emittance in the j-plane.

A typical set of luminosity beam parameters as obtained in 1992 and in 1993 are listed in table 1 and compared to the design values.

Table 1 Luminosity beam parameters

| Parameter | 1992 | 1993 | Design |
|---|---|---|---|
| $\sigma_{x,p} / \sigma_{y,p} (mm)$ | 0.27/0.09 | 0.14/0.05 | 0.29/0.07 |
| $\sigma_{x,e} / \sigma_{y,e} (mm)$ | 0.33/0.09 | 0.33/0.07 | 0.26/0.07 |
| $\beta_{x,p} / \beta_{y,p} (m)$ | 7/0.7 | 7/0.7 | 10/1.0 |
| $\beta_{x,e} / \beta_{y,e} (m)$ | 2.2/1.4 | 2.2/1.4 | 2/0.7 |
| $n_b^p$ | 9(10) | 84(90) | 210 |
| $n_b^e$ | 9(11) | 84(94) | 210 |
| Bunch spacing (ns) | 96 | 96 | 96 |
| $N_p$ | $2 \cdot 10^{10}$ | $2.9 \cdot 10^{10}$ | $10 \cdot 10^{10}$ |
| $N_e$ | $3 \cdot 10^{10}$ | $3 \cdot 10^{10}$ | $3.5 \cdot 10^{10}$ |
| $\Delta\nu_{x,p} / \Delta\nu_{y,p}$ | 0.0004/0.0002 | 0.0003/0.0002 | 0.001/0.0005 |
| $\Delta\nu_{x,e} / \Delta\nu_{y,e}$ | 0.006/0.01 | 0.008/0.017 | 0.021/0.019 |
| Lifetime | | | |
| Proton beam (h) | 50 | >100 | - |
| Electron beam (h) | 4 | 10 | - |
| Peak Luminosity ($cm^{-2}s^{-1}$) | $2 \cdot 10^{29}$ | $1.5 \cdot 10^{30}$ | $1.5 \cdot 10^{31}$ |

The transverse beam dimensions are derived from the measured values of the β-functions at the interaction point and the measured beam emittance. Due to careful matching between all the accelerators in the injection chain the proton invariant emittance at the start of a run in HERA is of order 10 πmm·mrad in both planes compared to the design value of 20 πmm·mrad.

The proton beam lifetime is strongly dependent on the transverse size of the electron beam at the interaction point and is maximized when the cross section of the proton beam is smaller or equal to that of the electron beam.

Under these conditions the proton lifetime is of order 100-300 hrs compared to electron beam lifetimes between 4 and 10 hrs. Thus in general the HERA proton ring is filled typically once every 24 hrs whereas electrons are dumped and reinjected every 5 to 10 hrs.

The stored proton and electron currents is plotted versus time in Fig. 5 for a well centered proton beam colliding with an electron beam of smaller or equal transverse size.

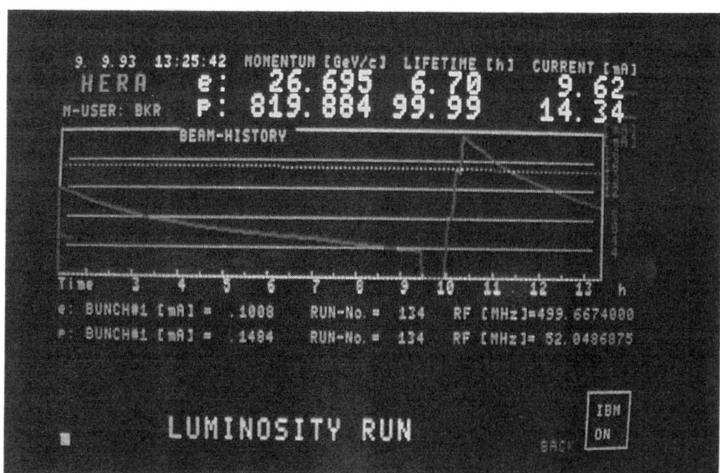


Fig. 5: Electron and proton beam currents versus storage time.

The number of electron bunches $n_b^e$ is limited by the maximum electron current which can be stored in the ring.

In 1992, the maximum electron current which could be stored with lifetimes on the order of a few hours was limited to 3mA by macroscopic particles trapped in the strong field of circulating beam. Thus a total of 11 electron bunches were stored of which 9 collided with proton bunches and 2 were used to measure the single beam background.

Using beam loss monitors the onset of the instability was traced back to particle emission from two of the vacuum chambers. Replacing these chambers in the 92/93 winter shutdown made it possible to raise the current to 20mA for the 1993 run. Thus in 1993 the HERA electron ring was filled with 94 electron bunches of which 84 was partnered with proton bunches. At the end of the 1993 run it was found that macroscopic particles are emitted by the distributed ion getter pumps. By reducing the voltage on the pumps and by switching off the main offenders a current of 33mA compared to the design current of 58mA could be stored at 26.7 GeV with a lifetime of several hours.

The number of proton bunches are listed in Table 1 and were chosen to match the number of electron bunches.

The proton bunch population in HERA is limited to some $2.9 \cdot 10^{10}$ protons/bunch compared to the design value of $10^{11}$ protons/bunch. This limit occurs both in the 7.5 GeV proton preinjector DESY III and in HERA during injection at 40 GeV and the first part of the ramp. In DESY III higher currents leads to a blow up of longitudinal beam emittance such that the beam do not fit into the R.F. bucket of the following accelerator PETRA. A longitudinal multibunch damping system is now under construction and will be installed in the 93/94 shutdown. The limitation arising in HERA is a single bunch transverse instability. Experiments have shown that this instability can be cured by a multibunch transverse feedback system. Such a system may become available towards the end of the 94 run.

A total of 33mA have been stored at 820 GeV in 180 bunches compared to the design value of 160mA in 210 bunches.

The luminosity and the specific luminosity as measured by the ZEUS experiment is plotted in Fig. 6 versus time.

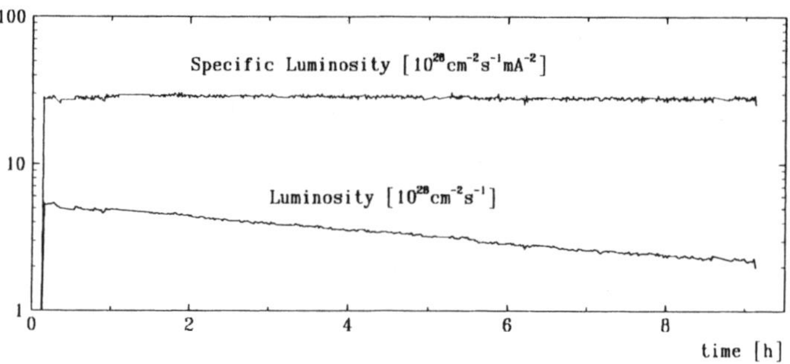


Fig. 6 The luminosity and the specific luminosity as measured by the ZEUS experiment.

The maximum luminosity was $1.5 \cdot 10^{30} cm^{-2} s^{-1}$ with 84 bunches colliding compared to the design luminosity of $1.5 \cdot 10^{31} cm^{-2} s^{-1}$ and 210 bunches. Also plotted in Fig. 6 is the specific luminosity obtained by normalizing the luminosity to the product of the stored currents.

The specific luminosity stays nearly constant over a fill and is now roughly 50% above its design value.

By the end of October an integrated luminosity of 1.1 $pb^{-1}$ had been delivered, or nearly a factor of 20 above the 1992 integrated luminosity of roughly 58$nb^{-1}$. Of the delivered luminosity roughly 60% is used by the experiments, the remainder being lost due to the time it takes to collimate the beams, to turn on the detectors, to detector trip offs and detector inefficiencies. The useful integrated luminosity collected by the H1 experiment is plotted as a function of calendar time in Fig. 7.

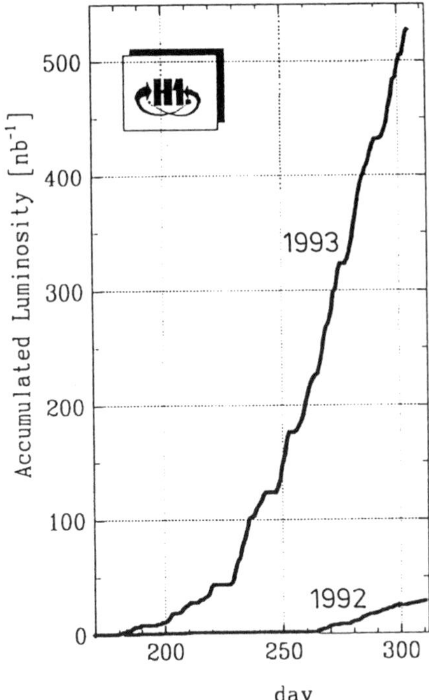


Fig. 7: Integrated luminosity versus time.

## 3. THE DETECTORS

### 3.1 Introduction

In a deep inelastic electron-proton collision as depicted in Fig. 8 the incomming electron interacts directly with one of the quarks in the proton by means of a spacelike current, charged or neutral. The kinematic variables $Q^2$ and $\nu$ or the scaled variables x and y, used to describe this process, are also defined in Fig. 8.

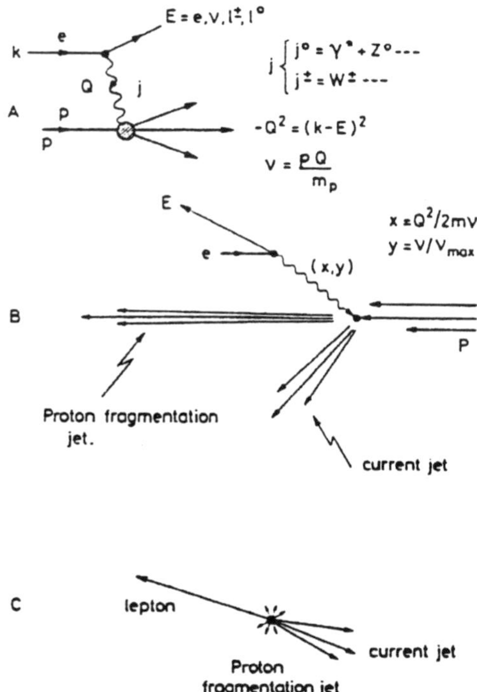


Fig. 8: Deep inelastic electron-proton interactions at HERA and the definition of kinematic variables.

A deep inelastic electron-proton collision in HERA yields a very simple final state topology. The struck quark will materialize as one or several jets of hadrons whose momentum components transverse to the beam axis are balanced by the transverse momentum of the final state lepton. The reminder of the proton will appear as a sharply collimated jet of hadrons travelling along the initial proton direction.

A simulated electron-proton neutral current event is shown in Fig. 9.

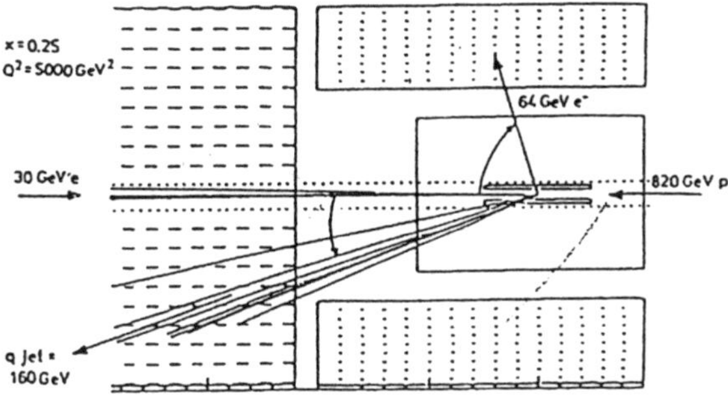


Fig. 9 The topology of a typical deep inelastic electron-proton event.

The kinematic variables $Q^2$ and $\nu$ or x and y can be determined from a measurement of energy and direction of either the final state electron or of the current jet or by combining electron and hadron data. Thus the basic requirement is that the detector can measure single electrons and hadrons with good resolution, both in space and energy, over nearly $4\pi$.

The energy of hadron jets vary between 820 GeV in the forward direction (incident proton direction) to roughly 30 GeV in the backward direction. It is thus clear that a high resolution, finely grained $4\pi$ calorimeter is the cornerstone of a general purpose detector at HERA. Since the ratio of neutral to charged pions vary from event to event, it is crucial that the calorimeter has the same response to showering as to non-showering particles. If not the data must be corrected off line.

New particles produced in the collision may reveal themselves through their semileptonic decay modes. The detector must thus be able to identify leptons travelling within the hadron jet and to determine the charge sign of the lepton.

The properties of various detector elements of H1 and ZEUS are listed in table 2 and a detailed description of the detectors and their performance can be found in references 6 and 7.

Table 2: Properties of the H1 and ZEUS detectors.

| | **H1** | **ZEUS** |
|---|---|---|
| Size | $12 \cdot 10 \cdot 15$ m$^3$ | $10.6 \cdot 10.8 \cdot 20$ m$^3$ |
| Polar angle | $0.7° - 175°$ | $10\sigma_{beam} - 176°$ |
| Solenoid | | |
| Field | 1.2 T | 1.8 T |
| Diameter | 5.6 m | 1.72 m |
| Length | 5.75 m | 2.8 m |
| Central Tracking | $11 < r < 85$ cm | $10 < r < 85$ cm |
| Resolution $\sigma(p)/p$ | 0.003 p | $0.002 \cdot p \oplus 0.002$ |
| Electromagnetic calorimeter | lead plates 2.4 mm<br>liquid Argon 2.5 mm | Uranium plates 3.3 mm<br>Scintillator 2.6 mm |
| Thickness | $20 - 30$ $X_0$ | $27 - 32$ $X_0$ |
| Tower size | $3 \cdot 3 - 8 \cdot 8$ cm$^2$ | $5 \cdot 20$ cm$^2$ |
| Long.segmentation | $3 - 4$ | 1 |
| Resolution | $(0.12/\sqrt{E}) \oplus 0.02$ | $(0.18/\sqrt{E}) \oplus 0.01$ |
| Hadronic calorimeter | steel plates 16 mm<br>liquid Argon $2 \cdot 2.5$ mm | Uranium plates 3.3 mm<br>Scintillator 2.6 mm |
| Response e/h | 1, by weighing | 1 inherent |
| Thickness | $(4.7 - 8)\lambda_{abs}$ | $(4 - 7.1)\lambda_{abs}$ |
| Size | $8 \cdot 8 - 20 \cdot 20$ cm$^2$ | $20 \cdot 20$ cm$^2$ |
| Long. segmentation | $4 - 6$ | 2 |
| Resolution | $(0.50/\sqrt{E}) \oplus 0.02$ | $(0.35/\sqrt{E}) \oplus 0.01$ |
| Tail catcher | $4.5\lambda_{abs}$ | $4\lambda_{abs}$ |
| Electron-pion separation | $10^4 - 10^5$ | $10^4 - 10^5$ with silicon pads |
| Myon-pion separation | $10^3$ | $10^3$ |

$\oplus$: errors added in quadrature

## 3.2 THE H1 DETECTOR

An isometric view of the H1 detector is shown in Fig. 10.

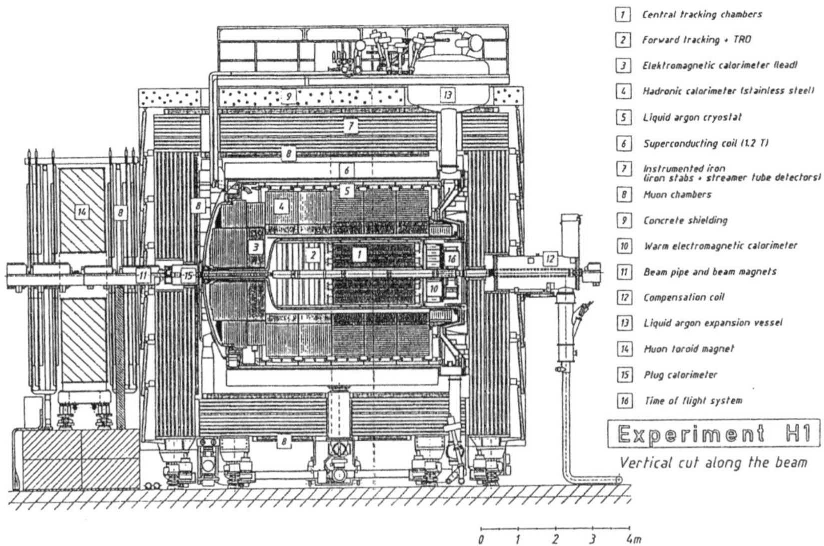


Fig. 10: An isometric view of the H1 detector.

The H1 electromagnetic calorimeter is made of 2.4 mm thick lead plates spaced 2.5 mm apart and the hadronic calorimeter is made of 16 mm thick stainless steel plates with gaps of 2 x 2.5 mm. Both calorimeters are read out using liquid Argon and are installed in a common cryostat. The liquid Argon read-out allows naturally for a fine segmentation both transversely and longitudinally. Transversely, the tower size varies between 3 x 3 and 8 x 8 cm² in the electromagnetic calorimeter and between 8 x 8 and 20 x 20 cm² in the hadronic calorimeter. Longitudinally, the electromagnetic shower is sampled 3 to 4 times, the hadronic shower 4 to 6 times. There is a total of 45000 independent segments in the detector.

A liquid Argon calorimeter has inherently a different response to showering and non-showering particles as shown in Fig. 11. However, an electron or neutral pion will deposit its energy over a short distance compared to the 6 to 8 absorption lengths needed to contain the energy of a charged pion. Since the liquid Argon calorimeter is finely segmented also in the longitudinal direction, the neutral pion content in a jet can be identified by looking for "hot spots" in the longitudinal energy loss profile. By weighting the measured longitudinal energy loss profile, the H1 group has achieved the same response to electrons and charged pions and hence an improved energy resolution. The energy resolution for electrons and hadrons (after weighting) is plotted in Fig. 11 as a function of energy.

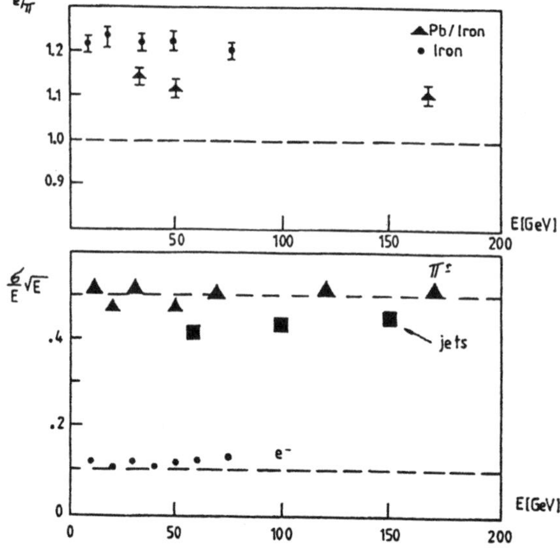


Fig. 11: a) The measured signal ratio of e to π for the lead and iron calorimeter modules as a function of energy.
b) The electron and hadron energy resolution after weighting.

The calorimetry is completed by a forward hadron calorimeter made of copper absorber plates with silicon read-out and a backward electromagnetic lead scintillator sandwich calorimeter.

The calorimeter has been operated since April 1991. The energy calibration has been stable to better than 0.2% and less than $10^{-3}$ of the read-out channels are not operational. Also the relative energy calibration between various units in the central calorimeter is quite good. However, more ep data are needed to reduce the uncertainty on the absolute energy calibration to a value below 5%. The measured energy resolution of the backward calorimeter is between 2% and 3% for electrons between 10 GeV and 30 GeV. An ultimate resolution of 1% can be reached using ep data.

The liquid Argon read-out can be operated in a magnetic field and the H1 group has chosen to install the calorimeter inside the large superconducting coil, which produces a 1.2 T field over a length of 5.75 m and a diameter of 5.2 m.

In the central region charged particles are being tracked using two jet chambers, augmented by two z-drift chambers and two multiwire proportional chambers. In the forward direction the trajectories are measured with three sets of radial and planar drift chambers. Electrons are identified using transition radiation detectors interleaved with tracking chambers. Trajectories of particles travelling in the backward direction are measured using multiwire proportional chambers. In the forward direction, a measured resolution ($\sigma$) of 150 μm has been achieved compared to the central chamber values of 190μm in r, Θ and 350 μm in z.

The calorimeters and the tracking detectors are enclosed and supported by an iron structure of 2200 tons which serves both as the return yoke of the magnet and as a hadronic backing calorimeter and muon-filter. The iron yoke is made of 7.5 cm thick iron plates interleaved with streamer tube chambers. Muons are identified and measured using three layers of chambers which are installed on both sides and in the middle of the iron structure. In the forward direction muons are detected in a muon spectrometer made of a magnetized iron toroid and six layers of drift chambers. The forward spectrometer covers production angles between 3° and 17° and is designed to detect muons in the range 5 GeV to 200 GeV.

The H1 group is now in the midst of a major detector upgrade. The aim is to extend the reach of the detector to lower values of x by installing spagetti type electromagnetic and hadronic calorimeters augmented by silicon trackers and drift chambers mounted along the direction of the incident electron. They are also constructing a central silicon detector to be used as a vertex detector.

## 3.3 THE ZEUS DETECTOR

A vertical cut of the ZEUS detector along the beam is shown in Fig. 12. It was the aim of the ZEUS Collaboration to construct an inherently compensating calorimeter with the best hadronic energy resolution possible. Compensation can be achieved by using a high-Z material as an absorber and judiciously choosing the thickness of the absorber plates and the scintillator plates. The ZEUS calorimeter is made of 3.3 mm thick, 20 cm wide and up to 4.6 m long depleted uranium plates, cladded with stainless steel, which are interleaved with 2.6 mm thick scintillator plates. The tower size, defined by the scintillator plates is 5x20 $cm^2$ in the electromagnetic part and 20x20 $cm^2$ in the hadronic calorimeter.

The scintillator plates are read out by means of electromagnetic wavelength shifters, light guides and photo-multipliers. Longitudinally the calorimeter is segmented into an electromagnetic and two hadronic sections. The calorimeter is made of three parts; the central region surrounded by the barrel (BCAL) calorimeter, the forward (FCAL) and the rear (RCAL) calorimeters cover the proton, respectively the electron direction. The total coverage is 99.8% of $4\pi$.

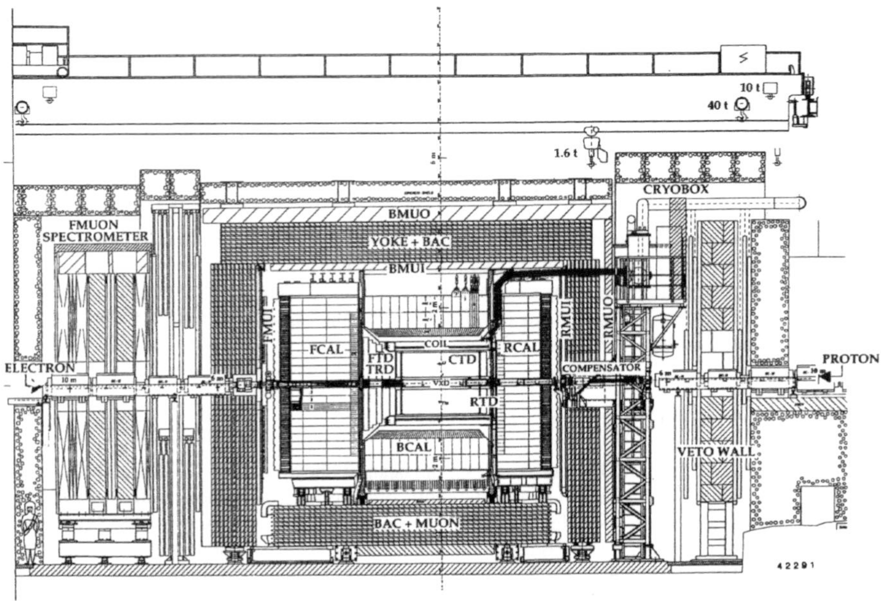


Fig. 12: Schematic view of the ZEUS detector. The component labels are defined in the text.

A prototype calorimeter module has been extensively tested and some of the results are shown in Fig. 13. Note that the compensation indeed works over the whole energy range tested and that hadrons are measured with excellent resolution.

The natural radioactivity of the uranium is used to monitor the calibration of the calorimeter to better than 0.2%. The noise caused by the uranium decay is on the order of 12-30 MeV depending on cell size. The total noise, integrated over an area of 80x80 cm$^2$ which corresponds to the area needed to contain a normal jet, it of order 200 MeV. The calorimeter has a time resolution of better than 1 ns for particle energies above 3 GeV. In order to be able to identify electrons within a jet, the ZEUS Collaboration has porposed to insert two layers of silicon pads at depths of 3 and 5 radiation lengths.

Photomultipliers are sensitive to magnetic fields and the calorimeters are hence installed outside the superconducting coil. The solenoid, 1.72 m in diameter and 2.8 m long, is 0.9 radiation lengths thick and produces a field of 1.8 T.

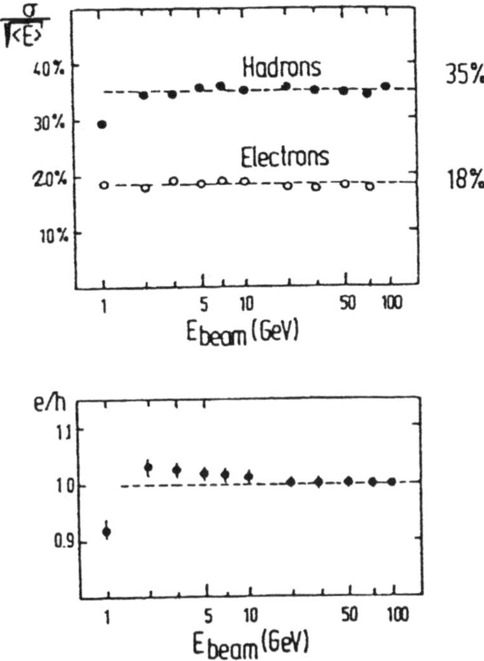


Fig. 13: a) The energy resolution observed with the uranium calorimeter module for incident hadrons and electrons as a function of energy.
b) The ratio of energy deposits by electrons and charged pions as a function of the particle energy.

The inner tracking detector is located in the magnetic volume. In the central part, particles are tracked with a time expansion vertex chamber (VXD) and a vector drift chamber (CTD). The vertex chamber has a measured resolution of 35 μm. The drift chamber has 9 superlayers each with 8 layers of sense wires, yielding a total of 4608 sense wires. A resolution of 120μm was achieved in a test beam. In the superlayers 1, 3 and 5 a z read-out is available with a resolution of 40 mm. The forward tracking detector (FDET) contains three planar drift chambers and four transition radiation detectors to identify particles within the jet. Particles emitted along the incident electron direction are measured in the rear tracking detector (RTD) which is made of planar drift chambers. The planar drift chambers have an expected spatial resolution of order 120 μm and the energy loss of a minimum ionizing particle is measured with an uncertainty of 6-7%.

The iron yoke of 2000 tons is segmented into 7.5 cm thick iron plates, intercleaved with proportional tube chambers /BAC) which serve to measure the energy leaking out of the hadronic calorimeter and to identify muons. Four layers of limited streamer tube chambers are installed on both sides of the barrel part (BMUI, BMUO) and the rear part (RMIZU, RMUO) of the iron yoke. In the forward direction high energy muons are detected using a toroidal spectrometer equipped with planar drift chambers and limited streamer tube chambers.

Protons scattered under small angles (~ 0.2 mrad) can be measured to within 10 $\sigma$ of the proton beam size by an elaborate set of six Roman pots, installed between 23 m and 90 m downstream of the interaction point.

The material in front of the Uranium-scintillator calorimeter of the ZEUS detector leads to a degradiation of the calorimeter energy measurement. The ZEUS group has proposed to install a presampler consisting of a layer of scintillator tiles in front of the forward and rear calorimeter sections. By combining on an event by event basis the information from the presampler and the calorimeter, the energy scale and the energy resolution of the calorimeter is recovered.

The ZEUS group is also installing a small angle tracker and a forward neutron calorimeter.

PHYSICS RESULTS

4.1 Introduction

The kinematic region in $Q^2$ and $\nu$ which has been liberated by the turn on of HERA is shown in Fig. 14 and compared to the region which can be investigated with a 600 GeV muon beam interacting with protons at rest.

The present maximum values of $Q^2$=600 GeV$^2$ and $\nu$=400 GeV from fixed target experiments can be extended by nearly two orders of magnitude to $Q^2$ and $\nu$ values of order 30000 GeV$^2$ respectively 40000 GeV. Furthermore, measurements of the cross section in the scaling region can be extended down to x-values on the order of a few $10^{-5}$ compared to the present lower limit of roughly $10^{-2}$.

Photoproduction and other low $Q^2$ processes can be measured up to c.m. energies of 250 GeV. Previous to HERA photoproduction data were available only up to 18 GeV.

Also the range available to search for new physics in electron-proton interactions is extended considerably. Leptoquarks and electron like new leptons can be searched for nearly up to the kinematic limit of 280 GeV - 300 GeV. Righthanded currents can be observed for a propagator mass of order 1 TeV.

A complete up-to-date discussion on HERA physics can be found in reference [8].

The data reported here are based on the 1992 integrated luminosity of 25 nb$^{-1}$ and they yield first information on photoproduction, deep inelastic scattering at low values of Bjorken x and the search for new particles.

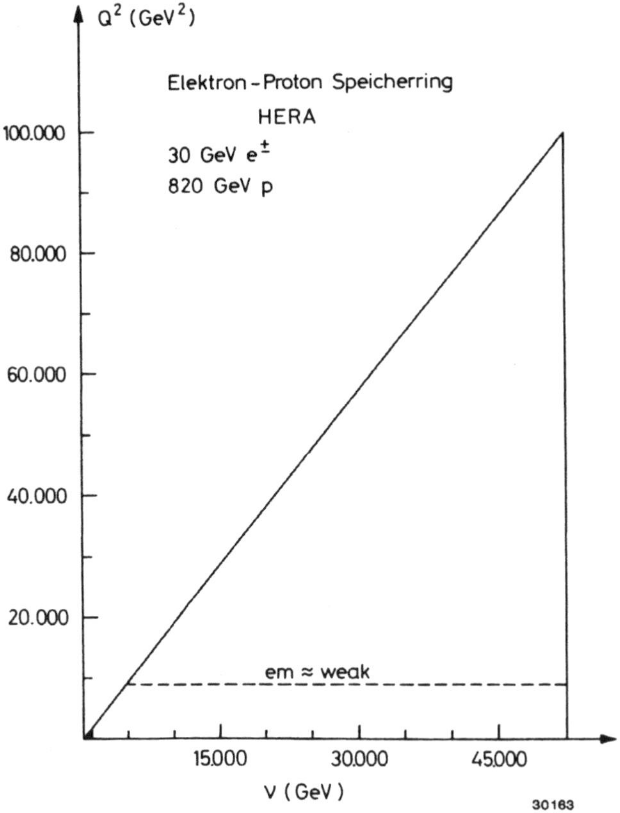


Fig. 14: The kinematic region in $Q^2$ and ν at HERA. The black area at the left hand corner corresponds to a 600 GeV muon beam incident on protons at rest.

## 4.2. PHOTOPRODUCTION

The incident electron beam for very small values of $Q^2$ is equivalent to a well collimated beam of quasi real photons. In this case the electron scattering cross section is related to the photoproduction cross section as follows:

$$\frac{d\sigma(ep)}{dy} = \frac{\alpha}{2\pi}\frac{1+(1-y)^2}{y}\cdot\sigma(\gamma p)\cdot\ln\frac{Q^2_{max}(y)}{Q^2_{min}(y)}$$

The photon energy and $Q^2$ can be determined from a measurement of energy and angle of the scattered electron. Both experiments use the luminosity counter to tag the scattered electron.

An electron which has emitted a photon at a small angle is deflected away from the circulating electron beam in the off-axis, low β-quadrupoles and in the downstream magnets, exits the accelerator vacuum through a thin window, and impinges on the luminosity electron counter.

The ZEUS luminosity monitor accepts electrons with scattering angles below 6 mrad and energies between 0.2E and 0.4 E. This corresponds to a range in $Q^2$ of the quasi real photons from $10^{-7}$ GeV² to $2 \times 10^{-2}$ GeV² and center of mass energies between 186 GeV and 220 GeV.

The H1 tagger accepts electrons scattered at angles below 5 mrad and energies in the range of 6 GeV to 19 GeV corresponding to a $Q^2$-range between $10^{-7}$ GeV² and 0.015 GeV² and a mean c.m. energy of 197 GeV.

The photon is a particle with unique properties. It is on one hand a fundamental gauge boson with well defined couplings to elementary fermions and other gauge bosons as shown in Fig. 15a. On the other hand the photon may fluctuate into a vector boson and then it behaves like a hadron (Fig. 15b). For large values of $p_\perp$, the transverse momentum in the event, the photon interacts dominantly via its hadronic constituents, quarks and gluons, yielding resolved photon events as shown in Fig. 15c. Note that direct and resolved photon events populates different final state topologies as discussed below.

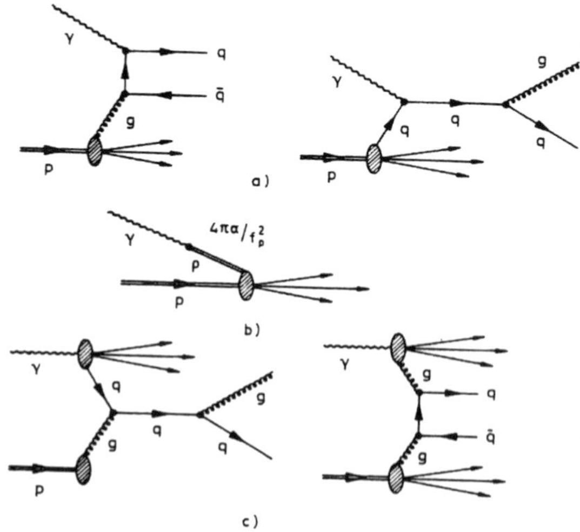


Fig. 15: Photon-hadron interaction.
a) Examples of the pointlike coupling of photons.
b) The vector dominance model.
c) "Resolved Photon" events induced by the gluon and quark content of the photon.

Both groups reports data on photoproduction. The data analysis and the event selection criteria are disussed in detail in reference 9 and 10.

The total photon-proton cross section has previously been studied for c.m. energies up to 18 GeV. At HERA these data have now been extended by an order of magnitude in c.m. energy.
H1 reports at a mean c.m. energy of 197 GeV: $\sigma_T(\gamma p) = (156 \pm 2 \pm 19)\mu b$.
ZEUS reports at a mean c.m. energy of 181 GeV: $\sigma_T(\gamma p) = (133 \pm 2.5 \pm 16)\mu b$.

Quoted are statistical and systematic errors. The systematic errors combines the uncertainty in the luminosity determination with the uncertainties resulting from acceptance and efficiency, added in quadrature.

The total photoproduction cross section at 200 GeV in c.m., compared to its value of roughly 118 μb measured at 18 GeV rises slowly with energy in agreement with Regge Model predictions.

The energy dependence of the photoproduction cross section is also similar to the rise observed in the total $\bar{p}p$ cross section. This is expected in a Vector Dominance Model in which the photon behaves like a hadron.

Direct and resolved photon processes can be separated by observing the final state jet topology. A pointlike interaction between the quark and the photon will in general yield two jets of hadrons from quark fragmentation. If the photon interacts via one of its hadronic constituents (Fig. 15c)) then, in addition to the two hadron jets at large angles there will also be a hadron jet along the incident electron direction resulting from the fragmentation of the remains of the photon.

At the high c.m. energies available at HERA, quark and gluon fragmentation jets can be readily identified and reconstructed. Two examples of reconstructed photoproduction events [11] are shown in Fig. 16.

Plotted in Fig. 16 are the energies deposited in the event as a function of azimuthal angle φ and pseudorapidity $\eta = -\ln tg\,\theta/2$ with the positive z-direction along the proton direction. Event 16a shows a clear two jet structure and is a candidate for a direct photon event. Event 16b shows, in addition to the two hadron jets at large angles, a third jet of hadron along the incident direction of the electron and is hence a candidate for a resolved photon event.

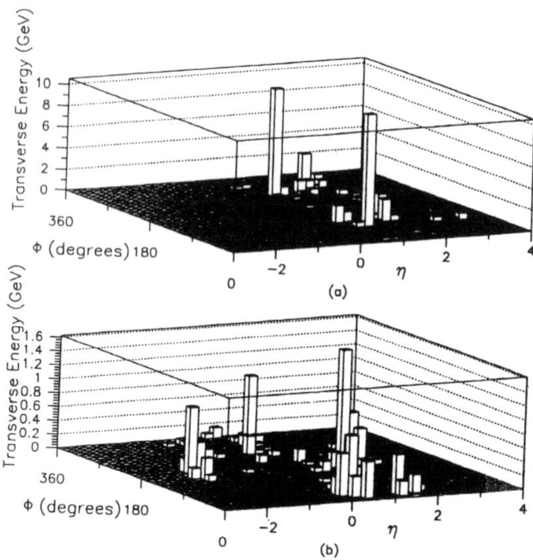


Fig. 16: Deposited energy in photoproduction events as a function of φ and $\eta = -\ln \mathrm{tg}\, \theta/2$

Since we are dealing with a twobody scattering process, the relative momenta $x_\gamma$ and $x_p$ of the partons in the photon, respectively the proton can be calculated from a measurement of energies and momenta of the final state partons.

The energies and momenta of the final state partons are estimated from a measurement of the energies and momenta of the hadronic jets observed in the detector. This yields [11]:

$$x_p^{meas} = \frac{\sum_{jets}(E+p_z)_{jet}}{2E_p}$$

$$x_\gamma^{meas} = \frac{\sum_{jets}(E-p_z)_{jet}}{\sum_i (E-p_z)_i}$$

where the sum in denominator runs over 4π.

The measured $x_\gamma^{meas}$ distribution is plotted in Fig. 17 and shows a clear peak at high values of $x_\gamma$ and a broad maximum around $x_\gamma \approx 0.2-0.3$.

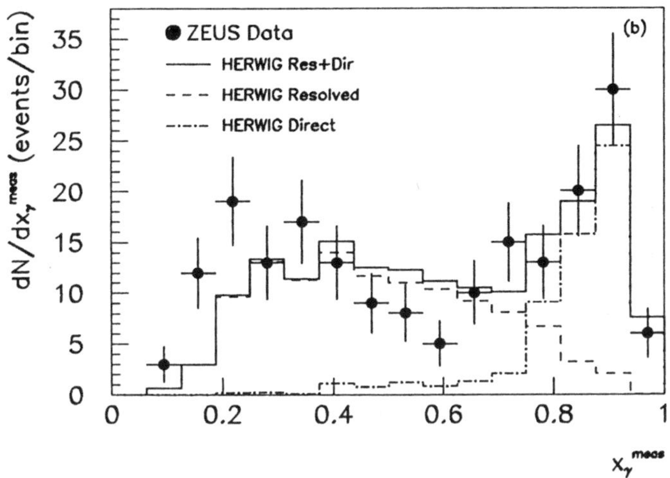


Fig. 17: The $x_\gamma$ distribution as reconstructed from a measurement of final state hadron jets. The contributions expected from direct and resolved photon events are also indicated. This analysis is from the ZEUS group and is based on the 1992 data.

The peak at high values of $x_\gamma$ can be identified with the direct process which predict $x_\gamma$=1. A Monte Carlo calculation of the direct contribution is shown as the dash-dotted line and is in good agreement with the data at high values of $x_\gamma$ and predicts only a small number of events with $x_\gamma$<0.7.

The broad distribution in $x_\gamma$ can be ascribed to resolved photon events where a partonic constituent of the photon interacts with the proton. The Monte Carlo prediction for resolved photon events is shown as the dashed histogram in Fig. 17.

Similar results have also been obtained [9] by the H1 group which has extended the ZEUS data down to lower values of $x_\gamma$. They find that at low $x_\gamma$ the photon, in addition to its quark content also has a sizeable gluon component.

### 4.3 Deep inelastic neutral current events

The kinematic region in $1/x$ and $Q^2$ available at HERA and at a 600 GeV muon beam incident on a proton at rest ist plotted in Fig. 18.

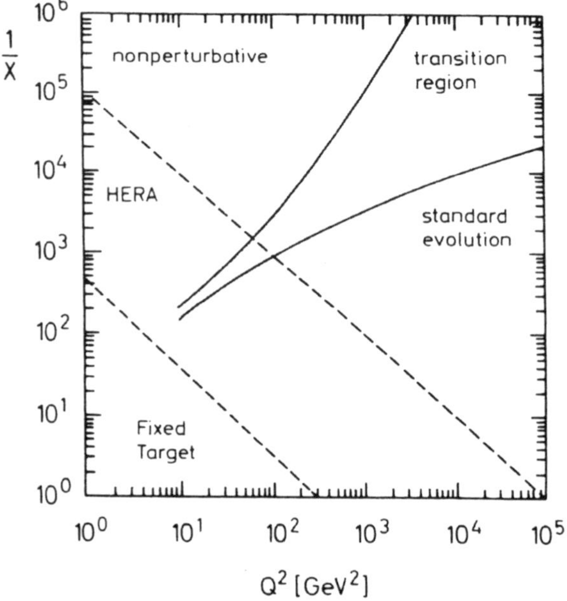


Fig. 18: The kinematic region in $1/x$ and $Q^2$ available at HERA and with a 600 GeV muon beam incident on protons at rest. The perturbative and non-perturbative domains are separated by a transition region.

The $1/x$, $Q^2$ plane can be divided into three main areas: a QCD perturbative region located in the lower right hand corner at moderate x and large $Q^2$, a non-perturbative region in the left hand upper corner and a transition region in between.

In the classic region of QCD, the $Q^2$ evolution of the structure functions can be computed from the Gribov-Lipatov-Altarelli-Parisi [12,13] equations. These equations are defined in terms of the parton splitting diagrams as indicated in Fig. 19.

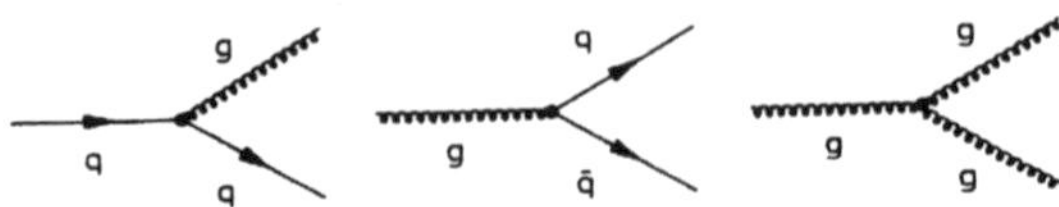


Fig. 19: The relevant parton splitting diagrams.

At moderate values of $Q^2$ but low values of x the structure functions are dominated by the gluon distribution function $xG(x,Q^2)$. Since the gluon density increases with 1/x, the gluon-gluon interaction can no longer be neglected although the gluon-gluon coupling is still weak. In this transition region one may be able to use the parton language and the behaviour of the structure functions may be described [14] by adding a recombination term onto the perturbative evolution equations.

A further extrapolation in $1/x$ yields a very dense partonic system. Although $\alpha_s(Q^2)$ ist still weak the effective interactions - due to the high parton density - are so strong that the parton picture may no longer be applicable.

The kinematic of the scattered electron is given in Fig. 20. Plotted are curves of constant energy and constant scattering angle of the final state electron in the x, $Q^2$ plane. The coordinate system is such that positive y is pointing upward and positive z is along the incident proton direction.

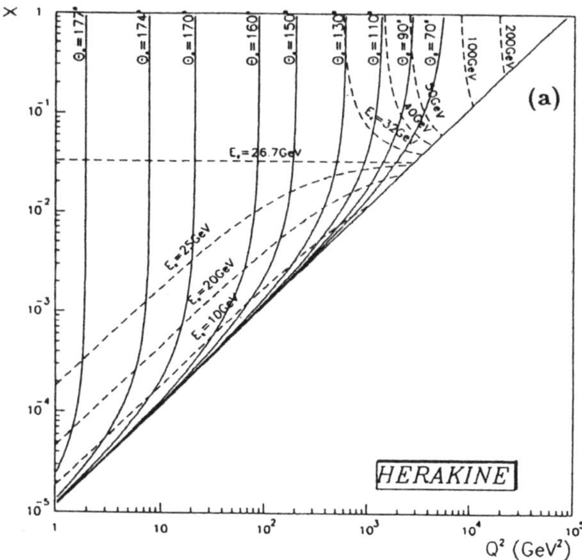


Fig. 20: Deep inelastic electron-proton kinematics

Note that low x events correspond to final state electrons travelling in the backward direction, i. e. scattered through small angles.

Both experiments collected data using a minimum bias trigger.

The H1 experiment [15] demanded that at least 4 GeV were deposited in the backward calorimeter (BEMC) which covers production angles between 160° and 172.5°. This resulted in a total of $5 \cdot 10^5$ events.

The ZEUS experiment required a threshold of 1 GeV in the electromagnetic calorimeter towers, or 10 GeV in the ring of towers which covers the region around the RCAL (downstream) beam pipe. A total of $4.2 \cdot 10^6$ events were recorded.

The bulk of the recorded events resulted from beam-gas interactions and photoproduction. A detailed discussion of the criteriea used to reject background events and to separate deep inelastic and photoproduction events can be found in references [15,16].

After the cuts H1 retains 1026 events with $E_0 > 10.4$ GeV ($y<0.6$) and 5 GeV$^2 < Q^2 < 80$ GeV$^2$. The background due to hadronic events is negligable as confirmed by measurements using non-colliding pilot bunches. The remaining photoproduction background is found to be less than 30% at the highest y and negligable at $y<0.4$.

A total of 1299 events including an estimated 35 background events satisfy the ZEUS selection criteria. The data cover 10 GeV$^2$ $< Q^2 <$ 640 GeV$^2$ and values of x between $4.2 \cdot 10^{-4}$ and $3.2 \cdot 10^{-2}$.

An example of a deep inelastic neutral current event is shown in Fig. 21.

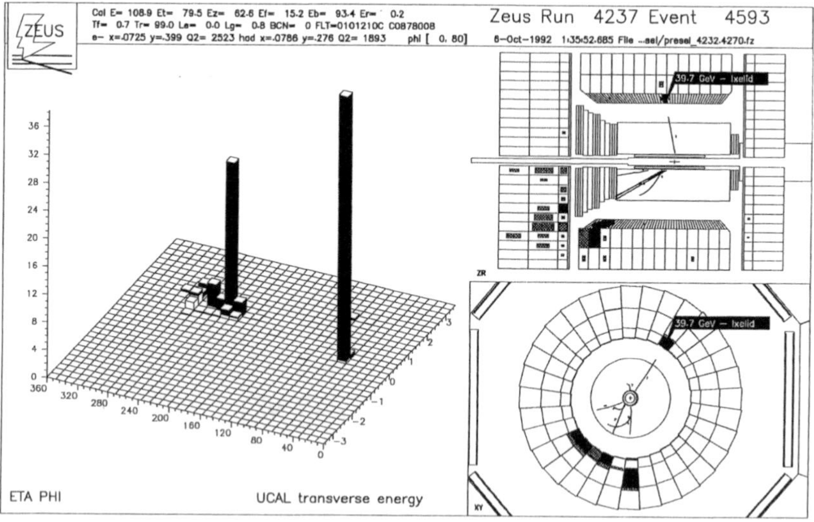


Fig 21: Deep inelastic neutral current event measured by the ZEUS collaboration.

Note that the event is clean with a well defined jet of hadrons. The transverse momentum components of the jet are balanced by an electron travelling at the opposite azimuthal angle.

In the present analysis H1 determines [15] $Q^2$ from a measurement of the energy and the angle of the scattered electron. The value of x is determined either from a measurement of the electron kinematics or from a combination of electron and final state hadron data, i. e. $x_m = Q_e^2/(sy_h)$, where $y_h$ is determined using the Jacquet-Blondel relation

$$y_h = \frac{\sum_h (E - P_z)_h}{2E_e}$$

where E is energy of the hadron and $P_z$ its momentum component along the incident proton direction. $s = 4E_e \cdot E_p$ is the center of mass energy squared.

ZEUS determines [16] the event variables x, $Q^2$ from a measurement of the angle of the final state electron and the angle $\gamma_h$ of the massless object balancing the momentum vector of the electron. It is determined from the hadronic energy flow measured in the detector.

$$\cos\gamma_h = \frac{\left(\sum_n P_x\right)^2 + \left(\sum_n P_y\right)^2 - \left(\sum_n (E - P_z)\right)^2}{\left(\sum_n P_x\right)^2 + \left(\sum_n P_y\right)^2 + \left(\sum_n (E - P_z)\right)^2}$$

The sums runs over all calorimeter cells n except those which have been assigned to the electron.

In the double angle method x,$Q^2$ can then be expressed as:

$$Q^2 = 4E_e^2 \frac{\sin\gamma_h (1 + \cos\Theta_e)}{\sin\gamma_h + \sin\Theta_e - \sin(\Theta_e + \gamma_h)}$$

$$x = \left(\frac{E_e}{E_p}\right)\left(\frac{\sin\gamma_h + \sin\Theta_e + \sin(\Theta_e + \gamma_h)}{\sin\gamma_h + \sin\Theta_e - \sin(\Theta_e + \gamma_h)}\right)$$

A scatter plot in the x, $Q^2$ plane of the events which met the selection criteria is shown in Fig. 22 for ZEUS and in Fig. 23 for H1. Note that both experiments observe events with x-values down to $10^{-4}$ - an extension by a factor of 100 compared to present fixed target data.

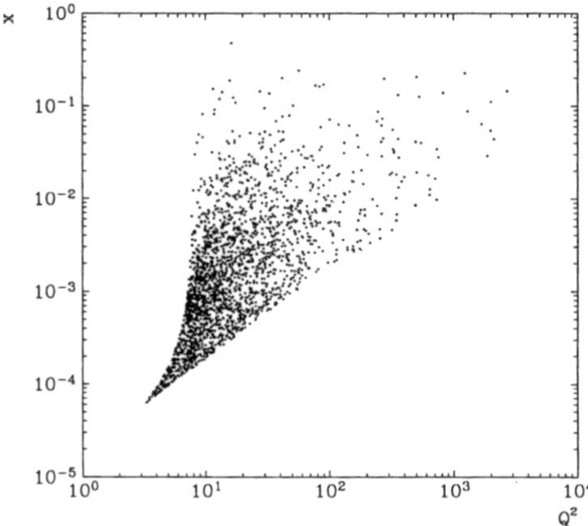


Fig. 22: A scatter plot of deep inelastic neutral current events in the x, $Q^2$ plane as measured by the H1 collaboration.

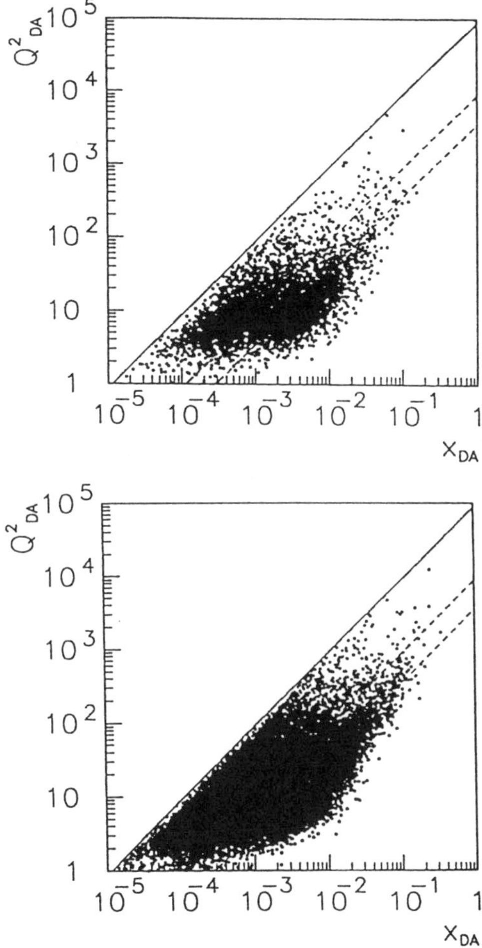


Fig. 23: A scatter plot of deep inelastic neutral current events in the x, $Q^2$ plane measured by the ZEUS collaboration. The 1992 and 1993 data are shown in the upper, respectively the lower figure.

The Born cross section for deep inelastic electron scattering of a proton can be expressed in terms of the structure function $F_2$ and $\sigma_L / \sigma_T$ , the photoproduction cross section of longitudinally and transversely polarized photons:

$$\frac{d^2\sigma}{dx\, d\, Q^2} = \frac{2\pi\alpha^2}{Q^4 x}\left(2(1-y)+\frac{y}{1+R}\right)F_2\left(x,Q^2\right)$$

This expression is valid for $Q^2 << M_z^2$.

The correction due to R has been evaluated [15 ] within the QCD model. The correction is rather small amounting to at most 8% for the H1 data.

The values of $F_2(x,Q^2)$ determined by the H1 group using either the electron data $(x_e,Q_e^2)$ or mixed electron-hadron data $(x_m,Q_e^2)$ are plotted in Fig. 24 versus x for $Q^2$=15 $GeV^2$ and $Q^2$=30 $GeV^2$. The formfactors evaluated by the two independent methods are in good agreement.

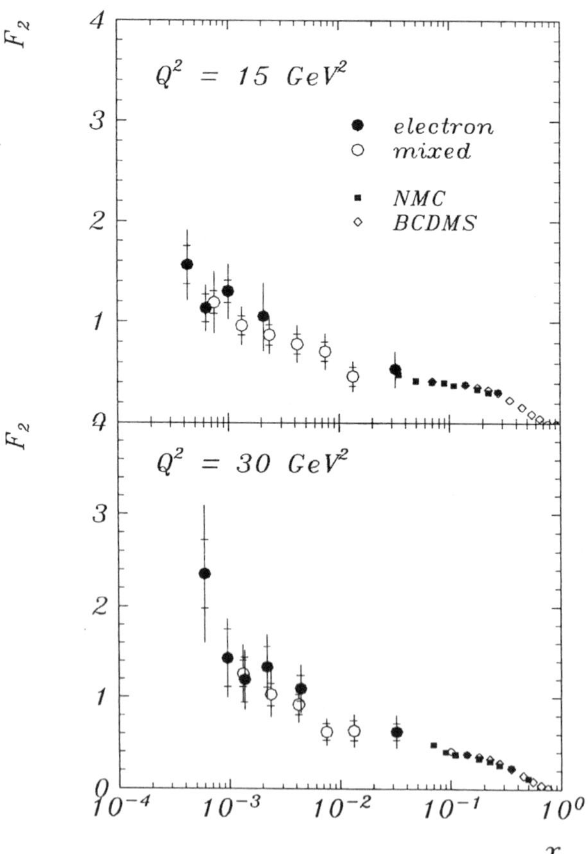


Fig. 24: Measurement of $F_2(x,Q^2)$ for two values of $Q^2$. The full circles identify points where x, $Q^2$ were extracted using electron data only, the points represented by the full circles were evaluated using mixed electron-hadron data.

The error bars show statistical and total errors obtained by adding the statistical and the systematic errors in quadrature. In addition all points have a normalization uncertainty of 8%. Data points of the fixed target muon proton scattering experiments NMC [17] and BCDMS [18] are shown for comparison.

The systematic errors for the data points varies between 15% and 22 %. They do not include an 8% global normalization uncertainty nor the effect due to the uncertainty in R.

H1 assign all events with y<0.06 to a single bin. Events in this region have x values in the range $6 \cdot 10^{-3} < x < 1$. These data overlaps to a large extent with fixed target muon data. H1 use the measured shape of $F_2(x,Q^2)$ from these data to determine the average x value of this bin. This high x-point agrees well with the lower energy data providing an independent cross-check of the absolute normalization with an acuracy of order 20%.

The $F_2(x,Q^2)$ form factor measured by H1 and ZEUS is plotted in Fig. 25 versus x for values of $Q^2$ between 8.5 GeV² and 1000 GeV².

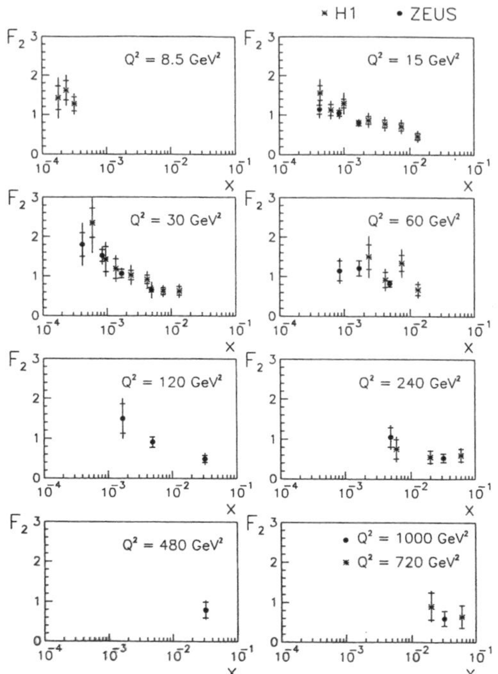


Fig. 25: $F_2(x,Q^2)$ as measured by the H1 and ZEUS collaborations plotted versus x for constant values of $Q^2$ between 8.5 GeV² and 1000 GeV².

The two sets of data are in excellent agreement and show a clear rise of the structure function for decreasing values of x.

A QCD evolution equation derived by Kuraev, Fadin and Lipatov [19] predict a $x^{-\Lambda}$ behaviour of the gluon density at low values of x with $\Lambda$ of order 0.5. The data are not yet precise enough to determine the value of $\Lambda$. However, the observed strong increase in $F_2(x,Q^2)$ with decreasing values of x indicate that interesting effects like screening and saturation may become detectable at HERA.

The ZEUS data [15] are plotted in Fig. 26 versus $\ln Q^2$ for fixed values of x.

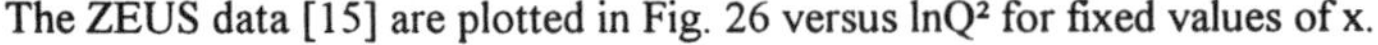

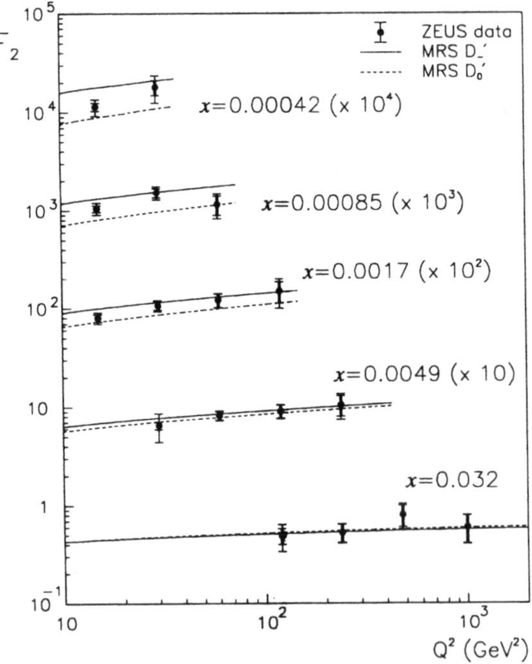


Fig. 26: The structure function $F_2(x,Q^2)$ as a function of $\ln Q^2$ for different values of x. The inner error bar is the statistical error, and the outer bar shows the systematic error added in quadrature.

The data are consistent with the QCD logarithmic scaling violation. Fits using two different parton distributions are also shown. The parton distribution MRSD′ has a singularity $x^{-1/2}$ for small x whereas the distribution $MRSD_0$ approaches a constant value [20].

The observed rise of $F_2(x,Q^2)$ at low values of x is, within the context of QCD due to the rise of the gluon density function $xG(x,Q^2)$. The $Q^2$ dependence of $F_2(x,Q^2)$ is thus a measure of $xG(x,Q^2)$. At small x the gluon term in the evaluation equation is expected to dominate leading to the approximate relation:

$$\frac{\partial F_2(x,Q^2)}{\partial \ln Q^2} \cong \frac{5\alpha_s(Q^2)}{9\pi}\int_x^1 \left(w^2+(1-w)^2\right)\cdot G\left(x/_w, Q^2\right)dw$$

This equation can be approximated [21] by:

$$\frac{\partial F_2(x,Q^2)}{\partial \ln Q^2} \cong \frac{10\alpha_s(Q^2)}{27\pi}\cdot G\left(2x, Q^2\right)$$

H1 has used this relationship to extract [22] $G(2x,Q^2)$. In the calculations they assume $\alpha_s = 0.24$ corresponding to four quark flavours and $\Lambda = 200$ MeV.

The data has been scaled to $Q^2 = 20$ GeV$^2$ using the observed linear dependence of $F_2(x,Q^2)$ on $\ln Q^2$.

The gluon distribution function extracted [22] from the data is plotted in Fig. 27 versus x and compared to various model predictions.

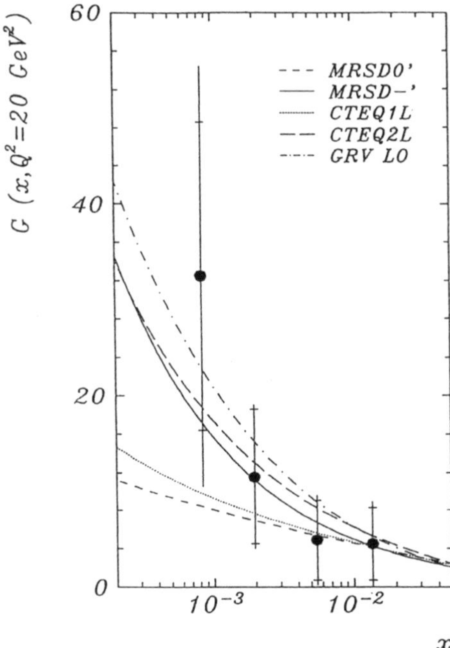


Fig. 27: The gluon distribution function $G(2x,Q^2)$ at $Q^2 = 20$ GeV$^2$. The inner error bars are the statistical errors, the full errors represents statistical and systematic errors added in quadrature.

The error bars do not yet allow to discriminate among the various models. However, the data indicate a finite value of $G(2x,Q^2)$, rising towards smaller values of x.

We will now briefly discuss the properties of the final state in deep inelastic events. In the naive parton model the struck quark will materialize as a well collimated jet of hadrons travelling along the initial direction of the struck quark with the remenants of the proton disappearing down the beam pipe.

QCD predicts a more complex final state as indicated in Fig. 28 a. The spacelike virtual photon interacts with one of the coloured quarks in the proton and the colour transfer between the quark and the proton remenant will lead to hadrons populating the rapidity gap between the current jet and the forward direction.

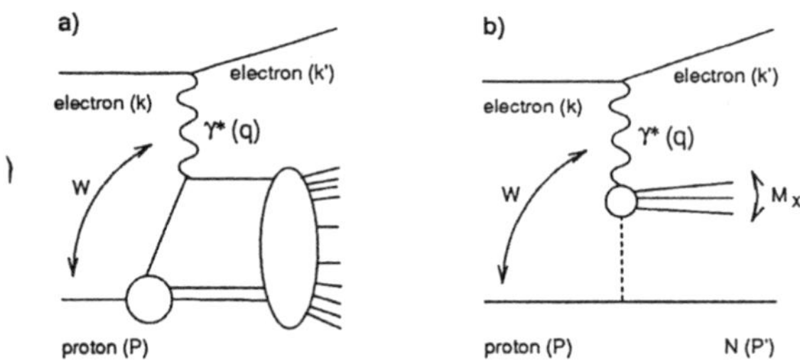


Fig. 28: a) Schematic diagram describing particle production in deep inelastic electron-proton scattering in which the photon interacts with a coloured quark.
b) Schematic diagram describing particle production by diffractive dissociation in deep inelastic electron-proton interactions. W is the center of mass energy of the $\gamma^*p$ system and $M_X$ is the invariant mass of the hadronic system measured in the detector. N represents the proton or a low mass nucleon system.

Furthermore, higher order processes in $\alpha_s$ will lead to multijet events. To first order in $\alpha_s$ three processes will contribute, gluon radiation by the struck quark $q \rightarrow g \cdot q$, boson-gluon fusion. $\gamma \cdot q \rightarrow q \cdot q$ and QCD Compton effect $\gamma \cdot q \rightarrow q \cdot g$.

Measurements of the hadronic final state in deep inelastic electron scattering are in qualitativ agreement with these QCD predictions.

Examples of final state event topologies obtained [23] by ZEUS group are shown in Fig. 29.

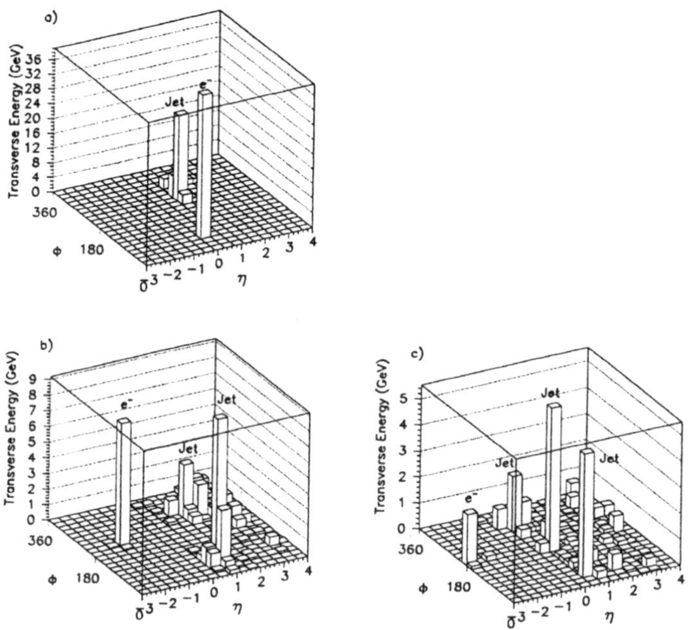


Fig. 29: Transverse energy distributions of events in the (η, φ) plane.
a) one jet event.
b) two jet events.
c) three jet events.

Also the H1 group observes multijet events and they have extracted [24] the topological multijet cross sections using the JADE algorithm [25].

The fraction of events with the current jet and the jet resulting from the remains of the proton is denoted by $R_{1+1}$, events with one additional jet by $R_{2+1}$ and so on. With $y_{cut} = 0.02$ the $R_{1+1}$ and $R_{2+1}$ topologies dominate with $R_{2+1}$ contributing on the order of 10% to 20% to the total cross-section. The authors showed that hadronization and detector effects are small using the JADE algorithm. Data were compared to the MEPS model [26] which is based on an evaluation of the matrix elements to order $\alpha_s$ and an approximate treatment of higher order effects by leading log hadron showers. This model represents the data rather well and shows that photon-gluon fusion makes the largest contribution to the $R_{2+1}$ topology.

The matrix element for jet production decreases with increasing values of $Q^2$ because of the $Q^2$ dependence of the coupling constant $\alpha_s$. However, this decrease is masked by the increase in phase space for jet production with $Q^2$, given the present cuts and the jet definition of $R_{2+1}$.

The measured values of $R_{2+1}$ extracted for $y_{cut} = 0.02$ are plotted versus $Q^2$ in Fig. 30 and compared to various theoretical predictions.

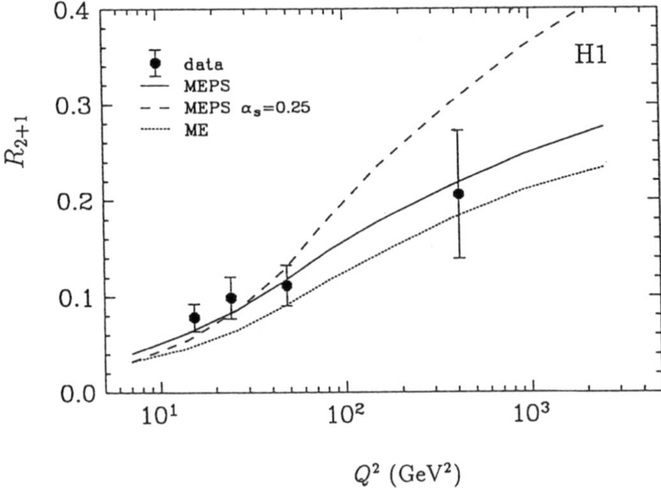


Fig. 30: The jet fraction $R_{2+1}$ versus $Q^2$ for $W^2 > 5000$ GeV$^2$ and $y < 0.5$. The data are compared to the MEPS model with $\alpha_s$ running. The data are also compared to the matrix element prediction with $\alpha_s$ running, neglecting the parton shower contribution.

The data clearly prefer the MEPS model with $\alpha_s$ running. However, better statistics are needed to make a quantitative statement.

The ZEUS collaboration has measured [27] the rapidity distribution of hadrons produced in deep inelastic neutral-current interactions at a center of mass energy of 296 GeV.

The hadronic energy flow for events with $x < 10^{-3}$ is shown as a function of pseudo rapidity in Fig. 31.

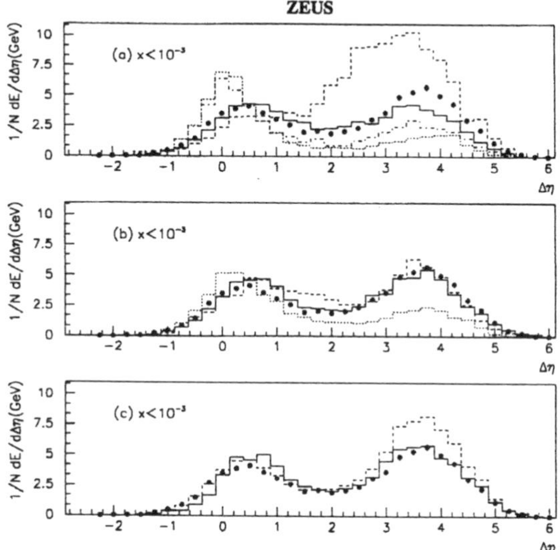


Fig. 31: The energy weighted pseudo rapidity difference, $\Delta\eta$, of the hadronic system calorimeter cells with respect to the struck quark from the quark parton model. The ZEUS data points are shown as dots. In (a) the full histogram is MEPS, the dashed histogram PS($W^2$), the dotted histogram PS($Q^2$) and the dash-dotted histogram ME. In (b) the full histogram is CDM+BGF, the dashed histogram CDM and the dotted histogram PS($Q^2$) (1-$\alpha$). In (c) the full histogram is based on HERWIG and the dashed histogram on HERWIG including SUE.

The pseudo rapidity $\eta$ is defined as $\eta$=-ln(tg $\Theta$/2). The plot shows the energy weighted pseudo rapidity differences between the calorimeter cells and that of the hadron angle $\gamma_h$ as defined above, i. e. $\Delta\eta = \eta - \eta_h$.

It is striking, and in agreement with QCD, that nearly all the energy is deposited at positive values of $\Delta\eta$ - i. e. at angles between the direction of the current jet and the remenant proton.

Also note that there is a small shift in the peak at low values of $\Delta\eta$ from the value 0 predicted by the quark parton model towards positive $\Delta\eta$ values.

The data show that the Lund model (MEPS), in which the first order matrix elements are combined with parton showers describes the hadronic energy flow reasonably well. The predictions of the colour dipole model (CDM) [28] and the HERWIG parton shower model [29] also reproduce the overall features of data.

Predictions of a model which is based on first order matrix elements (ME) fails to reproduce the energy distribution. Also the LUND parton shower model (PS) [30] does not fit the data irrespective of scale. HERWIG including a soft underlaying event (HERWIG + SUE) overestimate the energy deposited at large values of $\Delta\eta$.

The ZEUS collaboration shows that these differences are not due to the choice of fragmentation parameters, model parameters or the choice of parton density distribution functions.

The ZEUS collaboration has observed [31] a new class of events characterized by a large gap in rapidity between the remenant proton and the hadron energy deposited closest to the proton direction. An event of this type is shown in Fig. 32.

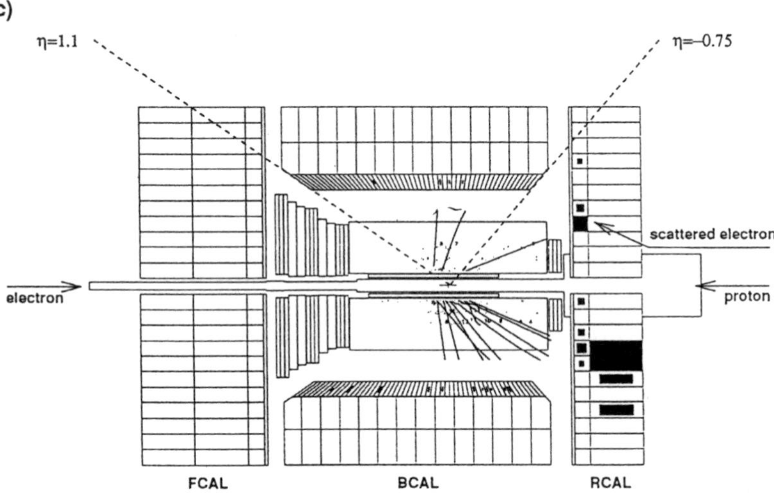


Fig. 32: Schematic view of the ZEUS calorimeter and central tracking. Overlaid is an event with a large rapidity gap.

In Fig. 33 the events are plotted versus $\eta_{max}$ which is the pseudo rapidity of the hadron cluster which is nearest to the remenant proton direction. At least 400 MeV must be deposited at the cluster.

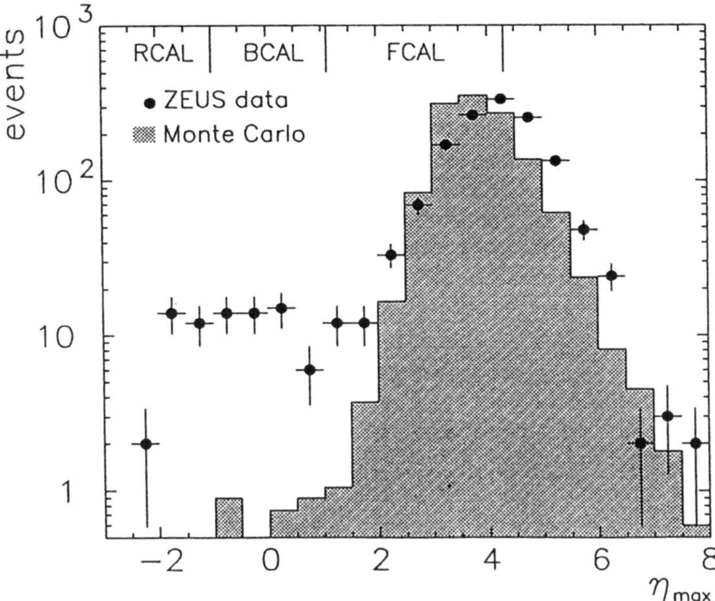


Fig. 33: The distribution of $\eta_{max}$ where $\eta_{max}$ is defined as the maximum pseudo rapidity of an hadronic energy cluster in an event. The data are compared to a Monte Carlo simulation of the $\eta_{max}$ distribution expected for deep inelastic neutral current events. The boundaries of various calorimeter segments are also indicated.

The distribution shows a clear peak at $\eta_{max}$ about 4 and then a flat distribution for $\eta_{max}$ between -2 and 2. Whereas the peak can be accounted for by the deep inelastic neutral current events, the observed $\eta_{max}$ distribution beyond $\eta_{max} < 1.5$ cannot.

89 events out of a total event sample of 1441 events have $\eta_{max} < 1.5$. Deep inelastic electron-proton scattering is expected to contribute 4.1 ± 1 event and the simulated electron gas background yields 7 ± 3 events in this region. Thus an excess of 78 ± 10 events with $\eta_{max} < 1.5$ remains corresponding to 5.4% of the total event sample.

The properties of events with $\eta_{max} < 1.5$ are displayed in Fig. 34 and compared to events with $\eta_{max} > 1.5$.

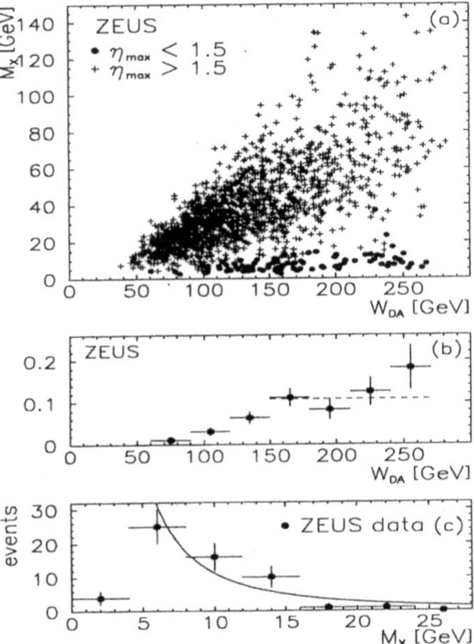


Fig. 34: Correlation between the invariant mass $M_x$ of the hadrons observed in the calorimeter and the invariant mass $W_{DA}$ of the $\gamma^*p$ system. Events with $\eta_{max} < 1.5$ are shown as the solid dots.

b) Fraction of events with $\eta_{max} < 1.5$ as a function of $W_{DA}$. The dashed line indicates the region in which acceptance corrections are independent of $W_{DA}$.

c) $M_x$ distribution of events with $\eta_{max} < 1.5$ and $W_{DA} > 150$ GeV. The solid line shows a $1/M_x^2$ dependence.

The large rapidity events have the following properties:

a) The invariant mass $M_x$ of the hadronic system observed in the detector is typically less than 10 GeV and nearly independent of the ($\gamma^*p$) invariant mass.

b) Acceptance corrections are nearly independent of W for $W_{DA} > 150$ GeV. In this region the fraction of large rapidity gap events is nearly constant.

c) The observed $M_x$ distribution is consistent with a $1/M_x^2$ distribution within the uncertainties due to acceptance and the rather large statistical errors. Note that the first point is strongly affected by acceptance and detector smearing.

The fraction of events with a large rapidity gap is plotted in Fig. 35 versus $Q^2_{DA}$ for two selected bins of $x_{DA}$.

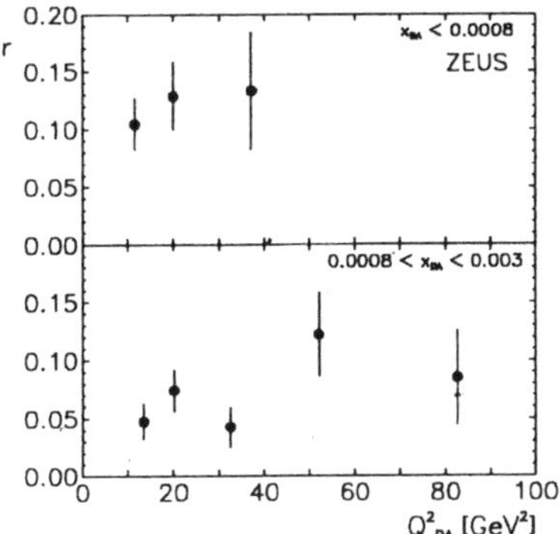


Fig. 35: Fraction r of events with $\eta_{max} < 1.5$ as a function of $Q^2_{DA}$ for two ranges of $x_{DA}$. No acceptance corrections have been applied.

Note that the fraction remains constant indicating that the production mechanism responsible for the large rapidity gap events is a leading twist effect.

The large gap in rapidity indicates that the photon interacts with a non-coloured object as shown in Fig. 33b. The general characteristics of these events shows that they are compatible with those expected from a diffractive dissociation involving pomeron exchange. Indeed, it has been suggested that the pomeron has a partonic structure which can be probed at HERA using virtual photons.

### 3.4 Search for exotic particles

The search for new forms of matter is a prime goal for any new accelerator facility. HERA as an ep collider is particularly well suited to search for excited electron like leptons or for leptoquarks, heavy particles with mixed electron-baryon quantum mumbers.

A composite electron or neutrino will have excited states. In electron-proton collisions excited electrons are predominantely produced by t-channel exchange of low $Q^2$-photons whereas excited electron like neutrinos are produced at much higher values of $Q^2$ by the interchange of a W-boson.

The HERA center of mass energy of 296 GeV makes it possible to explore a mass range, where for the first time the produced e* and ν* may decay into a heavy gauge bosons (W, Z°) and a light fermion.

The H1 experiment reports [32] the result of a search for e* and ν* and ZEUS reports [33] a search for e*. The experiments search in the channels:

| | |
|---|---|
| $e^* \to e + \gamma$ | $\nu^* \to \nu + \gamma$ |
| $e^* \to e + Z^\circ$ | $\nu^* \to \nu + Z^\circ$ |
| $e^* \to e + \nu + W$ | $\nu^* \to e + \nu + W$ |

The selection criteria used to extract the events are discussed elsewhere [32,33]. The number of events which meet the selection criteria are consistent with the expected background. The limits on heavy electron like leptons are quoted in terms of the three unknown parameters:

- Λ, the overall compositness scale
- C(eγe*) and C(eWν*), the production coupling constants
- B, the decay branching radius

The 95% confidence limits on $C(e\gamma e^*)\cdot B^{1/2}/\Lambda$ set by the ZEUS collaboration are plotted in Fig. 36 versus the mass of the heavy electron.

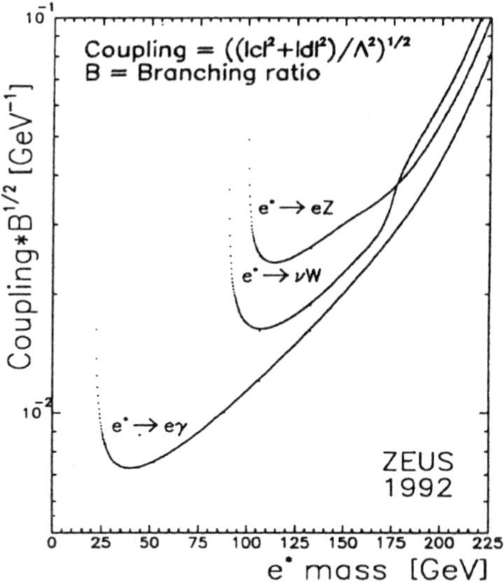


Fig. 36: 95% confidence level upper limits on the product of the coupling and the square root of the branching ratio $B^{1/2}$ divided by an overall scale parameter Λ.

Experiments at LEP reports [34] data on $e^* \to e\gamma$ and they find
$C(e\gamma e^*)\cdot B^{1/2}/\Lambda < 2\cdot 10^{-6}\ \text{GeV}^{-1}$ for $m_{e^*} = 90$ GeV

From a measurement of $e^+e^- \to \gamma\gamma$ the LEP experiments extract [35] indirect limits on $e^*\to e\gamma$. They find $C(e\gamma e^*)\cdot B^{1/2}/\Lambda < 10^{-2}\ \text{GeV}^{-1}$ for an e* mass less than 127 GeV.

Leptoquarks are massive coloured triplet bosons with spin = 0 or 1, with lepton and baryon numbers and with fractional charge. Leptoquarks will occur in grand unified theories [36] technicolour models [37], composite models [38] and in supersymmetric models [39].

Leptogluons are coloured octet states with lepton numbers. Leptogluons may exist in certain composite models.

Any state which is made of an electron and a proton constituent can be produced at HERA as an s-channel resonnance.

Assume a scalar leptoquark of mass $M_{Lq}$ made as a bound state of an electron and a quark q. This particle is produced as a Breit-Wigner resonnance with a peak at $\hat{s} = M_{LQ}^2$, where $\hat{s}$ is the square of the center of mass energy between the incident electron and the struck quark. The fractional momentum of the quark is given by the Bjorken x variable, i. e. $\hat{s} = x \cdot s$ where s is the ep center of mass energy.

The leptoquark may decay into an electron and a quark which will materialize as a jet of hadrons $e + q \to M_{Lq} \to e + X$.

The production of a leptoquark will thus show up in the x-distribution of neutral current events at $x_0 = M_{Lq}^2 / s$ .

The cross section for the s channel production of an (Lq) scalar leptoquark of mass $M_{Lq}$ can be written in the narrow width approximation as:

$$\sigma_0 = \frac{\pi}{4s} \cdot g^2 \cdot q(x_0, \mu).$$

Here $q(x_0, \mu)$ is the probability density of finding a quark with $x = x_0$ of the proton momentum. The scale $\mu$ is set equal to $M_{Lq}$.

The coupling $g = \sqrt{g_R^2 + g_L^2}$ is composed of lefthanded and righthanded couplings at the electron-quark-leptoquark vertex. These couplings strengths are not known, but they are usually set equal to the strength of the electroweak coupling, i. e. $g = \sqrt{4\pi\alpha_{EW}} \approx 0.31$.

Flavour-conserving leptoquarks of lefthanded couplings and appropriate charge may also populate the charged current final state $\nu X$.

$$e^- + p \to M_{Lq} \to \nu + X.$$

Note that scalar ($e^-p$) leptoquarks with righthanded couplings, or with lefthanded couplings and electric charge different from $-1/3$ can not decay into a charged current final state.

The selection criteria used by H1 and ZEUS in their search for leptoquarks are discussed in references [32,33]. None of the experiments observe a signal and the ZEUS 95% upper confidence limit on the coupling strength for various types of leptoquarks are plotted in Fig. 37. Similar limits are also reported [32] by H1.

Experiments at LEP limit [40] the mass of an elementary leptoquark to 44 GeV. The limit on leptoquarks extracted from experiments at the hadron colliders depends on the branching fraction, b, for decay into e + jet. UA2 quote [41] a mass limit of 79 GeV for b=1 and CDF reports [42] a limit from 45 GeV to 113 GeV for b from 0.1 to 1.

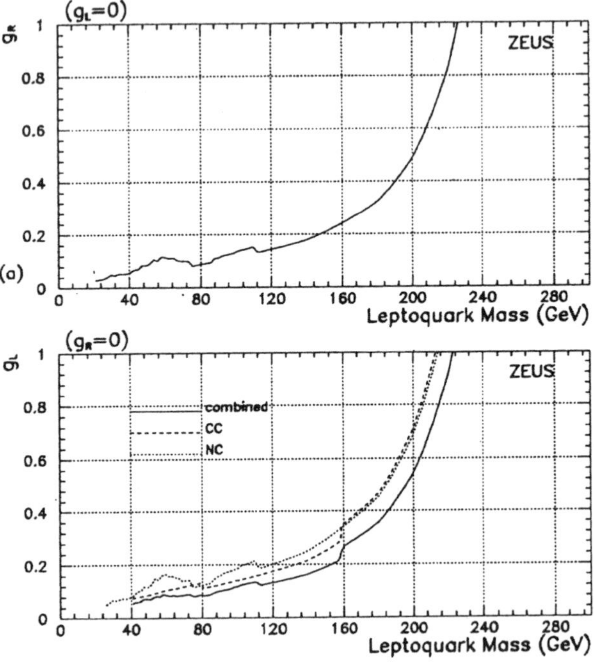


Fig. 37: The 95% confidence upper limits on the couplings of scalar leptoquarks with zero weak isospin and Fermion numer F = -2 versus the leptoquark mass in GeV.
a)The righthanded coupling limit assuming that the leptoquark decays only into $e^-X$.
b) The lefthanded coupling limit assuming equal decay probabilities with $e^-X$ and $\nu X$. The dotted curve is from $e^-X$ events, the dashed curve from $\nu X$ events and the limit from the combined sample is shown as the solid curve.

At HERA a leptogluon would be produced as an s-channel resonance through electron-gluon fusion. To first order the cross section depends on the gluon density at $x = M_{eg}^2/s$ where $M_{eg}^2$ is the mass of leptogluon and on $\left(M_{eg}^2/\Lambda\right)$ where $\Lambda$ is the unknown compositness scale parameter.

The H1 collaboration has searched [32] for leptogluons in neutral current events and have extracted the 95% upper confidence limit on $\Lambda^{-1}$ as a function of $M_{eg}$. The results are plotted in Fig. 38.

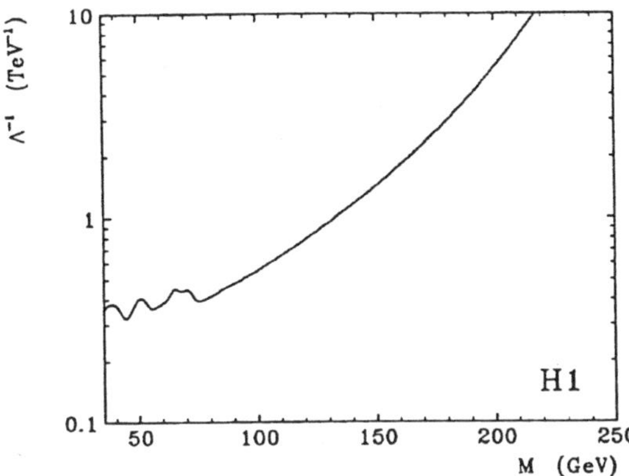


Fig. 38: 95% upper confidence limit on the inverse scale parameter Λ versus $M_{eg}$.

## 5. PHYSICS WITH INTERNAL TARGETS

The approved HERMES experiment plans to use the longitudinally polarized electron beam incident on a polarized H, D or $He^3$ gas jet target to investigate the nucleon spin structure. The experiment can also measure the scattered electron in coincidence with the hadronic final state.

The ARGUS collaboration is investigating the possibility of doing fixed target b-physics at HERA by positioning a thin wire target in the halo of the circulating proton beam. A broad range of b-physics experiments can be carried out with this facility in addition to the primary goal which is to measure the CP-violating parameter sin (2β) in the $B^0(\overline{B}^0) \to {}^{J}\!/\!_{\Psi_s} K^0_x$ channel.

### The HERMES Experiment

The spin-dependent part of the cross section for the scattering of polarized electrons on a nucleon target can be written in terms of two structure functions $g_1(x,Q^2)$ and $g_2(x,Q^2)$. These structure functions can be separated by performing two experiments with different orientation of the target polarization.

The structure function $g_1(x,Q^2)$, which has a very transparent interpretation within the quark model, can be written as:

$$g_1(x,Q^2) = \frac{1}{2}\Sigma e_f^2\left[q_f^+(x,Q^2) - q_f^-(x,Q^2)\right].$$

In this formula $q_f^{\pm}(x,Q^2)$ are the probabilities to find a quark of flavour f and charge $e_f$ in the infinite momentum frame with momentum fraction x and a helicity which is the same (+) or opposite (-) to that of the parent nucleon.

Such a transparent interpretation does not exist for $g_2(x,Q^2)$. This form factor arises only for heavy quarks and is sensitive to quark-gluon correlations.

Whereas $g_1(x,Q^2)$ has been measured in several experiments there are no data available on the $g_2(x,Q^2)$ structure function.

Measurements of the $g_1(x,Q^2)$ structure functions for protons and neutrons can be used to evaluate the fundamental Bjorken sum rule [43]. In the scaling limit, including first order QCD corrections this sum rule can be written as:

$$\int_0^1 \left[g_1^p(x,Q^2) - g_1^n(x,Q^2)\right] dx = g_A / 6g_V(1-\alpha_s/\pi)$$

where $g_A$ and $g_V$ are the weak axial and vector coupling constants as measured in Gamow-Teller nucleon beta decays.

In 1988 the EMC group at CERN reported [44] the first data on $g_1(x,Q^2)$ on a proton target. These data were then used to evaluate the Ellis-Jaffe sum rule [45]:

$$I_1^p = \int_0^1 g_1^p(x,Q^2)dx = 0.126 \pm 0.010 \pm 0.015$$

where the first error bar represents the statistical and the second the systematic uncertainties.

The mean z-component of the proton spin carried by each quark flavour can be extracted from these data assuming SU(3) flavour symmetry. This analysis yielded the surprising result that only a small fraction of the proton spin is carried by quarks and that the strange sea is making a large contribution of opposite sign. The results were:

$$< s_z > q = 0.060 \pm 0.047 \pm 0.069$$
$$< s_z > s-\text{quarks} = -0.095 \pm 0.016 \pm 0.023.$$

These results triggered a flurry of experimental activities and recently two groups reported new results on the $g_1(x,Q^2)$ structure function.

The SMC collaboration at CERN used a deuterated ammonia target and they find [46]:

$$I_1^d = 0.023 \pm 0.020 \pm 0.015.$$

The E142 collaboration at SLAC used a polarized $He^3$ target. In this case the contribution from the two protons with opposite spin alignment cancel, and the neutron $I_1^n(x,Q^2)$ can be extracted directly. They find [47]:

$$I_1^n = -0.022 \pm 0.011$$

The data are summarized [49] in Fig. 39 and compared to the predictions of the Bjorken and Jaffe-Ellis sum rule.

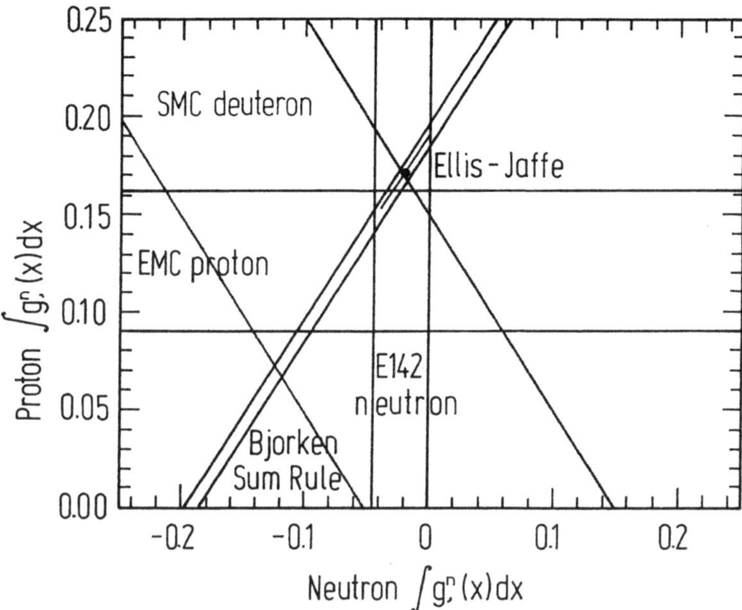


Fig. 39: Data on $I_1^p(x,Q^2)$, $I_1^D(x,Q^2)$ and $I_1^n(x,Q^2)$ are compared to the prediction of the Bjorken sum rule and the Jaffe-Ellis sum rule. Plotted are $2\sigma$ limits.

The data are in agreement within the rather large error bars and they are also consistent with the Bjorken and Jaffe-Ellis sum rule. Clearly higher precision data are needed.

The HERMES experiment [49] should indeed be able to provide such high precision data.

A layout of the HERMES experiment is shown in Fig. 40.

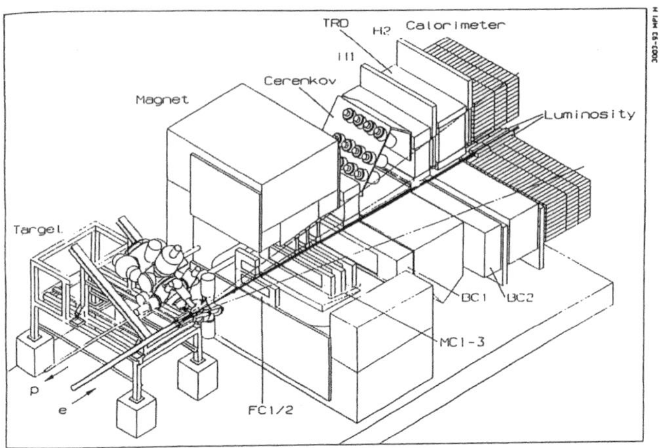


Fig. 40: A layout of the HERMES experiment.

The longitudinally polarized HERA electron beam is passed through a polarized gas jet target and electrons scattered between 40 mrad and 220 mrad are analyzed using a downstream magnetic spectrometer.

The HERMES group will use polarized hydrogen, deuterium and $He^3$ targets. These targets are pure atomic species targets with the atoms stored in a 40 cm long windowless target cell. Thus scattering occurs only from polarized atoms and there is no target dilution factor. The hydrogen and deuterium atoms are polarized using an atomic beam method, the $He^3$ target is polarized using optical pumping.

These targets have been constructed and successfully tested. The hydrogen and deuterium targets are 90% polarized at a target density of order $10^{10}$ atoms/cm$^2$. The $He^3$ target polarization is above 50% at a target density above $10^{15}$ atoms/cm$^2$.

The magnet is divided into two symmetrical parts by a horizontal iron plate. The electron and proton beams are spaced by 71.4 cm and traverse the plate through a bore. Trajectories of scattered particles are measured by drift chambers in the front and rear of the magnet and by proportional chambers installed within the magnetic volume. Particles traversing the magnet are identified by Cerenkov counters, transition radiation detectors and by a calorimeter.

With this detector the HERMES group can measure the polarized structure functions for x-values above 0.02 and for values of $Q^2$ between 1 $GeV^2$ and 12 $GeV^2$. Final state hadrons can be measured and identified in coincidence with the scattered electron.

Compared to other experiments HERMES should have smaller systematic errors. Also the statistical errors should be rather small given the high target polarizations and the fact that the experiment can run in parallel with the collider program. They will also be able to measure the tensor spin structure functions of the deuterium for the first time.

Test of the longitudinal spin rotators and various detector components will be carried out in 1994. The experiment will be installed in the 1994/95 winter shutdown and the first data on the polarized structure functions are expected in 1995.

**B-Physics at HERA**

In addition to its primary goal of providing information on CP-violation in the $b\bar{b}$-system, the proposed experiment will also yield high precision data on $B_S\overline{B}_S$ mixing and on the strength of the Cabbibo, Kabayashi and Maskawa matrix elements $V_{cb}$ and $V_{ub}$. Here we will briefly discuss the CP violation experiment and remark on the detector.

The CKM matrix V can be written in terms of three mixing angles and one complex phase. In the Wolfenstein representation [50] with four real parameters $\Lambda = \sin\theta_c$, A, $\rho$ and $\eta$ this matrix is given by:

$$V = \begin{pmatrix} V_{ud} & V_{as} & V_{ub} \\ V_{cd} & V_{cs} & V_{cb} \\ V_{td} & V_{ts} & V_{tb} \end{pmatrix}$$

$$= \begin{pmatrix} 1-\frac{\Lambda^2}{2} & \Lambda & \Lambda^3 A(\rho - i\eta) \\ \Lambda & 1-\frac{\Lambda^2}{2} & \Lambda^2 A \\ \Lambda^3 A(1-\rho-i\eta) & -\Lambda A & A \end{pmatrix}$$

The CKM matrix is unitary and one of the unitary conditions can be written as:

$$V_{ud} \cdot V_{ub}^* + V_{cd} \cdot V_{cb}^* + V_{td} \cdot V_{tb}^* = 0$$

This relation can be represented by the triangle in the $\rho$, $\eta$ plane as shown in Fig. 41.

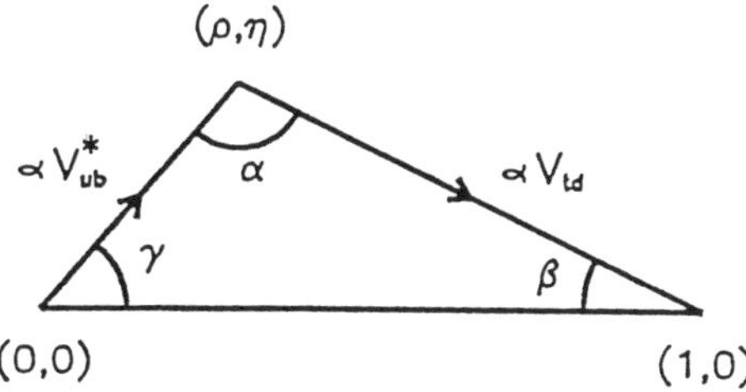


Fig. 41: Unitarity triangle in the $\rho$, $\eta$ plane.

Note that CP-violation occurs for finite values of $\eta$.

Measurements of decay modes in which the B° meson decay into CP eigenstates permits a direct determination of the angles in the unitarity triangle.

The HERA B-group has proposed [51] to measure the decay $B^\circ(\overline{B}^\circ) \rightarrow J/\Psi K^\circ_s$ and to determine the asymmetry:

$$A = \frac{B^\circ \rightarrow J/\Psi \cdot K^\circ_s - \overline{B}^\circ \rightarrow J/\Psi \cdot K^\circ_s}{B^\circ \rightarrow J/\Psi \cdot K^\circ_s + \overline{B}^\circ \rightarrow J/\Psi \cdot K^\circ_s}$$

This asymmetry is directly related to the angle β as defined in the unitarity triangle by:

$$A = \sin(\chi t / \tau) \cdot \sin(2\beta)$$

The mixing parameter $\chi$ is defined as $\chi = \Delta M/\Gamma = 0.67$, where $\Delta M$ is the mass difference of the mass eigenstates and $\Gamma$ is the inverse lifetime and t is measured in units of the B-lifetime. Thus the B°-meson must have time to transform into a $\overline{B}^\circ$-meson before a significant asymmetry can develop.

The predicted value [52] of sin(2β) is plotted in Fig. 42 versus the B-meson decay constant $f_B$.

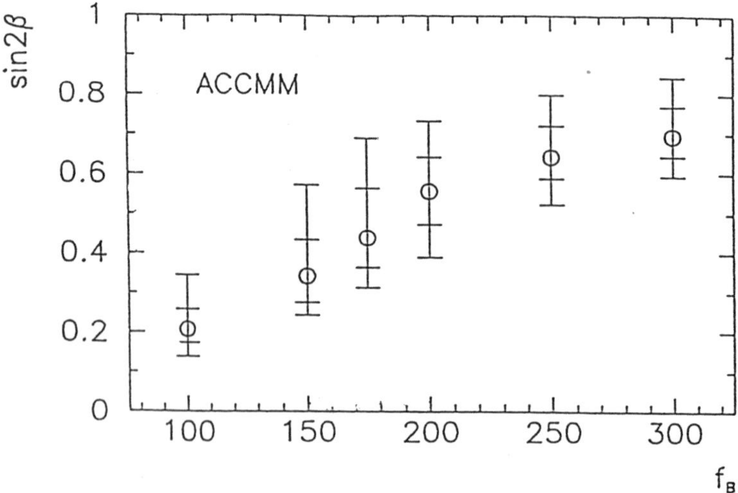


Fig. 42: Best fit results of sin(2β) as a function of $f_B$.

A total of some $10^{15}$ proton-nucleon interactions are needed to yield 3000 fully reconstructed $B^\circ\overline{B}^\circ$ events. With a total data taking time of 3 years, corresponding to $3 \cdot 10^7$ sec and a collision rate of 10 MHz this requires on the average 3 interactions per bunch crossing. This high rate imposes stringent requirements on the target, the detector, trigger and the data acquisition system.

A schematic layout of the proposed detector [51] is shown in Fig. 44.

The preferred value of $f_B$ yields sin(2β)=0.6 such that a measurement of the asymmetry with an error of 0.05 would allow the experiment to cover the complete parameter range.

The typical features [53] of a B-hadron event produced in a proton-nucleon collision at an incident proton energy of 1 TeV is shown in Fig. 43.

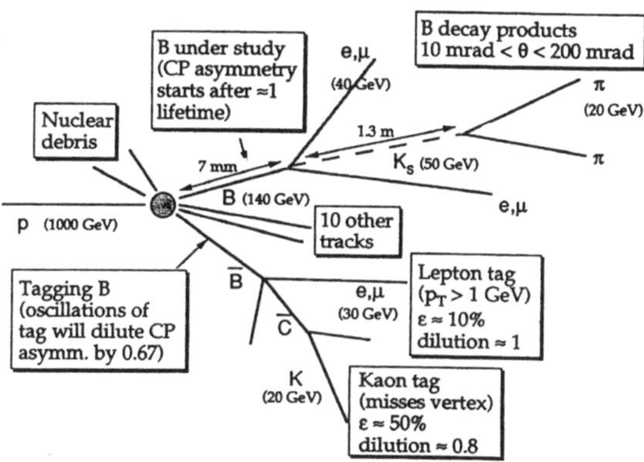


Fig. 43: Characteristics features of a B-hadron event produced in a proton-nucleon collision at an incident energy of 1 TeV. Typical mementa and decay angles are indicated.

The B°-meson is identified via its decay mode $B^\circ \rightarrow J/\Psi \cdot K_s^\circ$ with $J/\Psi \rightarrow e^+e^-$, $\mu^+\mu^-$ and $K_s^\circ \rightarrow \pi^+\pi^-$. The initial flavour of the decaying B°-meson is determined by tagging the flavour of its partner B-meson. This is done by observing the second B-meson decay into a final state containing either a charged lepton or a charged kaon and determine their charge. Mistagged events will dilute the observed CP asymmetry. A second dilution factor results from flavour mixing of the second B-meson before tagging.

The statistical error in the determination of sin(2β) is thus given by:

$$\Delta \sin(2\beta) = \frac{1}{D}\left(\frac{K}{N}\right)^{\frac{1}{2}}$$

N is the number of fully identified events which survive the vertex cut and D=0.5 is the net dilution factor. To suppress the background there must be a minimum separation between the primary production vertex and the decay vertex and this results in a dilution of statistical precision denoted by K which is of order K=1.5.

Therefore to measure sin(2β) with a statistical error of 0.05 requires a total of 3000 fully reconstructed events.

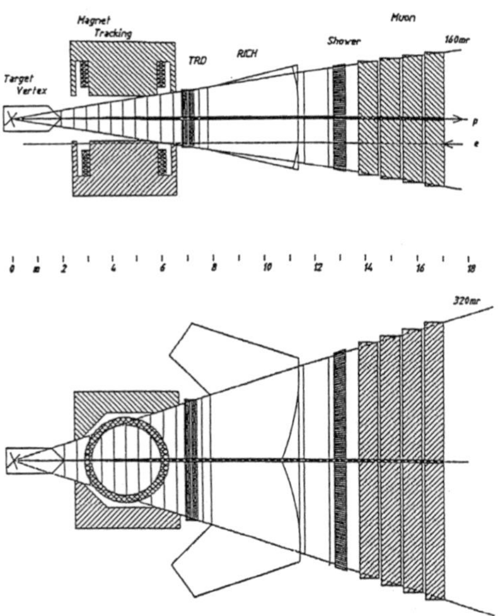


Fig. 44: The layout of the proposed B-experiment.

The set up consist of a wire target followed by a 2 m long silicon vertex detector assembly and a dipole magnet packed with tracking chambers. The dipole magnet covers roughly 90% of 4π in center of mass.

Downstream of the magnet there is a transition radiation detector for electron identification, a ring imaging Cerenkov counter for kaon identification, an electromagnetic calorimeter and a muon detector.

The readout and trigger system is a multilevel system which is based on systems now being designed for use in the LHC detectors.

A simplified wire target has been successfully tested in the beam. By using a feedback system to position the target in the beam halo the required rate can be obtained independent of the proton-beam lifetime.

Prototypes of various detector elements are being designed and tested. The experiment clearly benefits from the impressive development work underway for the LHC detectors.

The experiment is very rewarding but also very difficult. Compared to the LHC experiment the requirements to the detector and the data axquisition system are more relaxed with the exception of particle fluxes and radiation damage suffered by the silicon vertex detector.

The proposal is now being evaluated by the DESY experimental programme committee and a final recommendation is expected by the end of May 1994. Thus a final decision can be taken in the summer of 1994. If positive, we expect the first data to become available in 1998.

REFERENCES

[1] HERA, A Proposal for a Large Electron-Proton Colliding Beam Facility at DESY, DESY, HERA 81-10
B. H. Wiik, Proceedings of the IEEE Particle Accelerator Conference, May 1993, Washington.

[2] A. A. Sokolov and I. M. Ternov, Sov. Phys. Doklady 8 (1964), 1203.

[3] D. P. Barber et al., DESY Preprint 93-038.

[4] D. Kisielewska et al., DESY HERA 85-25.

[5] S. V. Levonian et al., DESY H1-TR113, 1987.

[6] H1 Collaboration, I. Abt et al. The H1 detector at HERA, DESY Preprint 93-103 (1993).

[7] D. Caldwell, Proceedings of the 26th International Conference on High Energy Physics, Dallas, J. R. Sanford, Editor (1992).

[8] Physics at HERA, Proceedings of the Workshop (Hamburg, October 24-30, 1991) Edited by W. Buchmüller and G. Ingelmann.

[9] H1 Collaboration, T. Ahmed et al., Phys. Lett. B297 (1992), 205, Phys. Lett. B299 (1993), 374, DESY Report 93-100.

[10] ZEUS Collaboration, M. Derrick et al., Phys. Lett. B293 (1992) 465.

[11] ZEUS Collaboration, M. Derrick et al., Phys. Lett. B297 (1992), 404, DESY Report 93-151.

[12] V. N. Gribov and L. N. Lipatov, Sov. Journ. Nucl. Phys. 15 (1972), 438 and 675.

[13] G. Altarelli and G. Parisi, Nucl. Phys. 126, (1977), 297.

[14] L. V. Gribov, E. M. Levin and M. G. Ryskin, Phys. Rep. 100 (1982), 1.

[15]H1 Collaboration, T. Ahmet et al., Phys. Lett. B298 (1993), 469, Phys. Lett. B299 (1993), 385.
I. Abt et al., Nucl. Phys. B407 (1993), 515.

[16]ZEUS Collaboration, M. Derrick et al., Phys. Lett. B303, (1993), 183, Phys. Lett. B316 (1993), 412.

[17]MMC Collaboration, P. Amaudruz et al., Phys. Lett. B295 (1992), 159 and Erratum to CERN-PPE/92-124.

[18]BCDMS Collaboration, A. C. Benvenuti et al., Phys. Lett. B223 (1989), 485.

[19]E. Kuraev, L. Lipatov and V. Fadin, Sov. Phys. JETP 45, (1977), 199.

[20]A. D. Martin, R. G. Roberts and W. J. Stirling, Phys. Lett. 306B, (1993), 145, Phys. Lett. B309 (1993), 492.

[21]K. Prytz, Phys. Lett. 311B, (1993), 286.

[22]H1 Collaboration, I. Abt et al., DESY Report 93-146.

[23]ZEUS Collaboration, M. Derrick et al., Phys. Lett. 306B, (1993), 158.

[24]H1 Collaboration, I. Abt et al., DESY Report 93-137.

[25]JADE Collaboration, W. Bartel et al., Z. Phys. C33, (1986), 23.

[26]G. Ingelman in Reference 8.

[27]ZEUS Collaboration, M. Derrick et al., Z. Phys. C59 (1993), 231.

[28]G. Gustafson, Phys.Lett B175 (1986), 453.
G. Gustafson and U. Peterson, Nucl. Phys. B306 (1988), 746.

[29]G. Marchesini et al., Comp. Phys. Comm. 67 (1992), 465.

[30]M. Bengtson, G. Ingelman and T. Sjöstrand, Nucl. Phys. B301 (1988), 554.
M. Bengtson and T. Sjöstrand, Z. Phys. C37 (1988), 465.
G. Marchesini and B. R. Webber, Nucl. Phys. B310 (1988), 461.

[31]ZEUS Collaboration, M. Derrick et al., Phys. Lett. B315 (1993), 481.

[32]H1 Collaboration, I. Abt et al., Nuclear Physics B396, (1993), 3.

[33]ZEUS Collaboration, M. Derrick et al., Phys. Lett. B306 (1993), 173, Phys. Lett. B316 (1993) 207.

[34]Opal Collaboration, M. Z. Akrawy et al., Phys. Lett. B257 (1991), 531.
L3 Collaboration, O. Adriani et al., Phys. Lett. B288 (1992), 404.
DELPHI Collaboration, P. Abreau et al., Z. Phys. C53 (1992), 41.
ALEPH Collaboration, D. De Camp et al., Phys. Rep. 216 (1992), 253.

[35] ALEPH Collaboration, D. De Camp et al., Phys. Rep. 216 (1992), 253.
L3 Collaboration, O. Adriani et al., Phys. Lett. B288 (1992), 404.
Opal Collaboration, M. Z. Akrawy et al., Phys. Lett. B257 (1991), 531.

[36] J. C. Pati and A. Salam, Phys. Rev. D10, (1974), 275.
H. Georgi and S. L. Glashow, Phys. Rev. Lett. 32, (1974), 438.

[37] S. Dinopoulos and L. Susskind, Nucl. Phys. B155, (1979), 237.

[38] B. Schremp and F. Schremp, Phys. Lett. B153, (1985), 101.

[39] For a review of MSSM see H. D. Niles, Phys. Rep. 110 (1984), 1.

[40] Opal Collaboration, M. Z. Akrawy et al., Phys. Lett. B263 (1991), 123.
L3 Collaboration, B. Adeva et al., Phys. Lett. B261 (1991), 169.
DELPHI Collaboration, P. Abreau et al., Phys. Lett. B275 (1991), 222.
ALEPH Collaboration, D. De Camp et al., Phys. Rep. 216 (1992), 253

[41] UA2 Collaboration, J. Alitti et al., Phys. Lett. B274 (1992), 507.

[42] CDF Collaboration, S. Moulding et al., invited paper given at the Seventh Meeting of the American Physical Society (DPF), Fermilab, November 1992.

[43] J. D. Bjorken, Phys. Rev. 148 (1966), 1467 and D1 (1971, 1376.

[44] EMC Collaboration, J. Ashman et al., Phys. Rev. Lett. B206 (1988), 364,
Nucl. Phys. B328 (1989), 1.

[45] J. Ellis and R. J. Jaffe, Phys. Rev. D9 (1974), 1444, Erratum D10 (1993), 533.

[46] SMC Collaboration, B. Adeva et al., Phys. Lett. B302 (1993), 533.

[47] The E142 Collaboration, P. L. Anthony et al., SLAC PUB-6101.

[48] E. Hughes, Workshop on Physics with Internal Targets at HERA, Hamburg, September 1993.

[49] Hermes Collaboration, Technical Design Report, DESY PRC 93/06, July 1993, MPIH-V20-1993.

[50] L. Wolfenstein, Phys. Rev. Lett. 51 (1983), 1945.

[51] H. Albrecht et al., "An Experiment to study CP Violation in the B System Using an Internal Target at the HERA Proton Ring", Letter of Intent, DESY PRC 92/04 (1992).

[52] M. Schmidtler and K. R. Schubert, Z. Phys. C53 (1992), 347.

[53] W. Hofmann, DESY Report 93-026.

## CHAIRMAN: B. Wiik

*Scientific Secretaries: G. Abu Leil, J.R. Forshaw, D. Kawall*

## DISCUSSION

– *Lenti:*

What is the HERA potential to study CP violation ? How many $b-\bar{b}$ events per year are necessary for such a study ?

– *Wiik:*

I did not bribe you to ask this question but I just happen to have a couple of transparencies (transparencies show 3000 identified $b-\bar{b}$ events are needed for $\Delta\sin(2\Phi) = 0.05$, requiring $10^{15}$ raw events - see lecture notes).

– *Forshaw:*

I'd first like to comment on the possibility of identifying the perturbative QCD pomeron at HERA. In my opinion, deep inelastic scattering is not going to prove an ideal test due to our inability to factorize out the soft physics. I think this makes the detection of processes such as DIS plus an associated high $P_T$ jet and hard diffractive processes (e.g. $p\gamma \rightarrow VX$ ) all the more important. So my question is to ask : What are the prospects for measuring such processes at HERA - where typically there is a problem associated with detecting a jet close to the beam pipe in the proton direction so that we have large rapidity ?

– *Wiik:*

What you are talking about is basically this process (see lecture notes for DIS event with an associated jet). I think the chances for measuring this process are very, very good. By the time we get $100pb^{-1}$ or so we should see it. This difference between the shadowing and no shadowing scenarios is significant. The hard diffractive process is also expected to be visible. The H1 and ZEUS groups have simulated this process and they conclude that the observation of the forward jet seems possible.

– *Kawall:*

Does the HERMES (polarized electron experiment) collaboration have any intention of measuring the asymmetry in the hadronic backgound? The semi-inclusive pion and kaon asymmetries can provide independent data on the nuclear structure functions.

– *Wiik:*

In addition to measuring g1 proton, g1 neutron and testing the Bjorken sum rule to the 5-8 % level, HERMES will also measure g2 proton, g2 neutron as well as the hadronic asymmetries. I can see no reason why they should not measure the asymmetry in the background rate.

– *Kawall:*

Do you know the highest $Q^2$ value for this experiment? - I'm wondering if it would be feasible to do colour transparency experiments as part of the fixed target program.

– *Wiik:*

As far as I know it is around 6 or 8 GeV$^2$ so I guess you could try but I don't know how meaningful it would be.

– *Lu:*

On the subject of B physics, what are the advantages and disadvantages of HERA compared to CLEO and the possible future B factory ?

– *Wiik:*

CLEO has basically no chance since the target is not moving. For a $b - \bar{b}$ factory with a luminosity of $10^{33} cm^{-2} s^{-1}$ , running for $3 \times 10^7$ seconds the statistical error is a factor of 2 worse. The $e^+e^-$ machine can do a lot of things that this hadron machine can't do. It can look into other decay modes and so on. But because there is such a clean signature I think the hadron machines will win. The ultimate machine will be the SSC/LHC - they will do the definitive experiments on CP violation - not the $e^+e^-$ machine simply because of the rate.

– *AbuLeil:*

How does shadowing affect $F_2$, do you expect to see it experimentally ?

– *Wiik:*

The observability of shadowing depends upon whether the gluons are distributed over the whole hadron or localized in "hot spots" around the valence quarks. In the latter case one should clearly see something. Perhaps the best way to see shadowing is not in the proton but in the pomeron because the pomeron is tightly bound and so has a smaller radius, leading to a higher gluon density.

– *McPherson:*

Does the event rate estimate in the B experiment include full background estimates ? FNAL E791 just wrote 40 Terabytes of data with no B signal. Will the background dominate ?

– *Wiik:*

Yes. Detailed simulations have been done. You should attend the workshop this fall.

– *McPherson:*

At the SSC/LHC, is a dedicated experiment needed, or can it be retrofitted into a conventional high-$P_T$ detector ?

– *Wiik:*

The importance of this physics justifies a dedicated experiment.

– *Montgomery:*

With respect to the experiments at the Tevatron - with conventional rapidity range and in $J/\Psi$ Ks then CDF/D0 would get $\sin(2\beta) = 0.015$. for 1000 $pb^{-1}$ . The need for major differences comes when you want $\sin(2\alpha)$ then you need pion, kaon and D. I think then you would need a dedicated experiment.

– *Wiik:*

I think the Tevatron should concentrate on top physics.

– *Montgomery:*

If you were staying until Saturday you would see some rather pretty B physics from the Tevatron.

– *Wiik:*

That is very encouraging.

– *Peccei:*

Was the dilution factor in this B experiment just computed by using kaon and lepton tagging ?

– *Wiik:*

Yes - both leptons and kaons.

– *Peccei:*

It may also be possible to use excited $B^{**} \to \pi^* B$ as a convenient tag, reducing the dilution factor.

– *Wiik:*

You also should come to the workshop !

– *Gibilisco :*

Have you at HERA the possibility and the program to test also the Bjorken sum rule? What about the EMC effect ?

– *Wiik:*

The EMC effect can be tested - in fact it is an important point from the theoretical point of view because it may indicate a violation of the Ellis-Jaffe sum rule. The Bjorken sum rule is more fundamental. We need of course precise proton and neutron data to check this rule.

– *Forshaw:*

Relating to the EMC, SMC and E142 experiments I believe a recent analysis by Close and Roberts suggests that there may in fact be no inconsistency. They attempt to compare "like with like" - meaning the results are scaled to the same $Q^2$ and higher twist effects are included amongst other things.

– *Wiik:*

I am not aware of that work - there is also the paper by Karliner and Ellis. I think it would be nice to get a consensus amongst the theorists and experimentalists. I agree with you, I think there are sufficient uncertainties in this business that you are not quite sure if there is a real disagreement.

– *Kawall:*

If I can comment on that - within the error bars the SMC and E142 measurements are entirely consistent. It's only when you do the integral (first moment) that you arrive at different conclusions.

– *Forshaw:*

Am I right in saying that the experiments are in agreement at the level of the asymmetry ?

– *Kawall:*

Yes.

– *Wiik:*

O.K. On the experimental side you need to know the structure functions and R values, whilst on the theoretical side there is the low $Q^2$ uncertainty for example due to higher twist.

– *Montgomery:*

At what electron energy did you get your polarization of 60 % ? Clearly the low $x$ limit of the HERMES experiment is controlled by the beam energy - it would be nice to have that polarization at 35 GeV !

– *Wiik:*

I could not agree with your last sentence more. The 60 % polarization is at the canonical limit of 27 GeV. There is no reason why it shouldn't be slightly better at 30 GeV. We are now on a different system than we used to be and I expect the value to go up to 70-80 %. On turning this transverse polarization into longitudinal polarization we lose a little but I still expect with a little luck to have about 70 %. Since the proposal calls for 50% I think we are in pretty good shape.

– *Peccei:*

You are still somewhat away from the theoretical limit on polarization of 92%. Why is that ?

– *Wiik:*

Once you start to bend out of the horizontal plane you dilute the 92 % theoretical value. There is thus a reduction in polarization due to the need for vertical bending.

# QCD FESTIVAL

*R. Arnowitt*

## QCD and Chiral Lagrangians

Chiral symmetry breaking is clearly a property of QCD. The chiral phase transition is seen in lattice gauge calculations (which are becoming increasingly reliable) and account naturally for the smallness of the pion mass and the largeness of the nucleon mass. However, the effective Lagrangians of chiral symmetry are only "weakly" consequences of QCD (and indeed were derived prior to the invention of QCD). Thus one derivation of the meson chiral Lagrangian makes use of the existence of some strong force binding quarks into mesons and vector bosons, the standard theorems of quantum field theory (LSZ, freedom to choose interpolating fields etc.) and the chiral current algebra. One expands the S–matrix in low order powers of momenta, and chooses as interpolating fields $\pi_a \equiv (\partial_\mu A^\mu_a)\ F_\pi m^2_\pi$, $\rho^\mu_a \equiv V^\mu_a/g_\rho$ (where the interpolating constants $F_\pi, g_\rho$ are defined by $\langle 0|\partial_\mu A^\mu_a|\pi_b, p\rangle = \delta_{ab} F_\pi m^2_\pi$, $\langle 0|V^\mu_a|\rho_b, p^\mu\rangle = \delta_{ab} g_\rho p^\mu$ etc.).

One then constructs an effective Lagrangian that correctly reproduces the S matrix to low order in momenta, that is obtained from the current algebra and the interpolating field choices relating the meson states to the currents. Even the Weinberg formula relating meson $(\text{mass})^2$ to quark masses, is only weakly dependent on QCD as the gluon interaction cancels out when one explicitly calculates $\partial_\mu A^\mu_a$ ( with $A^\mu_a = \bar{q}(\lambda_a/2)\gamma^\mu\gamma^5 q$).

The effective Lagrangians imply a number of approximations: the low order expansion of S in momenta, neglect of scattering states, assumption that interpolating field matrix elements are slowly varying etc. Whether these are good or bad approximations depends on the detailed dynamics, and thus one needs QCD to test these questions explicitly. One generally expects the approximation to be good because $m_\pi$ is small, which is a consequence of chiral symmetry breaking.

The true reason why we still consider models supposed to be consequences of QCD is simply due to our inability to perform credible calculations in the non-perturbative sector of QCD. In particular a convincing derivation of the confinement does not exist at the level of a mathematical proof.

Models like the quark model and the chiral model existed before QCD and they were and still are relatively successful. Their derivation from QCD cannot be a rigorous one and they cannot be considered as equivalent to QCD. They may be useful in some well defined circumstances but they are models with free parameters to be adjusted from experiment and even if they can help towards a better understanding of data they will never give the final explanation for this data. For instance, the so-called chiral perturbation theory, which is not a theory but a model is a non-renormalisable theory. Counter terms have to be introduced in order to cancel the appearing infinities and going to higher and higher orders $O(p^2), O(p^4), O(p^6), O(p^9)$ we must introduce more and more counter terms. This unavoidable feature which is absent in QCD reduces the power of prediction of the model.

Concerning Lattice Gauge theories, they are an ambitious attempt to formulate the gauge theory using a lattice. That means replacing continuous variables by a discrete set. Of course approximations have to be made in order to have a discretisation of the theory. Therefore in this case again we have a model with in addition, the problem of the lattice spacing and the necessity of defining a physical continuous limit before comparing the result of the computation with experimentally measured quantities like, for instance, the hadron mass spectrum. In the domain of lattice computations we can hope, for the future in technical improvements with more powerful computers. However, probably physical improvements for the formulations of gauge theories or a lattice would be necessary.

We must keep in mind that progress on the predictions of unmeasured physical quantities like $B_K, f_D, f_{D_S}, f_{B_d}, f_{B_S}$ are very slow and considerable uncertainties still exist. In my opinion a still open problem for decades, that of the $|\vec{\Delta I}| = 1/2$ rule in $K \to 2\pi$ decay, has not received a good quantitative explanation either from chiral perturbation theory or from lattice computations. However, it must have its solution in a calculation within QCD but we are still far from being able to do such a calculation.

I wish to make clear the distinction between a statement and its proof. In Professor Morpurgo's presentation the $\pi$ meson mass is the difference of two contributions $A - 3B$ using an exact parametrisation. The claims made by Guido

Altarelli and Roberto Peccei is that, in the zero quark limit, A becomes equal to $3B$, the $\pi$ meson mass vanishes and we have chiral symmetry. This is what I call a statement without proof. It is clear that in any model with a symmetry, if the basic masses are zero, the symmetry is chiral as a consequence of $\gamma_5$ invariance. That is trivial. What is less trivial is to give a proof e.g. to compute the coefficients $A$ and $B$ as functions of the basic QCD parameters and to study their limit when the quark masses go to zero. If in this limit we have $A = 3B$ then the statement is proven.

Incidentally we must keep in mind that the $\pi$ meson mass is not zero and therefore the chiral symmetry is broken. However the $\pi$ meson mass is small and the breaking is soft.

## Chiral Lagrangians

There is considerable experimental evidence that the strong interactions possess an approximate spontaneously broken chiral symmetry. Furthermore, QCD is chirally symmetric in the limit of neglecting the light quark masses, and there is strong evidence from lattice calculations that the symmetry is spontaneously broken, so that an effective chiral Lagrangian should be a useful description of the interactions of pions and kaons at low energy. I will briefly sketch how chiral symmetry is believed to work in QCD.

The QCD Lagrangian is

$$L_{QCD} = -\frac{1}{4} F_{\mu\nu}^i F^{\mu\nu i} + \sum_{\alpha=u,d,s,\cdots} \bar{q}_\alpha (i \not{D} - m_\alpha) q_\alpha \tag{1}$$

where $\alpha$ is the flavor label, the color index is suppressed, and $m_\alpha$ is the current mass, which is a bare mass as far as QCD is concerned. It is generated by the Higgs mechanism in the electroweak sector of the Standard model. The color interactions are independent of flavor. If $N$ of the $m_\alpha$ are degenerate then QCD possesses a global $SU(N)$ flavor symmetry under the (vector) transformations

$$q_\alpha \to U_{\alpha\beta} q_\beta \tag{2}$$

$$\alpha, \beta = 1, ...., N$$

e.g., $SU(2)$ of isospin for $m_u = m_d$, or flavor $SU(3)$ for $m_u = m_d = m_s$. If $N$ of the $m_\alpha$ vanish then one has the larger global $SU(N)_L \times SU(N)_R$ chiral symmetry, under which the left and right chiral projections of the quark fields transform independently,

$$q_{\alpha L} \to U_{L\alpha\beta} q_{\beta L}, \; q_{\alpha R} \to U_{R\alpha\beta} q_{\beta R} \tag{3}$$

Even before the development of QCD there was strong evidence that $SU(2)_L \times SU(2)_R$ is an excellent approximate symmetry of the strong interactions, and there are also remnants of $SU(3)_L \times SU(3)_R$, although that is more badly broken. In particular, the chiral symmetry must be spontaneously broken because of the large value of the nucleon mass. The pion is a pseudo-Goldstone boson of chiral $SU(2)$, with a small mass from explicit breaking in the Hamiltonian (the more massive $K$ and $\eta$ mesons are pseudo-Goldstone bosons of chiral $SU(3)$). The smallness of $m_\pi^2$ compared to other hadronic scales strongly suggests this picture, and it is further supported that the $\pi$ is a pseudo Goldstone boson by PCAC ( $\partial A^i = F_\pi m_\pi^2 \phi_\pi^i$, where $A^i, i = 1, 2, 3$ is the $i$th axial current, $\phi_\pi^i$ is the $i$th pion field, and $F_\pi$ is the pion decay constant, associated with the spontaneous breaking), the associated soft pion theorems, effective Lagrangians for low energy pion interactions (e.g., $\pi\pi \to \pi\pi$), Adler-Weisberger, and Goldberger-Treiman relations. These all test the underlying chiral symmetry and would not be expected to hold if the pions were merely accidentally light.

The fact that chiral symmetry is an automatic consequence of QCD for small current masses is one of the more compelling arguments for QCD. The basic scenario is that in the limit $m_u = m_d = m_s = 0$ (the $c$, $b$ and $t$ are so heavy that there is little point in discussing the chiral limit) the chiral $SU(3)_L \times SU(3)_R$ is spontaneously broken by a $\bar{q}q$ condensate (i.e., $< 0|\bar{q}q|0 > \neq 0$) to the ordinary flavor (vector) $SU(3)$ symmetry of Gell-Mann and Neeman. Since the eight axial generators that are spontaneously broken are global, there must be eight pseudoscalar Goldstone bosons formed as quark-antiquark bound states. These are identified with the pseudoscalar octet, $\pi, K, \eta$. The spontaneous breaking also generates a dynamical or effective quark mass $M$ of order $\Lambda_{QCD}$ (e.g., $M \sim M_N/3 \sim 300$ MeV), which gives masses to the nucleon, vector mesons and the hadrons other than the Goldstones.

In the second stage of the scenario one turns on current quark masses $m_s \gg m_d \gtrsim m_u \gtrsim 0$, which explicitly break the chiral symmetry. These give mass to the pseudo Goldstones ($0 \lesssim m_\pi^2 \ll m_{K,\eta}^2$), break the flavor $SU(3)(m_s \neq \frac{m_d+m_u}{2})$, and give a small breaking of isospin $SU(2)(m_d \neq m_u)$ that is of the same magnitude as but separate from electromagnetic breaking.

The spontaneous breaking is a complicated dynamical problem which has only been convincingly demonstrated by lattice calculations. However, the idea may be illustrated schematically as follows: In the chiral limit ($m_\alpha = 0$) the ground state of the theory is believed to be a chiral condensate, $< 0|\bar{q}_\alpha q_\alpha|0 > \neq 0$, with a value independent of $\alpha$. In a simple ladder approximation, $< \bar{q}q > \neq 0$ may be obtained by solving the homogenous Schwinger-Dyson equation:

Figure 1

where $G$ is a gluon and the blob represents $<\bar{q}q>$. If a solution exists, then the quark self-energy function $\Sigma(p^2)$ also has a non-zero solution (it satisfies the same equation) so that there is a non-zero effective or dynamical mass $M$ of order $\Lambda_{QCD}$, given by $(p^2 - \Sigma(p^2))_{p^2=M^2} = 0$. It must be emphasized that $\Sigma(p^2)$ is the solution of a homogeneous equation, and is not the result of solving an inhomogeneous equation with a bare mass driving term.

The $<\bar{q}q>\neq 0$ solution guarantees that there are also bound state solutions for an octet of pseudoscalar bound states with quantum numbers of $\bar{q}\lambda^i\gamma_5 q$ and wave functions $\Gamma_\pi(P,' P)$, where $P'$ and $P$ are the four-momenta of the $\bar{q}$ and $q$, because the chiral symmetry implies that $<\bar{q}q>$ and $\Gamma_\pi(P,P)$ satisfy the same Schwinger-Dyson equation (i.e., Figure (1), where the blob now represents $\Gamma_\pi(P,P)$). This is just the manifestation of the Goldstone theorem in the present case.

The directions $\bar{q}q$ of the condensate and $\bar{q}\lambda^i\gamma_5 q$ of the Goldstone bosons are equivalent by the chiral symmetry. The choice of the direction of the breaking (i.e., $<\bar{q}q>\neq 0$ rather than say, $<\bar{q}\lambda^4\gamma_5 q>\neq 0$) is one of convention, analogous to choosing $<\phi^0>\neq 0$ and $<\phi^\pm>=0$ in the electroweak theory or a convenient z-axis in a rotationally invariant theory. (Professor Morpurgo's assertion that $\lambda^1, \lambda^2, \lambda^4$ and $\lambda^5$ interactions of Goldstone bosons could never be generated in QCD was based on a confusion between explicit symmetry breaking, which is driven by powers of a term in the Hamiltonian, and spontaneous symmetry breaking).

One can also show from the Schwinger-Dyson equations that the pion wave function is related in a specific way to the vertex function $\Gamma^5_\mu(p+q,p)$ of the axial current:

$$\Gamma^5_\mu(p+q,p) = F_\pi q_\mu \Gamma_\pi(p+q,p)/q^2 \tag{4}$$

This is essentially the PCAC relation (i.e., its analogue in the chiral limit) and implies the dynamical soft-pion results.

Finally, when one turns back on the current quark masses, the Goldstones acquire masses $m_\pi^2 \propto (m_u + m_d)\Lambda^2_{QCD}/F_\pi$, $m_K^2 \propto m_s\Lambda^2_{QCD}/F_K$, etc. (the decay constants are also proportional to $\Lambda_{QCD}$), and the quarks obtain constituent masses $M_\alpha = M + m_\alpha$. From $m_\pi^2/m_K^2$, $SU(3)$ breaking terms like $m_{K^*} - m_\rho$ and $m_\Sigma - m_N$, and the non-electromagnetic parts of $m_n - m_p, m_{K^+} - m_{K^0}, \rho - \omega$ mixing, etc., one can estimate the current quark masses

$$\begin{aligned} m_s &\sim 150\, MeV \\ m_d &\sim 10\, MeV \\ m_u &\sim 5\, MeV \end{aligned} \tag{5}$$

We see that $m_s$ is sufficiently large compared to $\Lambda_{QCD}$ and (and $m_N/3$) that chiral $SU(3)$ symmetry is only marginally valid. However, $m_d$ and $m_u$ are so small compared to other hadronic scales (i.e., $m_\pi^2 \ll m_\rho^2$ ) that chiral $SU(2)$ is excellent. Furthermore, the approximate $SU(2)$ isospin symmetry is due more to the smallness of $m_d$ and $m_u$ than their near degeneracy (there is no reason to expect $m_d = m_u$ in the standard model).

The pion (and to a lesser extent the $K$ and $\eta$) play a dual role in QCD. In the quark model the pion is simply a quark-antiquark bound state, which is similar to the $\rho$ and pseudoscalar excitations, except that it is (apparently accidentally) very light. On the other hand, it can also be thought of as a pseudo Goldstone boson. In first approximation, it can still be considered a quark-antiquark bound state in this case, but now its small mass is not an accident. Furthermore, its wave function has a very specific relation to the axial current, which manifests the chiral symmetry and distinguishes it from the other mesons, and the chiral symmetry enforces relations between the parameters in an effective quark model.

These two roles are not necessarily in conflict. However, the chiral properties have never been fully implemented in any quark model as far as I am aware. A better wedding of these two viewpoints would be very useful for both chiral dynamics and the quark models.

G. Morpurgo

## Non Relativistic Quark Model (NRQM) from QCD and Effective quark-Goldstone-boson Lagrangians

The main point of my lectures was to show how one can derive rigorously from QCD (or from QCD-like relativistic field theories) parametrised expressions for many quantities. Such parametrised expressions have a structure very similar to that of the non relativistic quark model. This leads to a number of consequences, verified experimentally, that I have discussed in my lectures and will not report here.

Here I wish to focus only on one point, namely the relationship between QCD and several classes of effective Lagrangians, in particular chiral Lagrangians of the Georgi-Manohar type. For this I must first recall two points that emerge from my treatment.

The first point is this: From QCD one can derive a parametrised expression for the masses of the lowest pseudoscalar and vector mesons, namely:

$$M_{meson} = A + B\underline{\sigma}_1 \cdot \underline{\sigma}_2 + C(P_1^\lambda + P_2^\lambda) + D(\underline{\sigma}_1 \cdot \underline{\sigma}_2)(P_1^\lambda + P_2^\lambda)$$

I underline that this is a rigorous consequence of QCD although it has the appearance of a NRQM result. Although we cannot say how the 4 coefficients A,B,C D depend on the quark masses and on the quark-gluon coupling, nevertheless the above equation has an interesting consequence. From it: $M(\pi) = A - 3B$, $M(\rho) = A + B$. Thus from the experimental values of the masses $M(\pi) = 140, M(\rho) = 770$ MeV we get $A = 610, B = 160$ MeV. Suppose now that the quark-gluon coupling changed a little so that, for instance, B becomes 110 MeV and A becomes 660 MeV, the mass of the $\rho$ would remain the same and that of the $\pi$ would become 330 MeV, not so small. Should a pion of that mass still be identified with a quasi-Goldstone boson?

The second point is the following: The QCD Lagrangian with e.m. interactions contains only the flavor matrices $\lambda_3$ and $\lambda_8$. Because they commute, you cannot produce from them any such matrices as $\lambda_1, \lambda_2, \lambda_4, \lambda_5$ etc. Therefore in the calculation of any quantity from QCD $\lambda_1, \lambda_2, \lambda_4, \lambda_5$ cannot arise, no matter how complicated are the calculations of that quantity from QCD.

Both these points lead one to conclude that effective Lagrangians in which pions and kaons are introduced as separate degrees of freedom, <u>in addition to quarks</u>, and in particular, chiral Lagrangians of the Georgi-Manohar type (with pions and kaons identified as quasi-Goldstone bosons) cannot be mathematically equivalent, even in a narrow sector, to QCD. Indeed such chiral Lagrangians usually have an explicit dependence on $\lambda_1, \lambda_2, \lambda_4, \lambda_5$ etc.

My conclusion is the assertion often made and followed by many theorists working in this field that:

QCD $\rightarrow$ Spontaneously broken QCD $\rightarrow$ Effective quark Lagrangian with $\pi$ and $K$ local fields as quasi Goldstone bosons (where the arrows stand for unavoidable logical steps), this assertion, I repeat, is nothing but a " Dogma". The origin of this "Dogma" is that chiral dynamics is somehow regarded as compulsory to explain the great classical successes of PCAC + current algebra. But these successes (that were derived much earlier than QCD) depend only on the empirical fact that the pion mass is small (and in fact there are not similar successes for Kaon related phenomena). But we just saw that the smallness of the pion mass $[m(\pi) = A - 3B]$ could be an accidental fact unrelated to the notion of Goldstone bosons. As a matter of fact we did show how the NRQM can be derived from QCD.

We have a completely self-consistent and predictive picture that does not agree with the point of view generally taken for granted and advocated here particularly forcefully by Professor Langacker; what is still missing is to understand the relationship between $m_s/m$ for current and constituent quarks. We hope to do this soon.

## CLOSING CEREMONY

The closing ceremony took place on Sunday 11th July 1993.

The Prizes and Scholarships were awarded as specified below.

## PRIZES AND SCHOLARSHIPS

• Prize for **Best Student** awarded to:

*Martin BENEKE*, Max-Planck-Institut für Physik, Germany.

• Prize for **Best Scientific Secretary** awarded ex-equo to:

*Massimo GIOVANNI*, University of Turin, Italy

*Robert McPHERSON*, Princeton University, USA.

• Twelve **Scholarships** were open for competition among the participants. They were awarded as follows:

• **J.S. BELL** Scholarship to:
*Martin BENEKE*, Max-Planck-Institut für Physik, Germany.

• **Patrick M.S. BLACKETT** Scholarship to:
*Konstantinos SKENDERIS*, SUNY at Stonybrook, USA.

• **James CHADWICK** Scholarship to:
*Ming LU*, Caltech, USA.

• **Amos DE-SHALIT** Scholarship to:
*André HOANG*,University of Karlsrühe, Germany

• **Paul A.M. DIRAC** Scholarship to:
*Massimo GIOVANNINI*, University of Turin, Italy

• **Isidor I. RABI** Scholarship to:
*David M. KAWALL*, Stanford University, USA.

• **Robert HOFSTADTER** Scholarship to:
*Robert McPHERSON*, Princeton University, USA.

• **Andreij D. SAKHAROV** Scholarship to:
*Ioannis GIANNAKIS*, Houston Advanced Research Center, USA.

• **Jun John SAKURAI** Scholarship to:
*Lyubov VASSILEVSKAYA*, Yaroslavl State University, Russia.

• **Gunnar KÄLLEN** Scholarship to:
*Jeffrey FORSHAW*, Rutherford Appleton Laboratory, UK.

• **André LAGARRIGUE** Scholarship to:
*Carlo ACERBI*, University of Padua, Italy.

• **Giulio RACAH** Scholarship to:
*Pawel WEGRZYN*, Jagellonian University, Poland.

• One **EPS** Scholarship was awarded to:
*Robert BUDZYNSKI*, Poland

• Two **WFS** Scholarships were awarded to:
*Michail STOILOV*, Bulgaria.
*Ali YILDIZ*, Turkey - in honour of Feza GÜRSEY.

**The following participants gave their collaboration in the Scientific Secretarial work:**

| | |
|---|---|
| *Ghadir ABU LEIL* | *Robert McPHERSON* |
| *Carlo ACERBI* | *Alessandro PAPA* |
| *Robert BUDZYNSKI* | *Horia PETRACHE* |
| *Marco CAVAGLIA'* | *Fulvio PICCININI* |
| *Jeffrey FORSHAW* | *Ofelia PISANTI* |
| *Marina GIBILISCO* | *Nikolaos SARLIS* |
| *Massimo GIOVANNINI* | *Konstantinos SKENDERIS* |
| *André HOANG* | *Michail STOILOV* |
| *Aldo IANNI* | *Zoltan TROCSANYI* |
| *David M. KAWALL* | *Pawel WEGRZYN* |
| *Ming LU* | *Bo ZHENG* |

PARTICIPANTS

| | |
|---|---|
| Ghadir ABU LEIL<br>Palestina | Department of Physics<br>Science Laboratories<br>South Road<br>DURHAM DH1 3LE, UK |
| Carlo ACERBI<br>Italy | Dipartimento di Fisica "G. Galilei"<br>Università<br>Via Marzolo, 8<br>35131 PADOVA, Italy |
| Guido ALTARELLI<br>Italy | CERN<br>TH Division<br>1211 GENEVE 23, Switzerland |
| Alexei ANSELM<br>Russia | Petersburg Nuclear Physics<br>Institute<br>Gatchina<br>ST. PETERSBURG, 188350, Russia |
| Richard ARNOWITT<br>USA | Department of Physics<br>Texas A & M University<br>COLLEGE STATION, TX 77843,<br>USA |
| Iakov AZIMOV<br>Russia | World Laboratory<br>Eloisatron Project<br>Ettore Majorana Centre - EMCSC<br>Via Guarnotta, 26<br>91016 ERICE, Italy |
| Martin BENEKE<br>Germany | Max-Planck-Institut fur Physik<br>Föhringer Ring 6<br>8000 MUNCHEN, Germany |
| Paolo BRUNI<br>Italy | INFN<br>Via Irnerio, 46<br>40126 BOLOGNA, Italy |

| | |
|---|---|
| Robert BUDZYNSKI<br>Poland | Institute for Theoretical Physics<br>Hoza 69<br>00-681 WARSAW, Poland |
| Mariano CADONI<br>Italy | I N F N<br>Via A. Negri, 18<br>09127 CAGLIARI, Italy |
| Marco CAVAGLIA'<br>Italy | S I S S A<br>School of Astrophysics<br>34100 TRIESTE, Italy |
| Chungming CHU<br>Taiwan | Department of Physics<br>University of Michigan<br>ANN ARBOR, MI 48109, USA |
| Marco D'ATTANASIO<br>Italy | Dipartimento di Fisica<br>Università<br>Via delle Scienze<br>43100 PARMA, Italy |
| Yuri DOKSHITZER<br>Russia | Department of Theoretical Physics<br>Lund University<br>Solvegatan 14A<br>223 62 LUND, Sweden |
| Robert ELIA<br>USA | S L A C - MS 78<br>P.O. Box 4349<br>STANFORD, CA 94305, USA |
| John ELLIS<br>UK | C E R N<br>TH Division<br>1211 GENEVE 23, Switzerland |
| Niclas ENGBERG<br>Sweden | Inst. of Theoretical Physics<br>Chalmers Univ. of Technology<br>41296 GÖTEBORG, Sweden |

Alessandra FILIPPI
Italy

I N F N
Via Pietro Giuria, 1
0125 TORINO, Italy

Nicolao FORNENGO
Italy

Dipartimento di Fisica Teorica
Università
Via Pietro Giuria, 1
10125 TORINO, Italy

Jeffrey FORSHAW
UK

Theory Division
Rutherford Appleton Laboratory
CHILTON, Didcot, OX11 0QX, UK

Ioannis GIANNAKIS
Greece

Astroparticle Physics Group
Houston Advanced Res. Center
4800 Research Forest Drive
THE WOODLANDS, TX 77381, USA

Marina GIBILISCO
Italy

Dipartimento di Fisica
Università
Via Celoria, 16
20133 MILANO, Italy

Massimo GIOVANNINI
Italy

Dipartimento di Fisica Teorica
Università
Via Pietro Giuria, 1
10125 TORINO, Italy

Sheldon L. GLASHOW
USA

Department of Physics
Harvard University
CAMBRIDGE, MA 02138, USA

Michel GOURDIN
France

Departement de Physique
Universite' P. et M. Curie
Tour 16 - 1er Etage
4 Place Jussieu
75252 PARIS, France

Alec HABIG
USA

Gruppo MACRO
Laboratori Naz. del Gran Sasso
SS 17 bis - Km. 18,910
67010 ASSERGI, Italy

Arthur HEBECKER
Germany

D E S Y
Theory Group
Notkestrasse 85
2000 HAMBURG 52, Germany

Peter HIGGS
UK

Department of Physics
The University
Mayfield Road
EDINBURGH EH9 3JZ, UK

André HOANG
Germany

Theoretische Teilchenphysik
Universität
Kaiserstrasse 12 - Postfach 6980
7500 KARLSRUHE 1, Germany

Timothy James HOLLOWOOD
UK

TH Division
C E R N
1211 GENEVE 23, Switzerland

Aldo IANNI
Italy

Dipartimento di Fisica
Università degli Studi
06100 PERUGIA, Italy

Vadim S. KAPLUNOVSKY
Israel

Centre for Particle Physics
Department of Physics
R.L. More Hall
University of Texas
AUSTIN, TX 78712, USA

Michael KARSTENSEN
Denmark

Niels Bohr Institute
University of Copenhagen
Blegdamsvej 17
2100 COPENHAGEN, Denmark

David Michael KAWALL
Canada

Department of Physics
Stanford University
STANFORD, CA 94305-4060, USA

Valery KHOZE
Russia

Department of Physics
University
South Road
DURHAM DH1 3LE, UK

Athanasios B. LAHANAS
Greece

Nuclear & Particle Physics
Section
University of Athens
ATHENS, Greece

Paul LANGACKER
USA

Department of Physics
University of Pennsylvania
PHILADELPHIA, PA 19104, USA

Arne LARSEN
Denmark

Nordita
Blegdamsvej 17
2100 COPENHAGEN, Denmark

Massimo LENTI
Italy

I N F N
Largo E. Fermi, 2
50125 FIRENZE, Italy

Jorge L. LOPEZ
Peru

Center for Theoretical Physics
Texas A & M University
COLLEGE STATION, TX 77843,
USA

Vladimir LOUGOVOI
Russia

World Laboratory
Eloisatron Project
Ettore Majorana Centre - EMCSC
Via Guarnotta, 26
91016 ERICE, Italy

Ming LU
China

CALTECH 452-48
Lauritsen Laboratory of HEP
PASADENA, CA 91125, USA

Nikolaos MAVROMATOS
Greece

CERN
TH Division
1211 GENEVE 23, Switzerland

Giulio MIGNOLA
Italy

Dipartimento di Fisica Teorica
Università
Via Pietro Giuria, 1
10125 TORINO,Italy

Hugh MONTGOMERY
UK

FNAL
P.O. Box 500 - Wilson Road
BATAVIA, IL 65510, USA

Giacomo MORPURGO
Italy

Istituto di Fisica
Università
Via Dodecaneso, 33
16146 GENOVA, Italy

Robert McPHERSON
Canada

Physics Department
Princeton University
P.O. Box 708
PRINCETON, NJ 08544, USA

Dimitri NANOPOULOS
Greece

Center for Theoretical Physics
Physics Department
Texas A & M University
COLLEGE STATION, TX 77843, USA

and

CERN
TH Division
1211 GENEVE 23, Switzerland

Johnny Ng
USA

DESY
Notkestrasse 85
2000 HAMBURG 52, Germany

Marco PALLAVICINI
Italy

Dipartimento di Fisica
Università
Via Dodecaneso 33
16146 GENOVA, Italy

Alessandro PAPA
Italy

Dipartimento di Fisica
Universita'
Piazza Torricelli, 2
56100 PISA, Italy

Roberto PECCEI
Italy

Department of Physics
University of California
LOS ANGELES, CA 90024, USA

Horia PETRACHE
Romania

World Laboratory
Eloisatron Project
Ettore Majorana Centre - EMCSC
Via Guarnotta, 26
91016 ERICE, Italy

Fulvio PICCININI
Italy

Dip. di Fisica Nucleare e Teorica
Università
Via L. Bassi, 6
27100 PAVIA, Italy

Ofelia PISANTI
Italy

Dipartimento di Scienze Fisiche
Università
Mostra d'Oltremare - Pad. 20
80125 NAPOLI, Italy

Nikolaos SARLIS
Greece

Physics Department
University of Athens
Solonos 104
11634 ATHENS, Greece

William SCHMIDKE
USA

Max-Plank-Institut fur Physik
Fohringer Ring, 6
8000 MUNCHEN 40, Germany

Yuly SHABELSKI
Russia

World Laboratory
Eloisatron Project
Ettore Majorana Centre - EMCSC
Via Guarnotta, 26
91016 ERICE, Italy

Serguei SIVOKLOKOV
Russia

World Laboratory
Eloisatron Project
Ettore Majorana Centre - EMCSC
Via Guarnotta, 26
91016 ERICE, Italy

Konstatinos SKENDERIS
Greece

Physics Department
S U N Y
STONY BROOK, NY 11794, USA

Arild SKJOLD
Norway

Department of Physics
University
Allegaten 55
N-5007 BERGEN, Norway

Vassilios SPANOS
Greece

Physics Department
Nuclear & Particle Physics Section
Panepistimiopolis
ATHENS 15771, Greece

Michail STEPANOV
Russia

World Laboratory
Eloisatron Project
Ettore Majorana Centre - EMCSC
Via Guarnotta, 26
91016 ERICE, Italy

Michail STOILOV
Bulgaria

Institute for Nuclear Research
and Nuclear Energy
Tsarigradsko Chaussée Blvd. 72
1784 SOFIA, Bulgaria

Olga STROGOVA
Russia

World Laboratory
Eloisatron Project
Ettore Majorana Centre - EMCSC
Via Guarnotta, 26
91016 ERICE, Italy

Vittorino TALAMINI
Italy

Dipartimento di Fisica
Università
Via Marzolo, 8
35131 PADOVA, Italy

Zoltan TROCSANYI
Hungary

Theoretische Physik
E T H
Honggerberg
8093 ZURICH, Switzerland

Henrik URSIN
Denmark

Niels Bohr Institute
Blegdamsvej 17
DK-2100 COPENHAGEN, Denmark

Timo VAN RITBERGEN
Netherland

NIKHEF/H
Postbox 41882
1009 DB AMSTERDAM, Netherland

Lyubov VASSILEVSKAYA
Russia

Department of Physics
Yaroslavl State University
YAROSLAVL 150000, Russia

Lorenzo VITALE
Italy

I N F N
Lab. AREA
Padriciano 99
34012 TRIESTE, Italy

Mikhail VYSOTSKY
Russia

I T E P
B. Cheremunshkinskaya 89
117259 MOSCOW, Russia

Pawel WEGRZYN
Poland

Institute of Physics
Jagellonian University
Reymonta 4
KRAKOW, Poland

Stefan WERNER
Germany

Institute fur Hochenergiephysik
Schroderstr. 90
6900 HEIDELBERG, Germany

Bjorn WIIK
Germany

D E S Y
Notkestrasse 85
2000 HAMBURG 52, Germany

Ali YILDIZ
Turkey

Physics Department
Bogazici University
80815 BEBEK, Istanbul, Turkey

Kajia YUAN
China

Astroparticle Physics Group
Houston Advanced Res. Center
4800 Research Forest Drive
THE WOODLANDS, TX 77381, USA

Dominique YVON
France

Commissariat à l'Energie Atomique
DAPNIA/SPP - CEN Saclay
91191 GIF-SUR-YVETTE, France

Bo ZHENG
China

Fachbereich Physik
Universitat
Postfach 10 12 40
5900 SIEGEN, Germany